Topics in Environmental Physiology and Medicine

Edited by Karl E. Schaefer

Oxygen and Living Processes

An Interdisciplinary Approach

Edited by

Daniel L. Gilbert

With 105 Figures

Springer-Verlag
New York Heidelberg Berlin

Karl E. Schaefer, M.D.
Submarine Medical Research Laboratory
Groton, Connecticut 06340, U.S.A.

Daniel L. Gilbert, Ph.D.
Marine Biological Laboratory
Woods Hole, Massachusetts 02543, U.S.A.
and
Laboratory of Biophysics, NINCDS
National Institutes of Health
Bethesda, Maryland 20205, U.S.A.

Sponsoring Editor: Larry W. Carter
Production: William J. Gabello

Library of Congress Cataloging in Publication Data
Oxygen and living processes.

(Topics in environmental physiology and medicine)
Bibliography: p.
Includes index.
1. Oxygen in the body. 2. Oxygen—Physiological effect. 3. Oxygen. I. Gilbert, Daniel L. II. Series. [DNLM: 1. Oxygen. 2. Physiology. QV312 0955]
QP535.0109 574.19'214 81-8899
AACR2

Softcover reprint of the hardcover 1st edition 1981

Printed in the United States of America

9 8 7 6 5 4 3 2 1

ISBN-13: 978-1-4612-5892-6 e-ISBN-13: 978-1-4612-5890-2
DOI: 10.1007/978-1-4612-5890-2

Contents

Part I: General Aspects of Oxygen

Part III: Human Aspects of Oxygen

Part IV: Concluding Remarks

Preface

The field of oxygen study is immense. No single work on the subject can be comprehensive, and this volume makes no such claim. Indeed, coverage here is selective and the selection is somewhat personal. However, the choice of topics is vast. There are chapters on the history of oxygen, oxygen in the universe, the biochemistry of oxygen, and clinical uses of oxygen. An alternate title could have been, "Some things you always wanted to know about oxygen, but didn't know where to find them easily." Some information in this wide-ranging work can not be found elsewhere.

This book is intended not only for specialists, but also for nonspecialists engaged in or curious about any field of oxygen study, particularly if they wish to know more about other fields of oxygen. Thus, those who are interested in oxygen and are historians, astronomers, chemists, geochemists, evolutionists, biochemists, physiologists, pathologists, or clinicians will find here much of extreme value. It is intended to be read and understood at the graduate or advanced undergraduate level.

This volume is divided into four parts. The first constitutes the background for Parts II and III, and the last integrates the preceding material with an overall perspective on oxygen in living organisms.

Part I begins by putting oxygen into its historical context, including quotes from original sources; this leads up to the presentation of the Gerschman "free radical theory" of oxygen action in biological systems. The chemistry of the different reactive forms of oxygen and the history of this reactive substance in the entire universe and on earth are then presented. The present oxygen cycle in nature, with its photosynthetic production and its removal by metazoan life, is described.

Part II begins a description of the biology of oxygen, with the mechanism of photosynthetic oxygen production. The reader is introduced first to the harmful effects of oxygen in unicellular organisms and then to the mechanisms of oxygen transport to the tissues of multicellular organisms. The reaction of oxygen with cytochrome oxidase in the cell is the beginning of the biochemical events that eventually lead to the utilization of energy released by oxygen reduction in the cell. The uncontrolled release of oxidative energy produces harmful effects in the multicellular organisms, and description of these effects and the mechanisms of control in living organisms, by such naturally occurring antioxidants as superoxide dismutase, complete this part of the book.

Part III deals with the clinical and practical uses of oxygen. Since the lung is normally exposed to higher oxygen tensions than other organs, the detrimental effects of oxygen in the lung are discussed first. With the advent of manned exploration of undersea, high-altitude, and space environments, the optimal supply of oxygen to humans is of paramount importance, and this receives extensive treatment. The last chapters in this section deal with the clinical use of oxygen; of special mention is retrolental fibroplasia, a disease causing blindness in premature babies who have been treated by excess oxygen-inhalation therapy.

Part IV integrates the entire subject of oxygen in living organisms. It becomes apparent here that a multidisciplinary approach is vital to grasp a topic of such vast proportions and consequences. The contributors to this volume are recognized authorities in their various fields, ranging from microcosm to macrocosm. Their approaches are blended here to provide a rich contribution to our understanding of this gas in living organisms.

The reader will encounter, running through the chapters, the theme of the dualism of oxygen. Oxygen provides the energy that nourishes most of life on earth, but at the same time it destroys life. Aerobic life requires a proper balance between these opposing forces in order to maintain the optimal functioning of a biological system.

Many individuals have contributed to my understanding of oxygen over the years. Dr. R. Gerschman's fascinating seminar on oxygen initiated my interest, which led to an active collaboration between the two of us. Dr. W. O. Fenn, my Ph.D. advisor, supported this collaboration and was a source of inspiration to me. Dr. H. Rahn broadened my understanding of respiration physiology. Dr. K. Schaefer, editor of this series, invited me to be the editor of this volume and encouraged me throughout its preparation. He also invited me to present publicly for the first time in 1958 my thoughts on oxygen and evolution.

Besides these individuals, I wish to thank my wife, Dr. Claire Gilbert, for acting as my editorial assistant in the preparation of this volume, and my son, Raymond Gilbert, for being extremely patient with me while this book was in preparation. Finally, I wish to thank both my parents without whose support in my formative years this volume would never have been undertaken.

I would also like to take this opportunity to thank the staff of Springer-Verlag, whose endeavors have made this volume possible.

Daniel L. Gilbert

Contributors

Britton Chance, Ph.D., D.Sc. (Chapter 10) Johnson Research Foundation, Department of Biophysics, School of Medicine, University of Pennsylvania, Philadelphia, Pennsylvania, U.S.A.

David B. Drath, Ph.D. (Chapter 14) Departments of Medicine and Biological Chemistry, Harvard Medical School, Boston, Massachusetts, and The William B. Castle Laboratory, Mount Auburn Hospital, Cambridge, Massachusetts, U.S.A.

Aron B. Fisher, M.D. (Chapter 12) Department of Physiology, University of Pennsylvania, School of Medicine, Philadelphia, Pennsylvania, U.S.A.

Robert W. Flower, B.A. (Chapter 17) The Wilmer Ophthalmological Institute and The Applied Physics Laboratory, The Johns Hopkins University and Hospital, Baltimore, Maryland, U.S.A.

Henry J. Forman, Ph.D. (Chapter 12) Department of Physiology, University of Pennsylvania, School of Medicine, Philadelphia, Pennsylvania, U.S.A.

Irwin Fridovich, Ph.D. (Chapter 13) Department of Biochemistry, Duke University Medical Center, Durham, North Carolina, U.S.A.

Rebeca Gerschman, Ph.D. (Chapter 2) Departamento de Fisiologia Humana, Facultad Farm. y Bioquimica, Universitad de Buenos Aires, Buenos Aires, Argentina

Daniel L. Gilbert, Ph.D. (Chapters 1, 5, 18) Marine Biological Laboratory, Woods Hole, Massachusetts, U.S.A., and Laboratory of Biophysics, NINCDS, National Institutes of Health, Bethesda, Maryland, U.S.A.

Sheldon F. Gottlieb, Ph.D. (Chapter 7) Dean of the Graduate School and Director of Research, University of South Alabama, Mobile, Alabama, U.S.A.

Niels Haugaard, Ph.D. (Chapter 11) Department of Pharmacology, School of Medicine, University of Pennsylvania, Philadelphia, Pennsylvania, U.S.A.

George H. Herbig, Ph.D. (Chapter 4) Lick Observatory and Board of Studies in Astronomy and Astrophysics, University of California, Santa Cruz, California, U.S.A.

Gary L. Huber, M.D. (Chapter 14) Tobacco and Health Research Institute, University of Kentucky, Lexington, Kentucky, U.S.A.

Sheldon Kanfer, Ph.D. (Chapter 3) Department of Chemistry, Columbia University, New York, New York, U.S.A.

Gyula B. Kovachich, Ph.D. (Chapter 11) Institute for Environmental Medicine, University of Pennsylvania, Medical Center, Philadelphia, Pennsylvania, U.S.A.

David C. Mauzerall, Ph.D. (Chapter 6) The Rockefeller University, New York, New York, U.S.A.

Arnall Patz, M.D. (Chapter 17) The Wilmer Ophthalmological Institute and The Applied Physics Laboratory, The Johns Hopkins University and Hospital, Baltimore, Maryland, U.S.A.

Richard G. Piccioni, Ph.D. (Chapter 6) The Rockefeller University, New York, New York, U.S.A.

Johannes Piiper, M.D. (Chapter 8) Abteilung Physiologie, Max-Planck-Institut für Experimentelle Medizin, Göttingen, Federal Republic of Germany

Karl E. Schaefer, M.D. (Chapter 15) Submarine Medical Research Laboratory, Groton, Connecticut, U.S.A.

Peter Scheid, M.D. (Chapter 8) Abteilung Physiologie, Max-Planck-Institut für Experimentelle Medizin, Göttingen, Federal Republic of Germany

I. A. Silver, D.V.M. (Chapter 16) Department of Pathology, University of Bristol, Medical School, Bristol, England, U.K.

Nicholas J. Turro, Ph.D. (Chapter 3) Department of Chemistry, Columbia University, New York, New York, U.S.A.

Beatrice A. Wittenberg, Ph.D. (Chapter 9) Department of Physiology, Albert Einstein College of Medicine, New York, New York, U.S.A.

Jonathan B. Wittenberg, Ph.D. (Chapter 9) Department of Physiology, Albert Einstein College of Medicine, New York, New York, U.S.A.

1

Perspective on the History of Oxygen and Life

DANIEL L. GILBERT

The underlying theme of this chapter is to illustrate how the idea of the effects of oxygen on life, both beneficial and dangerous, was developed.

The notion that part of the atmosphere matters greatly to life is extremely old. It is first expressed as the "breath of life." Of course, that element in the atmosphere was later found out to be oxygen. Likewise, the analogy between fire and life was recognized by the ancients. Both Priestley and Scheele, the independent discoverers of oxygen, were aware of this analogy and recognized that oxygen as well as fire could be not only beneficial, but also dangerous.

According to the now discredited phlogiston theory of Stahl, some material named phlogiston was lost from combustible substances during the process of burning. This idea is analagous to our understanding of dehydrogenation reactions, in which hydrogen is lost from reduced compounds during the process of oxidation. The oxygen theory of Lavoisier overturned the phlogiston theory. Oxygen was shown to react with compounds directly. However, in biological oxidations, both removal of hydrogen and/or addition of oxygen occurs. The early workers suggested a balance in living systems between hydrogen and oxygen. This view is still essentially correct.

Shortly after the discovery of oxygen, some physicians began using oxygen as a cure-all for many illnesses. Within a few years after the introduction of oxygen therapy, the medical profession became extremely skeptical of this treatment. In fact, at the beginning of the nineteenth century, the medical profession was generally aware that oxygen can be toxic.

Paul Bert showed in his pioneering studies, a century after the discovery of oxygen, that all living organisms are subject to oxygen toxicity. Shortly before, Louis Pasteur had shown that the mere presence of oxygen could kill certain organisms. Since Bert, 100 years ago, there have been numerous studies on oxygen toxicity. Although he was not the first one to demonstrate that oxygen is toxic, he was the first one to study the effects of oxygen on living organisms so extensively. Both the beneficial and harmful effects were recognized by Bert. This duality of oxygen's nature in life processes was now put on a firm experimental basis.

We will see in this chapter that new ideas must ripen before they are accepted; they have to be born at the right time and the right place. Kuhn (1962) has stated, "To the historian, discovery is seldom a unit event attributable to some particular man, time, and place." Finally, as Fruton (1972) wrote, "The euphoria that attends great advances, some of which have been labeled 'revolutions,' is understandable; a study of the historical record suggests, however, more continuity and complexity than are implied by such dramatic words."

I. Ideas Preceding the Discovery of Oxygen

The Matter of the Sun, or of Light, the Phlogiston, Fire, the Sulphureous Principle, the Inflammable Matter, are all of them names by which the Element of Fire is usually denoted.

Macquer (1777b)

Fire and life have been linked as a theme persisting throughout much of recorded history. The atmosphere, hydrosphere, and lithosphere constitute the environment of living organisms, and are intimately related to this theme.

Anaximander (fl. 565 B.C.) believed that the fundamental qualities of the world were hot, cold, wet, and dry (Abel, 1973). Fire was the primordial element according to Heraclites (fl. 500 B.C.) (Able, 1973). Parmenides (fl. 475 B.C.) believed that fire within living organisms results in vitality (Hall, 1975a). The elements chosen by Empedocles (fl. 465 B.C.) were earth, water, air, and fire (Partington, 1970). Today, we recognize that the earth's crust is composed of the lithosphere, hydrosphere, atmosphere, and biosphere. Note the similarity between the elements of Empedocles and the components of the earth's crust. If we correlate fire with the biosphere, then there is a perfect correlation between these elements and the earth's crust. There may have been Iranian or Indian influence in choosing these elements (Partington, 1970). In India, the Bhuddists also had essentially the same four elements (Needham, 1969). Fire was also one of the five elements of Tsou Yen (fl. 300 B.C.), who influenced Chinese culture (Needham, 1969).

Burning is accelerated by a gentle wind, which replenishes the oxygen at the site of a fire. Devices were used to promote this wind in the ancient civilizations. For example, an Egyptian mural of the Eighteenth Dynasty (about 1500 B.C.) depicts a bellows used for fanning a fire. A bellows is also mentioned in the Bible. Blowing tubes for maintaining fires were used in Assyria, Syria, and the Indus Valley. Bird wings for fanning a fire are mentioned in the Indian Hindu *Rig Veda* (Ganzenmüller, 1943).

Empedocles also believed that "innate heat" was closely identified with the soul (Perkins, 1964). Earlier, Anaximenes (fl. 545 B.C.) discussed "pneuma," a vital principle, which is the same in the world and in living bodies. This "pneuma" is taken up by living bodies in breathing (Smith, 1976). Skinner (1970) mentions that "pneuma" was an attempt to explain the role of oxygen in living processes.

This ancient concept of breath of life has been documented in the writings of the Sumerians, who invented the first written language about 3000 B.C. The Sumerian creation myth, although not the original one, was that man was created for the service of the gods. This myth states: "For the sake of the good things in their pure sheepfolds Man was given breath" (Woolley, 1965).

Contenau (1938) refers to an inscription on a statue of Prince Gudéa with the words "generously endowed with the breath of life." Gudéa lived about 2200 B.C. (Woolley, 1965). Some cylinders also dating from this period refer to the breath of life given to the Prince Gudéa by the Dunshagge (Contenau, 1938).

The heretic Egyptian pharoah, Akhenaton (1350 B.C.), who worshipped the sun, has the following lines in his Hymn to the Sun: "Who givest breath to animate every one that he maketh" (Woolley, 1965). The ancient Egyptians looked upon life as a mere episode. This philosophy is reflected in *The Physician's Secret: Knowledge of the Heart's*

Movement and Knowledge of the Heart, written about 1600–1550 B.C., as illustrated by this citation: "The breath of life enters into the right ear, and the breath of death enters into the left ear" (Sigerist, 1967).

Thornton (1803) and Smith (1870) point out that this concept is mentioned in the Bible. In the Old Testament Genesis, chapter 2, verse 7, "God Yahweh formed man from clods in the soil and blew into his nostrils the breath of life. Thus man became a living being" (Speiser, 1964). This verse is from the J, or "Jehovah" biblical source, which presumably dates from about the tenth century B.C. (Speiser, 1964).

The Indian Hindu *Rig Veda* (1500–1000 B.C.) is the oldest surviving form of Sanskrit (Basham, 1959), and it contains the following phrase: "Giver of vital breath" (Eliade, 1974). This "breath of life" concept, represented by "chhi" (Needham, 1969), also occurs in the Chinese culture (Perkins, 1964).

Diogenes (fl. 435 B.C.) revived the "pneuma" concept of Anaximenes (Rothschuh, 1973). Democritus (fl. 420 B.C.) believed that fire atoms are inhaled by living organisms (Hall, 1975a). Plato (427–347 B.C.) in his "Timaeus" writes that fire and air enter the lungs and stomach during inspiration (Hall, 1975a). Aristotle (384–322 B.C.) believed that the heart was the hottest part of the body (Hall, 1975a), and stated, "Hence, of necessity, life must be coincident with the maintenance of heat, and what we call death is its destruction" (Aristotle, 1955). Erasistratus (fl. 260 B.C.), the founding father of physiology, believed that the lungs take up "pneuma" from the atmosphere. He called the vital spirit in the arteries "pneuma zotikon," and animal spirit in the nerves "pneuma psychikon." He postulated that the "pneuma psychikon" is derived from "pneuma zotikon" in the brain (Smith, 1976).

Three words which meant "breath" in ancient Greek were "anemos," "pneuma," and "psyche." "Pneuma" has been defined as air, breath of life, vital spirit, vital force, spirit, soul, and innate heat. "Psyche" has also been defined as breath of life, spirit, soul, and finally mind. The Latin word "anima" is derived from "anemos" and means breath or soul. It is similar in meaning to "pneuma." Interestingly, the word animal is derived from this source (Skinner, 1970).

McKie (1953) mentions that the Romans in the first century B.C. were well aware of the vital flame, or "flamma vitalis" ("flammula vitae"). Athenaeus (about 69 A.D.) established the Pneumatist sect, who adhered to the belief that the innate heat was the pneuma, and the heart was the center of the system (Skinner, 1970). Galen (130–200 A.D.) believed that the "pneuma zoticon" derived from the air was responsible for the source of the vital fire in the left ventricle of the heart. In many ways, this pneuma has many of the properties of oxygen (Smith, 1976). Galen not only believed in a "pneuma zoticon" or "spiritus vitalis," but also in the pneuma associated with the nervous system, the "pneuma psychicon" or "spiritus animalis" (Rothschuh, 1973). Spiritus means breath in Latin and has the same connotation as "pneuma" (Skinner, 1970). It is of interest to note that expire means either to exhale or to terminate life (Smith, 1976), and that inspire means either inhale or infuse an idea into an individual.

An inflammable substance, named Greek or marine fire, was discovered about 650 A.D. by Kallinikos of Heliopolis; another explosive mixture, named "Huo Yao" or fire chemical (gunpowder) (Leicester, 1956) by 850 A.D. was apparently used in China (Smith, 1976). The first written records of explosives appear in a Chinese work in 1044 (Needham et al., 1976). Both Greek fire and fire chemical were used for warfare (Leicester, 1956). Greek fire was probably an incendiary containing petroleum (Partington, 1965). Gunpowder is a mixture of potassium nitrate (niter or saltpeter) with carbon and sulfur. When potassium nitrate is heated alone between 650° to 750°C, the nitrite is formed with the liberation of oxygen (Brasted, 1961):

$$2KNO_3 \rightarrow 2KNO_2 + O_2 \qquad (1)$$

The nitrite decomposes at 1000°C, liberating the brown nitrogen dioxide (NO_2), nitric oxide (NO), and nitrogen (N_2) (Suttle, 1957). When the nitrate is heated in the gunpowder mixture, oxygen is released from the nitrate in the explosive theoretical reaction:

$$2KNO_3 + S + 3C \rightarrow K_2S + N_2 + 3CO_2 \quad (2)$$

Side reactions also produce the carbonate and sulfate salts of potassium (Parkes, 1952). These discoveries reinforced the importance of fire in living processes (Florkin, 1972).

Paracelsus (1490–1541) thought of life as fire (Hall, 1975a). He believed that air is required for the burning of wood and fuel. According to him, the principle of combustibility was sulfur (Partington, 1961). Aristotle (1953) considered sulfur as sacred. The burning of sulfur produces sulfur dioxide (Parkes, 1952):

$$S + O_2 \rightarrow SO_2 \quad (3)$$

This flammability was known before Paracelsus and was mentioned in Homer (Parkes, 1952) and in the Bible (Weeks, 1956). Other combustible substances were often designated as sulfur (Weeks, 1956). Fernel (1497–1558) believed that both vital heat and heat produced by combustion require air. Life, like a flame, will eventually die, according to Francis Bacon (1561–1626). Descartes (1596–1650) thought of life as fire-without-light. The idea that life is analagous to fire was prevalent at this time (Hall, 1975a).

II. Events Leading up to the Discovery of Oxygen

The Chinese knew how to oxidize mercury by heating it to produce mercuric oxide by 640 A.D. (Needham et al., 1976). In fact, Ko Hung (4th century A.D.) reported that heating cinnabar (mercuric sulfide) produced mercury:

$$HgS + O_2 \rightarrow Hg + SO_2 \quad (4)$$

It was claimed that cinnabar could be produced again by further heating (Partington, 1965). If the temperature is above 426°C in air, then mercury is oxidized to mercuric oxide (Mellor, 1922):

$$2Hg + O_2 \rightarrow 2HgO \quad (5)$$

Ko Hung could possibly have produced the red form of mercuric oxide, instead of the red cinnabar. Perhaps Zosimus in the 4th century A.D. released oxygen by heating mercuric oxide (Mellor, 1922). This reaction occurs when the temperature is below 426°C in air:

$$2HgO \rightarrow 2Hg + O_2 \quad (6)$$

In 1810, Henrich Julius Klaproth, the well known Orientalist, reported that he had translated a Chinese book by Maò hhóa, written in 756 A.D. Maò hhóa writes that the Yin of the air is not pure, and that it can be obtained by heating potassium nitrate (Mellor, 1922; Weeks, 1956; Needham et al., 1976). Obviously the Yin of the air refers to oxygen, which means that the discovery of oxygen by the Chinese predated by one thousand years the European rediscovery. However, there is some doubt about the authenticity of this Chinese work (Weeks, 1956; Needham et al., 1976). Although it is not at all likely that Klaproth perpetuated a hoax, it is possible that at the end of the 18th century, a false manuscript was given to Klaproth. Needham et al. (1976) further point out that at this time, the concept of positive and negative electricity was well known, and that this dualism could easily fit into the Chinese Yin-Yang philosophy of opposing forces (Needham, 1969). And further, "and could... one ... have translated hydrogen and oxygen into terms of Yang and Yin?" (Needham et al., 1976).

Jābir Ibn Hayyān (Geber) (b. ca. 721 A.D.) heated mercury and obtained a red solid (mercuric oxide) (Mellor, 1922; Holmyard, 1955). Masloma al-Majriti (d. ca. 1007 A.D.), a Moorish astronomer, observed that the weight of mercury was equal to the weight of mercuric oxide produced. Obviously, the measurement was not sensitive enough to show the increase in weight due to the addition of oxygen to the mercury (Leicester, 1956).

Leonardo da Vinci (1452–1519) recognized that air is composed of at least two components. He also stated that air which is unfit for fire is also unfit for life (Weeks, 1956).

Perhaps Paul Eck of Sulzbach in 1489 found that mercuric oxide was heavier than mercury (Partington, 1961). He may have also formed oxygen by heating mercuric oxide. Cardan in 1557 released oxygen by heating potassium nitrate (Mellor, 1922).

The magnale in the air of Van Helmont (1577–1644) is related to fire and therefore shows some similarity to oxygen. Incidentally, he coined the word "gas" from the Greek "chaos" (Partington, 1961). "Chaos" is defined to mean empty space (Jaeger, 1944). Paracelsus had earlier used "chaos" but in a different sense (Partington, 1961). Both Paracelsus and Van Helmont were iatrochemists, or medical chemists (Partington, 1965).

Drebbel in 1608 decomposed potassium nitrate by heating it. Oxygen was released by Boyle in 1673 when he heated lead oxide. He also heated mercury and obtained the red form of mercuric oxide, which reverted to mercury again with further heating (Partington, 1961). According to Fulton (1932), "BOYLE approached the discovery of oxygen more closely than anyone before PRIESTLEY and LAVOISIER."

Experiments were also conducted in the seventeenth century to show that when metals are heated, their weight is increased. The heated metal was called the "calx" and the process was termed "calcination." Today, we refer to the calx as an oxide and the process as an oxidation. In 1618 Poppius heated antimony and measured a weight increase (Rey, 1630; Partington, 1961). Rey (1630) pointed out that the weight increases of tin and lead upon heating were due to air. Le Fèvre confirmed the observation of Poppius about 1600 (Partington, 1961). Brouncker heated both lead and copper; the data (Sprat, 1667) clearly showed a gain in weight for lead. Boyle also noted that metals increased in weight when heated. In 1679, Thomas Henshaw observed again the increase in weight of antimony when heated (Partington, 1961).

Boyle also found that subjecting an animal and a lighted candle to a partial vacuum caused the animal to die and the candle to be extinguished. However, gunpowder burned when subjected to a partial vacuum (Partington, 1961). Thus, both the ordinary flame and the "flamma vitalis" depend upon some component in the air (Hall, 1975a).

Previously, Acosta (1590a) had described altitude sickness in the earliest documented source known to us, in recounting crossing the Peruvian Andes: "it is not only... the Pariacaca mountain pass, which produces this effect [of altitude sickness], but also... the entire mountain range,... and much more for those who ascend from the seacoast to the mountain, than for those who return from the mountain to the plains." The original Spanish for the latter part of this quote is "y mucho mas [más] a los q̃ [que] suben de la çosta de la mar a la sierra, q̃ [que] no en los q̃ [que] bueluen [vuelven] de la sierra a los llanos."[1]

This part of the sentence was incorrectly rendered in the 1604 English translation (Acosta, 1590d) as: "you shall finde strange intemperatures [altitude sickness], . . . rather to those which mount from the sea, then from the plaines." A French translation (Acosta, 1590b) originally published in 1598 also altered the original meaning: "this strange intemperature [altitude sickness] is felt... more by those who ascend from the side of the sea than by those who come from the side of the plains." Bert (1878) quotes a later 1606 revised edition of this French translation (Acosta, 1590c), but mistakenly identified it as a 1596 edition. However, this quotation was not changed.

[1]The material in square brackets here represents modern Spanish.

On the other hand, an Italian translation (Acosta, 1590e) did not alter the original meaning of this section: "much more [altitude sickness] for those that ascend from the seacoast and go to the mountain, than for those who return from the mountain to the plains."

If the French and English translators meant the plains to be the high altiplano, then they realized that acclimatization occurred in the high plains. Earlier in the sixteenth century, writings reflected a knowledge of acclimatization (Monge, 1948). However, it is more likely that the translators simply erred; if an alternate Spanish text exists which justifies their versions, it is unknown to the present author. Monge (1948) quoted Acosta from a 1608 Spanish edition (Acosta, 1590f) and no essential difference in the text (Acosta, 1590a) can be observed.

Boyle probably read the English translation (Acosta, 1590d), in which Acosta describes graphically cases of altitude sickness: "Some... demaunded confession, thinking verily to die,... I beheld one that did beate himselfe against the earth, crying out for the rage and griefe which this passage of *Pariacaca* hadded caused." Acosta suggested that "the aire is there so subtile [thin] and delicate, as it is not proportionable with the breathing of man, which requires a more gross [dense] and temperate aire, and I believe it is the cause that doth so much alter the stomacke, & trouble all the disposition." The name is derived from "paria" meaning sparrow; Pariacaca means "rock of the sparrow" (Tauro, 1967).

Although the name "Pariacaca... has disappeared in Peru as well as in Ecuador" (Bert, 1878), we think that we have located it. At least two sites in Peru with that spelling have been noted earlier in the twentieth century (Espasa, 1920).

One is a town about 300 km (200 miles) north of Lima. Today it is known as Pariacancha (9°12′ S, 76°57′ W) and is at an elevation of about 3000 m (9800 ft) (Map of Peru, 1974). It is to the east of both the Cordillera Blanca and Cordillera Negra ranges, and is not at all near the sea. Because of this, and because it is at a relatively low altitude, it is probably not near the Pariacaca mountain mentioned by Acosta.

The other site is a ranch by a mountain known today as Cerro Pariachaca (12°01.4′ S, 76°00.5′ W) at 5100 m (16,700 ft). Lake Pariachaca, at about 4400 m (14,500 ft), lies at the base of this mountain and about a minute west of it (Map of Peru, 1972a). A minute latitude or longitude equals about 1.85 km (1.15 miles) at this latitude. A town, Pariachaca, is 65 km (40 miles) away, at 11°41′ S, 76°29′ W (U.S. Office of Geography, 1955). On a 1775 map (Map of Peru, 1775), this town was spelled Pariacaca, and was on the shortest road from Lima over the Andean divide at that time. Cerro Pariachaca is not the highest peak in the vicinity. There is a peak two minutes north at 5768 m (18,924 ft) (Map of Peru, 1965). Therefore the Pariacaca mountain which Acosta described is undoubtedly one in the same vicinity, most probably Cerro Pariachaca.

Additional support for the conclusion that the Pariacaca mountain is in this general vicinity comes from an Indian legend. It seems that Pariacaca, a god, destroyed the mountain of Wallallo, a rival fire deity, with snow, hail and rain, near the Rimac valley (Tauro, 1967). This is the valley in which a highway and a railroad are located (Maguiña, 1965; Maps of Peru, 1965, 1979). The highway is between Lima (12°3′ S, 77°3′ W) and Morococha (11°37′ S, 76°9′ W)(U.S. Office of Geography, 1955); it goes over the Andean crest at the Anticona pass (11°35′ S, 76°11′ W)(Map of Peru, 1979). Morococha is the location of a high altitude laboratory at 4540 m (14,900 ft)(Heath and Williams, 1977). The town of Pariachaca is not on this road.

From an examination of the region in the vicinity of Cerro Pariachaca (Maps of Peru, 1972a, b, 1973, 1979), it appears that the elevation of any pass near here across the Andean divide is about the same as the one at Anticona, which is at 4843 m (15,890 ft) (Maguiña, 1965). This altitude is even greater than Mont Blanc (15,771 ft or 4807 m) on the border of France and Italy (Delury, 1978). Bert (1878) mentions that no snow was mentioned by Acosta, and therefore the altitude of the described pass was less than 14,000 ft (4300 m). However, the snow line on mountains depends upon latitude. From personal observation, there is not always snow on the road at the Anticona pass.

Hence, it seems that the elevation of the pass by Pariacaca mountain which Acosta described is the same as that at Anticona Pass, or close to it. He was correct when he noted (Acosta, 1590d) that it is "one of the highest parts of land in the worlde." The Andes are the second highest mountain range in the world, being surpassed only by those in central Asia. Marco Polo (1299?) wrote about the Hindu Kush and Pamir mountains located in Afghanistan, the Soviet Union, and China, as being "the highest place in the world.... No birds fly here because of the height and cold."

Boyle (1670) noted that other travelers besides Acosta, whom he mentions, observed altitude sickness at high elevations. He also thought that the rarity of the air was the cause of this sickness, but did not rule out natural pollution of the air as being a factor. He writes, "the [altitude] Sickness, if not also the Difficulty of breathing, that some have been abnoxious to in the uppermost parts of *Pariacacha* [sic], and perhaps some other high Mountains, may not be imputed not so precisely to the Thinness and Rarity of the Air in places so remote from the lowermost part of the Atmosphere, as to exclude certain steams of a peculiar nature, which in some places the Air may be imbued with?" Shortly before, in 1648, Périer, at the suggestion of Pascal, noted that the measured barometric pressure decreased when the altitude was increased (Bert, 1878).

The effects of high altitude were definitely recognized by the residents of the high Andes of South America before and after Acosta (Monge, 1948). The ancients must have been aware of the effects of altitude (Ward, 1975). Herodotus (fl. 450 B.C.) writes about legends concerning the inhabitants in the high mountains in Asia, possibly the Hindu Kush range, as being "inhabited by men with goats' feet" (Herodotus, 1928), but he adds that he doesn't believe these legends. Alexander the Great in 329 B.C. crossed the Hindu Kush mountains through Khawak pass, Afghanistan, at an altitude of 11,600 ft (3540 m). Perhaps he also crossed the Kaoshan pass, also in Afghanistan at the very high altitude of 14,300 ft (4360 m), but this is very doubtful (Tarn, 1956). Livy (59 B.C.–17 A.D.) in his description of Hannibal's crossing of the Alps from France into Italy in 218 B.C. mentions that "surely no lands touched the sky or were impassable to man" (Livy, 1929). Hannibal probably crossed at an altitude of 2000 to 2700 m (6500 to 9000 ft) (Bert, 1878). A first century B.C. Chinese report mentions the Headache Mountains, where one suffers from mountain sickness, characterized by headaches, dizziness, and vomiting. According to Needham (1965), these mountains probably refer to the Tibetan plateau. Marco Polo (1299?) writes about mountains of immense height in northern Afghanistan. He mentions "On the mountain tops the air is so pure and so salubrious that if a man... falls sick of a fever,... he has only to go up into the mountains, and a few days rest will banish the malady and restore him to health." However, he also writes about "lofty mountains [in the present provence of Yunnan, China] which no one may visit in summer at any price, because the air in summer is so unwholesome and pestilent that it is death to any foreigner."

In addition, techniques were now available to compress air, which could cause the oxygen concentration to be increased above its natural level. Animals were exposed to compressed air by Hooke in 1664, who observed that a mouse was healthy at a pressure of eight bar: "Hence it was inferred, that a man may breathe under water at the depth of 200 fathoms [1200 feet. The pressure at this depth is 37 bar and not 8 bar (Gilbert, 1981b)], if he can breathe and live with as thick air, as a mouse can do" (Birch,

1756). In 1665 Hooke found that a lamp would burn longer when the air was compressed (Patterson, 1931). Boyle observed that a mouse lived longer in air compressed to two bar, in 1667 (McKie, 1953).

Compressed air, as a therapeutic agent, was advocated by Nathaniel Henshaw (1664). He suggested building air chambers next to homes for this purpose. The pressure within the chambers could be maintained between 0.5 and 3 bar. According to Henshaw, diseases could be divided into two types. One type was the cold or chronic disease, such as scurvy, rickets, arthritis, and probably gout. The other type was the hot or acute disease, such as fevers and inflammations. "Where the disease seems to depend upon a deficient Fermentation of the humours [chronic disease]...the Patient...shall... force the air out [of the chamber]; till having considerably alter'd the tone [pressure], and rarefied what remains.... On the other side, if the disease be Acute, and seem to depend upon the too violent Fermentation of the humours, then it is necessary that the Chamber be well charged with air." His idea was that increasing the air density impeded movement; thus compressing the air decreases the movement of the humors. He suggests patients with chronic diseases should be in the rarefied atmosphere two or three hours; whereas those with acute diseases should be in the compressed air environment "during the whole course of the disease."

About 1650, Glauber showed that potassium nitrate is good for plants (Hogben, 1951). As stated, it is a major component of gunpowder, which remains effective in the absence of air.

Sulfur was still considered to be the combustible principle. This idea came from the Greeks and Chinese. Aristotle considered sulfur to be a constituent of the smoky exhalation; the Chinese considered it as Yang, the active fiery male essence (Mason, 1953). Geber, a leader of Muslim Alchemy, wrote "For in fire, all burnable sulfurous is destroyed" (Ganzenmüller, 1943). Paracelsus took up this idea by naming sulfur the "combustible principle" (Partington, 1961). In addition, sulfur is an ingredient of gunpowder, which was at this time believed to be involved in electrical storms and earthquakes (Leicester, 1956).

With this background, Hooke (1665) concluded "*that the dissolution* [fire] of sulphureous bodies [combustible materials] is made by a substance inherent, and mixt with the Air, that is like, if not the very same, with that which is fixt in *Salt-peter* [potassium nitrate]." Of course, oxygen is in potassium nitrate, so Hooke was on the right path. McKie (1953) remarked, "Hooke's theory marks a great advance; it is the first step towards the modern theory of combustion."

Thomas Henshaw in 1667 also believed that air contained particles of niter (potassium nitrate). De Locques in 1664 mentions a substance which is volatile, all over, a congealed air in the atmosphere, in the sea, and not vulgar niter. Oxygen fits this description (Partington, 1962). Willis in 1670 emphasized the similarity between a flame and the "flamma vitalis" (Hall, 1975a).

Hooke performed an experiment by cutting open the thorax of a dog that was then kept alive by blowing with a bellows into the lungs which were perforated. He concluded that "as the *bare* Motion of the Lungs *without fresh Air* contributes nothing to the life of the Animal, he being found to survive as well when they were not mov'd, as when they were; so it was not the subsiding or movelessness of the Lungs that was the immediate cause of Death, or the stopping the Circulation of the Blood through the Lungs, but the *want* of a sufficient *supply of fresh Air*" (Hook, 1667).

Mayow (1674) in his discussion on respiration stated, "With respect, then, to the use of respiration, it may be affirmed that an aërial something essential to life, whatever it may be, passes into the mass of the blood. And thus air driven out of the lungs, these vital particles having been drained from it, is not longer fit for breathing again... it is probable that nitro-aërial spirit, mixed with the saline-sulphureous particles of the blood, excites in it the neccessary fermentation."

Foster (1901) points out "By his nitro-aereal, or igneo-aerial particles, Mayow evidently meant what we now call oxygen. He saw that this formed only a part of the atmosphere, that it was essential for burning, that it was essential for all the chemical changes on which life depends, that it was absorbed into the blood from the lungs, carried by the blood to the tissues, and in the tissues was the pivot, the essential factor of the chemical changes by which the vital activities of this or that tissue are manifested." However, Mayow did not envision his nitro-aerial particles forming a stable union with the combustible material; rather, these particles escaped in the form of heat or light (Hall, 1975a). Patterson (1931) thinks that Mayow has been given too much credit for his role in the events leading to the discovery of oxygen.

Borrichius (or Borch) in 1678 heated potassium nitrate and released oxygen (Weeks, 1956). A short time before, Becher in 1667 theorized that combustible substances contained "terra pinguis," fatty earth (Leicester, 1956), or "brennbare Erde," combustible earth (Rothschuh, 1973). This fatty combustible earth corresponded to the sulfur of Paracelsus and was lost during burning. This idea of losing something in a fire seemed reasonable, since fires can destroy large objects and leave only a small visible residue (Leicester, 1956). Stahl (1697) popularized this idea of Becker when he renamed this fatty earth: "But also, according to a reasonable manner of speaking, it is the corporeal fire, the essential fire material, the true basis of fire movement in all inflammable compounds. . . . From all these various conditions, therefore, I have believed that it should be given a name, as the first, unique, basic, inflammable principle. . . . And therefore I have chosen the Greek name phlogiston, in German *Brennlich* [inflammable]. . . .it is chiefly found in the fatty materials. . . .coal and bitumin are full of it; sulfur, not indeed in weight, but in the number of its finest particles, is completely possessed with it. Not less is it found in all inflammable, incomplete, and so-called 'unripe' metals."

Earlier, Hopelius (Raphael Eglin) in 1606 had used the word phlogiston (Partington, 1961). Perhaps phlogiston can be defined as "fire-stuff," since "phlox" in Greek means flame (McKenzie, 1960). Although it was known that an oxidized metal is heavier than the metal, Stahl suggested that phlogiston decreases the weight of the metal (Partington, 1965).

We now know that this theory is not correct. However, it was a very successful theory, which lasted nearly a century. Ihde (1964) states: "The potential comprehensiveness of the phlogiston theory proved amazingly good in a world in which chemistry still held a qualitative attitude towards matter." The opinion of Rothschuh (1973) is that: "Stahl's theory was important for the development of inorganic, organic, and even physiological chemistry. For the first time... the reversibility of chemical actions... became a general principle." Leicester (1956) has written: "the concept is one of the transfer of something from one substance to another. It was essentially this concept... that made the theory so useful.... It was thus the first great unifying principle in chemistry. Its success accounted for the importance it assumed for eighteenth century chemists."

Geoffrey in 1717 heated potassium nitrate and found it lost weight. However, he missed the production of oxygen (Partington, 1962). Hales in 1727, without realizing it, collected oxygen from heating potassium nitrate (Mellor, 1922). Bayen in February, 1774, heated mercuric oxide and released oxygen, but identified this gas as carbon dioxide (Conant, 1970). Cadet-Gassicourt in September, 1774, heated mercuric oxide and released oxygen without knowing it. A commission, consisting of Sage, Brisson, and Lavoisier, repeated Cadet's experiment, and they too failed to realize that oxygen was liberated. The report of this commission was presented on November 19, 1774 (Partington, 1962).

Joseph Priestley (Fig. 1A) a liberal minister, became interested in the "doctrine of air" (Priestley, 1795). He reported: "on the 17th of August, 1771, I put a sprig of mint

A

B

C

Figure 1. The discoverers of oxygen and of the mechanism of oxidation. **A.** Joseph Priestley (1733–1804), the codiscoverer of oxygen. **B.** Carl Wilhelm Scheele (1742–1786), the independent codiscoverer of oxygen. **C.** Antoine Laurent Lavoisier (1743–1794), the discoverer of the mechanism of oxidation. (Priestley, Scheele, and Lavoisier photographs courtesy of Library of Congress, Washington, D.C.)

into a quantity of air, in which a wax candle had burned out, and found that, on the 27th of the same month, another candle burned perfectly well in it... Experiments made in the year 1772, abundantly confirmed my conclusion concerning the restoration of air, in which candles had burned out, by plants growing in it." He then described the effects of plants on animal respiration: "I took a quantity of air, made thoroughly noxious, by mice breathing and dying in it, and divided it into two parts; one of which I put into a phial immersed in water; and to the other... I put a sprig of mint. This was about the beginning of August, 1771, and after eight or nine days, I found that a mouse lived perfectly well in that part of the air, in which the sprig of mint had grown, but died the moment it was put into the other part of the same original quantity of air; and which I had kept in the very same exposure, but without any plant growing in it" (Priestley, 1772).

This is the first demonstration that plants can restore the air, so that combustion and animal respiration can once again occur. Franklin, in commenting on this discovery, states, "I hope this will give some check to the rage of destroying trees that grow near houses, which has accompanied our late improvements in gardening, from an opinion of their being unwholesome. I am certain, from long observation, that there is nothing unhealthy in the air of woods; for we Americans have every where our country habitations in the midst of woods, and no people on earth enjoy better health, or are more prolific" (Priestley, 1772). Perhaps this is the origin of our common idea that fresh air is good for you.

Toward the end of this extensive paper, Priestley described an experiment in which he isolated oxygen from potassium nitrate and noticed that in it "a candle burned in it just as in common air. In one quantity... a candle not only burned, but the flame was increased.... This series of facts, relating to air extracted from nitre [potassium nitrate], appear to me to be very extraordinary and important, and, in able hands, may lead to considerable discoveries." And they did!

III. The Discovery of Oxygen and Its Significance

Joseph Priestley (1775b) wrote: "on the 1st of August, 1774, I endeavoured to extract air from *mercurius calcinatus per se* [HgO]; and I presently found that, by means of this lens, air was expelled from it very readily.... But what surprized me more than I can well express, was, that a candle burned in this air with a remarkably vigorous flame, very much like that enlarged flame with which a candle burns in nitrous air, exposed to iron." In his terminology, nitrous air was nitric oxide (NO) and nitrous air exposed to iron was nitrous oxide (N_2O) or laughing gas. The reaction proceeds (Parkes, 1952):

$$2NO + H_2O + Fe \rightarrow N_2O + Fe(OH)_2 \quad \textbf{(7)}$$

Nitrous oxide, discovered by Priestley (1772), supports combustion by releasing oxygen and forming nitrogen and is somewhat soluble in water (Parkes, 1952). Thus, heating copper in nitrous oxide gives (Parkes, 1952):

$$Cu + N_2O \rightarrow CuO + N_2 \quad \textbf{(8)}$$

Priestley (1775b) continues, "I observed now at this time (Nov. 19), and which surprized me no less than the fact I had discovered before, was, that, whereas a few moments agitation in water will deprive the modified nitrous air [nitrous oxide] of its property of admitting a candle to burn in it; yet, after more than ten times as much agitation as would be sufficient to produce this alternation in the nitrous air [nitrous oxide], no sensible change was produced in this.... after two days,... I agitated it violently in water about five minutes, and found that a candle still burned in it as well as in common air.... These facts fully convinced me, that there must be a very material difference between the constitution of the air from mercurius calcinatus [mercuric oxide], and that of phlogisticated nitrous air [nitrous oxide]." Priestley suspected that he had isolated nitrous oxide on August 1, 1774, but it was not until Saturday, November 19, 1774, that he began his significant experiment. The following Monday he realized that he had indeed isolated a new gas. Therefore, we feel that November 21, 1774, is the real date that Priestley discovered oxygen. Ironically, Dr. Priestley might have made this discovery a day earlier, if he had not been occupied as a minister that Sunday.

Earlier, Priestley (1772) devised the nitrous air (nitric oxide) test for determining the purity of air for respiration. Nitric oxide is introduced into the air sample, and the new volume is determined (Conant, 1970). The reaction is:

$$2NO + O_2 \rightarrow 2NO_2 \rightleftharpoons N_2O_4 \quad \textbf{(9)}$$

The oxygen in the sample reacts with the nitric oxide, resulting in the formation of nitrogen peroxide, which is a mixture of a red gas, nitrogen dioxide (NO_2) with nitrogen tetroxide (N_2O_4). Nitrogen peroxide reacts with water (Parkes, 1952) and is removed from the system. On March 1, 1775, Priestley performed his nitric oxide test on the newly discovered gas. The next day, he introduced a candle into the unknown gas which had not reacted with the nitric oxide, and observed that it burned "even better than in common air." On March 8, 1775, he found that a mouse could breathe the new gas (Priestley, 1775b).

Priestley submitted a letter to the Royal Society of London on March 15, 1775. This letter was read March 23, 1775, and on this date was the first public announcement of the discovery of oxygen (McKie, 1952). In this publication, he named the newly discovered gas, which we now call oxygen, "dephlogisticated air." He stated, "a quantity of this air [oxygen] required about five times as much nitrous air [nitric oxide] to saturate it, as common air requires" (Priestley, 1775a).

Carl Wilhelm Scheele (Fig. 1B), a pharmacist in Sweden, actually discovered oxygen before Priestley. Sometime between 1770 and 1773, he called this new gas in a manuscript "vitriol air." He changed the name to "fire air" in 1775. He described experiments with oxygen in his *On Air and Fire*, which he sent to the printer in December, 1775. There was a delay in printing, and the book was not published until July 13 and August 22, 1777. Forster in his translation of this work used "empyreal air" instead of "fire air" for oxygen (Partington, 1962). Scheele states, in Section 29: "Since this air is necessarily required for the origination of fire, and makes up about the third part of our common air, I shall call it after this, for the sake of shortness, Fire Air" (Scheele, 1777). He noted that fire air supports respiration as well as combustion.

On September 30, 1774, Scheele wrote Antoine Lavoisier (Fig. 1C) a thank you note for a book which he sent. At the end of this letter, Scheele writes, "Since I have not a large burning-glass, I beg of you to make an experiment with yours in the following manner.... reduce it [silver carbonate] by means of the burning-glass in your apparatus,... a little quick lime [calcium oxide] should be put into the water in which the bell-glass has been placed, in order that... fixed air [carbon dioxide] may unite rapidly with the lime. It is by this means that I hope you will see how much air is produced during this reduction and whether a lighted candle can carry on its flame, and animals live in it" (Scheele, 1774). Thus, Scheele suggested to Lavoisier a way of obtaining oxygen by heating silver carbonate. Silver oxide is obtained upon heating silver carbonate, and if the temperature is about 300°C, then the silver oxide is decomposed into silver and oxygen (Parkes, 1952):

$$Ag_2CO_3 \rightarrow Ag_2O + CO_2 \qquad \textbf{(10)}$$

$$2Ag_2O \rightarrow 4Ag + O_2 \qquad \textbf{(11)}$$

In his *Chemical Treatise on Air and Fire*, Scheele writes in Section 38: "I then placed this calx of silver [it was not a calx, but silver carbonate instead] in a small glass retort on the open fire for reduction, and fastened an empty bladder to the neck.... [carbon dioxide] was necessarily present also in the bladder. This acid was removed from it by milk of lime [calcium hydroxide] and there remained behind one half of pure fire air [oxygen]" (Scheele, 1777). Scheele did not tell Lavoisier of his discovery of oxygen in his letter (Scheele, 1774); there is evidence that Lavoisier received this letter (Ihde, 1980). Scheele later became a member of the Swedish Royal Academy of Sciences in 1775 and declined a position offered by Frederick the Great in 1777; he died in 1786 (Partington, 1962).

The next month, October 1774, Priestley visited Lavoisier in Paris (Partington, 1962). Thus, about the same time, Lavoisier heard about oxygen in a vague way from both independent codiscoverers of oxygen. In addition, Lavoisier had been part of the commission that heated mercuric oxide and released oxygen during the autumn of 1774 (see above).

Lavoisier read to the French Academy of Science on April 26, 1775, and also on August 8, 1778, his famous papers on calcination (oxidation) of metals. Both papers have the publication date of 1775. The complete texts of the two French papers, given years apart with the same title, were translated into English by Conant (1970). In the earlier version, Lavoisier first collected carbon dioxide by heating mercuric oxide in the presence of charcoal, and then collected oxygen by heating mercuric oxide by itself. The weight of the metal was less than that of the oxide. In addition, the oxygen which he collected could transform the metal back into an oxide. When he applied the nitric oxide test to his sample of gas, he could not distinguish between common air and his oxygen gas. This was due to the amount of nitric oxide chosen, since practically all the oxygen in his common air sample reacted with the nitric oxide. Thus, the minimum amount of oxygen in the gas sample, as determined by this test, was found to be the same as the amount in common air by Lavoisier. In the first version, he did not realize this, and implied that the gas sample was the same as common air; however, he does point out that this gas is more combustible and more pure than common air (Conant, 1970). This implies that Lavoisier thought he had isolated a common air which was purer than commonly observed, and not that he had isolated a new gas. Between the first and second versions, he recognized that indeed he had isolated a new gas only after he had read about Priestley's discovery (Conant, 1970; Ihde, 1980). In the second version, without mentioning Priestley, he refers to this gas as the "healthiest and purest part of the air"; he names this new gas "air éminement respirable", or eminently respirable air (Lavoisier, 1775).

Lavoisier (1777a) removed the oxygen from the air and found that this deoxygenated air could not support respiration. Then he added oxygen back to this air and made common air. He attacked Stahl's theory of phlogiston (Lavoisier, 1777b). He knew that oxides were heavier than their respective metals, since they contained oxygen, which he now called "air pur" or pure air. He states: "The existence of fire matter, of phlogiston, in metals, sulfur, etc., in fact is only a hypothesis, a supposition; it is true that once this is granted, it explains some of the phenomena of calcination [oxidation] and of combustion. But, if I show that these same phenomena can be explained by the opposite hypothesis in an equally natural way, that is without supposing that fire matter or phlogiston exist in the material called *combustible*, then the system of Stahl will be shaken to its very foundations." Lavoisier then concludes that "pure, dephlogisticated air [oxygen] of Mr. Priestley is therefore, in this opinion, the true combustible material."

This report was followed by other papers of Lavoisier which attacked the phlogiston theory, until hardly anyone believed in phlogiston. Lavoisier realized the full significance of the fact that metals gain in weight when heated. Although he did not discover oxygen, he did discover that the process of combustion involved the addition of oxygen to the metal. In other words, he discovered that combustion is an oxidation process. We can note that later in the eighteenth century, hot sulfur was found to be another oxidizing agent (Ihde, 1980).

Both Priestley and Scheele, the independent codiscoverers of oxygen, however, never gave up on the phlogiston theory (Weeks, 1956). In fact, Priestley reduced metallic oxides to metals by hydrogen and at first his conclusion was that hydrogen was phlogiston; later when he found that water was also produced, his revised conclusion was that hydrogen was water plus phlogiston (Toulmin, 1957). Macquer (1777a) believed that phlogiston was light, whereas Cavendish (1766) speculated that "Phlogiston flies off... and forms the inflammable air [hydrogen]." Earlier, phlogiston was supposed to be sulfur, the principle of combustion, but now, according to Cavendish, it was hydrogen, the inflammable air.

The phlogiston theory had served its purpose, and now it was time to lay it to rest. Interestingly, Davy in 1808 attempted to devise a theory in which both phlogiston and oxygen participated (Ihde, 1980). The major difficulty in the phlogiston theory, even as formulated by Stahl, was that it implied a negative weight for phlogiston. This theory belongs to a class of dualistic theories, in which there exists a balance between two principles or states. It is possible to explain a change of state either by an addition of one principle or by a subtraction of the opposing principle. Thus, excluding the implication of a negative weight for phlogiston, the process of metal oxidation can be explained by either a loss of phlogiston or a gain of oxygen. Unfortunately, the loss of phlogiston theory was wrong. Lavoisier showed that the oxidation he studied occurred through an oxygen effect. If hydrogen is substituted for phlogiston, then the phlogiston theory is essentially a dehydrogenation theory.

The electrical theory of Franklin (1750) is another example of such a dualistic theory in which the wrong alternative was chosen; he chose to call the electricity carrier a positive, instead of a negative, charge. (Cations, with a positive charge, were found later to carry electricity, as well.) The nomenclature of Franklin still persists in defining electrical current flow, while the nomenclature of Stahl has disappeared.

Now, who was the discoverer of oxygen? Kuhn (1962) has pointed out that discovery is usually a complex series of events, instead of a unitary occurrence. By the discoverer of oxygen, we mean here the principal person involved in its discovery. Bayliss (1915) considered Mayow as its discoverer, but his ideas on combustion were wrong. Scheele was the first one to isolate and characterize oxygen. Although he did not submit his oxygen research for publication until after Priestley announced his discovery, he certainly has a claim to being called the discoverer.

Priestley also acted independently; even though he did his research some years later than Scheele, he did publish it first. It should be added that although Priestley in 1772 first reported on the isolation of oxygen and on the observation that a flame was increased in it; he didn't realize that he had isolated a new gas. Therefore, both Priestley and Scheele should be considered as the independent codiscoverers of oxygen. Lavoisier only recognized that he had isolated oxygen after he had read about Priestley's discovery. His contribution was to realize the chemical significance of the action of oxygen, which Scheele and Priestley never did. At the 1974 Bicentennial of the Discovery of Oxygen meeting, Ihde (1980) called Priestley "the qualitative scientist *par excellence*," whereas Lavoisier had more of a "quantitative bent." In spite of this, Lavoisier still had some erroneous ideas concerning oxygen. First, he believed that oxygen was an acidifying principle (Kuhn, 1962; Partington, 1962; Le Grand, 1972). Second, he believed that oxygen gas was formed when the element of oxygen combined with caloric, the matter of heat or fire (Kuhn, 1962; Partington, 1962). I do not consider him to be the discoverer, although Hemmeter (1921), McKie (1952), Singer (1962), and Foster (1901) take a more positive view of Lavoisier.

IV. Oxygen Terminology

In a manuscript received September 5, 1777, but read November 23, 1779, Lavoisier (1778) changed the name of dephlogisticated air or eminently respirable air to "the name of *principe acidifiant*, or, if the same meaning is preferred in the guise of a Greek word, the name of *principe oxygine*." He believed that all acids contain oxygen. To him, this mistaken belief was more important than the role of oxygen in combustion (Le Grand, 1972). Lavoisier (1789) in his *Traité Elémentaire de Chimie* mentions that "*air vital*" was also used for oxygen. Condorcet thought of this term, according to Lavoisier in 1782 (Partington, 1965). Lavoisier used this term after he coined the word "oxygine." Thus, he certainly thought that a most important role of oxygen is for respiration.

In his *Traité*, Lavoisier (1789) gives the derivation for oxygen from the Greek: "We gave to the basic breathable part of air the name of oxygen [spelled oxygène], deriving it from two Greek words [transliterated from the Greek according to the rules given by Jaeger (1944)]: oxys, *acid* [acide],[2] ginomae, *I produce* [j'engendre]." However, "ginomae" actually means "I am born (Jones and McKenzie, 1940), or "je nais" in French.

[2]The material in square brackets here represents the original French words used by Lavoisier.

Partington (1962) notes that "gennao" should have been used instead of "ginomae." The original spelling, "oxygine," was probably from the Greek form "ginomae." "Oxygine" was then changed to "oxygène," possibly due to the fact that the suffix "gine" does not occur in French for Greek derivatives, whereas the suffix "gène" does occur. The spelling was changed to "oxigène" in 1787, and reverted to "oxygène" in 1835 (Oxford University Press, 1971).

The introduction of the terminology of the element into the various languages generally followed two different paths (Table 1). The Romance languages, part of the Indo-European family of languages (Bodmer, 1944), and comprising French, Spanish, Italian, Romanian, and Portuguese, all adopted the Greek root for acid, as put forward by Lavoisier. In the Teutonic and Slavonic languages, also part of the Indo-European family, the root for acid was generally taken from their own linguistic system. For example, "Sauerstoff" is the word for oxygen in German; "Saüre" means acid and "Stoff" means matter. This method of deriving the word for oxygen is found in the Teutonic languages German, Dutch, Swedish, and Norwegian, as well as in the Slavonic languages Russian, Czech, Slovene, and Serbo-Croatian. The English language, although Teutonic, does not follow this rule because the English were greatly influenced by the French chemical revolution. For Danish, another Teutonic language, and Polish, a Slavonic language, the words for oxygen are unrelated to their words for acid (see Table 1).

Latin, the ancestor of Romance languages, in its modern version, and modern Greek, another Indo-European language, both follow the French system. Hungarian, part of the Finno-Ugrian family of languages, and not part of the Aryan or Indo-European family, follows the French system; yet Finnish in the same family follows the German system. Two languages in the Semitic family split also; Arabic has the French system, while Hebrew has the German system. In Asian languages, Vietnamese follows the French system, and Indonesian and Japanese adopted the German system. The Chinese did not derive their word for oxygen from acid.

Table 1. *Roots for Oxygen in Various Languages*[a]

Greek root for acid used		Root in the language for acid used			Root unrelated to acid used		
Language	Oxygen	Language	Oxygen	Acid	Language	Oxygen	Acid
French	oxygène	German	Sauerstoff	Säure	Danish	Ilt	Syre
Spanish	oxígeno	Dutch	zuurstof	zuur	Polish	tlen	kwas
Italian	ossigeno	Swedish	syre	syra	Chinese[2,11]	yang[b]	suan
Romanian	oxigen	Norwegian[6]	surstoff	sur			
Portugese[1]	oxigénio	Russian	kislorod	kislota			
Latin[2]	oxygenium	Czech	kyslík	kyselina			
English	oxygen	Slovene[7]	kisik	kisel			
Greek[3]	oxygonon	Serbo-Croatian[8]	kisik	kiseo			
Hungarian	oxigén	Finnish	happi	happo			
Arabic[4]	oksizen	Hebrew[9]	hamtsan	hoomtsah			
Vietnamese[5]	oxygen	Indonesian[10]	zat asam	asam			
		Japanese	sanso	san			

[a]Unless otherwise noted, the words are taken from Reid (1970). 1, Langenscheidt (1961); 2, Commercial Press (1961); 3, Divry (1974); 4, Cowan (1971); 5, U.S. Dept. of Commerce (1966); 6, Haugen (1967); 7, Grad et al. (1967); 8, Filipović et al. (1963); 9, Ben-Yehuda and Weinstein (1961); 10, Pino and Wittermans (1955); 11, Needham et al. (1976).

[b]The ideograph for oxygen is a single character which is derived from the ideograph for gas (chhi). Previously, oxygen was represented by two characters, yang chhi, meaning nourishing gas.[11]

Mellor (1922) suggested that "oxys," the Greek word for acid or sour, is derived from the Greek word for vinegar, which is "oxos" (Jones and McKenzie, 1940). "Oxycrat" and "oxymel" are two words derived from oxos, which have existed prior to the eighteenth century in French (Godefroy, 1888) and English (Oxford University Press, 1971). They mean, respectively, vinegar mixed with water, and vinegar mixed with honey. Thus, these two words are related to the more recent word oxygen.

In 1787, Berthollet showed that hydrogen cyanide was an acid which did not contain oxygen (Partington, 1962). However, Lavoisier's acidifying principle was too well entrenched by this French father of modern chemistry to be shaken at this time. Some twenty years later in 1810, Davy pointed out that hydrochloric acid did not contain oxygen. Davy had destroyed Lavoisier's acid theory, just like Lavoisier had destroyed Stahl's phlogiston theory (Partington, 1964). Later in 1838 Liebig showed that hydrogen in acids can be replaced by metals (Partington, 1964; Skinner, 1970). Therefore, as Skinner points out, "Hydrogen therefore should be called oxygen, as it more truly deserves the title of acid producer."

In three languages, or just 10% of the 26 languages listed in Table 1, the root for oxygen seems derived from appropriate terminology. The Danish word for oxygen, "Ilt," is derived from the Danish word "Ild" meaning fire (Bodmer, 1944). The Polish, "tlen," is derived from "tlić sie," the Polish verb meaning to smother (Stanislawski, 1968). The Chinese word for oxygen originated from the ideographs meaning nourishing gas (Needham et al., 1976).

However, Haldane (1947) did report that 6 bar of oxygen had a sweet and sour (acid) taste. Certainly, the use of the vital air nomenclature is more appropriate than our present term, oxygen. This vital air terminology did persist in English (Riadore, 1845), in French (Demarquay and Leconte, 1864), and in German (Gmelin, 1799) where the term was "Lebensluft" (De Vries, 1959).

V. Early Biological and Medical Research with Oxygen

If the atmospheric air were perfectly pure, the life of animals breathing it, would be much more energetic, better, and more pleasant in many ways; but at the same time it might be proportionately shortened, and being rapidly consumed by such active air, they might live only one quarter of the time that they live in the ordinary air of our atmosphere, impure though it may be.

Macquer (1777a)

Man began to breathe higher than normal partial pressures of oxygen when he began to use diving bells in which the air was compressed. Aristotle (1953) wrote about divers who can "respire equally well by letting down a cauldron [used as a diving bell]; for this does not fill with water, but retains the air, for it is forced down straight into the water." Alexander the Great was reported to have one by two authors, Neckam (1157–1217) and Roger Bacon (1214–1292) (Partington, 1961). In 1531, de Lorena designed one which was used in Lake Nemi, adjacent to Rome, and in 1538, two Greeks designed one which was used in Toledo, Spain. In the seventeenth century, a German used one in 1616; one was used at Cadaqués, Spain, in 1677; and about 1689 two were independently designed, one by Papin, a French physicist, and the other by Halley, the English astronomer (Marx, 1971). Drebbel reportedly used a submarine which had "air within being freshened by a subtle spirit [oxygen?] which he had extracted from the atmosphere" (Thorndike, 1958). According to Kerr (1779), a physician (Partington, 1962), the compressed air pressure inside the diving bells sometimes reached nine bar.

The modern therapeutic use of gases began with the utilization of carbon dioxide. In 1775, Black had shown that carbon dioxide was a constituent of carbonated alkalis (Black, 1755). Priestley prepared carbonated water and noted the similarity between this water and natural mineral water. Since many mineral waters were supposed to possess medicinal effects, he concluded that water charged with carbon dioxide also had medicinal value (Priestley, 1772). Priestley had so much influence that he persuaded the British Lords of the Admiralty in 1772 to equip Captain Cook's ships with devices to produce carbonated water (Florkin, 1977). However, his house in Birmingham, England, was attacked in 1791 due to enemies who did not like his liberal views (Partington, 1962).

Fire and life have intrigued man for over 2500 years. In the second half of the eighteenth century, it is not suprising that the two independent discoverers, Priestley and Scheele, performed experiments with this newly isolated oxygen on both combustion and respiration. Priestley (1775b) compared these two processes, stating: "[Oxygen] might not be so proper for us in the usual healthy state of the body: for, as a candle burns out much faster in dephlogisticated [oxygenated] than in common air, so we might, as may be said, *live out too fast*, and the animal powers be too soon exhausted in this pure kind of air. A moralist, at least, may say, that the air which nature has provided for us is as good as we deserve."

Lavoisier and his collaborators have also pointed out the similarities between combustion and respiration. In respiration, they found that oxygen was taken up and carbon dioxide and water were given off. Seguin and Lavoisier (1789) wrote: "respiration is but a slow combustion of carbon and hydrogen, that it is exactly like the combustion functioning in a lit candle,... animals that breathe are truly combustible bodies that burn and are consumed." However, they believed that "blood is the combustible material." They also proclaimed that: "This fire from the heavens, this Promethean torch, is not only the embodiment of an ingeniously poetic idea, but the very faithful portrait of nature's operations, at least for the breathing animals: we can therefore say, along with the ancients, that the torch of life is lit at the moment the infant first breathes, and it is extinguished only at the moment of his death." In this article, both terminologies, oxygen and vital air, are used.

Priestley became the principal advocate for suggesting the therapeutic use of artificially obtained gases (Duncum, 1947), the "factitious airs" of Boyle (McKenzie, 1960). When he discovered oxygen, he wrote: "From the greater strength and vivacity of the flame of a candle, in this pure air [oxygen], it may be conjectured, that it might be peculiarly salutary to the lungs in certain morbid cases, when the common air would not be sufficient to carry off the phlogistic putrid effluvium fast enough.... perhaps, we may also infer from these experiments, that... pure dephlogisticated air [oxygen] might be very useful as a *medicine*." (Priestley, 1775b).

A couple of years later, Macquer (1777a) (Fig. 2A) proposed oxygen therapy for the treatment of "asphyxias by suffocation" in the second edition of his classic *Dictionnaire de Chymie*, which first appeared without an index; it later appeared with an index in 1778 (Partington, 1962). Interestingly, the biological and medical dissussions of oxygen appeared at the end of the article, "Volatilité" in Volume 3 in his treatment of asphyxias. He states that the discussions are a supplement to his main one on oxygen appearing in Volume 2 under the article "Gas ou Air Dephlogistiqué." Perhaps Macquer's discussion was inserted as an afterthought.

Another advocate for the therapeutic use of factitious gases was a physician, Ingenhousz (Fig. 2B). His great contribution was his demonstration that oxygen is given off by plants exposed to sunlight (Ingen-Housz, 1779). In his book on photosynthesis, he states, "we touch at the happy moment, at which a very easy and little expensive method of producing this beneficial fluid [oxygen] in any quantity wanted, will be produced for the cure of many diseases." The title of the book refers to "purifying the common air in the sun-shine," for his main interest was the health profession. He recommended the use of at least 1000 to 1200 in^3 (16,000 to 20,000 cm^3) of oxygen per day as a therapeutic agent (Soc. Phil. Exp. Batavia, 1782).

A

B

C

D

E

Figure 2. Some of the early proponents of oxygen therapy. **A.** Pierre Joseph Macquer (1718–1784), a chemist who proposed oxygen therapy for asphyxias caused by suffocation in 1777. **B.** Jan Ingenhousz (1730–1799), a physician who recommended oxygen therapy and who observed that oxygen is liberated by plants exposed to sunlight. **C.** Thomas Beddoes (1760–1808), the leading eighteenth century advocate of the therapeutic use of gases including oxygen in England. **D.** Antoine François de Fourcroy (1755–1809), the leading eighteenth century advocate of the medical use of oxygen in France. **E.** Robert John Thornton (1768?–1837), a physician and botanist who reported more than anyone else on the use of oxygen therapy. (Fourcroy photograph courtesy of Library of Congress, Washington, D.C.; Macquer, Ingenhousz, Beddoes, and Thornton photographs courtesy of History of Medicine Division, National Library of Medicine, Bethesda, Maryland.)

Fothergill, another early proponent of oxygen therapy, suggested its use in lung diseases, according to Reid (1782).

As soon as oxygen was discovered, experimenters began to breath this gas. Both Scheele and Priestley experienced breathing pure oxygen. Scheele wrote, in Section 90 of his Treatise (1777): "I began to respire air [oxygen] from this bladder. This proceeded very well, and I was able to make forty inspirations before it became difficult for me." Priestley (1775b) described his experience as "The feeling of it [oxygen] to my lungs was not sensibly different from that of common air; but I fancied that my breast felt peculiarly light and easy for some time afterwards. Who can tell but that, in time, this pure air may become a fashionable article in luxury. Hitherto only two mice and myself have had the privilege of breathing it."

In 1782, it was reported that "Mr. Ingenhousz tried upon himself the medicinal properties of this [oxygen]. After having breathed a certain amount, he felt more gay, more robust, and with a better appetite: his sleep was more gentle and more refreshing than ordinary" (Soc. Phil. Exp. Batavia, 1782). Beddoes (Fig. 2C) the leading exponent of factitious air therapy, in England, also breathed oxygen. "To my own lungs, it [oxygen] feels like ardent spirits applied to the palate" (Beddoes, 1794). Sir Humphrey Davy, when he was in the employ of Beddoes, stated: "In respiring eight or ten quarts [of oxygen], for the first two or three minutes I could perceive no effects" (Davy, 1800).

The first report of oxygen toxicity was given by Scheele. He mentioned in Section 92 of this Treatise that, "Plants [peas], however, will not grow noticeably in pure fire air [oxygen]" (Scheele, 1777). This report on oxygen toxicity in peas was noted by Ingenhousz (1779). More recent observations show that oxygen toxicity occurs in peas exposed to oxygen pressures of 1 bar (Galston and Siegel, 1954) and of 5 bar (Turner and Quartley, 1956). Leonard and Pinckard (1946) studied the influence of oxygen and carbon dioxide on cotton plants and found that a lack of carbon dioxide had hardly any effect whereas 1 bar oxygen inhibited the plants. Hence, it appears that the effects observed by Scheele were not due to a lack of carbon dioxide, but rather to an excess of oxygen.

In one experiment, Priestley exposed two mint plants for a month, one to oxygen and the other to common air. He "found the plant in the dephlogisticated air [oxygen] quite dead and black, and the other partially so, but the uppermost leaves were still alive" (Priestley 1777). Priestley (1781) in Section 15 writes, "I found that mice would not live in dephlogisticated air [oxygen] till they had completely phlogisticated [reduced] it, though they lived longer in it than, in proportion to its purity, with respect to common air." He described an experiment in which the exhaled carbon dioxide was absorbed and obtained the same result. Puzzled by this result, Priestley continued in Section 33, and repeated this observation with a mouse. The measured oxygen concentration was considerably more than 67% when the mouse died. He "then put another young mouse into the remainder of the air, and it also continued at its ease two or three hours; but then seemed to be expiring,... But bringing it near the fire,... it lived several hours longer, and when it died the air was as completely phlogisticated [deoxygenated] as common air is generally found to be when mice have died in it. This experiment fully satisfied me, that it was nothing in the dephlogisticated air [oxygen] itself that was the reason that mice could not live in it." Mice can survive in continuously flowing pure oxygen for some days. The average survival time of 347 female mice was 111.3 ± 1.4 hours (Gerschman et al., 1958).

Lavoisier (1782–1783) read a paper on February 15, 1785, in which he extended Priestley's study on oxygen toxicity. He stated: "In the experiment performed on the guinea-pig enclosed in vital air [oxygen], which I have just reported upon, I noticed that this animal suffered considerably at the end of the experiment. However, it was seen that only a very small part of the air was vitiated [spoiled], that is to say converted into fixed air [carbon dioxide], and that there still remained more vital air [oxygen] than would be required to constitute healthy air. This circumstance was already observed by Mr. Priestley." Lavoisier, along with Bucquet, autopsied the guinea-pigs, after they died in oxygen, and found that all "of them appeared to have died of a burning fever and of an inflammatory sickness. Upon inspection, the flesh was very red; the heart was livid and full with blood, especially in the ventricle and the right auricle. The lung was very flaccid, but very red, even outside, and much gorged with blood. Healthy air is therefore composed of a good proportion between vital air [oxygen] and atmospheric moffete [nitrogen],... when there is an excess of vital air [oxygen], the animal only undergoes a severe illness; when it is lacking, death is almost instantaneous."

Lavoisier also believed that carbon and hydrogen react with oxygen in the lung to produce caloric, i.e., fire matter or heat, carbon dioxide, and water. The source of carbon and hydrogen is the ingested food. Inflammation is augmented by a decrease of carbon and hydrogen in the blood (Seguin and Lavoisier, 1789). Thus, according to Lavoisier, either an increase of oxygen or a decrease of hydrogen produces an inflammatory condition. On the other hand, let us recall that during the previous century, compressed air, with an increase in oxygen, was advocated for treatment of inflammations (Henshaw, 1664). Since, according to Seguin and Lavoisier, respiration controls oxygen and digestion controls carbon and hydrogen, both systems must be in the proper balance to maintain a healthy state.

Seguin and Lavoisier (1789) emphasized the importance of removing carbon dioxide. They report guinea pigs surviving days exposed to 1 bar of oxygen, providing alkali was present to absorb the carbon dioxide. Some other investigators were not so careful. Above 0.02 bar, carbon dioxide begins to have some effect at normal oxygen tensions (Graham, 1962).

Beddoes (1794) in commenting on oxygen toxicity, writes: "Dr. Priestley and Mr. Lavoisier found animals either to die, or to become exceedingly ill in such air, while it continues more oxygenated than the atmosphere.... The heart and arteries pulsate more quickly and forcibly; the eyes grow red and seem to protrude; the heat of the body is said considerably to increase, sweat to break out over the whole body, and fatal mortification of the lungs to come on.... The existence of inflammation is fully established by dissection." Beddoes then describes the autopsy of a kitten exposed to oxygen, and found results similar to Lavoisier's description.

Another example of an early medical view of the effects of oxygen is given in the following quotation: "When breathed in a larger proportion that that in which it naturally occurs in the atmosphere, it raises the spirits, promotes the circulation, increases the heat of the body, and heightens the colour of the blood. If it be breathed alone by healthy animals, in large quantities, and for any length of time, its stimulant effects are carried to excess, and febrile and inflammatory symptoms ensue" (Pearson, 1795).

On the other hand, oxygen was seen as beneficial, both in nature and in its use in a wide range of therapeutic applications. Priestley (1781) demonstrated that "air contained in water... is as necessary to the life of fishes, as air... is to that of land animals." The Count de Morrozzo noted in 1784 that birds were more lively in oxygen (Demarquay, 1866).

Caillens (1783) reported oxygen was successful for the treatment of phthisis (consumption or tuberculosis). He wrote, "Besides, the patients always breathe this air with great pleasure, and if they are in a totally despairing state, it prolongs their life;... They would desire to breathe it continually, so much good do they derive from it, although it must not be breathed more than occasionally." Chaussier (1780–1781) reported that in 1783 he confirmed Caillens' reported successful treatment of tuberculosis. Chaussier also recommended "that physicians have on hand dephlogisticated air [oxygen] for every time following a difficult childbirth an infant is born without giving signs of life" and that it be given as "first aid to victims of drowning and asphyxiation." Poulle (1785) thought it was good for certain asthmatics, suffocation cases, and elderly men. Jurine (1789) of Geneva also reported on the use of oxygen in the treatment of tuberculosis.

The last decade of the eighteenth century saw the rise of the treatment of all kinds of diseases with factitious gases (Duncum, 1947; Cartwright, 1952). In France, Chaptal (1790) wrote that it was effective in a case of humid asthma. Chaptal also warned of mercury contamination in oxygen, when the gas was produced from heating mercuric oxide. Fourcroy (1790) (Fig. 2D) used oxygen therapy for the treatment of tuberculosis and found it to be detrimental; he reasoned that since oxygen generates heat, it should not be used in any inflammatory disease, such as tuberculosis. However, he believed where generation of heat is desirable in an illness, then oxygen therapy is indicated. Fourcroy was a leading advocate in France for oxygen therapy (Demarquay, 1866). However, the French Revolution was now in progress, and Fourcroy was occupied with politics (Smeaton, 1962). Hence, there was little concern with oxygen therapy in France for this decade. In Germany, Girtanner, as well as Stoll and Ferro, advocated oxygen therapy (Demarquay, 1866) during this period. In a review, Gmelin (1799) pointed out that inflammatory conditions such as fevers and burns can be aggravated by an excess of oxygen. In addition, he believed that an elevated oxygen was responsible for some fevers, such as yellow fever.

Beddoes was the principal figure in the therapeutic use of factitious airs, not only in his native England, but also in other countries as well (Demarquay, 1866; Cartwright, 1952). Oxygen, carbon dioxide, hydrogen, nitrogen, and carbon monoxide, as well as ether, were investigated for their therapeutic potential during the last decade of the eighteenth century in England (Cartwright, 1952). Beddoes (1793a) expected "essential benefit from oxygene [sic] air in a considerable variety of diseases." He thought that scurvy was caused by "want of air sufficiently furnished with oxygene [sic]" (Beddoes, 1793b). Barr, Thornton, and Carmichael initiated oxygen therapy about this time (Cartwright, 1952). However, Beddoes (1793b) did not advocate the use of oxygen in tuberculosis.

James Watt, the famous engineer, designed an apparatus for producing factitious gases (Cartwright, 1952). Watt (1794) wrote a description of this apparatus and it was published together with a contribution by Beddoes (1794). Their joint publication was so popular that two years later, a third edition was published (Beddoes and Watt, 1796a). Many letters were included from others, proclaiming the effective use of the various gases in the treatment of a variety of diseases. Still more letters made up Part III (Beddoes and Watt, 1796b). The last joint publication of Beddoes and Watt (1796c) also includes many more letters; a supplement to the description of Watt's apparatus; and some comments made by critics, one of which is "*The whole* SEEMS *a flimsy fiction*." There was growing disillusionment about the efficacy of these factitious airs in the treatment of diseases. Beddoes (1797) exhibited some scepticism when he wrote, "I have now no chemical theory of any one disease.... I started conjectures to be compared with facts; and now I think all these conjectures are shewn to be erroneous by facts." In spite of this attitude, Beddoes opened up the Pneumatic Insititution at Bristol, England, in 1799, and hired Humphry Davy as its superintendent (Cartwright, 1952).

Davy became interested in the effects of oxygen and wrote "I hope to be able to ascertain why pure oxygene [sic] is incapable of supporting life" (Davy, 1800). Davy believed that oxygen gas was combined with light, and for a while used the term "phosoxygen" to represent this combination (Partington, 1962). During his two year employ at the Pneumatic Institution, Davy did study the effects of nitrous oxide, the "laughing gas" (Cartwright, 1952). Beddoes (1799) believed that "*the newly tried gas* [nitrous oxide] *may be regarded as a more powerful form of oxygen gas.*"

Like other scientists of his day, Beddoes

was intrigued by the similarity of respiration and fire. In his "Conjectures on Explosive Composition" (in Stock, 1811), he wrote about Greek fire and gun powder. He concluded that Greek fire contained potassium nitrate.

In a book written for clinicians of 1801, it is stated that currently, there is "a *revival* of chemistry, in consequence of the late discoveries in Pneumatic chemistry, the facts of which are now introduced as forming the basis of Medical theory, and of course much influencing its practice" (Nisbet, 1801).

Erasmus Darwin (1796) obtained mixed results with oxygen therapy on his patients. He noted, "The necessity of perpetual respiration shews, that the oxygen of the atmosphere supplies the source of the spirit of animation." An ardent supporter of pneumatic medicine, Hill (1800) cured diseases with oxygen therapy. He breathed oxygen for several weeks and noticed an improvement in his health.

Trotter (1792) noted that scurvy can be cured by fresh vegetables and especially those containing an acid, such as citric acid. He thought (Trotter, 1795): "scurvy is produced, 'from a deficiency of recent vegetable matter *alone*,' and that the oxygen is by this means abstracted from the body."

Rollo (1797) believed that respiration controls oxygen, and food controls hydrogen. He stated: "in proportion to the quantity of food received into the stomach abounding with hydrogene [sic], the system covets oxygene [sic], taking up a greater quantity of it by respiration from the atmospheric air. Therefore, when food of an opposite quality is taken... the system will covet hydrogene [sic], or at least, it will not solicit oxygene [sic]... in order to preserve the healthful oxygenated balance. Thus, under certain circumstances of predisposition these foods... must produce the two opposite states of system [system of oxidation states], and lay the foundation of two different morbid affections." Earlier, Seguin and Lavoisier (1789), as mentioned before, expressed similar ideas.

Rollo thought that diabetes melitus and tuberculosis were caused by the body being in a hyperoxygenated state, whereas scurvy was caused by the body being in a deoxygenated state. So treatment for hyperoxygenated disease states consists of conditions which remove oxygen. Such conditions are confinement for the patient and giving animal food which contain hydrogen. Hepatified ammonia and kali sulphuratum were also recommended. On the other hand, treatment for deoxygenated disease states consists of conditions which release oxygen. Such conditions are exercise for the patient and giving vegetables, with citric acid and mercuric oxide also recommended; acids were thought to contain oxygen then. Today, we know that citric fruits contain ascorbic acid, an antioxidant (Forman and Fisher, 1981). Thus, Rollo deemphasized the role of inhalation therapy.

Townsend (1802) also believed in a balance between respiration and digestion. Vinegar was a good source of oxygen. He believed that "animal food, and all the articles of diet, which abound with hydrogen, evidently contribute to increase the vital heat." Therefore, "in the inflammatory fever,... we have little inclination for animal food."

Lavoisier and Fourcroy, although scientists, also engaged in politics during this turbulent decade in France. Lavoisier was a tax collector and Commissioner for Gunpowder. Marat, a journal editor, wrote in 1780 a book about fire, in which it was claimed that the element of fire was made visible. Since Lavoisier did not seem to be sympathetic to his book, Marat was upset; in 1791, Marat attacked Lavoisier ferociously in his journal. Finally, the Reign of Terror under the leadership of Robespierre claimed the life of Lavoisier on Thursday, May 8, 1794 (McKie, 1952). It is of interest to note that Priestley on April 8, 1794, left his native England for a two month sea voyage to New York. He settled in Northumberland, Pennsylvania, where he died in 1804. His mansion in Northumberland is now open to the public (Willeford, 1979).

The Reign of Terror lasted for only a

couple of months more and ended when Robespierre was executed on July 28, 1794 (McKie, 1952). Although Fourcroy had been criticized by Marat (McKie, 1952), Marat supported him for public office (Smeaton, 1962) in 1792. When Marat was assassinated on July 13, 1793. Fourcroy filled his seat in the National Convention. From 1793 to 1797 he was either on the Committee of Public Instruction or the Committee of Public Safety. Hence, Fourcroy did not have time to devote himself to his main field, chemistry, until 1797 (Smeaton, 1962).

At the time, Fourcroy extended his ideas concerning oxygen therapy. He (1798b) enthusiastically supported the views of Rollo (1797) in a French translation. In a lengthy article, Fourcroy (1798a) advocated the use of chemicals instead of gases as therapeutic agents. When more oxygen is indicated, metallic oxides can be given; when less oxygen is indicated, powdered iron can be administered. The rationale for this treatment was that iron can remove oxygen by combining with it.

In 1795, he began a book on oxygen, *Oxigènologie*, which was never completed (Smeaton, 1962). It would have been the first comprehensive book on oxygen.

VI. Measurements of Atmospheric Oxygen and High Altitude Effects

Priestley (1772) devised the nitric oxide test as previously described for determining the fitness of common air for respiration. Two volumes of nitric oxide react for each volume of oxygen present in the original gas sample, and so a total of three volumes of gas are removed from the system. If the ratio of added nitric oxide to the initial gas volume is designated by N, and if the ratio of the final gas volume to the initial gas volume by G, it can be shown that the fraction of oxygen in the initial dry gas volume is $(N + 1 - G)/3$. This calculation is valid only when sufficient nitric oxide is added to react with all the oxygen. Other factors such as temperature and purity of chemicals are also important. Priestley noticed that when the ratio N was equal to or greater than 0.5 in testing common air, then $(N - G)$ was constant. Thus, if this condition is met, the calculation is valid. Priestley (1772) wrote, "For if one measure of nitrous air [nitric oxide] be put to two measures of common air, in a few minutes... there will want about one ninth of the original two measures." Thus, the value of N equaled $\frac{1}{2}$ and G equaled $\frac{8}{9}$; the first determination of the fraction of oxygen in dry common air ever made was $\frac{11}{54}$ or 0.204. The modern dry value is 0.2095 (Gilbert, 1981a). Priestley (1775b) found with later experiments that when N was set to 0.5, then G was 0.8, which corresponds to an oxygen fraction of 0.23.

Priestley (1779) did not find any measurable difference in the oxygen concentration of air at different times and places. He cautioned about the dangers of phlogisticated matter produced by people in crowded dining rooms. Priestley (1775b) theorized about atmospheric composition with these words: "Whether the air of the atmosphere was, in remote times, or will be in future time, better or worse than it is at present, is a curious speculation; but I have no theory to enable me to throw any light upon it."

Scheele (1779) determined the fraction of oxygen of dry air from January 1 to December 31, 1778, and found it to be constant and equaled $\frac{9}{33}$ or 0.27. The measurements of Ingenhousz (1780) indicated that the oxygen pressure was about 0.3 bar. Ingenhousz obtained small differences in the air samples he collected and concluded, "the air at sea and close to it is in general purer and fitter for animal life than the air on the land." Townsend (1802) reported that "Dr. Ingenhousz discovered... that the atmosphere at Vienna contains a greater proportion of vital air [oxygen] than in Holland, and to this he attributes the remarkable increase of appetite felt by strangers on their arrival at Vienna."

Lavoisier (1777a) measured how much the volume of air was diminished when mercury was oxidized to mercuric oxide. He found by this procedure that air contained 17% oxygen. Later, he revised this figure to 28% (Lavoisier, 1782–1783). An analysis of a hospital ward indicated that the oxygen was only 18.5% down at the floor level, and was 25% up at the ceiling level of the rooms. Air collected at the top of a crowded theater measured 21% oxygen; he therefore stated that society should be alerted to these health hazards. Later, Seguin and Lavoisier (1789) stated that air contains 25% oxygen.

When this paper was written, there was social unrest as France began its revolution. The authors point out that "the air belongs equally to all," and they ask, "why... does the rich man profit from an abundance which is not physically necessary for him, and which seemed to be destined for the laborer?" They conclude with their hopes that working wages will be augmented, to equalize the fortunes of men. Lavoisier (1862), in a work published posthumously, speculated that water in the atmosphere was decomposed into oxygen and hydrogen, which would combine to produce water again. Hydrogen would be in the uppermost part of the earth's atmosphere.

Cavendish (1783) noted that the purity of air values obtained by the nitric oxide test was subject to the modus operandi. During the last half of 1781, he measured the purity of air on 60 different days and "found no difference that I could be sure of." He also could find no difference between city and country air. Therefore, he suggested that common air be used as a standard in comparing purity of gas mixtures. He found that the oxygen concentration was 4.8 times greater in pure oxygen than in common air, or in other words the fraction of oxygen in common air is $\frac{1}{4.8}$ or 0.208.

Davy (1801) reported that the percentage of oxygen in the atmosphere was 21%, and that places exposed to winds have the same percentage. The measurements of Dalton (1805) using hydrogen gas instead of nitrous air indicated that there is 20% oxygen in the atmosphere. He also pointed out that the ratio of oxygen to nitrogen in the atmosphere is relatively independent of elevation.

Many others measured the purity of air (Partington, 1962). Some measurements were not the best. For example, Davidson (Hosack, 1797) found in 1796 that there was 67% oxygen in the atmosphere on Martinique.

Boussingault (1841) found no significant effect on the percentage of oxygen in the atmosphere at different altitudes. Thus, the oxygen compositions at the altitudes of 1798, 4340, and 8671 ft were, respectively, 20.77%, 20.7%, and 20.65%.

Soroche, or high altitude sickness, in the high Andes of South America, played a role in the history and customs of the region (Monge, 1948). Humboldt (1838) compared the symptoms of altitude sickness from his own personal account in the Andes with that given by Acosta (1590a). After comparing it to being seasick, Humboldt noted that "the air seems as rich in oxygen in these high regions as in inferior regions; but in this rarified air, the barometric pressure was less than half the level to which we are normally exposed in the plains, a lower quantity of oxygen is taken up by the blood at each breath."

Charles Darwin (1860) gives a lively account of it in his 1835 crossing of the Andes: "The short breathing from the rarefied atmosphere is called by the Chilenos 'puna.'... The only sensation I experienced was a slight tightness across the head and chest, like that felt on leaving a warm room and running quickly in frosty weather. There was some imagination even in this; for upon finding fossil shells on the highest ridge, I entirely forgot the puna in my delight."

In the nineteenth century, many mountaineering expeditions and balloon ascensions experienced the deleterious effects of the lack of oxygen in the rarefied atmosphere of high altitude (Bert, 1878). Payerne (1851) noted that even when the barometric pressure was reduced to 0.5 bar, the oxygen content of the air was sufficient for oxygen extraction by respiration. Jourdanet (1861) correlated "the famous mountain sickness" with a "lack of the normal quantity of oxygen in the circulation of arterial blood."

VII. Developments in the Nineteenth Century

Although the use of oxygen was almost abandoned by 1800 as a therapeutic agent due to its observed lack of effectiveness in curing disease (Cartwright, 1952), some reports on oxygen therapy still continued by the original promoters (Smith, 1870). Thornton (Fig. 2E) was one of these and he commenced using oxygen therapy on his patients in 1792 (Thornton, 1803). In the volumes edited by Beddoes and Watt (1796a,b,c) on pneumatic medicine, there were many cases presented by Thornton using oxygen therapy.

After Beddoes became disillusioned with pneumatic medicine (Cartwright, 1952), Thornton was still an extremely strong advocate for oxygen therapy, as evidenced by his many reports in the *Philosophical Magazine*. From 1799 to 1807, he published 29 reports describing 31 conditions for which oxygen therapy was used. In 1799 alone, ten appeared, and seven in 1806. There was an attempt to rationalize the results by some of the authors of pneumatic medicine studies. According to Thornton (1813), Baumé classified diseases into five classes, two of which are diseases of oxygenation and the contrasting diseases of hydrogenisation. Accordingly, not all diseases could be cured with oxygen; some diseases could be treated with other gases. Thus, hydrogen was used in the treatment of tuberculosis (Thornton, 1806b). Thornton (1803) believed in the "balance of principles" in the body, that is the balance between oxygen and hydrogen. He stated: "blood by possessing *hydrogen* becomes more attractive of the *oxygen*" (Thornton, 1806c). This idea of the proper balance probably goes back to Seguin and Lavoisier (1789). Note the similarity to Yin and Yang, the fundamental dual principles of the Chinese (Bernal, 1974) and the harmony between them (Needham, 1969). In both philosophies the dual elements or principles produce healthy or unhealthy states.

The philosophy of the oxygen and hydrogen balance in the body is essentially correct. Hydrogen is stored in reduced compounds, and antioxidant mechanisms regulate its availability (Gilbert, 1960). A proper balance of oxygen is required for the body. Energy requirements of the body are not met with too little oxygen and toxicity is evident with too much oxygen. Due to this dual nature of oxygen, an optimum oxygen concentration exists for the body (Gilbert, 1972).

There must have been a considerable amount of resistance to the therapeutic use of oxygen. For example, Thornton (1805) mentions a Mr. Morton who ridiculed the use of oxygen in the theater: "Mr. Morton, in his School of Reform, [introduces] one DR. OXYGEN, who gives his patient, by mistake, instead of a certificate of Cures, the bills of *Mortality*!" Thornton could have been called Dr. Oxygen, since he was the leading advocate of oxygen therapy. Thornton (1806a) also mentions that a Dr. Rowley calls him, "*vital air* [oxygen] *and gas mad*." The conditions reported in these papers to be treated with oxygen ranged from dyspepsia (Thornton, 1799), discharge of blood from the breasts (Thornton, 1805), laudanum or opium poisoning (Thornton, 1806a), and suspended animation (Thornton, 1806c), to asthma (Thornton, 1807). After 1807, reports from him ceased except for a single one in 1821 (Thornton, 1821).

Oxygen was probably never obtained for therapeutic use in the pure state. Besides, it was common to mix oxygen with air. Thus, Loane (1800) administered 18 quarts of 47% oxygen daily to a patient. There was only one other report on the use of oxygen in the early years of the 19th century, which appeared in the *Philosophical Magazine* (Ince, 1805). In France, a report on its use for the cure of tetanus was also reported (Sarasin, 1802). Bean (1945) pointed out that oxygen therapy slackened off after the first decade of the 19th century. Probably, this was due to the ineffectiveness of oxygen to cure the multitude of illnesses.

In addition, there were warnings about using oxygen as a therapeutic agent. Brizé-Fradin wrote in 1808 that pure oxygen does not maintain life, but wears it out instead (Bert, 1878). Around 1830, Broughton (1829, 1830) reported on the toxicity of oxygen. De Lapasse (1846) wrote that oxygen must be considered as dangerous. Demarquay and Leconte (1864) warned of using oxygen in inflammatory conditions.

Due to a cholera epidemic of 1832 in Europe, there was a renewed interest in oxygen therapy (Hahn, 1899; Demarquay, 1866; Smith, 1870) in a desperate attempt to treat this disease. It was suggested that it could be treated by oxygen inhalation (Stein, 1831). Earlier, oxygen therapy had been used in the cholera epidemics in India (1819–1821) and in Russia (1830–1831) (Hahn, 1899).

At about this time, there was an interest in compressed air therapy (Jacobson et al., 1965). Tabarié, one of the leaders of this therapy, presented a sealed packet to the French Academy of Science entitled, "On the therapeutic and healthful use of oxygen," after his publication on the therapeutic use of compressed air (Tabarié, 1840). Pereira (1836), who had great influence on the use of drugs, wrote the following about oxygen gas: "no very obvious effects result from a few inhalations; that an animal will live longer in a given volume of this gas than in the same volume of atmospheric air, but that the continued use of it causes death."

In 1842, Liebig (Fig. 3A) pointed out that carbohydrates and fat combine in the organism with the inhaled oxygen to produce heat (Rothschuh, 1973). His studies caused a resurgence in the use of oxygen. Liebig (1842) himself believed that oxygen was capable of burning up all the tissues. In starvation, the food does not remove the oxygen and "the particles of the brain begin to undergo the process of oxidation, and delirium, mania, and death close the scene; that is to say, all resistance to the oxidizing power of the atmospheric oxygen ceases." He believed that one function of the carbon and hydrogen in the food was to act like antioxidants by reacting with oxygen; thus it would prevent the oxygen from destroying the tissue. He wrote, "hardly a doubt can be entertained, that this excess of carbon alone, or of carbon and hydrogen, is expended in the production of animal heat, and serves to protect the organism from the action of the atmospheric oxygen."

Riadore (1845), an enthusiastic supporter of oxygen therapy, did not think that there were any dangers in this therapy; he wrote, "with regard to these remedies [gas and/or electrical therapy], I may state, that in no instance have they, in the various diseases in which I have employed them, proved injurious." Bouchardat (1851) used oxygen inhalation for glycosuria. On the other hand, Savory (1857) wrote that the inhalation of oxygen is worthless.

The discoveries of nitrous oxide in 1844, ether in 1846, and chloroform in 1847 as anesthetic agents had a profound effect upon the medical field (Raper, 1945). Reports on the use of oxygen in removing the anesthetic agents from the patient appeared soon after (Gardner, 1847; Jackson, 1847; Robinson, 1847; Cattell, 1847; Ozanam, 1860). Other developments with reference to oxygen included the discovery of hydrogen peroxide by Thenard in 1818 and Schönbein's discovery of ozone in 1840. Because of its odor, Schönbein named it after the Greek "ozin," meaning to smell. In 1860, Andrews and Tait showed that ozone is a modification of oxygen (Partington, 1964). In their classical studies, Regnault and Reiset (1849) showed that the ratio of the carbon dioxide output to the oxygen intake of organisms depended upon the type of food ingested, but was generally less than unity.

Birch, an ardent user of oxygen therapy, wrote in his book on it, "Oxygen gas may be 'too much' whenever the food digested and assimilated by the patient does not supply the necessary proportion of hydrogen and carbon to enter into combination with it" (Birch, 1857). Seguin and Lavoisier (1789), as mentioned earlier, had similar ideas. The review of this book in *Lancet* (Book Review, 1857) was quite critical, ending with the statement, "Speculations so vague and unreasonable as these will add little to his credit as an author or physician." Birch (1869) also thought oxygen could be added to foods. He ad-

A

B

C

Figure 3. Some of the nineteenth century scientists who emphasized the toxic effects of oxygen. **A.** Justus von Liebig (1803–1873), an influential chemist who believed that carbon and hydrogen in food act as antioxidants by removing the toxic oxygen from the organism. **B.** Louis Pasteur (1822–1895), the first one to demonstrate that the mere presence of oxygen can be lethal to certain microorganisms. **C.** Paul Bert (1833–1886), a physiologist who demonstrated that oxygen is toxic to all living matter. (Liebig and Pasteur photographs courtesy of Library of Congress, Washington, D.C.; Bert photograph courtesy of History of Medicine Division, National Library of Medicine, Bethesda, Maryland.)

vocated oxygenated bread as a therapy. In addition, he advocated the therapeutic oral use of oxygenated water where "pure oxygen being made to take the place of all the atmospheric air in the water." This treatment was better than giving hydrogen peroxide, which was "found to be a disagreeable medicine by most persons, and is certainly more apt to cause internal discomfort and flatulence than the oxygenated water." Ozanam (1861) also used oxygenated water as a medicine.

Goolden (1866) described the therapy he ordered for a patient suffering from double pneumonia as follows: "His mouth was washed with a solution of Condy's ozone water constantly, He took one drachm of binoxide of hydrogen [hydrogen peroxide] in two ounces of water three times daily, and inhaled from the oxygen bag twice daily." The patient recovered. Laugier (1863) noted that oxygen baths were successful in treating gangrene. Trousseau (Wolff, 1865) reported that oxygen inhalation acts as an irritant when used in the unsuccessful treatment of tuberculosis. Tamin-Despalle (1875) used oxygen inhalation instead of leeches for the successful treatment of a cerebral stroke.

Demarquay (1866) and Smith (1870) were both well-respected advocates of oxygen therapy.

As noted before, compressed air therapy was also investigated. This type of therapy, which increases the pressure of oxygen, was intensified after Triger in 1841 developed the caisson (Bert, 1878; Jacobson et al., 1965). Compressed air or hyperbaric chambers for therapeutic purposes were built in several places in Western Europe, but by 1885 this interest in compressed air theory had subsided.

The medical profession was sceptical of many of the wild, unbelievable claims of the oxygen therapists. The actual concentration of oxygen used in therapy was usually less than 0.5 bar and the duration of this therapy per day was minimal. Placebo effects undoubtedly played a major role.

The dangers of oxygen were recognized as soon as oxygen was discovered. However, it was Paul Bert (1878) who put oxygen toxicity on a firm experimental basis. Bert wrote that he approved of the "charming mystification" of the space novel written by Jules Verne. In a translated version of Verne's science fiction novel *All Around the Moon* (1876), the travelers were accidentally exposed to an excess of oxygen. Verne writes, "A few minutes more and it [oxygen] would have killed the travellers, not like carbonic acid by smothering them, but by burning them up, as a strong draught burns up the coak in a stove." After the voyagers recovered, one of them said, "Would it not be worth some enterprising fellow's while to establish a sanatorium provided with oxygen chambers, where people of a debilitated state of health could enjoy a few hours of intensely active existence!"

VIII. Pasteur and Later Developments

Pouchet was an influential and important adversary of Pasteur on the issue of spontaneous generation (Farley and Geison, 1974; Roll-Hansen, 1979). Pouchet (1858) claimed that spontaneous generation occurred in pure oxygen; in fact a new species, *Aspergillus Pouchetii*, was supposedly generated. Concerning the possible role of oxygen in spontaneous generation, Louis Pasteur (1862) (Fig. 3B) wrote, "I thought that a choice of two possibilities must be made: the essential material of fermentations being organized are of spontaneous generation, if oxygen alone, as oxygen, gives rise to them, by its contact with the azote [nitrogen] material, if these fermentation products are not spontaneous, then oxygen does not intervene in their formation as oxygen alone, but as the exciting agent of a germ that is brought in at the same time as it, or exists in the azote [nitrogen] materials or fermentable 'matter.'" Pasteur continued his studies on spontaneous generation and "virtually convinced the French in the 1860's that spontaneous generation was a dead issue" (Farley, 1977).

In addition, Pasteur (1861a) found that some unicellular organisms were so sensitive to oxygen poisoning that "the air kills them." Pasteur recognized the significance of this discovery and wrote: "This is, I think, the first known example of animal ferments, and also of animals living without free oxygen gas." Two years later, he wrote, "I propose with every type of scruples, these new words *aėrobies* [aerobes] and *anaėrobies* [anaerobes], to indicate the existence of two classes of inferior beings; the ones are incapable of living outside of the presence of free oxygen gas, and the others can propagate themselves infinitely outside of any contact with this gas" (Pasteur, 1863). Moreover, he noted the obligate and faculative aerobes (Pasteur, 1861b). Comparing the facultative ones to the obligate ones, he wrote, "like them needing oxygen, but differing from them in that they apparently can, in the absence of free oxygen gas, breath with oxygen gas that is extracted from unstable compounds."

Claude Bernard was a slightly older contemporary of Pasteur's (Hall, 1975b). His ideas on oxidation and reduction were: "The animal was thus considered as an apparatus for *combustion, oxidation, analysis* or *destruction*, while the plant, to the contrary, was an apparatus of *reduction, formation*, and *synthesis*" (Bernard, 1878). In other words, photosynthesis uses energy in forming oxygen and respiration releases energy in consuming oxygen.

Paul Bert (Fig. 3C), an assistant of Claude Bernard, was the first person to study extensively the effects of barometric pressure. In addition, he was active in French politics. He was a member of the National Academy, the Minister of Education, and Governor-General of French Indo-China (Vietnam), where he died in 1886 (Kellogg, 1978). Not only did he study the effects of pressures below one bar (Bert, 1871a), but also above one bar (Bert, 1871b). In 1878, he wrote his monumental 1178 page book on *Barometric Pressure* (Bert, 1878). He gives an extensive history on the biological and medical effects of high altitude experienced by man on mountains and in balloon ascensions, showing that these effects were due to oxygen deprivation in this book. According to Macedo Dianderas (1961), V. Tschudi gave an excellent description of altitude sickness before Bert. A vivid drama of the fatal high altitude balloon ascent, at 28,200 ft, over Paris in 1875, is given by Bert (1878). Only one of the three who ascended in the *Zenith* survived. Bert had studied the two victims the year before in his decompression chamber.

In his book, Bert also discussed the effects of elevated pressures experienced by those in diving bells, diving suits, caissons, and compressed air therapeutic chambers. He exposed animals, plants, and microscopic organisms to various oxygen pressures under different total barometric pressures. His results show that "The whole aggregation of living beings, in a word, dies absolutely when the oxygen tension rises high enough." Thus, Bert extended Pasteur's observation on oxygen poisoning to include not only anaerobes but also aerobes. He observed that elevated oxygen pressures result in "violent convulsions displayed by the higher animals." This effect of oxygen is known as the Paul Bert effect" (Bean, 1945). Bert stated that: "The organisms at present existing in a natural state... are acclimated to the degree of oxygen tension in which they live: any decrease, any increase seems to be harmful to them when they are in a state of health." Concerning oxygen therapy, he suggested that it be "used only in threatening cases of asphyxia, poisoning by carbon monoxide or sewer gas, where the time for action is short."

Bert wrote in his book that "the tension of this gas [oxygen on earth] has probably been diminishing and no doubt will continue to diminish." He also expressed the thought that "microscopic organisms must have appeared first [on earth] and... they will disappear last, when life becomes extinct through insufficiency of oxygen tension." On the contrary, it is believed that life on earth began in a reducing environment (Gilbert, 1981a).

Bert's studies mark a milestone in our knowledge concerning the physiological effects of oxygen. They came a century after the discovery of oxygen, and mark the midpoint in time up to the publication of the present chapter.

Paul Regnard was an assistant and a disciple of Bert (Fenn, 1970). Regnard (1891) wrote his pioneering monograph on *La Vie Dans les Eaux (Life in the Waters)* in which he described measuring the oxygen uptake and carbon dioxide output of fish under carefully controlled conditions: He found that the carbon dioxide production was less than the oxygen consumption. Fenn (1970) points out that these were the "first reliable measurements of aquatic respiration."

Smith investigated the findings of Lavoisier (1782–1783) concerning inflammation of the lungs upon exposure to oxygen. Smith (1899) found that an increase in oxygen pressure produced an inflammatory condition of the lungs. This lung pathology is known as the "Lorraine Smith effect" (Bean, 1945).

IX. Intracellular Oxidation

The comparison of fire and life, which we saw to be an important theme in earlier times, has more recently been used in biochemistry to study intracellular oxidation processes. Dixon (1970) has written, "The living organism—the living cell—has sometimes been compared with a flame, and this comparison is instructive, for they have much in common." He points out that the two major differences are: First, a flame has a higher temperature than an organism. This difference is due to the presence of the biological catalysts, the enzymes. Second, in the flame, there are degradation reactions, whereas in the organism, there are many synthetic reactions. Of course, "if the production of heat is excluded as an end in itself, other processes must consume the greater part of the output of energy by the organism: these comprise growth, reproduction and maintenance" (Ogston, 1950).

It has been known since the time of Lavoisier that in biological oxidations, carbon and hydrogen from food react with oxygen to produce carbon dioxide and water.

$$nO_2 + (CH_2O)_n \rightarrow nCO_2 + nH_2O \quad \textbf{(12)}$$

The first theory developed in 1856 concerning intracellular oxidation was Schönbein's idea that oxygen had to first be activated to ozone (Dixon, 1970; Fruton, 1972; Florkin, 1975). Schönbein, as mentioned earlier, discovered ozone in 1840. He idolized oxygen. He felt, "oxygen was not an element like any other, but, to use his own expression, it was the king of elements, the Jupiter of the scientific Olympus. He spoke of oxygen as his hero, which he regarded as ominipotent" (Florkin, 1975). He believed that if substances were not present to convert oxygen into ozone, then living organisms would suffocate, even if the oxygen sup- was ample. This idea led to the "ozone craze" for several years. Daily ozone determinations were made in Paris in 1866–1867; and correlations were made between the weather forecasts in Britain and the Doctor's ozonometer (Dixon, 1970). In 1870, Lender was the strongest advocate of ozone inhalation therapy (Cohen, 1876).

Warburg in 1925 concluded that an iron containing respiratory ferment is responsible for the reduction of oxygen in living cells. However, according to the Wieland–Thunberg theory, certain hydrogen atoms of the substrate transfer to an acceptable hydrogen acceptor, such as oxygen (Florkin, 1975). Thus, we have two opposing theories, one with an oxygen activation, and the other with a hydrogen activation. In a certain sense, the hydrogen activation theory represents a phlogiston theory revisited. In other words, a dehydrogenated substrate is analagous to Priestley's dephlogisticated air; that is, the dehydrogenated substrate is more oxidized than the original substrate, just as dephlogisticted air, or oxygen, is more oxidized than common air.

However, unlike the phlogiston theory which Lavoisier had crushed, the dehydrogenation theory still exists. Keilin discovered the cellular pigment, or cytochrome (Keilin, 1925). Keilin showed that there are several cytochromes which are capable of hydrogen atom transfers, and only one of these reacts with oxygen. The one that reacts with oxygen is Warburg's respiratory ferment or cytochrome oxidase (Dixon, 1970). It is of interest to note that the cytochromes had been discovered earlier by MacNunn in 1884, who called them "myohaematin" and "histohaematin." However, contemporaries "met with serious and baffling difficulties in MacNunn's work, which they could not overcome" (Keilin, 1966). We think today of hydrogen atoms being transformed by a chain of carriers to cytochrome oxidase where they react with oxygen. Thus, oxidation in this system occurs not only by a direct oxidation by oxygen, by also by dehydrogenation reactions. For more details, the reader is referred to the review by Florkin (1975).

X. Conclusion

This chapter has been a biased one, with emphasis on the therapeutic and toxic effects of oxygen on living organisms. Oxygen has been known to promote life as well as destroy life for two centuries. This dual nature makes oxygen a double-edged sword.

As pointed out by Liebig (1842), antioxidants inhibit the destructive influence of oxygen in the body. More recently, Dufraisse (1935) writing about the "many strange features of the phenomenon of autoxidation," mentioned that "the most important of which is inhibition, or the antioxygen effect."

Many important topics were necessarily omitted, so that obviously, this chapter is not comprehensive.

There are several reviews in which some of the early literature is discussed (Demarquay, 1866; Smith, 1870; Cohen, 1875; Bert, 1878; Hahn, 1899; Campbell and Poulton, 1934; Bean, 1945; Sackner, 1974). References to oxygen therapy published between 1780 and 1802 were collected in 1902 by J. Pagel in his *Handbuch der Sauerstoff-Therapie* (Campbell and Poulton, 1934). The history of low oxygen effects on life has been discussed by Bert (1878), Monge (1948), Kellogg (1968), and Lenfant and Sullivan (1971).

Hardly any mention of the developments since Paul Bert in 1878 have been discussed, since other extensive reviews cover this area (Stadie et al., 1944; Bean, 1945). Clements (1921) gives some history of the toxic effects of oxygen on plants. Precautions in utilizing oxygen therapy have been emphasized in the classical book by Comroe and Dipps (1953).

Several books on various aspects of oxygen function in the living organism have appeared recently (Dickens and Neil, 1964; Hayaishi, 1974; Davis and Hunt, 1977; Hayaishi and Asada, 1977; Jöbsis, 1977; Caughey, 1979; Fitzsimons, 1979). Some review articles on this subject which are not included in these books are Gottlieb (1965), Haugaard (1968), Gottlieb (1971), Polistena (1973), Clark (1974), Smith and Shields (1975), Fridovich (1976), Chance and Boveris (1978), Fridovich (1978), and Lambertsen (1978).

The development of the Gerschman theory (Gerschman, 1964; Gilbert, 1972) of free radicals being responsible for at least some of the biological effects of oxygen appeared in 1954 (Gerschman et al., 1954). Fridovich (1975) in discussing this theory wrote, "Similarities between the lethality of oxygen and of ionizing radiation led, in 1954, to the theory that the undisciplined reactivities of free radicals were the root cause of oxygen toxicity [Gerschman et al., 1954]. This was a remarkable prescient theory considering the paucity of information concerning the generation and scavenging of specific free radicals in biological systems available at that time. The developments of recent years have provided a firm foundation for a reasonable discussion of the basis of oxygen toxicity." The next chapter (Gerschman, 1981) will deal with the history of this important theory.

Acknowledgments

Acknowledgment is given to Dr. Claire Gilbert for her helpful discussions, criticisms, extensive editing, and translation of all French, Spanish and Italian articles. Acknowledgments are also give to the following librarians who have been of invaluable assistance to me in the preparation of this chapter: Ms. Jane Fessenden of the Library of the Marine Biological Laboratory, Woods Hole, Massachusetts; Mrs. Dorothy Hanks and Mrs. Young Rhee, both of the History of Medicine Division of the National Library of Medicine, Bethesda, Maryland; Mme. Catherine Ludovici, Conservateur de Bibliothèque de l'Académie National de Médecine, Paris, France for supplying the article by Caillens; and to the Rare Book and Special Collections Division, Library of Congress, Washington, D.C. For their aid in obtaining information on the Pariacaca Mountain, Peru, I would like to thank Mr. Edward Szymanski, Defense Mapping Agency, Department of Defense, Washington, D.C.; Director, DMAHTC, Washington, D.C.; Dr. Tulio Velásquez, Loayza Hospital, Instituto de Bilogia Andina, Lima, Peru; Luis Gonzales

Cardenas, Brigadier General, Director of Instituto Geografico Militair, Lima, Peru; Mr. Gary L. Fitzpatrick, Reference Librarian, Geography and Map Division, Library of Congress, Washington, D.C.; Mr. Everett E. Larson, Reference Librarian, Hispanic Division, Library of Congress, Washington, D.C.; and Dr. Yolanda C. Oertel, Department of Pathology, George Washington University Hospital, Washington, D.C. I also wish to express my gratitude to Dr. Ralph H. Kellogg, Department of Physiology, University of California School of Medicine, San Francisco, California, for giving me copies of portions of two of the Acosta references (1590e, 1590f). Also, I am grateful to Dr. Harold L. Burstyn, Historian, U.S. Geological Survey, Reston, Virginia, and to Dr. A. J. Ihde, Department of Chemistry, University of Wisconsin, Madison, Wisconsin, for their helpful comments. In addition, for photographs showing scientists discussed in this chapter, acknowledgments are given to Ms. Lucinda H. Keister, History of Medicine Division of the National Library of Medicine, Bethesda, Maryland, and to Prints and Photographs Division, Library of Congress, Washington, D.C. Finally, acknowledgments are given to Dr. Mark Plunguian and Dr. Richard FitzHugh for their aid in translating the German articles.

References

Abel, E. L. (1973). Ancient Views on the Origins of Life. Rutherford, N. J.: Fairleigh Dickinson University Press.

Acosta, I. de (1590a). Historia Natvrall y Moral de las Indias, En qve se Tratan las Cosas Notables del Cielo, y Elementos, Metales, Plantas, y Animales Dellas: y los Ritos, y Ceremonias, Leyes, y Gouierno, y Guerras de los Indios. Seuilla: Iuan de Leon.

Acosta, I. (1590b). Histoire Natvrelle et Moralle des Indes, tant Orientalles qu'Occidentalles. Où il est traicté des choses remarquables du Ciel, des Elemens, Metaux, Plantes & Animaux qui sont Propres de ces Païs. Ensemble des Moeurs, Ceremonies, Loix, Gouuernemens & Guerres des mesmes Indiens (Translated from Spanish into French by R. R. Cauxois). Paris: Marc Orry. 1598.

Acosta, I. (1590c). Histoire Natvrelle et Moralle des Indes, tant Orientalles qu'Occidentalles. Où il est traitté des choses remarquables du Ciel, des Elemens, Metaux, Plantes & Animaux qui sont Propres de ce Païs. Ensemble des Moeurs, Ceremonies, Loix, Gouuernemens, & Guerres des mesmes Indiens (Translated from Spanish into French by R. R. Cauxois. Revised Edition). Paris: Marc Orry. 1606.

Acosta, J. (1590d). The Natvrall and Morall Histoire of the East and West Indies. Intreating of the Remarkeable Things of Heaven, of the Elements, Mettalls, Plants and Beasts which are proper to that Country: Together with the Manners, Ceremonies, Lawes, Governements, and Warres of the Indians (Translated from Spanish into English by E[dward] G[rimstone]). London: Edward Blount. 1604.

Acosta, G. di (1590e). Historia Natvralle, e Morale della Indie; Nellaquale si trattano le cose notabili del Cieolo, & de gli Elementi, Metalli, Piante, & Animali di quelle: i suoiriti, & Ceremonie: Leggi, & Gouerni, & Guerre degli Indiani. Venice: Bernardo Basa. 1606.

Acosta, I. de (1590f). Historia Natvral y Moral de las Indias, en qve se Tratan las Cosa Notables del Cielo, y Elementos, Metales, Plantas, y Animales dellas: y los Ritos, y Ceremonias, Leyes, y Gouierno, y Guerras de los Indio. Madrid: Alonso Martin. 1608.

Aristotle (1953). The Works of Aristotle Translated into English. W. D. Ross (Ed.) Vol. 7. Problemata by E. S. Forster. New York: Oxford University Press.

Aristotle (1955). The Works of Aristotle Translated into English. W. D. Ross (Ed.) Vol. 3. Parva Naturalia. New York: Oxford University Press.

Basham, A. L. (1959). The Wonder That Was India. A Survey of the Culture of the Indian Sub-continent before the Coming of the Muslims. New York: Grove Press.

Bayliss, W. M. (1915). Principles of General Physiology. New York: Longmans, Green, and Co.

Bean, J. W. (1945). Effects of oxygen at increased pressure. Physiol. Rev. 25:1–147.

Beddoes, T. (1793a). A Letter to Erasmus Darwin, M. D. on a New Method of Treating Pulmonary Consumption and some other Diseases hitherto found Incurable. London: J. Murray.

Beddoes, T. (1793b). Observations on the Nature

and Cure of Calculus, Sea Scurvy, Consumption, Catarrh, and Fever: Together with Conjectures upon Several Other Subjects of Physiology and Pathology. London: J. Murray.

Beddoes, T. (1794). Part I. A familiar exploration of the principles on which benefit may be expected from factitious airs in various diseases. In: Beddoes, T., and Watt, J. Considerations on the Medicinal Use of Factitious Airs, and on the Manner of Obtaining them in Large Quantities. Part I. Part II. London: J. Johnson, pp. 1–48.

Beddoes (1797). Letters to Dr. Rollo. In: Rollo, J. An Account of Two Cases of the Diabetes Mellitus: with Remarks as They Arose During the Progress of the Cure. To Which are Added, A General View of the Nature of the Disease and its Appropriate Treatment, Including Observations on Some Diseases Depending on Stomach Affection; and a Detail of the Communications Received on the Subject Since the Dispersion of the Notes on the First Case. With the Results of the Trials of Various Acids and Other Substances in the Treatment of the Lues Venerea; and some Observations on the Nature of Sugar, &c. By William Cruickshank. In Two Volumes. Vol. II. London: C. Dilly, pp. 6–8.

Beddoes, T. (1799). Notice of some Observations made at the Medical Pneumatic Institution. London: T. N. Longman and O. Rees.

Beddoes, T., and Watt, J. (1796a). Considerations on the Medicinal Use and on the Production of Factitious Airs. 3rd Ed. Part I. Experiments, Cautions, and Cases, tending to illustrate the Medicinal Use of Factious airs, and of other Substances, of which the Application to Medicine has been Suggested by Modern Philosophical Discoveries. By T. Beddoes. Part II. Description of a Pneumatic Apparatus, with directions for Procuring the Factitious Airs. By J. Watt. Addenda. Cases and Observations in Elucidation of the Medicinal Effects of Factitious Airs and their Production. London: J. Johnson.

Beddoes, T., and Watt, J. (1796b). Considerations on the Medicinal Use and Production of Factitious Airs. 2nd Ed. Part III. London: J. Johnson.

Beddoes, T., and Watt, J. (1796c). Medical Cases and Speculations; Including Parts IV and V of Considerations of the Medicinal Powers, and the Production of Factitious Airs. Part IV. Part V. Supplement to the description of a Pneumatic Apparatus, for preparing Factitious Airs; Containing a Description of a Simplified Apparatus, and of a Portable Apparatus. By J. Watt. London: J. Johnson.

Ben-Yehuda, E., and Weinstein, D. (1961). Ben-Yehuda's Pocket English-Hebrew Hebrew-English Dictionary. New York: Pocket Books.

Bernal, J. D. (1974). Science in History. Vol. I: The Emergence of Science. 3rd Ed. Cambridge, Massachusetts: M.I.T. Press.

Bernard, C. (1878). Leçons sur les Phénomènes de la Vie Communs aux Animaux et aux végétaux. (1st Ed. 1878. 2nd Ed. conforme à la Première Ed. 1885). vol. I. Paris: Baillière.

Bert, P. (1871a). Recherches expérimentales sur l'influence que les changements dans la pression barométrique exercent sur les phénomènes de la vie. Note. Comp. Rend. Acad. Sci. 73:213–216.

Bert, P. (1871b). Recherches expérimentales sur l'influence que les changements dans la pression barométrique exercent sur les phénomènes de la vie. 2e note. Comp. Rend. Acad. Sci. 73:503–507.

Bert, P. (1878). Barometric Pressure. Researches in Experimental Physiology. Translated by M. A. Hitchcock and F. A. Hitchcock. Columbus, Ohio: College Book Co. 1943.

Birch, S. B. (1857). On the Therapeutic Action of Oxygen, with Cases proving its singular Efficacy in various Intractable Diseases. London: H. Baillière.

Birch, S. B. (1869). Some remarks on the exhibition of oxygen as a therapeutic in connexion with a new, aggreeable, and easy form of administration by the stomach. Lancet 1:492–493.

Birch, T. (1756). The History of the Royal Society of London for Improving of Natural Knowledge, from its First Rise in Which the Most Considerable of those Papers Communicated to the Society, Which have Hitherto not been Published, are Inserted in their Proper Order, as a Supplement to The Philosophical Transactions. Vol. 1. London: A. Millar. Brussels: Culture et Civilisation, 1968.

Black, J. (1755). The Discovery of Carbonic Acid Gas. From Experiments upon Magnesia, Quicklime, and some other Alkaline Substances. In: Knickerbocker, W. S. (Ed.). Classics of Modern Science (Copernicus to Pasteur). Boston: Beacon Press, 1962, pp. 89–95.

Bodmer, F. (1944). The Loom of Language. Ed. by L. Hogben. New York: Norton and Co.

Book Review (1857). Birch, S. B. On the therapeutic value of oxygen; with cases proving its singular efficacy in various intractable diseases. Baillière, London. Lancet 2:553.

Bouchardat (1851). Emploi du gaz oxygène dans la glucosurie. Comp. Rend. Acad. Sci. 33: 543–544.

Boussingault (1841). Sur la composition de l'air qui se trouve dans les pores de la neige. Ann. Chim. Phys. Ser. 3 1:354–360.

Boyle, R. (1670). The continuation of the experiments concerning respiration. Phil. Trans. 5:2036–2056.

Brasted, R. C. (1961). Comprehensive Inorganic Chemistry. Vol. 8. Sulfur, Selenium, Tellurium, Polonium, and Oxygen. New York: D. Van Nostrand Co.

Broughton, S. D. (1829). An experimental inquiry into the physiological effects of oxygen gas upon the animal system. Paper read Mar. 26 Proc. Roy. Soc. In: Phil. Mag. Ser. 2 5:383.

Broughton (1830). On the poisonous effects of oxygen and some other gases on the animal body. Quart. J. Sci., Literature, and Art, April 1830. Cited in: Am. J. Med. Sci. Ser. 1 7:547–548.

Caillens (1783). Observations sur un nouveau moyen de remédier à la phthisie pulmonaire. Gazette de Santé, p. 38.

Campbell, A., and Poulton, E. P. (1934). Oxygen and Carbon Dioxide Therapy. London: Oxford University Press.

Cartwright, F. T. (1952). The English Pioneers of Anesthesia (Beddoes, Davy, and Hickman). London: Simpkin Marshall.

Cattell, T. (1847). On oxygen, as a corrective of the secondary effects of ether in surgical operations. Lancet 1:422.

Caughey, W. S. (Ed) (1979). Biochemical and Clinical Aspects of Oxygen. New York: Academic Press.

Cavendish, H. (1766). Three papers, containing experiments on factitious air. Phil. Trans. 56:141–184.

Cavendish (1783). An account of a new eudiometer. Phil. Trans. Roy. Soc. 73:106–135.

Chance, B., and Boveris, A. (1978). Hyperoxia and hydroperoxide metabolism. In: Robin, E. D. (Ed.). Extrapulmonary Manifestations of Respiratory Disease. New York: Marcel Dekker, pp. 185–237.

Chaptal (1790). Lettre de M. Chaptal, à M. Berthollet. Ann. Chim. 4:21–24.

Chaussier (1780–1781). Sur les moyens propres à déterminer la respiration dans les enfans qui naissent sans donner aucun signe de vie, & à rétablir cette fonction dans les asphyxiés; & sur les effets de l'air vital ou déphlogistiqué employé pour produire ces avantages. Hist. Soc. Roy. Méd. 4:346–354 (printed 1785).

Clark, J. M. (1974). The toxicity of oxygen. Am Rev. Respir. Dis. 110(Suppl.):40–50.

Clements, F. E. (1921). Aeration and Air-Content. The Rôle of Oxygen in Root Activity. Carnegie Institution of Washington Publication No. 315. Washington: Carnegie Institute of Washington.

Cohen, J. S. (1876). Inhalation in the Treatment of Disease: Its Therapeutics and Practice. A Treatise on the inhalation of gases, vapors, fumes, compressed and rarified air, nebulized fluids, and powders. 2nd Ed. Philadelphia: Lindsay and Blakiston.

Commercial Press (1961). Latin-English-Chinese Dictionary of Medical Terms. Hong Kong: Commercial Press Ltd.

Comroe, J. H., Jr., and Dripps, R. D. (1953). The Physiological Basis for Oxygen Therapy. 2nd Print. Springfield, Illinois: Charles C Thomas.

Conant, J. B. (1970). The overthrow of the phlogiston theory. The chemical revolution of 1775–1789. In: Conant, J. B. and Nash, L. K. (Eds.), Harvard Case Histories in Experimental Science. Cambridge, Massachussets: Harvard University Press, pp. 65–115.

Contenau, G. (1938). La Médecine en Assyrie et en Babylonie. Paris: Librairie Maloine.

Cowan, J. M. (Ed.) (1971). Hans Wehr. A Dictionary of Modern Written Arabic. 3rd Ed. Ithaca, New York: Spoken Language Services, Inc.

Dalton, J. (1805). Experimental enquiry into the proportion of the several gases or elastic fluids, constituting the atmosphere. Mem. Literary Phil. Soc. Manchester, Second Ser. 1:244–258. In: Alembic Club Reprints. No. 2. Foundations of the Atomic Theory comprising Papers and Extracts by John Dalton, William Hyde Wollaston, and Thomas Thomson (1802–1808). Edinburgh: E. and S. Livingston, 1948, pp. 5–15.

Darwin, C. (1860). The Voyage of the Beagle (Edited by L. Engel). Garden City, New York: Doubleday and Co., 1962.

Darwin, E. (1796). Zoonomia; or, the Laws of Organic Life. Vol. 2. London: J. Johnson.

Davis, J. C., and Hunt, T. K. (Eds.). (1977). Hyperbaric Oxygen Therapy. Bethesda, Maryland: Undersea Medical Society.

Davy, H. (1800). Researches, Chemical and Philosophical; chiefly concerning Nitrous Oxide, or Dephlogisticated Nitrous Air, and its Respiration. London: J. Johnson.

Davy (1801). An account of a new eudiometer. J. Roy. Inst., G. Brit. Vol. I. In: Phil. Mag. 10:56–58.

de Lapasse (1846). De l'action de l'oxygène sur les organes de l'homme, et des moyens de diriger convenablement cette action. Comp. Rend. Acad. Sci. 22:1055–1056.

Delury, G. E. (1978). The World Almanac & Book of Facts. 1979. New York: Newspaper Enterprise Association.

Demarquay, J. N. (1866). Essay on medical pneumatology: A physiological, clinical, and therapeutic investigation of the gases. Trans. by Wallian, S. S. Philadelphia: F. A. Davis, 1889.

Demarquay and Leconte (1864). Des indications et des contre-indications à l'emploi de l'oxygène. (Troisième Mémoire.) Comp. Rend. Acad. Sci. 58:463–465.

De Vries, L. (1959). German-English Science Dictionary for Students in Chemistry, Physics, Biology, Agriculture, and Related Sciences. 3rd Ed. New York: McGraw-Hill.

Dickens, F., and Neil, E. (Eds.) (1964). Oxygen in the Animal Organism. New York: Pergamon Press, Macmillan Co.

Divry, G. C. (1974). Divry's Modern English-Greek and Greek-English Desk Dictionary. New York: D. C. Divry, Inc.

Dixon, M. (1970). The history of enzymes and of biological oxidations. In: Needham, J. (Ed.). The Chemistry of Life. New York: Cambridge University Press, pp. 15–37.

Dufraisse, C. (1935). L'Absorption chimique réversible de l'oxygène libre par les corps organiques. In: Institute International de Chimie Solvay. Cinquiéme Conseil de Chimie tenu a L'Université de Bruxelles deu 3 au 8 Octobre 1934. Rapports et Discussions relatifs a L'Oxygène. Ses Réactions Chimiques et Biologiques. Paris: Gauthier-Villars, pp. 205–254.

Duncum, B. M. (1947). The Development of Inhalation Therapy with Special Reference to the Years 1846–1900. New York: Oxford University Press.

Eliade, M. (1974). Gods, Goddesses, and Myths of Creation. A Thematic Source Book of the History of Religions. Part 1 of From Primitive to Zen. New York: Harper and Row.

Espasa, H. de J. (1920). Enciclopedia Vniversal Ilvstrada Evropeo-Americana. Vol. 42. Pare-Peke. Barcelona: Espasa.

Farley, J. (1977). The Spontaneous Generation Controversy from Descartes to Oparin. Balitmore: Johns Hopkins University Press.

Farley, J., and Geison, G. L. (1974). Science, politics, and spontaneous generation in nineteenth-century France: The Pasteur-Pouchet debate. Bull. Hist. Med. 48:161–198.

Fenn, W. O. (1970). A study of aquatic life from the laboratory of Paul Bert. A review of "La Vie dans Les Eaux" by Paul Regnard, Paris 1891. Respir. Physiol. 9:95–107.

Filipović, R., Grgić, B., Cizelj, K., Mosković, V., Ratnik, V., Spalatin, L., Sovary, R., Tomljenović, B., and Urbany, M. (1963). English Croato-Serbian Dictionary. 3rd Ed. Zagreb: Zora.

Fitzsimons, D. W. (Ed.) (1979). Oxygen Free Radicals and Tissue Damage. Ciba Foundation Symposium 65 (New Series). New York: Elsevier/North-Holland, Inc.

Florkin, M. (1972). A History of Biochemistry. Part I. Proto-Biochemistry. Part II. From Proto-Biochemistry to Biochemistry. Vol. 30. In: Florkin, M., and Stotz, E. H. (Eds.). Comprehensive Biochemistry. New York: Elsevier Publishing Co.

Florkin. M. (1975). A History of Biochemistry. Part III. History of the Identification of the Sources of Free Energy in Organisms. Vol. 31. In: Florkin, M., and Stotz, E. H. (Eds.). Comprehensive Biochemistry. New York: Elsevier Publishing Co.

Florkin, M. (1977). A History of Biochemistry. Part IV. Early Studies on Biosynthesis. Vol. 32. In: Florkin, M., and Stotz, E. H. (Eds.). Comprehensive Biochemistry. New York: Elsevier Publishing Co.

Forman, H. J., and Fisher, A. B. (1981). Antioxidant defenses. This volume.

Foster, M. (1901). Lectures on the History of Physiology during the Sixteenth, Seventeenth and Eighteenth Centuries. New York: Dover Publishers, 1970.

Fourcroy (1790). Extrait d'un mémoire sur les propriétés médicinales de l'air vital. Ann. Chim. 4:83–93. Lu dans la séance publique de la Soc. Roy. Méd. 1789.

Fourcroy (1798a). Sur l'application de la chimie pneumatique à l'art de guérir, et sur les propriétés médicamenteuses des substances oxigénées. Ann. Chim. 28:225–281.

Fourcroy (Year 6)[1798b]. Notes. In: Rollo, J. Traité du Diabète Sucré des Affections Gastriques et des Maladies qui en Dépendent,

Suivi du Résultat des Essais des Acides et Autres Substances Oxigénées dans le Traitement de la Maladies Vénérienne; d'une Nouvelle Nosologie; d'un Traité de Quelques Poisons Morbifiques; de l'Analyse Chimique du Sucre, et de Plusiers Applications de la Chimie à la Médecine. Première Partie, Deuxième Partie. Paris: Moutardier, pp. 115–156 (in Part 2).

Franklin, B. (1750). The one fluid theory of electricity. June 1, 1747. From Phil. Trans. 45:98. In: Magie, W. F. (Ed.), A Source Book in Physics. Cambridge, Massachusetts: Harvard University Press, 1963, pp. 401–402.

Fridovich, I. (1975). Oxygen: Boon and bane. Am. Sci. 63:54–59.

Fridovich, I. (1976). Oxygen radicals, hydrogen peroxide, and oxygen toxicity. In: Pryor, W. (Ed.), Free Radicals in Biology. Vol. I. New York: Academic Press, pp. 239–277.

Fridovich, I. (1978). The biology of oxygen radicals. Science 201:875–880.

Fruton, J. S. (1972). Molecules and Life. Historical Essays on the Interplay of Chemistry and Biology. New York: Wiley-Interscience.

Fulton, J. F. (1932). Robert Boyle and his influence on thought in the Seventeenth Century. Isis 18:77–102.

Galston, A. W., and Siegel, S. M. (1954). Antiperoxidative action of the cobaltous ion and its consequences for plant growth. Science 120:1071–1072.

Ganzenmüller, W. (1943). Gmelins Handbuch der Anorganischen Chemie. System-Nummer 3: Sauerstoff. Lieferung 1. Geschichtliches. Ed. 8. Berlin: Verlag Chemie.

Gardner, J. (1847). Ether-vapour and oxygen. Lancet 1:395.

Gerschman, R. (1964). Biological effects of oxygen. In: Dickens, F., and Neil, E. (Eds.). Oxygen in the Animal Organism. New York: Pergamon Press, Macmillan Co., pp. 475–494.

Gerschman, R. (1981). Historical introduction to the "free radical theory" of oxygen toxicity. This volume.

Gerschman, R., Gilbert, D. L., Nye, S. W., Dwyer, P., and Fenn, W. O. (1954). Oxygen poisoning and x-irradiation: A mechanism in common. Science 119:623–626.

Gerschman, R., Gilbert, D. L., and Caccamise, D. (1958). Effect of various substances on survival times of mice exposed to different high oxygen tensions. Am. J. Physiol. 192:563–571.

Gilbert, D. L. (1960). Speculation on the relationship between organic and atmospheric evolution. Perspect. Biol. Med 4:58–71.

Gilbert, D. L. (1972). Introduction: Oxygen and life. Anesthesiology 37:100–111.

Gilbert, D. L. (1981a). Significance of oxygen on earth. This volume.

Gilbert, D. L. (1981b). Oxygen: An overall biological view. This volume.

Gmelin, J. F. (1799). Geschichte der Chemie seit dem Wiederaufleben der Wissenschaften bis an das Ende des achtzehnten Jahrhunderts. Vol. 3. Göttingen: Johann Georg Rosenbusch.

Godefroy, F. (1888). Dictionnaire de L'Ancienne Langue Française et de tous ses dialectes du IXe au XVe Siècle. Composé d'après le dépouillement de tous les plus importants documents, manuscrits ou imprimés qui se trouvent dans les grandes bibliothèques de la France et de l'Europe et dans les principales archives Departmentales, Municipales, Hospitalière ou Privées. Vol. 5. Liste-Parsomme. Paris: F. Vieweg.

Goolden, R. H. (1866). Treatment of disease by oxygen. Lancet 1:270–271.

Gottlieb, S. F. (1965). Hyperbaric oxygenation. Adv. Clin. Chem. 8:69–139.

Gottlieb, S. F. (1971). Effects of hyperbaric oxygen on microorganisms. Annu. Rev. Microbiol. 25:111–152.

Grad, A., Skerlj, R., and Vitorovic, N. (1967). English-Slovene Dictionary. Ljubljana: Drzavna Zalozba Slovenije.

Graham, J. D. P. (1962). The Diagnosis and Treatment of Acute Poisoning. New York: Oxford University Press.

Hahn, L. (1899). L'Oxygène et son emploi médical depuis sa découverte. Janus 4:6–13, 57–63.

Haldane, J. B. S. (1947). Life at high pressures. Science News No. 4:9–29, July. Penguin Books.

Hall, T. S. (1975a). History of General Physiology. 600 B.C. to A.D. 1900. Vol. 1. From Pre-Socratic Times to the Enlightenment. Chicago: University of Chicago Press.

Hall. T. S. (1975b). History of General Physiology. 600 B.C. to A.D. 1900. Vol. 2. From the Enlightenment to the End of the Nineteenth Century. Chicago: University of Chicago Press.

Haugaard, N. (1968). Cellular mechanisms of oxygen toxicity. Physiol. Rev. 48:311–373.

Haugen, E. (1967). Norwegian English

Dictionary. Madison, Wisconsin: University of Wisconsin Press.

Hayaishi, O. (Ed.) (1974). Molecular Oxygen in Biology. Topics in Molecular Oxygen Research. New York: American Elsevier Publishing Co.

Hayaishi, O., and Asada, K. (Eds.) (1977). Biochemical and Medical Aspects of Active Oxygen. Baltimore: University Park Press.

Heath, D., and Williams, D. R. (1977). Man at High Altitude. The Pathophysiology of Acclimatization and Adaptation. New York: Churchill Livingstone.

Hemmeter, J. C. (1921). Antoine-Laurent Lavoisier. Janus 25:1–22, 57–86.

Henshaw, N. (1664). Aero-chalinos: or, a Register for the Air; in Five Chapters. Dublin: Samuel Dancer.

Herodotus (1928). With an English translation [from the Greek] by A. D. Godley. In 4 Volumes. Volume II (Books III and IV). New York: G. P. Putnam's Sons, p. 225 (Paragraph 25, Book IV).

Hill, D. (1800). Practical Observations on the Use of Oxygen, or Vital Air in the Cure of Diseases: To which are added a few Experiments on the Vegetation of Plants. Vol. I. London: F. and C. Rivington.

Hogben, L. (1951). Science for the Citizen. A Self-Educator based on the Social background of Scientific Discovery. 4th Ed. New York: W. W. Norton and Co.

Holmyard, E. J. (1955). Alchemy in medieval Islam. Endeavour 14:117–125.

Hook, R. (1665). Extracts from Micrographia: or Some Physiological descriptions of Minute bodies made by magnifying Glasses with Observations and Inquiries thereupon. Alembic Club Reprints. No. 5. Edinburgh: Oliver and Boyd, Ltd., 1944.

Hook. (1667). An experiment of preserving animals alive, by blowing through their lungs with bellows. Phil. Trans. 2:539–540.

Hosack, A. (1797). An inaugural Essay on the Yellow Fever, as it Appeared in this City in 1795. New York: T. and J. Swords.

Humboldt, A. de (1838). Notice sur deux tentatives d'ascension du Chimborazo. Ann. Chim. Phys. Ser. 2 69:401–434.

Ihde, A. J. (1964). The Development of Modern Chemistry. New York: Harper and Row.

Ihde, A. J. (1980). Priestley and Lavoisier. In: Kieft, L. and Willeford, B. R. (Eds.). Joseph Priestley. Scientist, Theologian, and Metaphysician. Cranbury, New Jersey: Associated University Presses, Inc., pp. 62–91.

Ince, H. R. (1805). Communication from Mr. Ince, Surgeon, relative to pneumatic medicine. Phil. Mag. 21:128.

Ingen-Housz, J. (1779). Experiments upon vegetables, discovering their great power of purifying the common air in the sunshine, and of injuring it in the shade and at night. To which is joined, a new method of examining the accurate degree of salubrity of the atmosphere. London: P. Elmaly. In: Reed, H. S. Jan Ingenhousz. Chron. Bot. 11:285–393, 1949.

Ingen Housz, J. (1780). On the degree of salubrity of the common air at sea, compared with that of the sea-shore, and that of places far removed from the sea. Phil. Trans. Roy. Soc. 70:354–377.

Jackson, C. T. (1847). Extrait d'une lettre de M. le docteur Charles T. Jackson à M Élie de Beaumont. Comp. Rend. Acad. Sci. 24:492–494.

Jacobson, J. H. II, Morsch, J. H. C., and Randall-Baker, L. (1965). The historical perspective of hyperbaric therapy. Ann. N. Y. Acad. Sci. 117:651–670.

Jaeger, E. C. (1944). A Source-Book of Biological Names and Terms. Springield, Illinois: Charles C Thomas.

Jöbsis, F. J. (Ed.) (1977). Oxygen and Physiological Function. Dallas, Texas: Prof. Int. Library.

Jones, H. S., and McKenzie, R. (1940). A Greek-English Lexicon compiled by H. G. Liddell and R. Scott. Rev. Ed. New York: Oxford University Press.

Jourdanet, D. (1861). Les Altitudes de L'Amérique Tropicale Comparées au Niveau des Mers au Point de Vue de La Constitution Médicale. Paris: J.-B. Baillière et Fils.

Jurine (1789). Mémoire sur la question suivante proposée par la Société de Médecine: Déterminer quels avantages la Médecine peut retirer des découvertes modernes sur l'art de connoître la pureté de l'air par les différens eudiomètres. Soc. Roy. Med. Paris. Mém. Soc. Med. 10:19–99.

Keilin, D. (1925). On cytochrome C, a respiratory pigment, common to animals, yeast, and higher plants. Abridged from Proc. Roy. Soc. London 98B:312–329. In: Gabriel, M. L., and Fogel, S. (Eds.). Great Experiments in Biology. Englewood Cliffs, New Jersey: Prentice-Hall Inc., 1962, pp. 31–38.

Keilin, D. (1966). The History of Cell Respiration and Cytochrome. Prepared for publication by J. Keilin. New York: Cambridge University Press.

Kellogg, R. H. (1968). Altitude acclimatization, a historical introduction emphasizing the regulation of breathing. Physiologist 11:37–57.

Kellogg, R. H. (1978). La Pression Barométrique: Paul Bert's hypoxia theory and its critics. Respir. Physiol. 34:1–28.

Kerr, J. (1779). A treatise on the various kinds of permanently elastic fluids or gases. 2nd Edition. In: Macquer. Additions to the Dictionary of Chemistry. Translated by J. Kerr. London: T. Cadell.

Kuhn, T. S. (1962). Historical structure of scientific discovery. Science 136:760–764.

Lambertsen, C. J. (1978). Effects of hyperoxia on organs and their tissues. In: Robin, E. D. (Ed.). Extrapulmonary Manifestations of Respiratory Disease. New York: Marcel Dekker, Inc., pp. 239–303.

Langenscheidt (1961). Langenscheidt's Universal Dictionary. English-Portuguese. Portuguese-English. New York: Barnes and Noble, Inc.

Laugier (1863). Nouveaux faits concernant l'utilité des bains d'oxygène dans les cas de gangrène sénile. Comp. Rend. Acad. Sci. 56:1011–1014.

Lavoisier (1775). Mémoire sur la nature du principe qui se combine avec les métaux pendant leur calcination et qui en augmente le poids. Mém. Acad. Sci. p. 520. In: Oeuvres de Lavoisier. Vol. II. Mémoires de Chimie et de Physique. Paris: Imp. Impériale, 1862, pp. 122–128.

Lavoisier (1777a). Expériences sur la respiration des animaux et sur les changements qui arrivent à l'air en passant par leur poumon. Mém. Acad. Sci. p. 185. In: Oeuvres de Lavoisier. Vol. II. Mémoires de Chimie et de Physique. Paris: Imp. Impériale, 1862, pp. 174–183.

Lavoisier (1777b). Mémoire sur la combustion en général. Mém. Acad. Sci. p. 592. In: Oeuvres de Lavoisier. Vol. II. Mémoires de Chimie et de Physique. Paris: Imp. Impériale, 1862, pp. 225–233.

Lavoisier (1778). Considérations générales sur la nature des acides et sur les principes dont ils sont composés. Mém. Acad. Sci., pp. 535–547. (Received in 1777; Read in 1779.)

Lavoisier (1782–1783). Sur les altérations qui arrivent à l'air dans plusieurs circonstances où se trouvent les hommes réunis en société. Mémoires de Médecine. In: Hist. Soc. Méd. 5:569–582. (Read in 1785.)

Lavoisier (1789). Traité Elémentaire de Chimie. In: Oeuvres de Lavoisier. Vol. I. Paris: Imp. Impériale, 1864.

Lavoisier (1862). Vues générales sur la formation et la constitution de l'atmosphère de la terre. Rec. Mém. Chim. de Lavoisier. Vol. 2, p. 398. In: Oeuvres de Lavoisier. Vol. II. Mémoires de Chimie et de Physique. Paris: Imp. Impériale, pp. 804–811.

Le Grand, H. E. (1972). Lavoisier's oxygen theory of acidity. Ann. Sci. 29:1–18.

Leicester, H. M. (1956). The Historical Background of Chemistry. New York: John Wiley.

Lenfant, C., and Sullivan K. (1971). Adaptation to high altitude. New England J. Med. 284: 1298–1309.

Leonard, O. A., and Pinckard, J. A. (1946). Effect of various oxygen and carbon dioxide concentrations on cotton root development. Plant Physiol. 21:18–36.

Liebig, J. (1842). Animal Chemistry or Organic Chemistry in Physiology and Pathology edited from the Author's Manuscript by William Gregory. With Additions, Notes, and Corrections, by Dr. Gregory, and Others. By John W. Webster. 2nd Ed. Cambridge, Massachusetts: John Owen, 1843.

Livy (1929). With an English translation [from the Latin] by B. O. Foster. In 13 Volumes. Volume V (Books XXI–XXII). New York: G. P. Putnam's Sons, p. 87 (Chapter XXX, Book XXI).

Loane (1800). A communication from Dr. Loane, relative to pneumatic medicine. A case of atonic gout cured by vital air. Phil. Mag. 6:82–83.

Macedo Dianderas, J. (1961). Hipoxia de altitud: "Aspectos fisiologicos y clinicos." Rev. Assoc. Med. Prov. Yauli 6:63–70.

Macquer (1777a). Dictionnaire de Chymie, contenant la Théorie et la Pratique de cette Science, son Application à la Physique, à l'histoire naturelle, à la médecine, & aux arts dépandans de la chymie. 2nd Ed. 3 Vols. Paris: P. Fr. Didot jeune, Libraire de la Faculté de Médecine.

Macquer (1777b). Elements of the Theory and Practice of Chemistry. Translated from the French. 5th Ed. Edinburgh: Alex Donaldson.

Maguiña, J. D. [1965]. Guide to Peru. Lima, Cuzco, Machu Picchu, Arequipa, Puno, Iquitos, and Callejon de Huaylas. Lima: Maguiña.

Map of Peru (1775). Como y Olmedilla. Reprint-Barcelona, 1907. No. 1 [In Library of Congress, Washington, D.C.].

Map of Peru (1965). USAF Operational Navigation Chart. Scale 1,000,000. N-25. Bolivia, Brazil, Chile, Ecuador, Peru. 1st Ed. Washington, D.C.: Coast and Geodetic Survey. U.S. Dept. of Commerce.

Map of Peru (1972a). Huarochiri. Carta Nacional. Scale 1:100,000. Sheet 25-k. 1st Ed. Lima: Instituto Geographico Militar.

Map of Peru (1972b). La Oroya. Carta Nacional. Scale 1:100,000. Sheet 24-l. 1st Ed. Lima: Instituto Geographico Militar.

Map of Peru (1973). Yauyos. Carta Nacional. Scale 1:100,000. Sheet 25-l. 1st Ed. Lima: Instituto Geographico Militar.

Map of Peru (1974). Singa. Carta Nacional. Scale 1:100,000. Sheet 19-j. 1st Ed. Lima: Instituto Geographico Militar.

Map of Peru (1979). Matucana. Carta Nacional. Scale 1:100,000. Sheet 24-k. 1st Ed. Lima: Instituto Geographico Militar.

Marx, R. E. (1971). The early history of diving. Oceans Mag. 4(4):66–74; (5):24–34. In: Pirie, R. G. (Ed.). Oceanography. Contemporary Readings in Ocean Sciences. 2nd Ed. New York: Oxford University Press, 1977, pp. 3–23.

Mason, S. F. (1953). Main Currents of Scientific Thought. A History of the Sciences. New York: Henry Schuman.

Mayow, J. (1674). On Respiration. In: Mayow, J. Medico-Physical Works. Being a Translation of Tractatus quinque medico-physici. Edinburgh: The Alembic Club, 1907, pp. 183–210. In: Faulconer, A., and Keys, T. C. (Eds.). Foundations of Anesthesiology. Vol. I. Springfield, Illinois: Charles C Thomas, 1965, pp. 15–32.

McKenzie, A. E. E. (1960). The Major Achievements of Science. Vol. I. New York: Cambridge University Press.

McKie, D. (1952). Antoine Lavoisier. Scientist. Economist. Social Reformer. New York: Henry Schuman.

McKie, D. (1953). Fire and the Flamma Vitalis: Boyle, Hooke and Mayow. In: Underwood, E. A. (Ed.), Science Medicine and History. Essays on the Evolution of Scientific Thought and Medical Practice Written in Honour of Charles Singer. Vol. One. New York: Oxford University Press, pp. 469–488.

Mellor, J. W. (1922). A Comprehensive Treatise on Inorganic and Theoretical Chemistry. Vol. I. H, O. New York: Longmans, Green and Co.

Monge, C. (1948). Acclimatization in the Andes. Historical Confirmations of "Climatic Aggression" in the Development of Andean Man (translated by Donald F. Brown from Spanish). Baltimore: Johns Hopkins Press.

Needham, J. (1965). Science and Civilization in China. Vol. 1. Introductory Orientations. New York: Cambridge University Press.

Needham, J. (1969). Science and Civilization in China. Vol. 2. History of Scientific Thought. New York: Cambridge University Press.

Needham, J., Ping-Yü, Ho and Gwei-Djen, Lu (1976). Science and Civilization in China. Vol. 5. Chemistry and Chemical Technology. Part III: Spagyrical Discovery and Invention: Historical Survey, from Cinnabar Elixirs to Synthetic Insulin. New York: Cambridge University Press.

Nisbet, W. (1801). The Clinical Guide, or, a concise view of the leading facts, on the history, nature, and cure of diseases; to which is subjoined, a practical pharmacopoeia; in three parts: viz. materia medica, classification, and extemporaneous prescription: intended as a memorandum-book for practitioners. 4th Ed. Edinburgh: James Watson.

Ogston, A. G. (1950). Oxidation reactions and the exchange of energy in animals. In: Institut International de Chimie Solvay. Huitième Conseil de Chimie tenu à l'Université de Bruxelles, du 10 au 15 Septembre 1950. Le Mécanisme de L'Oxydation. Rapports et Discussions. Brussels: R. Stoops, pp. 369–400.

Oxford University Press (1971). The Compact Edition of the Oxford English Dictionary. Complete text reproduced micrographically. Vol. 1. A–O. New York: Oxford University Press.

Ozanam, C. (1860). Note sur l'oxygène employé comme antidote de l'éther et du chloroforme. Comp. Rend. Acad. Sci. 51:59–60.

Ozanam (1861). Sur la préparation et l'emploi en thérapeutique de l'eau oxygénatée. Comp. Rend. Acad. Sci. 53:791–792.

Parkes, G. D. (1952). Mellor's Modern Inorganic Chemistry. New York: Longmans, Green, and Co.

Partington, J. R. (1961). A History of Chemistry. Vol. 2. New York: St. Martin's Press.

Partington, J. R. (1962). A History of Chemistry. Vol. 3. New York: St. Martin's Press.

Partington, J. R. (1964). A History of Chemistry. Vol. 4. New York: St. Martin's Press.

Partington, J. R. (1965). A Short History of

Chemistry. 3rd Ed. New York: Macmillan and Co.

Partington, J. R. (1970). A History of Chemistry. Vol. 1. Part 1: Theoretical Background. New York: St. Martin's Press.

Pasteur, L. (1861a). Animalcules infusoires vivant sans gaz oxygène libre et déterminant des fermentations. Comp. Rend. Acad. Sci. 52: 344–347.

Pasteur, L. (1861b). Expériences et vues nouvelles sur la nature des fermentations. Comp. Rend. Acad. Sci. 52:1260–1264.

Pasteur (1862). Mémoire sur les corpuscles organisés qui existent dans l'atmosphère. Examen de la doctrine des générations spontanées. Ann. Chim. Phys. Ser. 3 64:5–110. In: Vallery-Radot, P. (Ed.). Oeuvres de Pasteur. Vol. 2. Fermentations et Générations dites Spontanées. Paris: Masson, 1922.

Pasteur, L. (1863). Recherches sur la putréfaction. Comp. Rend. Acad. Sci. 56:1189–1194.

Patterson, T. S. (1931). John Mayow in contemporary setting. A contribution to the history of respiration and combustion. Isis 15:47–96.

Payerne, M. (1851). Observations tendant à demontrer que, dans les ascensions sur les hautes montagnes, la lassitude et l'anhélation éprouvées par la plupart des explorateurs n'ont pas pour cause une insuffisance d'oxygène dans l'air respiré. Comp. Rend. Acad. Sci. 33:198–199.

Pearson, R. (1795). A Short Account of the nature and properties of different kinds of airs, so far as relates to their medicinal use, intended as an Introduction to the Pneumatic Method of treating diseases, with miscellaneous observations on certain remedies used in consumption. Birmingham: Thomas Pearson.

Pereira, J. (1836). Lectures on materia medica, or pharmacology, and general therapeutics. London Med. Gaz. 17:785–793.

Perkins, J. F., Jr. (1964). Historical development of respiratory physiology. In: Fenn, W. O., and Rahn, H. (Eds.) Handbook of Physiology–Section 3: Respiration. Vol. I. Washington, D.C.: American Physiological Society, pp. 1–62.

Pino, E., and Wittermans, T. (1955). Kamus Inggeris. Part 1: English-Indonesian. Part 2: Indonesian-English. Djakarta: J. B. Walters.

Polistena, A. (1973). Moderne vedute sulla tossicità dell'ossigeno. Riv. Med. Aeronaut Spaz. 36:205–243.

Polo, M. (1299?). The Travels of Marco Polo. Translated and with an Introduction by Ronald Latham. New York: Penguin Books, 1978.

Pouchet, F. (1858). Note sur des proto-organismes végétaux et animaux, nés spontanément dans de l'air artificiel et dans le gaz oxygène. Comp. Rend. Acad. Sci. 47:979–982.

Poulle, A. (1785). Positiones chemico-medicae de aëre vitali, feu dephlogisticato, tanquam novo sanitatis praesidio. (Thèses chimico-médicinales sur l'air vital déphlogistiqué, considéré comme un nouveau moyen de santé, soutenues à Montpellier par A. Poulle pour son baccalauréat.) Montpellier: J. F. Picot. In: J. Med. Chir. Pharm. 63:247–248.

Priestley, J. (1772). Observations on different kinds of air. Phil. Trans. 62:147–264.

Priestley, J. (1775a). An account of further discoveries in air. Phil. Trans. 65:384–394.

Priestley, J. (1775b). Experiments and observations on different kinds of air. Vol. II, Sections III–V, pp. 29–103. In: Priestley, J. The Discovery of Oxygen. Part I. Edinburgh: The Alembic Club, 1923. In: Faulconer, A., and Keys, T. C. (Eds.), Foundations of Anesthesiology. Vol. 1. Springfield, Illinois: Charles C Thomas, 1965, pp. 39–70.

Priestley, J. (1777). Experiments and Observations on Different Kinds of Air. Vol. III. London: J. Johnson.

Priestley, J. (1779). Experiments and Observations relating to various Branches of Natural Philosophy; with a Continuation of the Observations on Air. London: J. Johnson.

Priestley, J. (1781). Experiments and Observations relating to various Branches of Natural Philosophy; with a Continuation of the Observations on Air. Vol. Two. London: J. Johnson.

Priestley, J. (1795). Memoirs of the Rev. Dr. Joseph Priestley to the year 1795, written by himself. In: Brown, I. V. (Ed.) Joseph Priestley. Selections from his Writings. University Park, Pa.: Pennsylvania State University Press, 1962, pp. 1–75.

Raper, H. R. (1945). Man against Pain. The Epic of Anesthesia. New York: Prentice-Hall, Inc.

Regnard, P. (1891). Recherches Expérimentales sur les Conditions Physiques de la Vie Dans Les Eaux. Paris: Masson.

Regnault, V., and Reiset, J. (1849). Recherches chimiques sur la respiration des animaux des diverses classes. Ann. Chim. Phys. Ser. 3 26: 299–519.

Reid, E. E. (1970). Chemistry Through the Language Barrier. How to Scan Chemical articles in Foreign Languages with emphasis on Russian and Japanese. Baltimore: The Johns Hopkins Press.

Reid, T. (1782). An Essay on the Nature and Cure of the Phthisis Pulmonalis. London: T. Cadell.

Rey, J. (1630). Essays of Jean Rey, Doctor of Medicine, On an Enquiry into the Cause Wherefore Tin and Lead Increase in Weight on Calcination. Alembic Club Reprints. No. 11. Edinburgh: E. and S. Livingston, 1953.

Riadore, J. E. (1845). On the Remedial Influence of Oxygen or vital air, nitrous oxide, and other gases, electricity and galvanism, in restoring the healthy functions of the principal organs of the body, and the nerves supplying the respiratory, digestive and muscular systems. London: J. Churchill.

Robinson, J. (1847). Ether-vapour and oxygen. Lancet 1:422.

Roll-Hansen, N. (1979). Experimental method and spontaneous generation: The controversy between Pasteur and Pouchet, 1859–64. J. Hist. Med. 34:273–292.

Rollo, J. (1797). An Account of Two Cases of the Diabetes Mellitus: with Remarks as They Arose During the Progress of the Cure. To Which are Added, A General View of the Nature of the Disease and its Appropriate Treatment, Including Observations on Some Diseases Depending on Stomach Affection; and a Detail of the Communications Received on the Subject Since the Dispersion of the Notes on the First Case. With the Results of the Trials of Various Acids and Other Substances in the Treatment of the Lues Venerea; and some Observations on the Nature of Sugar, &c. By William Cruickshank. In Two volumes. Vol. I. London: C. Dilly.

Rothschuh, K. E. (1973). History of Physiology. Translated by B. G. Risse. Huntington, New York: Robert E. Krieger Publishing Co.

Sackner, M. A. (1974). Oxygen therapy. A history of oxygen usage in chronic obstructive pulmonary disease. Am. Rev. Respir. Dis. 110 (Suppl.):25–34.

Sarasin (1802). Observations et expériences sur l'emploi de l'oxigène dans la cure du tétanos. Ann. Chim. 42:43–50.

Savory, W. S. (1857). On the inhalation of oxygen. Lancet 2:44–45.

Scheele, C. W. (1774). Letter to Lavoisier, Sept. 30. Rev. Gen. Sci. Pure Appl. 1:1–2, 1890. In: Scheele, C. W. The Collected Papers of Carl Wilhelm Scheele. Translated by L. Dobbin. London: G. Bell and Sons, 1931. New York: Kraus Reprint Co., 1971, pp. 350–351.

Scheele, C. W. (1777). Chemische Abhandlung von der Luft und dem Feuer. Uppsala and Liepzig. Section 2. Chemical Treatise on Air and Fire. In: Scheele, C. W. The Collected Papers of Carl Wilhelm Scheele. Translated by L. Dobbs. London: G. Bell and Sons, 1931. New York: Kraus Reprint Co., 1971, pp. 85–178.

Scheele, C. W. (1779). Experiments on the quantity of pure air which is present day by day in our atmosphere. Kongl. Vet Acad. Hand. 40:50–55. In: Scheele, C. W. The Collected Papers of Carl Wilhelm Scheele. Translated by L. Dobbs. London: G. Bell and Sons, 1931. New York: Kraus Reprint Co., 1971, pp. 196–200.

Seguin and Lavoisier (1789). Premier mémoire sur la respiration des animaux. Mém. Acad. Sci, p. 185. In: Lavoisier, A.-L. Mémoires sur la Respiration et la Transpiration des Animaux. Paris: Gauthier-Villars, 1920, pp. 31–51.

Sigerist, H. E. (1967). A History of Medicine. I. Primitive and Archaic Medicine. New York: Oxford University Press.

Singer, C. (1962). A Short History of Scientific Ideas to 1900. New York: Oxford University Press.

Skinner, H. A. (1970). The Origin of Medical Terms. 2nd Ed. New York: Hafner Publishing Co.

Smeaton, W. A. (1962). Fourcroy. Chemist and Revolutionary. 1755–1809. Cambridge, England: W. Heffer and Sons Ltd.

Smith, A. H. (1870). Oyxgen Gas as a Remedy in Disease. New York: D. Appleton and Co.

Smith, C. U. M. (1976). The Problem of Life. An Essay in the Origins of Biological Thought. New York: John Wiley and Sons.

Smith, G., and Shields, T. G. (1975). Oxygen toxicity. Pharmacol. Ther. B 1:731–756.

Smith, J. L. (1899). The pathological effects due to increase of oxygen tension in the air breathed. J. Physiol. 24:19–35.

Société de Philosophie expérimentale de Batavia à Rotterdam (1782). Verhandelingen van het Bataafsch Genootschap, etc. C'est-à-dire, Mémoires de la Société de philosophie expérimentale de Batavia à Rotterdam, Vol. VI. In: J. Med. Chir. Pharm. 61:187–192, 1784.

Speiser, E. A. (1964). The Anchor Bible. Genesis. Introduction, translation, and notes. Garden City, New York: Doubleday and Co., Inc.

Sprat, T. (1667). The History of the Royal Society of London. St. Louis: Washington University Studies. 1959.

Stadie, W. C., Riggs, B. C., and Haugaard, N. (1944). Oxygen poisoning. Am J. Med. Sci. 207:84–114.

Stahl, G. E. (1697). Zymotechnia fundamentalis. Trans. In: Leicester, H. M., and Klickstein, H. S. (Eds.) A Sourcebook in Chemistry., Cambridge: Harvard Univ. Press, 1952. In: Schwartz, G., and Bishop, P. W. (Eds.). Moments of Discovery. Vol. 1. The Origins of Science. New York: Basic Books, 1958, pp. 209–211.

Stanislawski, J. (1968). The Great English-Polish Dictionary. Warsaw: Wiedza Powszechna.

Stein, J. H. (1831). Treatment of cholera by inhalation of oxygen. Lancet 1:364–365, 1831/32.

Stock, J. E. (1811). Memoirs of the life of Thomas Beddoes, M. D. with an Analytical Account of his Writing. London: J. Murray.

Suttle, J. F. (1957). The alkali metals. In: Sneed, M. C., and Brasted, R. C. (Eds.). Comprehensive Inorganic Chemistry. Vol. 6. The Alkali Metals. Hydrogen and Its Isotopes. New York: D. Van Nostrand Co. Inc., pp. 1–182.

Tabarié (1840). Sur l'action thérapeutique de l'air comprimé. Comp. Rend. Acad. Sci. 11:26–28.

Tamin-Despalle (1875). Sur les effects thérapeutiques de l'oxygène. Comp. Rend. Acad. Sci. 80:1031–1032.

Tarn, W. W. (1956). Alexander the Great. Boston: Beacon Press.

Tauro, A. (1967). Diccionario Enciclopédico del Peru. Three Vols. [Lima]: Mejia Baca.

Thorndike, L. (1958). A History of Magic and Experimental Science. Vols. VII and VIII. The Seventeenth Century. Vol. VII. New York: Columbia University Press.

Thornton (1799). Communication from Dr. Thornton, Physician to the General Dispensary, relative to trials made with the different factitious gases. Phil. Mag. 2:420.

Thornton, R. J. (1803). Tenth communication from Dr. Thornton. Phil. Mag. 17:171–178.

Thornton (1805). Twenty-first communication from Dr. Thornton, relative to pneumatic medicine. Phil. Mag. 21:126–128.

Thornton, R. J. (1806a). Twenty-sixth communication from Dr. Thornton, relative to pneumatic medicine. Phil. Mag. 24:47–49.

Thornton, R. J. (1806b). Thirtieth communication from Dr. Thornton, relative to pneumatic medicine. A case of consumption cured by the inhalation of hydrogen gas. Phil. Mag. 25:161–163.

Thornton (1806c). Thirty-third communication from Dr. Thornton, relative to pneumatic medicine. Suspended animation restored by vital air. Phil. Mag. 26:257–264.

Thornton, R. J. (1807). Thirty-fourth communication from Dr. Thornton, relative to pneumatic medicine. Phil. Mag. 27:234–236.

Thornton, R. J. (1813). The Philosophy of Medicine: Being Medical Extracts, on the nature and preservation of Health, and on the nature and removal of Disease. 5th Ed. Vol. 1. London: Sherwood, Neely, and Jones.

Thornton, R. J. (1821). On the cure of scrofula by means of vital air, and the use of the juice of sorrel. Phil. Mag. 57:351–353.

Toulmin, S. E. (1957). Crucial experiments: Priestley and Lavoisier. J. Hist. Ideas 18:205–220. In: Wiener, P. P., and Noland, A. (Eds.). Roots of Scientific Thought. New York: Basic Books, 1957, pp. 481–496.

Townsend, J. (1802). Elements of Therapeutics; or a Guide to Health; being Cautions and Directions in the Treatment of Diseases. Designed Chiefly for the Use of Students. 1st Am. Ed. Boston: Thomas and Andrews.

Trotter, T. (1792). Observations on the Scurvy; with a Review of the Opinions Lately Advanced on that Disease, and a New Theory Defended, on the Approved Method of Cure, and the Induction of Pneumatic Chemistry: Being an Attempt to Investigate that Principle in Recent Vegetable Matter, Which, Alone, has Been Found Effectual in the Treatment of this Singular Disease; and from thence to Deduce More Certain Means of Prevention than have been Adopted Hitherto. Ed. 2. London: T. Longman.

Trotter, T. (1795). Medical and Chemical Essays. London: J. S. Jordan.

Turner, E. R., and Quartley, C. E. (1956). Studies in the respiratory and carbohydrate metabolism of plant tissues. VIII. An inhibition of respiration in peas induced by "oxygen poisoning." J. Exp. Bot. 7:362–371.

U.S. Department of Commerce (1966). Vietnamese-English Dictionary. Vols. 1, 2. Washington, D.C.: U. S. Dept. Commerce.

U. S. Office of Geography (1955). NIS Gazetteer. Peru. Official Standard Names Approved by the United States Board on Geographic Names.

Washington, D. C.: Division of Geography, Dept. of the Interior.

Verne, J. (1876). All around the moon. In: Verne, J. Space Novels. From the Earth to the Moon. All Around the Moon. Translated by E. Roth (1874–1900). New York: Dover Publishers Inc., pp. 215–470.

Ward, M. (1975). Mountain Medicine. A Clinical Study of Cold and High Altitude. New York: Van Nostrand Reinhold Co.

Watt, J. (1794). Part II. Description of an air apparatus; with hints respecting the use and properties of different elastic fluids. In: Beddoes, T., and Watt, J. Considerations on the Medicinal Use of Factitious Airs, and on the manner of Obtaining them in Large Quantities. Part I. Part II. London: J. Johnson, pp. 1–32.

Weeks, M. E. (1956). Discovery of the Elements. 6th Ed. Easton, Pennsylvania: J. Chem. Educ.

Willeford, B. R. (1979). Das Portrait: Joseph Priestley (1733–1804). Chem. Unserer Zeit 13(4):111–117.

Wolff, J. R. (1865). Inhalation of oxygen in phthisis and anaemia. Med. Times Gaz., Nov. 25. In: Am. J. Med. Sci. New Ser. 51:249–250, 1866.

Woolley, L. (1965). History of Mankind. Cultural and Scientific Development. Vol. I, Part 2. The Beginnings of Civilization. New York: New American Library.

2

Historical Introduction to the "Free Radical Theory" of Oxygen Toxicity

Rebeca Gerschman

There is nothing more powerful than an idea whose time has come.

Victor Hugo

It all started in 1952 while I was a Research Associate in the Physiology Department of the School of Medicine at the University of Rochester. At that time, the adrenal cortex was known to be stimulated by many stress-producing conditions. Furthermore, the release of the adrenal cortical hormones was shown in stress situations, such as hypoxia, to aid the organism to resist the adverse effects of the stress. Previously, with Dr. Bernardo Houssay here in Buenos Aires, I had performed experiments with the adrenal gland and the hormones in the medulla, studying its effects on the plasma potassium concentration.

Since I had experience with adrenals, Dr. Wallace O. Fenn, Chairman of the Department at Rochester, suggested that I study hyperoxia stress on the adrenal system. Our hyperoxia experiments in rats showed, as expected, that high oxygen pressures activated the release of the adrenal cortical hormones, but the rats' survival times were significantly diminished. We were puzzled by the fact that adrenalectomy did not decrease but on the contrary clearly increased the rats' survival time during oxygen poisoning (Gerschman and Fenn, 1954).

Since sudden increases of oxygen pressure in the atmosphere do not exist, it was logical to think that the adrenal cortical hormones were not the right answer to the demands of a stress previously unknown to living matter. It became apparent that oxygen was in a class apart.

I began to prepare a Departmental Seminar that I was scheduled to give on "Adrenals and Hyperoxia." While I had read many papers on the effects of hyperbaric oxygen, describing its morphological and functional changes, I did not find adequate explanations for the observed toxic effects. I was confused, and needed more information to understand these surprising findings. The explanations given in the literature for the adverse effects of oxygen were not satisfying to me.

Dr. Hermann Rahn, then a member of the Department, suggested that if I did not like those explanations, then I should make up my own theory. I received different advice when I visited Dr. Otto Loewi in New York City. As a personal friend of mine, he sincerely suggested that I should not attempt to tackle the fundamental mechanism of oxygen poisoning, because it was so perplexing. I was already deeply involved in the problem, though, and so I did not take his advice.

In reading the numerous observations made by scientists in the hyperoxia field, I was inclined to think that the variety and variability of the oxygen effects might all be the result of the same basic, relatively simple principles. This and the surprising fact of the biologists' remarkable lack of interest in the intrinsic properties of oxygen led me to get acquainted with these properties: oxidation potential, electronic configuration, paramagnetism, sluggishness, activation, excited species, oxidizing free radicals, chain reactions, etc.

The unstable and metastable intermediates produced by oxygen were very often the same postulated for irradiation effects. This brought to my memory old reports by Ozorio de Almeida (1934) describing striking similarities between the histological changes in the testes produced by oxygen and by X-irradiation. An extensive review of the literature on the X-irradiation effects in living systems strengthened my feeling of a significant relation between these two processes. Since X-irradiation effects were thought to be due to free radicals, I felt it was also true for oxygen. According to Michaelis (1940), "every oxidation (or reduction) can proceed only in steps of univalent oxidations (or reductions)." When this theory is applied to the reduction of oxygen, free radicals are formed (Michaelis, 1946).

For my scheduled seminar, I succinctly reported my experimental results and devoted the rest of the time to the formulation of the free radical theory as the fundamental mechanism of oxygen toxicity and its implications. It was very well received; in fact, Dr. Daniel L. Gilbert, then a graduate student in the Department, was so enthusiastic that he became my principal collaborator.

Since we wanted to publish the free radical theory in *Science*, we prepared a short manuscript from the large amount of collected material. It was indeed gratifying to read the referee's report, recommending it for a sublead article because "We are here lifted to a higher plane of observation in which the similarity of the two effects is established, first by citation of the literature, secondly by the submission of new data showing cumulative effects." The editor agreed on "its scientific importance" and it was published as a sublead article (Gerschman et al., 1954).

The essential feature of this theory is that both oxygen poisoning and the biological effects of X-irradiation share a common mechanism of action, which are free radicals. There is no doubt that the oxidizing effects of oxygen are mediated by free radicals, as discussed in several chapters of this volume. Oxygen also enhances the biological effects of radiation as discussed for example in a current publication (Stone, 1979). We have found that many chemical agents which afford protection against the toxic effects of radiation, also protect against oxygen poisoning (Gerschman, 1964; Gerschman et al., 1979).

Implications of the free radical theory abound (Gerschman, 1964). For example, at least part of the radiation effects in biological systems can be considered to act merely as a catalyst for the toxic effects of oxygen (Gilbert, 1972).

Other implications lead us to believe that during evolution, living matter had to develop adequate antioxidant defenses in order to resist the destructive capability of the oxygen in the air. Antioxidant defenses inhibit oxygen toxicity but cannot prevent it completely. We adapted for a normal concentration of oxygen in the atmosphere by developing adequate antioxidant defenses. Uncontrolled oxidations from a continuous small "slipping" in the defending system could be an important factor in the process of aging and also a factor contributing to life span. In this sense, one could consider that for toxic effects of oxygen to appear, there is no threshold tension necessary. Paracelsus (1967), in the sixteenth century, wrote, "Poison is in everything, and no thing is without poison. The dosage makes it either a poison or a remedy."

"Knowledge is like a sphere," wrote Pascal, "the larger its volume, the larger is its contact with the unknown." Nature does not easily yield its secrets; what seems to be the answer to a research inquiry is only the beginning of scrutinizing new aspects of the problem. And so I see those exciting times of discovery for me as part of scientific endeavor, which progresses slowly and asymptotically toward the elusive essence of knowledge.

References

Gerschman, R. (1964). Biological effects of oxygen. In: Dickens, F., and Neil, E. (Eds.). Oxygen in the Animal Organism. New York: Pergamon Press, Macmillan Co., p. 475.

Gerschman, R., and Fenn, W. O. (1954). Ascorbic acid content of adrenal glands of rat in oxygen poisoning. Am. J. Physiol. 176:6.

Gerschman, R., Gilbert, D. L., Nye, S. W., Dwyer, P., and Fenn, W. O. (1954). Oxygen poisoning and x-irradiation: A mechanism in common. Science 119:623.

Gerschman, R., Diez, O., and Rojo, R. (1979). Protective effects of WR-2721 on survival time of mice exposed to high oxygen pressure. (Abst. A-31-4). Abst. 6th Int. Cong. Radiat. Res. Tokyo, p. 108.

Gilbert, D. L. (1972). Oxygen and life. Anesthesiology 37:100.

Michaelis, L. (1940). Occurrence and significance of semiquinone radicals. Ann. N. Y. Acad. Sci. 40:39.

Michaelis, L. (1946). Fundamentals of oxidation and respiration. Am. Sci. 34:573.

Ozorio de Almeida, A. (1934). Recherches sur l'action toxique des hautes pressions d'oxygène. Comp. Rend. Soc. Biol. 116:1225.

Paracelsus. (1967). Quotation from Time, March 9, 1959. In: Seldes, G. The Great Quotations. New York: Pocket Books, p. 733.

Stone, H. B. (1979). The role of oxygen in the radiation biology of tumors. In: Caughey, W. S. (Ed.). Biochemical and Clinical Aspects of Oxygen. New York: Academic Press, p. 811.

3

Reactive Forms of Oxygen

SHELDON KANFER AND NICHOLAS J. TURRO

Research into the chemical and physical properties of at least ten forms of "active oxygen" is frequently encountered in today's scientific literature (see Table 1). Much of this diverse effort is unified in its attempt to illuminate the fundamental, or primary, chemical processes attending activated oxygen reactions. Gross features of chemical reactivity have been delineated in the last decade for a number of these species, most prominent among them oxygen atoms, singlet molecular oxygen, and ozone. This wealth of experimental research has made possible inquiry of a more detailed nature into reaction mechanisms and dynamics of active forms of oxygen.

In this review the major reactions involved in activated oxygen chemistry and the methodology employed will be emphasized. Results of individual investigations have

Table 1. *Major Species of Active Oxygen*

Species	References to recently published articles or reviews
Atomic oxygen	
$O(^3P)$	28, discussed in text
$O(^1D)$	Discussed in text
$O(^1S)$	17, p. 152
Dioxygen or molecular oxygen	
$O_2(^3\Sigma_g^-)$	119
$O_2(^1\Delta_g)$	77–81, discussed in text
$O_2(^1\Sigma_g^+)$	17, p. 183, 120–122
Ozone	
O_3	17, p. 237, 88, discussed in text
Superoxide	
O_2^-	114–118, discussed in text
Ozonide	
O_3^-	114, 123, 124
Dioxygenyl cation	
O_2^+	125–127

been selected to represent larger bodies of research. Investigations into the chemistry of five active oxygen species—triplet and singlet atomic O, $O_2(^1\Delta_g)$, O_3, $O_2^{\overline{\cdot}}$—are the major subjects of this review.

I. Atomic Oxygen

We start our discussion with "active O," i.e., oxygen atoms. The ground electronic configuration of the oxygen atom (an eight electron system) is $1s^22s^22p^4$. Electric dipole transitions among the three electronic states to which this configuration gives rise are strongly forbidden (1), and as a result the excited states are metastable (see Table 2). Nearly all of the experimental work on oxygen atoms has been performed in the *vapor phase* and, unless specified differently, the following discussion will refer only to vapor phase results. In Table 3 are summarized the experimental techniques most commonly used to generate the ground state of atomic oxygen, $O(^3P)$. Table 4 lists common methods of $O(^3P)$ detection.

A. Chemistry of $O(^3P)$

Trends in research in $O(^3P)$ reactivity reflect the typical progression from identification and rationalization of products of reactions to detailed mechanistic inquiry. In the period 1967–1977 the results of product studies of the reactions of $O(^3P)$ with representative organic compounds were reported.

$O(^3P)$ has been observed to undergo H-abstraction reactions with alkanes and ethers

Table 2. *States of Atomic Oxygen from Ground Configuration* (2)

State	Designation	Electronic energy (kcal/mole)	Natural radiative lifetime (sec)
$2s^22p^4$	^3P	0.00	—
$2s^22p^4$	^1D	45.38	150
$2s^22p^4$	^1S	96.65	0.71

Table 3. *Techniques for Generating $O(^3P)$*

$O(^3P)$ generation[a]	Quantum or chemical yield
1. Dissociation of O_2 in a microwave discharge (3)	$O:O_2(^1\Delta):O_2(^3\Sigma) = 0.03:0.11:1.00$ (4)
2. Hg-sensitized photodecomposition of N_2O (5): $Hg(^1S_0) + h\nu(2537\ \text{Å}) \rightarrow Hg^*(^3P_1)$; $Hg^* + N_2O \rightarrow N_2 + O(^3P) + Hg$	$\Phi O(^3P) = 0.78$ (6)
3. $N + NO \rightarrow N_2 + O(^3P)$ (7,8)	Chemical yield = 1.0 (9)
4. $NO_2 + h\nu(>3000\ \text{Å}) \rightarrow NO + O(^3P)$ (10)	$\Phi O(^3P)$ at 313 nm = 1 (11–13)
5. Vacuum ultraviolet photolysis of O_2, NO: $(O_2 \text{ or } NO) + h\nu(\lambda > 1250\ \text{Å}) \rightarrow O(^3P)$ (14)	$\Phi_{O_2}O(^3P) = 1$ at 1470 Å, 1849 Å (15, 16); $\Phi_{NO}O(^3P)$ variable (17, p. 176)
6. $SO_2 + h\nu(\lambda > 165\text{ nm}) \rightarrow SO + O(^3P)$ (18)	$\Phi O(^3P) = 0.27$ (19, 20)
7. $O_3 + h\nu(4400 < \lambda < 8500) \rightarrow O(^3P) + {}^3O_2$	$\Phi = 1$ in this region, known as Chappuis band (21)
8. O_2 passed over Re or W filament at 2300°K produces atomic oxygen $O(^3P)$ (22)	

[a]References in this column are to representative experiments.

Table 4. *Methods of Detecting $O(^3P)$*

$O(^3P)$ detection	Comments
1. $NO + O(^3P) \rightarrow NO_2^* - NO_2 + h\nu$	NO_2 chemiluminescence in spectral region 3875 Å $< \lambda <$ 14,000 Å indicative of $O(^3P)$ (23)
2. $O(^3P_2) + h\nu(1300\ \text{Å}) \rightarrow O(3s^3S^0)$ $O(3s^3S^0) \rightarrow {}^3P_{2,1,0} + h\nu$(1303.2, 1304.9, 1306.0 Å, resp.)	Detection by resonant fluorescence (24)
3. $O(^3P) + NO_2 \rightarrow NO + O_2$ $O(^3P) + NO \rightarrow NO_2^*$ as in 1.	A standard method of titration for determining concentration of atomic oxygen (25)
4. $O(^3P) + O(^3P) \rightarrow O_2(A^3\Sigma_\mu^+) \rightarrow {}^3O_2 + h\nu$	Herzberg band emission in ultra-violet and blue; intensity of emission α $(O)^2$ (26)
5. ESR (27)	

(29,5), C—C bond cleavage reactions with strained hydrocarbons (30) and a variety of reactions with olefins (31–33). In all cases it must be remembered triplet–singlet interconversion must have occurred at some point along the reaction coordinate from $O(^3P)$ + singlet reactants to one isolated singlet ground state product.

The products of the reaction of $O(^3P)$ with allene (34), for example, could be accounted for by initial addition to a C=C bond followed by intra-molecular rearrangements of the type indicated in Scheme 1 (compounds **1–3** represent isolated products). The origin of products of the reaction of $O(^3P)$ with a cyclic olefin such as cyclohexene (30) could also be understood in terms of an initial addition followed by the radical fragmentation pattern shown in Scheme 2 (compounds **1–5** represent isolated products).

As a final example, stoichiometry of the reactions of $O(^3P)$ with acetylene and methylacetylene (27) could be best accounted for by the following elementary reaction

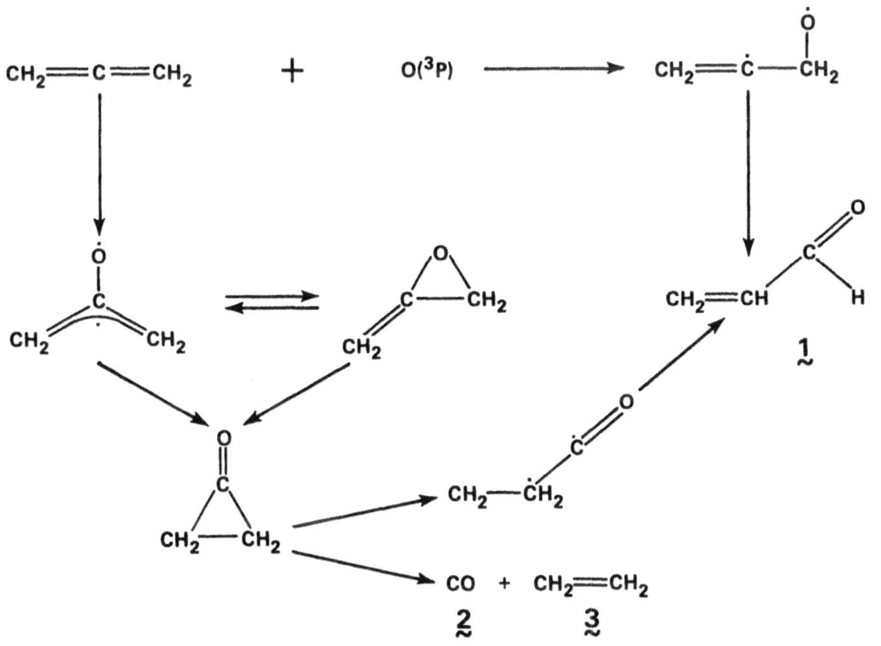

Scheme 1. Possible pathways in atomic oxygen plus allene reaction (35) (required intersystem crossing not shown explicitly in scheme).

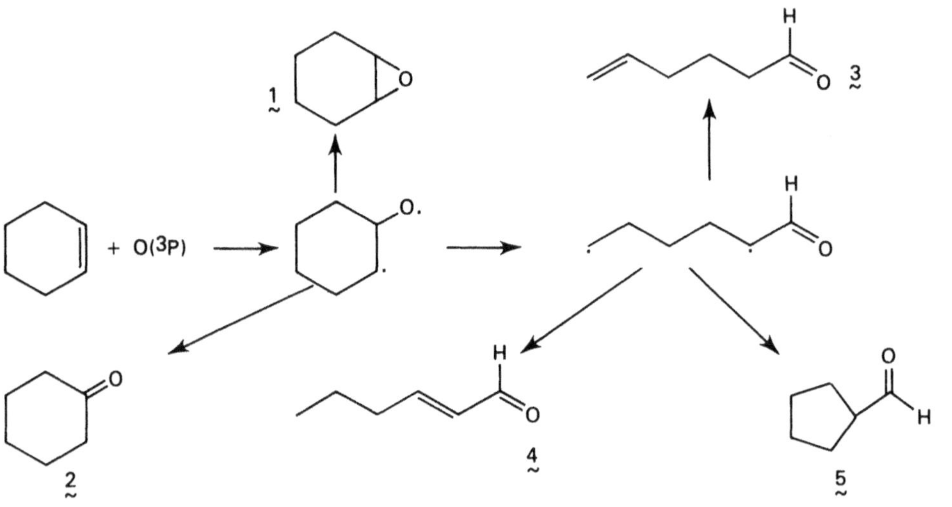

Scheme 2. Possible pathways in atomic oxygen plus cyclohexene reaction (31) (required intersystem crossing not shown explicitly in scheme).

steps:

$$O + CH{\equiv}CH \rightarrow CO + CH_2 \quad \Delta H^0 = -40 \text{ kcal/mole}$$

$$O + CH_3{-}C{\equiv}CH \rightarrow CO + CH_3CH$$

$$CH_3CH \rightarrow CH_2{=}CH_2^* \rightarrow HC{\equiv}CH + H_2$$

$$\searrow CH_2{=}CH_2$$

$$\Delta H^0 = -40 - -50 \text{ kcal/mole} \qquad \textbf{(1)}$$

One early dynamical study, of the reaction of $O(^3P)$ with cyclooctene, was published in 1973 (35). Some undetermined fraction of the initial biradical adduct is partitioned to cyclooctanone (analogous to product **2** in Scheme 2) in the collision-free environment of the experiment. Detailed information on the extent of deposition of reaction exothermicity, 118 kcal/mole, in the various vibrational modes of this product molecule was obtained from the infrared chemiluminescence emission spectrum. This spectrum revealed excitation of C—H stretching (2960 cm^{-1}), C=O stretching (1700 cm^{-1}), and CH_2 bending (1450 cm^{-1}) modes of vibration in cyclooctanone.

Most of the recent literature on $O(^3P)$ is devoted to questions of energy distribution in chemically activated $O(^3P)$-substrate adducts, i.e., of how the typically high reaction exothermicities are funneled into the molecular degrees of freedom of the various structures along the reaction coordinate. The results of an investigation of the reaction of $O(^3P)$ with carbon suboxide (10) are relevant to these questions:

$$O(^3P) + O{=}C{=}C{=}C{=}O \rightarrow 3CO \quad \Delta H^0 = -115 \text{ kcal/mole} \qquad \textbf{(2)}$$

The extent to which reaction exothermicity was channeled into vibrational excitation of CO was determined by flash absorption spectroscopy, and a random statistical model of energy distribution in the initial adduct accounted well for the observed CO vibrational state populations. The initial adduct was hypothesized to possess the symmetrical cyclotrione structure:

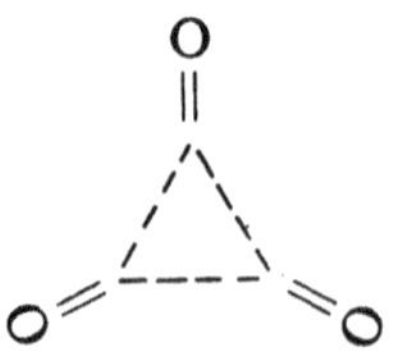

(3)

In this case, however, agreement between theory and experiment was revealed to be fortuitous. Further isotopic labeling studies of the reaction indicated that the CO molecule associated with the $O(^3P)$ reactant emerges from its precursor approximately 10 kcal/mole hotter in vibrational excitation than the two CO molecules associated with the termini of the C_3O_2 reactant. This result represents a significant contribution to our understanding of atomic oxygen reaction dynamics, for it is an example of molecular memory sustained in the course of triplet–singlet interconversion and possible geometrical loss of identity.

In dynamical studies of $O(^3P)$ reactions with allene and methylacetylene (7), reactions we recognize from earlier product studies, vibrational energy distributions of the CO product were measured and found to be in good agreement with those predicted from random statistical treatment of activated precursors:

$O(^3P) + CH_2{=}C{=}CH_2 \rightarrow$ 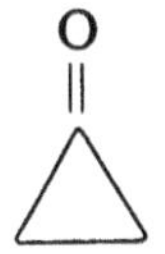 $\rightarrow$

chemically
activated
cyclopropanone

$CO + CH_2{=}CH_2$ **(4)**

$O(^3P) + CH_3{-}C{\equiv}CH \rightarrow$

$CH_3\backslash{=}C{=}O \rightarrow CO + CH_3CH$ (ethylidene)

chemically
activated
methylketene **(5)**

As a final example of $O(^3P)$ chemistry we consider a study of the reaction dynamics of $O(^3P)$ plus ^1CHF (18). Atomic H and F, HF, and CO are the observed products of the reaction, and as a guide to reaction energetics the energy level diagram of Hsu et al. is reproduced in Fig. 1.

If the approximately 192 kcal/mole of reaction exothermicity are distributed in a random fashion among the modes of the fluoroformaldehyde molecule, then agreement between theory and experiment for HF vibrational energy distribution is good. It is clear from Fig. 1, however, that CO may emerge from two energetically distinct product channels. A reaction exothermicity of 192 kcal/mole was supplied to HFCO evolving into the HF + CO exit channel and 56 kcal/mole for the H + F + CO channel (corresponding to dissociation energy of H—F bond), affording two sets of CO vibrational state populations. The CO vibrational state distribution determined experimentally was intermediate between pure HF and pure (H + F) exit channels, with HF:(H + F) = 2:3.

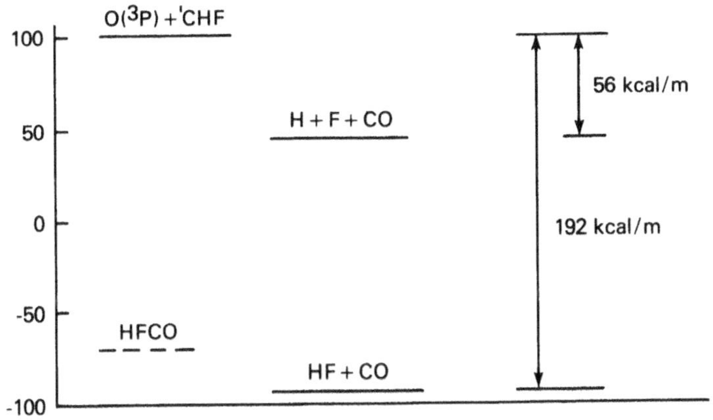

Figure 1. Energy level diagram for $O(^3P) + {^1CHF}$, adapted from Hsu et al. (18). There exists uncertainty in heat of formation of fluoroformaldehyde HFCO.

B. Chemistry of $O(^1D)$

Spin and energetic considerations prepare us for the different reactivity manifested by the lowest metastable state of atomic oxygen, $O(^1D)$. In Tables 5 and 6, respectively, are collected methods of generating and detecting this species.

A guide to the chemistry of $O(^3P)$ and of $O(^1D)$ atoms is available from the more familiar, isoelectronic species methylene (CH_2). The radical behavior of the ground state methylene triplet is analogous to the reactivity of $O(^3P)$. The more "concerted" reactivity patterns of the first excited state of methylene, like $O(^1D)$ a singlet, lead us to predict an analogous measure of concertedness in $O(^1D)$ reactions (48).

Singlet methylene inserts into C—H bonds of hydrocarbons forming (in the vapor phase) the corresponding methylated hydrocarbon in a state of chemical activation (49). We are thus inclined to expect similar insertion behavior of $O(^1D)$ leading, in this case, to chemically activated alcohols:

$$^1CH_2(\text{ref. } 50) + H_3C{-}H \rightarrow H_3C{-}CH_2{-}H$$

$$\Delta H^0 = -102 \text{ kcal/mole (ref. 51)} \quad \textbf{(6)}$$

$$O(^1D) + H_3C{-}H \rightarrow H_3C{-}O{-}H$$

$$\Delta H^0 = -130 \text{ kcal/mole (ref. 51)} \quad \textbf{(7)}$$

The classic work on $O(^1D)$ reactivity with organic substrates was published in 1964 (36). Product ratios of the reaction $O(^1D)$ plus propane were studied as a function of total system pressure and inert buffer gas. The following primary processes were deduced from the three types of products observed:

$O(^1D)$ + 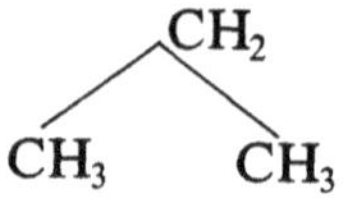 →

***Table 5.** Techniques for Generating $O(^1D)$*

$O(^1D)$ generation[a]	Quantum yield
1. $N_2O + h\nu(1849\text{Å}) \rightarrow O(^1D) + N_2$ (36)	$\Phi O(^1D) = 0.97$ at 2139 Å (37)
2. $O_3 + h\nu(248\text{ nm} < \lambda < 285\text{ nm}) \rightarrow O(^1D) + O_2(^1\Delta)$ (38)	$\Phi O(^1D) = 1$ throughout region (39–41)
3. $NO_2 + h\nu(\lambda < 2440\text{ Å}) \rightarrow O(^1D) + NO$ (42)	$\Phi O(^1D) = 0.5 \pm 0.1$ (43)
4. $CO_2 + h\nu(1200 < \lambda < 1670) \rightarrow O(^1D) + CO$	$\Phi O(^1D) = 1.0$ (44)
5. $O_2(^3\Sigma) + h\nu(\lambda < 1750\text{ Å}) \rightarrow O(^3P) + O(^1D)$ (42)	$\Phi O(^1D) = 1.0$ at 1470 Å (45)

[a]References in this column are to representative experiments.

***Table 6.** Methods of Detecting $O(^1D)$*

$O(^1D)$ detection	Comments
1. $O_3 + h\nu \rightarrow O(^1D) + O_2(^1\Delta)$ (see Table 4, 2) $O(^1D) + N_2O \rightarrow 2NO$ or $N_2 + O_2$ $NO + O_3 \rightarrow NO_2^* + O_2$ $NO_2^* \rightarrow NO_2 + h\nu(\lambda < 600\text{ nm})$ (41)	Method relies on detection of NO_2^* chemiluminescence, which is related to $O(^1D)$ concentration
2. Growth of $O(^3P)$ resonant absorption (Table 4, 2) evidence for depletion of $O(^1D)$, but $O(^1D)$ not observed directly (24)	
3. $O(^1D) \rightarrow O(^3P) + h\nu(6300\text{ Å})$ (46)	Detection of emission corresponding to forbidden transition
4. $O(^1D) + h\nu(1152\text{ Å}) \rightarrow O(3s^1D^0)$ (47)	Detection of allowed absorption

1. C—H Insertion Products

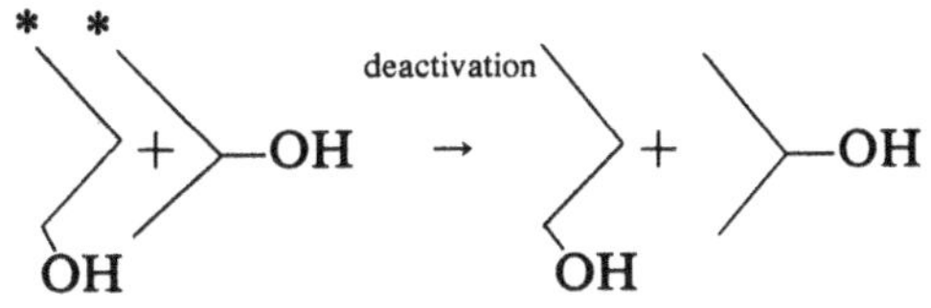

2. H-Abstraction Products

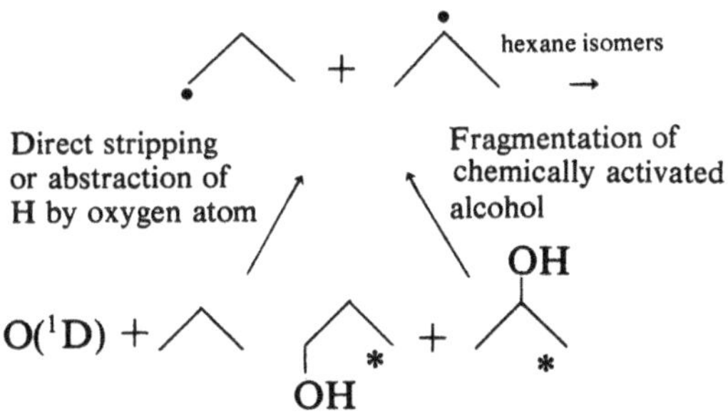

3. Fragmentation Products

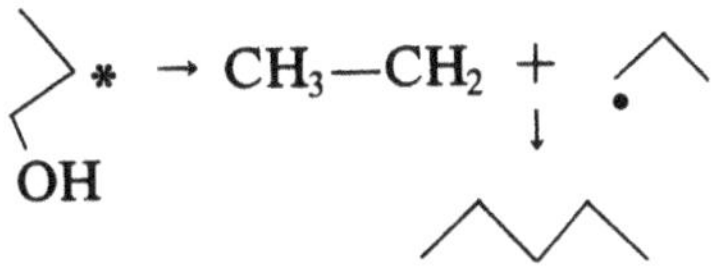

The pressure dependence of product type ratios is illustrated in Fig. 2. $O(^1D)$ was generated in this study by the 1849 Å photolysis of N_2O, and the overall reactant pressure was varied under conditions of fixed $p(N_2O)$:p(propane). The higher rate of stabilizing collisions at higher pressures facilitates alcohol formation (1) at the expense of fragmentation (3), whereas collision rate has little bearing on the rate of formation of type 2 products. Of significance is the observation that (rate of formation of primary alcohol): (rate of formation of secondary alcohol) = 3:1 = primary hydrogens:secondary hydrogens. This suggests $O(^1D)$ reactions with saturated hydrocarbons are regiochemically indiscriminate, and relative bond strengths are of little importance in determining the course of reaction.

Studies of excited state behavior confront the possibility of nonchemical degradation of energy stored in atoms and molecules. For this reason the effect of foreign gas on product-type ratio was a subject of inquiry in this work on $O(^1D)$ reactivity. Admission of sufficient pressures of foreign gases into the photolysis vessel caused the product spectrum to evolve continuously, as shown in

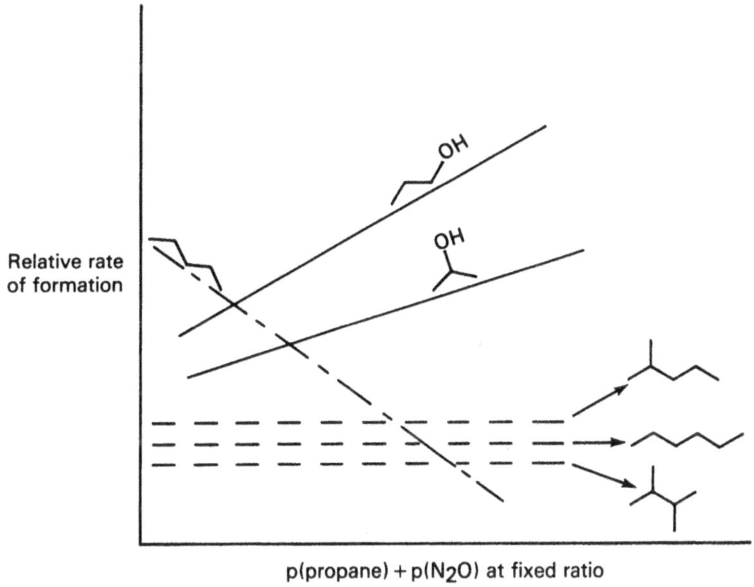

Figure 2. Reactant pressure dependence of product-type ratio.

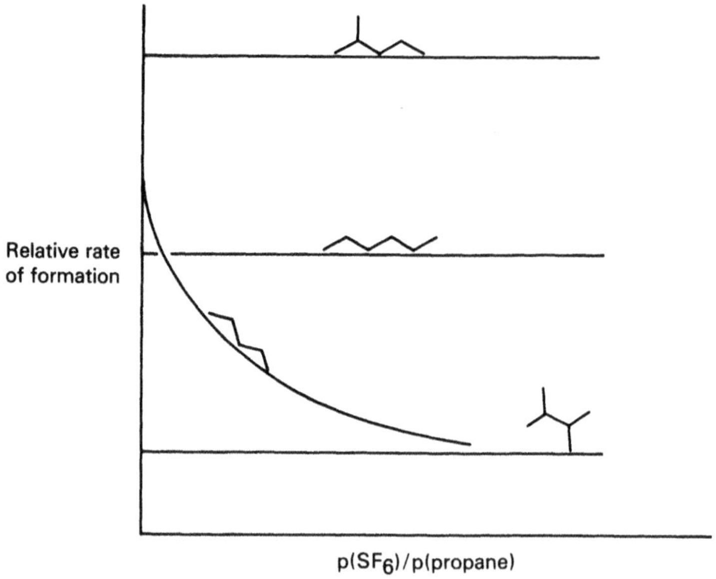

Figure 3. SF_6 pressure dependence of product-type ratio.

Figs. 3 and 4. SF_6 inhibits formation of type 3 fragmentation products without affecting relative rates of formation of hexanes. This observation suggests the primary reactive species $O(^1D)$ is not deactivated to the ground state $O(^3P)$, a species which abstracts hydrogen atoms, to a significant extent. Yet chemically activated intermediates capable of eventuating in pentane by fragmentation are stabilized by collision with SF_6. The trends depicted in Fig. 3 are identical with those of Fig. 2.

Xenon, on the other hand, modifies the entire product spectrum: hexane products reflecting secondary hydrogen abstraction by oxygen more successfully compete with primary hydrogen abstraction products. This more discriminate manner of attack, in con-

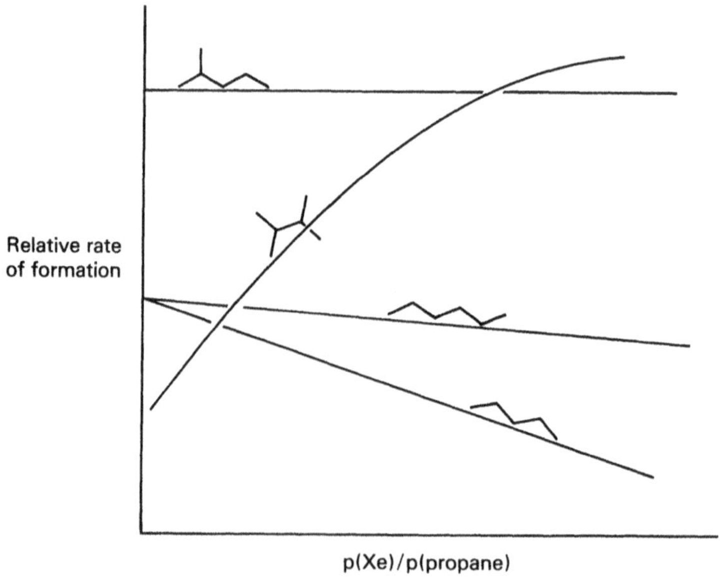

Figure 4. Xe pressure dependence of product-type ratio.

Table 7. O(^{1}D → ^{3}P) Quenching Rate Constants

Reactant	Product	Rate constant ($cm^3 molecule^{-1} sec^{-1}$)
Xe + O(^{1}D)	Xe + O(^{3}P)	$1.4 \pm 0.3 \times 10^{-10}$
N_2 + O(^{1}D)	N_2 + O(^{3}P)	$0.55 \pm 0.15 \times 10^{-10}$
CO + O(^{1}D)	CO + O(^{3}P)	$0.75 \pm 0.15 \times 10^{-10}$
NO + O(^{1}D)	NO + O(^{3}P)	$2.1 \pm 0.4 \times 10^{-10}$
CO_2 + O(^{1}D)	CO_2 + O(^{3}P)	$1.8 \pm 0.3 \times 10^{-10}$

trast to random attack, is characteristic of both O(^{3}P) and $^3CH_2$ (48), and suggests the heavy atom Xe supplies O(^{1}D) with a mechanism for singlet–triplet interconversion that cannot be supplied by SF_6, N_2O, or propane. Table 7 adapted from Okabe (52) lists O(^{1}D → ^{3}P) quenching rate constants for a variety of chemical species.

II. Dioxygen Species (Singlet Molecular Oxygen)

In the ensuing discussion of "active O_2" chemistry the relationship between electronic structure and reactivity of the type sought above with atomic oxygen will be of primary importance. The ground electronic configuration of the O_2 molecule, like that of the oxygen atom, gives rise to three electronic states, of which the upper two constitute significant active species. Optical transitions among states of the same electron configuration are strongly forbidden, rendering the $^1\Delta$ and $^1\Sigma$ states metastable (1) (see Table 8). In Tables 9 and 10, respectively, are listed prominent methods of generating and detecting $O_2(^1\Delta)$, the only active O_2 species to be treated in this review.

We will first highlight new trends in $O_2(^1\Delta)$ research and then proceed to recent fundamental mechanistic work on this species. The extensive singlet oxygen literature has received excellent treatment in a number of review articles and books (77–81).

Of particular interest in the literature since 1977 have been reports of research into the kinetic properties of singlet oxygen in aqueous micellar systems. Experiments of this type are conducted by sequestering organic substrates of established reactivity with singlet oxygen in cationic or anionic micelles. Aqueous singlet oxygen is generated by energy transfer to ground state oxygen from ionic or neutral photosensitizing dyes. Rate constants for quenching of singlet oxygen are determined spectroscopically by monitoring bleaching of absorption of the micellized singlet oxygen acceptor. In Table 11 a chronological abstract of these experiments is presented, which together indicate neither change of lifetime nor attenuation of reactivity of singlet molecular oxygen in micellar solution. A warning is in order here. Many workers assume that $O_2(^1\Delta_g)$ is the only reactive species generated in photosensitized oxidations. This need not be true and other species (e.g., O_2^-) may be involved in certain cases.

Singlet oxygen mechanistic questions are often investigated in a manner akin to those of atomic oxygen reactions. In the vapor phase $O_2(^1\Delta)$ reacts with ethylenes to produce

Table 8. States of Diatomic Oxygen Arising from Ground Configuration $(K)(K)(\sigma_g 2s)^2(\sigma_u 2s)^2(\sigma_g 2p)^2(\pi_u 2p)^4(\pi_g 2p)^2$ (53)

State designation	Electronic energy (kcal/mole)	Natural radiative lifetime
$O_2(X^3\Sigma_g^-)$	0.0	—
$O_2(a^1\Delta g)$	22.6 (54)	Radiative halflife = 2700 sec (55)
$O_2(b^1\Sigma_g^+)$	37.7 (54)	7–12 sec (56, 57)

Table 9. *Techniques for Generating $O_2(^1\Delta_g)$*

$O_2(^1\Delta_g)$ generation[a]	Quantum or chemical yield
1. Generation of $O_2(^1\Delta)$ by ground state absorption:	
A. $2\ O_2(^3\Sigma) + h\nu(1065\ nm) \rightarrow O_2(^1\Delta) + O_2(^3\Sigma)$ (58)	$\Phi(O_2(^1\Delta)) = 1.0$
B. $2\ O_2(^3\Sigma) + h\nu(632.8\ nm) \rightarrow 2\ O_2(^1\Delta)$ (59)	$\Phi(O_2(^1\Delta)) = 2.0$
2. Excitation of $O_2(^1\Delta)$ in microwave discharge (8,60)	Table 2, 1
3. Generation of $O_2(^1\Delta)$ by energy transfer from organic and inorganic donors (also known as photosensitization) (61): $D + h\nu(\lambda = 1065\ nm) \rightarrow D^*$ $D^* + O_2(^3\Sigma) \rightarrow D + O_2(^1\Delta)$	Efficiency of this diffusional quenching process yielding $O_2(^1\Delta)$ determined by energetic and spin properties of the particular system
4. Chemical generation of $O_2(^1\Delta)$:	
A. Thermal decomposition of phosphite ozonides (253°K) (62) $(PhO)_3PO_3 \rightarrow (PhO)_3PO + O_2(^1\Delta)$	Chemical yield unknown
B. Thermal decomposition of aromatic endoperoxides (63) 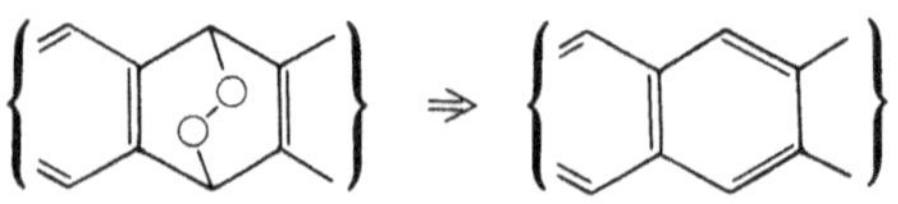 $+ O_2(^1\Delta)$	Chemical yield variable, up to (93 ± 2%) (64)
C. $H_2O_{2(aq)} + NaOCL_{(aq)} \rightarrow O_2(^1\Delta) +$ $NaCl + H_2O$ (65)	—
D. $ClFSO_3 + H_2O_2/OH^- \rightarrow O_2(^1\Delta)$ (66)	—
E. $2O_2^- + R(CO)OO(CO)R \rightarrow O_2(^1\Delta) +$ $2RCO_2^-$ (67)	Chemical yield unknown

[a]References in this column are to representative experiments.

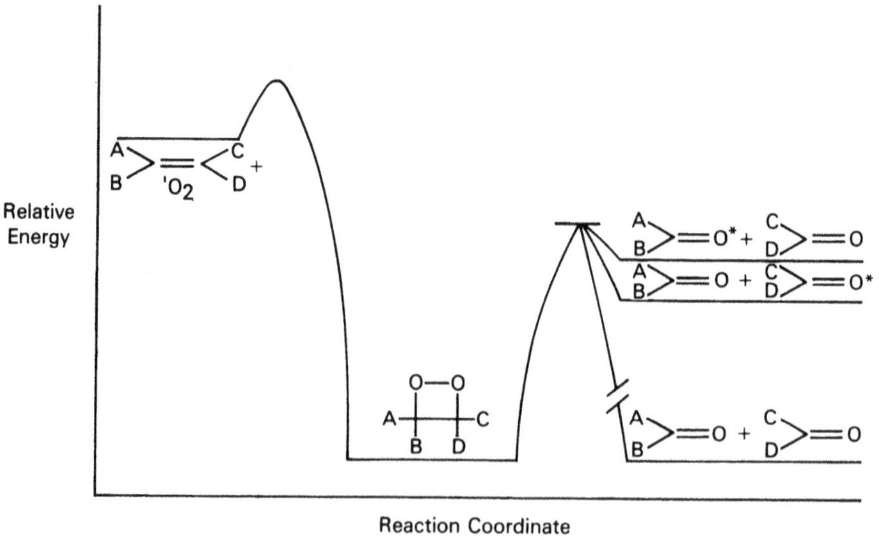

Figure 5. Energy level diagram for model dioxetane reaction, after Bogan et al. (86).

Table 10. *Methods for Detecting $O_2(^1\Delta)$*

$O_2(^1\Delta)$ detection	Comments
1. $(O_2(^1\Delta)\ O_2(^1\Delta)) \rightarrow (O_2(^3\Sigma)\ O_2(^3\Sigma)) + h\nu(634\ \text{nm},\ 701\ \text{nm})$ (65,68)	Detection of singlet oxygen "dimol emission," reverse of absorption process (Table 9, 1b)
2. $O_2(^1\Delta) \rightarrow O_2(^3\Sigma) + h\nu(1270\ \text{nm})$ (69)	Detection of parent singlet molecular oxygen emission
3. Detection of electron paramagnetic resonance absorption of $O_2(^1\Delta)$ (70)	
4. Detection of $O_2(^1\Delta)$ by absorption spectroscopy (71–73)	
5. Diagnostic tests for $O_2(^1\Delta)$:	
A. H_2O/D_2O lifetime effect (74)	
B. "ene" reaction with olefins possessing allylic hydrogen in plane of π-system (75) 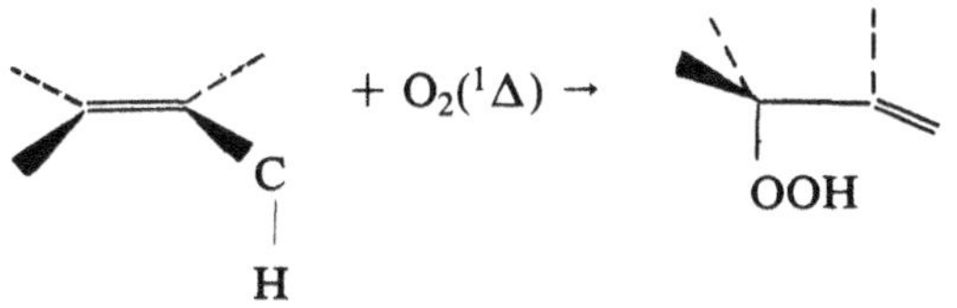	
C. Dioxetane reaction with electron-rich monoolefins (76) 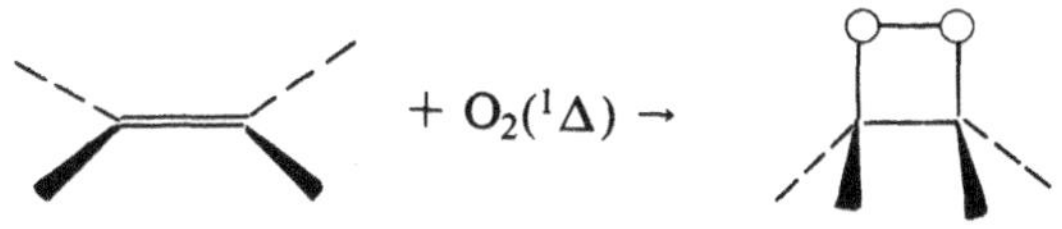	
D. Endoperoxide formation with isolated diene acene systems (75) 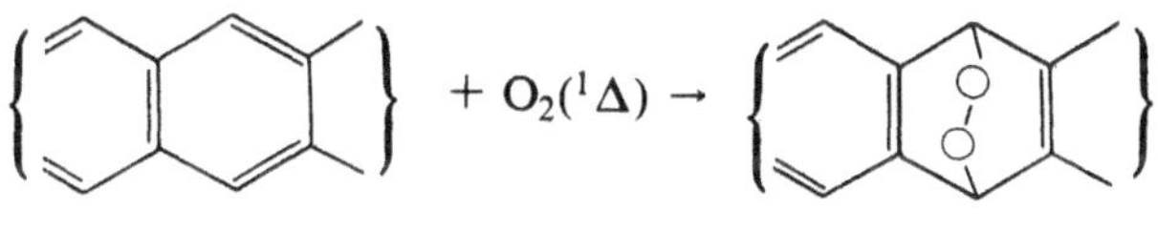	

dioxetanes that subsequently undergo decomposition to chemiluminescent carbonyl fragments. The chemiluminescent dioxetane reaction has been the subject of the most extensive mechanistic research for the wealth of dynamical information that can be obtained from emission spectroscopy of dioxetane decomposition fragments.

We consider first an energy level diagram for a model dioxetane reaction (see Fig. 5) analogous to the one modeling the $O(^3P)$ plus ^1CHF reaction discussed above. According to this model the initial $O_2(^1\Delta)$-olefin adduct is formed with excitation energy far in excess of the activation energy for decomposition of a "vibrationally cool" dioxetane. Electronically excited carbonyl products are formed on decomposition of this chemically activated dioxetane and, depending upon the energetics of the reaction at hand, these fragments are produced in varying degrees of vibrational and rotational excitation.

Branching ratios governing partitioning of excitation between the two carbonyl fragments are of fundamental mechanistic interest. An example of a similar reaction

Table 11. *Chronological Abstract of Singlet Oxygen-Micellized Substrate Kinetics Experiments*

Bulk solvent	Sensitizer	Surfactant	Substrate	Conclusions
H_2O (82)	MB[a]	SDS[b]	DPBF[c]	1. Kinetic results consistent with unhindered diffusion of singlet oxygen from aqueous to micellar phases. 2. Substrate found to possess comparable reactivity in micellar solutions and homogeneous solvent (CH_3OH)
H_2O/D_2O (83)	MB, eosin[d]	SDS, DDTBr[e]	DPBF, DMA[f]	1. In micellar systems composed of either surfactant the singlet oxygen lifetime determined by bulk solvent
D_2O (84)	MB	SDS, HDTBr[g]	DPBF	1. $\tau_{SDS}^{D_2O}(^1O_2) = 32 \pm 2\ \mu sec$ $\tau_{HDTBr}^{D_2O}(^1O_2) = 36 \pm 2\ \mu sec$ 2. Substrate found to possess comparable reactivity in micellar solution and homogeneous solvent
H_2O/D_2O (85)	2-AN[h]	SDS	DPBF	1. $\tau_{SDS}^{D_2O}(^1O_2) = 30\ \mu sec$ $\tau_{SDS}^{H_2O}(^1O_2) = 3\ \mu sec$

[a]Methylene blue, a cationic sensitizer.
[b]Sodium dodecyl sulfate, an anionic surfactant.
[c]1,3-Diphenylisobenzofuran.
[d]Eosin, an anionic sensitizer.
[e]Dodecyltrimethylammonium bromide, a cationic surfactant.
[f]9,10-Dimethylanthracene.
[g]Hexadecyltrimethylammonium bromide, a cationic surfactant.
[h]2-Acetonaphthone, a micellized sensitizer.

network branching out from a chemically activated intermediate was seen in the fragmentation of the $O(^1D)$-propane adduct above. Molecules such as ketene and allene (8) have been studied with an eye to the branching ratios of their respective dioxetanes.

Fundamental insight into the singlet oxygen dioxetane reaction has been provided at a different level by the molecular beam studies of Cross et al. (87). Observation of chemiluminescence from dioxetane decomposition fragments in crossed $O_2(^1\Delta)$-olefin beams is consistent with reaction occurring in a single bimolecular collision. Moreover, the transition state for dioxetane formation was shown to be promoted by activation energy in the relative translation coordinate. Reactant vibrational energy was less effective in producing chemiluminescence, indicating little change of reactant geometry at the transition state for dioxetane formation.

The singlet oxygen–acene reaction resulting in aromatic endoperoxide has also been the subject of recent mechanistic inquiry (64). Information on endoperoxide formation must be inferred in this case from the principle of microscopic reversibility, as the study cited is devoted to endoperoxide decomposition to acene and oxygen:

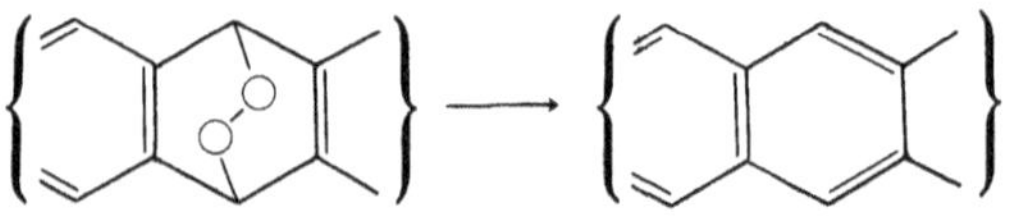

$$+ x\,^3O_2 + (1 - x)\,^1O_2 \qquad \mathbf{(8)}$$

Turro and Chow succeeded in correlating radical character in the transition state for endoperoxide decomposition (or activation enthalpy) in a series of endoperoxides with triplet oxygen yield on decomposition: high degree of radical character in the transition state (or high activation enthalpy) inevitably implied high $^3O_2$ yield. Further enhancement of $^3O_2$ yield, upon application of external magnetic fields up to 17 kG, was most dramatically demonstrated for those very endoperoxides possessing high activation enthalpy for decomposition. Together these observations provide a consistent picture of the mechanism of decomposition of aromatic endoperoxides (see Scheme 3).

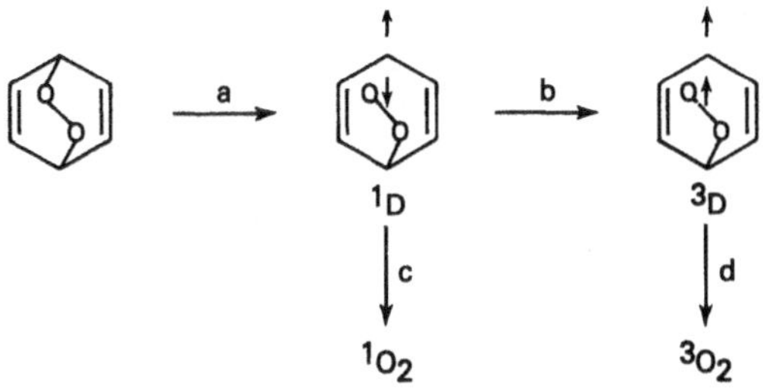

Scheme 3. Diradical mechanism for thermolysis of endoperoxides of aromatic compounds.

III. Trioxygen (Ozone)

We have chosen under the heading of O_3 chemistry to discuss recent advances in work on the mechanism of ozonolysis of olefins. Ozone ground state chemistry and photochemistry have been thoroughly discussed in review articles and books (88, 89).

The Criegee mechanism of ozonolysis of olefins (90), proposed in 1957, is highlighted by the following sequence of structures:

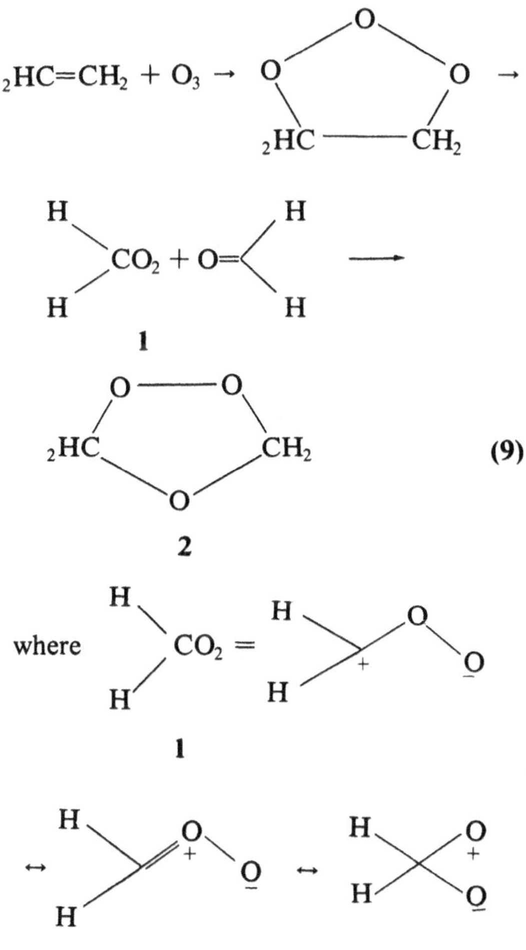

(9)

The nature of the parent species **1**, the so-called Criegee intermediate, is addressed in recently published mass spectroscopic (91) and microwave investigations (91,93) of the ozonolysis of ethylene. An ozone–ethylene mixture is permitted to warm from 77°K in a cell that can be adapted for either mass spectroscopic or microwave analysis of the equilibrium vapor. The temperature profile of the $m/e = 46$ peak (CH_2O_2) matched the profile of a group of microwave lines consistent with absorption by a cyclic peroxirane structure:

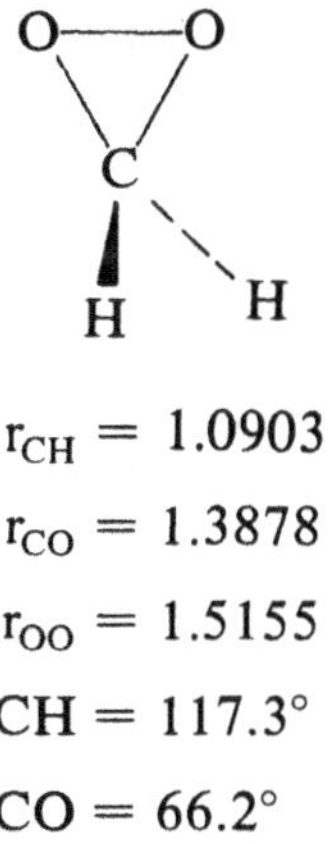

$$r_{CH} = 1.0903\ \text{Å}$$
$$r_{CO} = 1.3878\ \text{Å}$$
$$r_{OO} = 1.5155\ \text{Å}$$
$$HCH = 117.3°$$
$$OCO = 66.2° \qquad \textbf{(10)}$$

The $m/e = 46$ signal and the microwave lines were observed to persist in the temperature range −105°C to −60°C. At higher temperatures both analytical techniques reveal predominantly secondary ozonide **2** in vapor.

It is fascinating to note that the parent CH_2O_2 structure above was first invoked in 1954 (94), as an intermediate capable of accounting for product yields in ketene-O_2

photolysates. More recently this same structure was proposed as an intermediate in the $CH_2 + O_2 \rightarrow CO^*$ laser medium (95,96). In neither case, however, is the relevance of CH_2O_2 to the mechanism of ozonolysis of olefins explicitly recognized.

IV. Superoxide

The history of research into the monoanion of dioxygen, superoxide, is strangely episodic. The first spectroscopic identification of superoxide was reported as recently as 1959, in a publication which interpreted the ESR signal of alkali halide crystals heated in oxygen atmospheres in terms of an axially symmetric $O_2^{\cdot}$ crystal defect (97). A ten-year explosion of experimental effort devoted to the more complete spectroscopic characterization and alternative preparation of this molecule followed the initial report.

There can be little question that the implication of superoxide as an intermediate in certain enzyme-catalyzed dioxygen reductions in 1969 (100) had a strong influence on the direction of superoxide research. The subsequent discovery of superoxide dismutase introduced, from the chemical point of view, a sensitive nonspectroscopic method of detection of solution phase superoxide (101). This technique of chemical detection has and continues to play an important role in distinguishing singlet molecular oxygen-mediated from superoxide-mediated oxidations in dye-

Table 12. *Methods of Superoxide Generation and Detection, in Chronological Succession*

Generation	Phase	Detection	Reference
Alkali-halide crystals grown from melt, heated in oxygen atmosphere.	Crystal	ESR	97
Alkali-halide crystal grown from melt in oxygen atmosphere	Crystal	UV absorption, excitation, fluorescence, spectroscopy	107
Pulse radiolysis of oxygenated aqueous solutions	Aqueous	Transient UV absorption spectroscopy	98
1. e^- attachment to O_2 in He discharge 2. Charge transfer to O_2 in N_2O/Ar discharge	Vapor	Negative ion mass spectroscopy	99
γ irradiation of the molecular crystal H_2O_2—$CO(NH_2)_2$	Molecular crystal	ESR	108
1. Alkali metal/H_2O/O_2 or alkali metal/alcohol/O_2 matrix 2. γ irradiation of O_2-saturated alcohol matrix 3. Rapid cooling of aqueous or alcoholic supernatant of NaO_2 slurry	Molecular matrix	ESR	109
1. O_2 reduction by xanthine oxidase 2. $H_2O_2 + IO_4^-$ 3. $H_2O_2 + OCl^-$	Aqueous	Rapid-freeze ESR	100
Pulse radiolysis of oxygenated alkaline aqueous solution	Aqueous	Rapid-freeze ESR	110
Dye photosensitization	Liquid	Lactoperoxidase-superoxide absorption spectroscopy	101
Cation-promoted solubilization of superoxide salt	Liquid	Chemical	111
Crown ether-promoted solubilization of superoxide salt	Liquid	Chemical	112
O_2-reduction at surfactant-covered Hg electrodes	Aqueous	Electrochemical	113
Alkali superoxide-doped alkali halide crystals	Crystal	UV absorption spectroscopy	106

sensitized photooxidations of organic substrates (102).

In the course of this second generation of superoxide research reactivity patterns—superoxide as nucleophile (103), oxidant (104), reductant (104)—became manifest. Molecular beam electron-photodetachment experiments revealed fundamental information on electronic structure and molecular geometry of superoxide (105). The crystalline state spectroscopy of superoxide has again, more recently, been subject to investigation (106). In Table 12 are summarized, in chronological fashion, methods developed for superoxide generation and detection.

Still, our understanding of ground and excited state behavior of superoxide is nascent when compared with that of other species discussed in this review. Further exploration of these areas is assured a place in future superoxide research.

V. Summary

The prominence of active oxygen species as subjects of investigation throughout the spectrum of chemical research is reflected in the numerous review articles that have appeared and focused attention on the specific properties of these reactive species. In this more inclusive review of active oxygen chemistry five species—$O(^3P)$, $O(^1D)$, $O(^1\Delta_g)$, O_3, $O_2^{\overline{\cdot}}$—were discussed separately in detail, yet in a manner intended to reveal the interplay among the different research interests. The results of individual experiments were brought forward to highlight the problems addressed by workers in the field of active oxygen chemistry, and approaches to their elucidation were discussed.

References

1. Steinfeld, J.I., Molecules and Radiation, Harer and Row, New York, 1974, p. 50 (atomc transitions), p. 76 (diatomic transitions).
2. Moore, C.E., Atomic Energy Levels as Derived from Analyses of Optical Spectra, Vol. 1, Circular of the National Bureau of Standards 467, 1949, p. 45.
3. Macdonald, R.G., and Moore, C.B., J. Chem. Phys. 68:513 (1978).
4. Brown, R.L., J. Phys. Chem. 71:2492 (1967).
5. Takezaki, Y., Mori, S., and Kawasaki, H., Bull. Chem. Soc. Jpn., 29:1643 (1966).
6. Cvetanovic', R.J., J. Chem. Phys. 23:1203, 1208 (1955).
7. Lin, M.C., Shortridge, R.G., and Umstead, M.E., Chem. Phys. Lett. 37:279 (1976).
8. Bogan, D.J., Durant, J.C., Jr., Sheinson, R.S., and Williams, F.W., Photochem. Photobi. 30:3 (1979).
9. Bauch, D.C., Drysdale, D.D., and Horne, D.G., Evaluated Kinetic Data for High Temperature Reactions, Vol. II, The University, Leeds, 1973.
10. Hsu, D.S.Y., and Lin, M.C., J. Chem. Phys. 68:4347 (1978).
11. Pitts, J.N., Jr., Sharp, J.H., and Chan, S.I., J. Chem. Phys. 40:3655 (1964).
12. Ford, H.W., and Jaffe, S., J. Chem. Phys. 38:2935 (1963).
13. Jones, I.T.N., and Bayes, K.D., J. Chem. Phys. 59:4836 (1973).
14. Atkinson, R., and Pitts, J.N., Jr., J. Chem. Phys. 68:911 (1978).
15. Washida, N., Mori, Y., and Tanaka, I., J. Chem. Phys. 54:1119 (1971).
16. Sullivan, J.C., and Warneck, P., J. Chem. Phys. 46:953 (1967).
17. Okabe, H., Photochemistry of Small Molecules, John Wiley and Sons, New York, 1978.
18. Hsu, D.S.Y., Shortridge, R.G., and Lin, M.C., Chem. Phys. 38:285 (1979).
19. Driscoll, J.N., and Warneck, P., J. Phys. Chem. 72:3736 (1968).
20. Warneck, P., Marmo, F.F., and Sullivan, J.O., J. Chem. Phys. 40:1132 (1964).
21. Castellano, E., and Schumacher, H.J., Z. Phys. Chem NF 34:198 (1962).
22. Hughes, A.N., Scheer, M.D., and Klein, R., J. Phys. Chem. 70:798 (1966).
23. Fontijn, A., Meyer, C.B., and Schiff, H.I., J. Chem. Phys. 40:64 (1963).
24. Davis, D.D., Huie, R.E., Herron, J.T., Kurylo, M.J., and Braun, W., J. Chem. Phys. 56:4868 (1972).
25. Clyne, M.A.A., Physical Chemistry of Fast Reactions, ed. by B.P. Levitt, Plenum, London, 1973.

26. Young, R.A., and Sharpless, R.L., J. Chem. Phys. 39:1071 (1963).
27. Brown, J.M., and Thrush, B.A., Trans. Far. Soc. 63:630 (1967).
28. Cvetanonic, R.J., Advan. Photochem 1:115 (1963).
29. Marsh, G., and Heicklen, J., J. Phys. Chem. 71:250 (1967).
30. Havel, J.J., and Chan, K.H. J. Am. Chem. Soc. 97:5800 (1975).
31. Havel, J.J., and Chan, K.H., J. Org. Chem. 42:569 (1977).
32. Gaffney, J.S., Atkinson, R., and Pitts, J.N., Jr., J. Am. Chem. Soc. 97:5049 (1975).
33. Gaffney, J.S., Atkinson, R., and Pitts, J.N., Jr., J. Am. Chem. Soc. 98:1828 (1976).
34. Havel, J.J., J. Am. Chem. Soc. 96:530 (1974).
35. Moehlmann, J.G., and McDonald, J.D., J. Chem. Phys. 59:6683 (1973).
36. Yamazaki, H., and Cvetanovic, R.J., J. Chem. Phys. 41:3703 (1964).
37. Paraskevopoulos, G., and Cvetanovic, R.J., J. Am. Chem. Soc. 91:7572 (1969).
38. DeMOre, W., and Raper, O.F., J. Chem. Phys. 37:2048 (1962).
39. Amimoto, S.T., Force, A.P., and Wiesenfeld, J.R., Chem. Phys. Lett. 60:40 (1978).
40. Fairchild, P.W., and Lee, E.K.C., Chem. Phys. Lett. 60:36 (1978).
41. Philen, D.L., Watson, R.T., and Davis, D.D., J. Chem. Phys. 67:3316 (1977).
42. Ref. 17, p. 149.
43. Uselman, W.M., and Lee, E.K.C., Chem. Phys. Lett. 30:212 (1975).
44. Slanger, T.G., and Black, G., J. Chem. Phys. 68:1844 (1978).
45. Young, R. A., Black, G., and Slanger, T.G., J. Chem. Phys. 49:4758 (1968).
46. Gilpin, R., Schiff, H.I., and Welge, K.H., J. Chem. Phys. 55:1087 (1971).
47. Heidner, R.F., III, Husain, D., and Wiesenfeld, J.R., Chem. Phys. Lett. 16:530 (1972).
48. Kirmse, W., Carbene Chemistry, 2nd ed., Academic Press, 1971; pp. 213, 214 for all references cited.
49. Butler, J.N., and Kistiakowsky, G.B., J. Am. Chem. Soc., 82:759 (1960).
50. Moss, R.A., and Jones, M. Jr., eds., Carbenes, Vol. II, Ch. VI by P.P. Gaspar, G.S. Hammond, John Wiley and Sons, New York, 1975, p. 292.
51. Rossini, F.D., Wagman, D.D., Evans, W.H., Levine, S., and Jaffe, I., Selected Values of Chemical Thermodynamic Properties, Circular of the National Bureau of Standards 500, 1952.
52. Ref. 17, p. 153.
53. Herzberg, L., and Herzberg, G., Astrophys. J. 105:353 (1947).
54. Herzberg, G., Spectra of Diatomic Molecules, Van Nostrand, New York, 1950.
55. Badger, R.M., Wright, A.C., and Whitlock, R.F., J. Chem. Phys. 43:4345 (1965).
56. Childs, W.H.J., and Mecke, R., Z. Phys. 68:344 (1951).
57. Wallace, L., Hunten, D.M., J. Geophys. Rev. 73:4813 (1968).
58. Matheson, I.B.C., and Lee, J., Chem. Phys. Lett. 7:475 (1970).
59. Evans, D.F., Chem. Commun. p. 367 (1969).
60. Foner, S.N., and Hudson, R.L., J. Chem. Phys. 25:601 (1956).
61. Kearns, D.R., Khan, A.U., Duncan, C.K., and Maki, A.H., J.Am.Chem.Soc. 91:1039 (1969).
62. Wasserman, E., Murray, R.W., Kaplan, M.L., and Yager, W.A., J.Am.Chem.Soc. 90:4160 (1968).
63. Wasserman, H.H., and Scheffer, J.R., J.Am.Chem. Soc. 89:3073 (1967).
64. Turro, N.J., and Chow, M.F., J.Am. Chem.Soc. 101:3701 (1979).
65. Khan, A.U., and Kasha, M., J.Am.Chem. Soc. 92:3293 (1970).
66. Pritt, A.T., Jr., Coombe, R.D., Pilipvich, D., Wagner, R.I., Bernard, D., and Dymek, C., Appl. Phys. Lett. 31:745 (1977).
67. Danen, W.C., and Arudi, R.L., J.Am. Chem.Soc. 100:3944 (1978).
68. Benard, D.J., McDermott, W.E., and Pchelkin, N.R., Chem. Phys. Lett. 55:552 (1978).
69. Browne, R.J., and Ogryzlo, E.A., Proc. Chem. Soc. p. 117 (1964).
70. Falick, A.M., Mahan, B.H., and Myers, R.J., J. Chem. Phys. 42:1837 (1965).
71. Alberti, F., Ashby, R.A., and Douglas, A.S., Can. J. Phys. 46:337 (1968).
72. Collins, R.J., and Husain, D., J. Chem. Soc. Faraday II 69:145 (1973).
73. Ogawa, M. J. Chem. Phys. 53:3754 (1970).
74. Merkel, P.B., and Kearns, D.R., J.Am. Chem.Soc. 94:7244 (1972).
75. Foote, C.S., Acc. Chem. Res. 1:104 (1968).

76. Foote, C.S., and Lin, J.W.-P., Tetrahedron Lett. 29:3267 (1968).
77. Ranby, B., and Rabek, J.F., eds. Singlet Oxygen Reactions of Organic Compounds and Polymers, John Wiley and Sons, New York, 1978.
78. Schapp, P.A., ed., Singlet Molecular Oxygen, Dowden, Hutchinson and Mross, Inc., 1976.
79. Denny, R.W., and Nickon, A., Organic Reactions, Vol. XX, John Wiley and Sons, New York, 1973, p. 133.
80. Wayne, R.P., Adv. Photochem. 7:311 (1969).
81. Kearns, D.R., Chem. Rev. 71:395 (1971).
82. Gorman, A.A., Lovering, G., and Rodgers, M.A.J., Photochem. Photobiol. 23:399 (1976).
83. Usui, Y., Tsukada, M., and Nakamura, H., Bull. Chem. Soc. Jpn. 51:379 (1978).
84. Matheson, I.B.C., Lee, J., and King, A.D., Chem. Phys. Lett. 55:49 (1978).
85. Gorman, A.A., and Rodgers, M.A.J., Chem. Phys. Lett.: 55:52 (1978).
86. Bogan, D.J., Sheinson, R.S., and Williams, R.W., Am. Chem. Soc. Symp. Ser. 56:127 (1977).
87. Alben, K.T., Auerbach, A., Ollison, W.M., Weiner, J., and Cross, R.J., Jr., J.Am. Chem.Soc. 100:3274 (1978).
88. Blomquist, A.T., and Wisserman, H.H., eds., Organic Chemistry, Vol. 39. Academic Press, New York, 19; Bailey, P.S., Ozonation in Organic Chemistry, Vol. I, Olefinic Compounds, Academic Press, New York, 1978.
89. Ref. 17, p. 237.
90. Criegee, R., Rec. Chem. Prog. 18:111 (1957).
91. Martinez, R.I., Huie, R.E., and Herron, J.T., Chem. Phys. Lett. 51:457 (1977).
92. Lovas, F.J., and Suenram, R.D., Chem. Phys. Lett. 51:453 (1977).
93. Suenram, R.D., and Lovas, F.J., J.Am. Chem.Soc. 100:5117 (1978).
94. Strachan, A.N., and Noyes, W.A., Jr., J.Am.Chem.Soc. 76:3258 (1954).
95. Lin, M.C., Chemical lasers produced from $O(^3P)$ atom reactions. II. A mechanistic study of 5 μm CO laser emission from the O + C_2H_2 reaction. chemiluminescence and bioluminescence Proc. Int. Conf. Chemil. Eds. D. Hercules, J. Lee, and M.J. Cormier, Athens, Georgia, 1972.
96. Gordon, R.J., and Lin, M.C., Chem. Phys. Lett. 22:107 (1973).
97. Känzig, W., and Cohen, M.H., Phys. Rev. Lett. 3:509 (1959).
98. Czapski, G., and Dorfman, K.M., J. Phys. Chem.: 68:1169 (1964).
99. Fehsenfeld, F.C., Ferguson, E.E., and Schmeltekopf, A.L., J. Chem. Phys. 45: 1844 (1966).
100. Knowles, P.F., Gibson, J.F., Pick, F.M., and Bray, R.C., Biochem. J. 111:53 (1969).
101. Balny, C., and Douzou, P. Biochem. Biophys. Res. Commun. 56:386 (1974).
102. Rosenthal, I., Isr. J. Chem. 13:32 (1975).
103. Johnson, R.A., Nidy, E.G., and Merritt, M.V., J. Am. Chem. Soc. 100:7960 (1978).
104. Sawyer, D.T., Gibian, M.J., Morrison, M.M. and Seo, E.T., J. Am. Chem. Soc. 100:627 (1978).
105. Celotta, R.J., Bennett, R.A., Hall, J.L., and Siegel, M.W., J. Levine, Phys. Rev. A 6:631 (1972).
106. Rolfe, J., J. Chem. Phys. 70:2463 (1979).
107. Rolfe, J., Lipsett, F.R., and King, W.J., Phys. Rev. 123:447 (1961).
108. Ichikawa, T., and Iwaskai, M., J. Chem. Phys. 44:2979 (1966).
109. Bennett, J.E., Mile, B., and Thomas, A., Trans. Faraday Soc. 64:3200 (1968).
110. Nilsson, R., Pick, F.M., Bray, R.C., and Fielden, M., Acta Chem. Scand. 23:2554 (1969).
111. Peters, J.W., and Foote, C.S., J. Am. Chem. Soc. 98:873 (1976).
112. Valentine, J.S., and Curtis, A.B., J. Am. Chem. Soc. 97:224 (1975).
113. Divisek, J., and Kastening, B., J. Electroanal. Chem. 65:603 (1975).
114. Michelson, A.M., McCord, J.M., and Fridovich, I., Superoxide and Superoxide Dismutases, Academic Press, New York, 1971.
115. Bielski, B.H.J., and Gebicki, J.M., Adv. Radiat. Chem. 2:177 (1970).
116. Lee-Ruff, E., Chem. Soc. Rev. 6:195 (1977).
117. Fridovich, I., Adv. Enzymol. 41:35 (1974).
118. Sawyer, D.T., and Bibian, M.J., Tetrahedron 35:1471 (1979).
119. Hayaishi, O., Molecular Mechanisms of Oxygen Activation, Academic Press, New York, 1974.
120. Slanger, T.G., and Black, G., J. Chem. Phys. 70:3434 (1979).

121. Lawton, S.A., Novick, S.E., Broida, H.P., and Phelps, A.V., J. Chem. Phys. 66:1381 (1977).
122. Martin, L.R., Cohen, R.B., and Schat, J.F., Chem. Phys. Lett. 41:394 (1976).
123. Novick, S.E., Engelking, P.C., Jones, P.L., Futrell, J.H., and Lineberger, W.C., J. Chem. Phys. 70:2652 (1979).
124. Jacox, M.E., and Milligan, D.E., J. Mol. Spectrosc. 43:148 (1972).
125. Moseley, J.T., Cosby, P.C., Ozenne, J.-B., and Durup, J., J. Chem. Phys. 70:1474 (1979).
126. Tanaka, K., and Yoshimine, M., J. Chem. Phys. 70:1626 (1979).
127. Cotton, F.A., and Wilkinson, G., Adv. Inorg. Chem., 3rd ed. Wiley, New York, 1972, pp. 106, 414–418.

4

The Origin and Astronomical History of Terrestrial Oxygen

GEORGE H. HERBIG

The oxygen atoms available to biological processes in the atmosphere and the oceans and (to some extent) in the crust of the present earth have had a complex history. Astronomers, most of whom have an optimistic attitude toward the huge extrapolations and presumptions that are so often necessary on the frontiers of their profession, believe that one can now perceive the dim outlines of that history, although all readily admit that major gaps and uncertainties still exist. Let us sketch the outlines of this subject as it appears today.

I. The Astronomical Scenario

Our sun is one of slightly more than 10^{11} stars in our galaxy, or Milky Way system. This total population figure comes not from an actual census, but essentially by dividing the total mass of the galaxy, which is well determined from the dynamics of its rotation, by the mass of the average star. The galaxy has the form of a large, flattened disk in which the space density of stars is enormously higher (by a factor of over 10^8) very near the center than in the disk at a radius of 30,000 light years,[1] where the sun circulates about the center in a slightly noncircular orbit with a period of about 200 million (2×10^8) years. At this distance, the galaxy does not rotate as a solid body, but differentially, approximately as do the planets in the solar system, with more distant stars having longer orbital periods than those nearer the center. This galactic disk is surrounded by a faint, roughly spherical halo of faint, very much older stars, which is believed to outline the shape of the primeval pregalactic gas cloud when, in its collapse to form the disk, it gave birth to the first generation of stars about 15×10^9 years ago.

The idea that stars are *born*, not in some remote era of which we know nothing, but under circumstances that prevail today in the galaxy, was recognized in the 1940s. Once the nuclear sources of stellar energy were clearly identified, and the amount of fuel accessible to these sources was recognized to be (at most) the hydrogen content of a star, then it was apparent that the most luminous

[1] The light year is the distance traveled by light in 1 year, or 5.88×10^{12} miles $= 9.46 \times 10^{13}$ km; it is used here rather than the technical unit of distance, the parsec, which is 3.26 l.y.

stars were converting hydrogen to helium at rates that unarguably limited the duration of that phase to a few million years. We now recognize that there are short-lived stages in such stars during which other energy sources can be tapped, but there is no escape from the conclusion that such stars have very short luminous lifetimes (on an astronomical time scale). The fact of their existence requires a continuing source of new stars that operates today, and which must be located very near the present locations of these very luminous stars, because at typical velocities, there is simply no time for them to travel very far from their birthplaces.

Such luminous stars have masses up to 50–60 times that of the sun (whose mass is 2.0×10^{33} g), and at the present time they frequently occur in clusters having a total mass of the order of 1000 solar masses. In the outer parts of our galaxy today, there is only one known source of uncommitted material in such quantity: the clouds of interstellar gas and dust which one observes on photographs of our galaxy and of others like it, usually as vast, inchoate obscuring masses that block the light of stars beyond. That these interstellar clouds must indeed be the stellar birth sites is supported by the fact that the youngest, shortest-lived stars are always found imbedded in, or very near, such clouds.

The study not only of young stars of all luminosities but of the properties of interstellar material has come far in the past 35 years. There has sprung up, in addition, a considerable body of theory upon the processes of star formation, a theory, which although still imperfect, is an essential guide to the early stages which are (at the present time) not directly detectable because of the initially low temperatures, and the fact that they take place in sites heavily obscured by local interstellar dust. The fundamental imperative of star formation is gravitational collapse: once a cool, dense fragment of an interstellar cloud wanders or is pushed below a certain critical threshold of size, specified largely by its mass and temperature, self-gravitation prevails over internal forces and the cloud begins to collapse, irreversibly, toward stellar dimensions.

The details of this collapse—which for stars like the sun probably transpires in less than a million years—are currently a subject of intense investigation, which there is no need to discuss here. What is important is that astronomers are now able to recognize stars having masses about that of our sun which have recently emerged from this collapse stage, and are now very nearly stable. Many hundreds of these very young stars are now known, scattered through interstellar clouds like raisins in a pudding; they are called "T Tauri stars" after one of the best-known examples.

There is no reason to doubt that the sun was a T Tauri star at the beginning of its life, and there is today much interest in the possibility that studies of the early solar system as revealed in the surface of the moon and in the meteorites can be connected to the properties of today's T Tauri stars. We know from the T Tauri stars, all of which fluctuate erratically in their light output, sometimes on time scales of a few hours, and are subject to high-speed ejection of gas from their surfaces, that the young sun was still not quite the stable star that supports terrestrial life today. We recognize also that these stars are still quite cool in their interiors, and are still subject to a slow shrinkage (by typical factors of 2–5 in radius) until the interior temperature has risen to the point that the nuclear "burning" of hydrogen into helium can begin, and assure long-term stability. For a star having about the mass of our sun, this slow contraction from the beginning of the T Tauri stage to the onset of H burning requires 50 to 100 million years, but so high is the efficiency of the thermonuclear conversion of H to He that the H-burning lifetime (the so-called "main sequence" lifetime) of such a star is about 10×10^9 years. The time since the solidification of the oldest meteorites, generally believed to be by-products of the formation of the sun and the planetary system, is near 4.6×10^9 years. If this is also the lifetime of the sun essentially as we know it now, one sees that there need be no concern for any cutback in the solar luminosity for the next 5×10^9 years. In fact, the sun will *brighten* very slowly over that time.

The belief that the sun and its planetary system are very nearly contemporaneous is plausible, but is as yet not proved. As mentioned, the ages of meteorites (determined by radiochemical methods) run up to, but do not exceed, 4.6×10^9 years. The ages of the oldest samples returned from the lunar surface do not exceed 4.0 or 4.1×10^9 years, although there are hints of a somewhat older component as well. Theoretical evolutionary models can be fitted to the radius and surface temperature of the present sun if an age of 4.6×10^9 years is assumed, but the evolutionary changes are now so slow that model fitting is unable to *determine* the age precisely enough to be useful in this connection. One argument that the major planets originated in the same process that formed the sun is that the sun's axial rotation is in the same direction and its equator is in very nearly the same plane as that defined by the orbits of the major planets. This correspondence could be explained either if the planets were produced as a direct by-product of the original condensation of the sun, or through a later, close encounter of the primitive sun with another protostar. However, it seems unlikely that any event that took place very long after the formation of the sun would establish such an alignment. Consequently, most current models of the early history of the sun strive to produce a flattened disk as a product of the later stages of collapse, from which emerge a central sun (sometimes more than one) and a coplanar arrangement of residual material from which, it is supposed, the planets in turn condense.

Somewhat belatedly, let us consider how our oxygen came to be present in the interstellar clouds in the first place. Current belief is that the primeval, very massive gas cloud whose collapse produced first the galactic halo and then the galactic disk was composed almost entirely of hydrogen and helium, with at most a very small admixture of heavier elements. Such a composition, with the heavier elements having considerably less than 1% of their abundance (relative to H) in the sun, is actually observed today in the very old halo stars. There is much evidence, from the compositions of stars of all ages, that the heavy element content of stars has risen since the first few generations were born, steeply at first and then more gradually. There seems to have been only a small increase, if any, since the formation of the sun, some 5×10^9 years ago. If since the beginning, stars have always been formed from interstellar material, clearly the composition of the parent interstellar clouds must have changed systematically. There are two possible sources of such progressive heavy-element contamination: the debris of massive stars which is ejected suddenly back into the interstellar medium when such stars explode as supernovae; and the infall of extragalactic material into the galactic disk.

As already mentioned, the energy-generation rates of massive stars are so high that these objects rush through their H-burning evolution very quickly, and as their interior temperatures subsequently rise, they proceed successively through the alternate nuclear fuels that are available. It is obvious that this course cannot continue: the amounts of helium, carbon, etc., are limited, the region of the star where the temperature is sufficiently high for ignition is relatively small, and the yield of heavier-element reactions diminishes as the process proceeds to heavier nuclei. The final event is an explosive ignition of one of these elements, or a catastrophic collapse of the core (Arnett, 1973). Either has the consequence that the outer layers of the star, heavily contaminated with the heavy-element end products of earlier and concurrent nuclear burning episodes, are ejected into space at high velocity. It is this material, slowed down and mixed with ordinary interstellar material that it has swept up, that eventually finds its way back into interstellar clouds, and thence into the next generation of young stars.

Prediction of the amounts of various heavy elements that are produced in supernova explosions depends upon detailed modeling of the evolution of these massive stars, as well as on the theorist's ability to follow the complex sequence of events that take place very rapidly near the star's center just before the explosion and in the outer layers as a consequence of the detonation. The last word has not been said on this subject; its present state is described by Arnett (1978a). Furthermore, the composition of an interstellar cloud will reflect not only the recent incidence of supernovae in its immediate neighborhood but also the relative frequency of supernova explosions in stars of different mass, because the model calculations show that both the properties and amounts of heavy elements in the ejecta are strongly stellar-mass dependent. For example, the calculations of Arnett (1978b) indicate that the oxygen/magnesium ratio in the ejecta of stars of 50 solar masses is 7 times higher than at 12 solar masses, while the total mass of O ejected is 70 times greater for the more massive star. Thus one must know how the proportions of stars of various masses has changed over galactic history. One would therefore not be surprised to find local variations superimposed upon the general trend due to the overall aging of the galaxy. Of course we know nothing about the cloud from which the sun formed. That cloud has long since been dissipated, but there are a few clues (to be discussed later) that provide a hint that there was supernova activity in that vicinity just prior to the solidification of the meteorites.

Our knowledge of the constituents of interstellar clouds has multiplied remarkably in recent years as the result of earth-based spectroscopy at millimeter and centimeter wavelengths, and of far ultraviolet spectroscopy from space. One interesting result has been that there are apparent depletions of certain elements in the gas with respect to their normal cosmic abundance. The depletion of oxygen is small, but in the case of elements such as Ca, Al, or Ti it can amount to factors of 1000 or more. It is presumed that these elements have not been removed from the clouds, for if they were, newly formed stars would be similarly depleted, which is not observed. Rather it is assumed that the missing element fractions have condensed out as part of the dust grains which are mixed with the gas. The distribution of oxygen between its various carriers in the cloud gas has been modeled by Mitchell et al. (1978). In one typical cloud, at low total densities (~ 10 cm^{-3}) the oxygen is overwhelmingly in the neutral atomic form, but at high densities ($\sim 10^6$ cm^{-3}) it is divided mainly between neutral atoms and O_2, SO_2, and CO molecules.

The subject of the chemical evolution of our galaxy is an active one; a review of the situation will be found in Audouze and Tinsley (1976). An overall view of the problems of nucleosynthesis in stars, as well as a discussion of elemental abundances, has been presented by Trimble (1975).

The possibility that infall of intergalactic matter into the galactic disk may make a significant contribution to this picture is under discussion (for example, see Reeves and Meyer, 1978). It is supposed that such material represents gas that predated the formation of our galaxy, and thus is composed predominantly of hydrogen and helium. If so, its contribution would not be of interest in the present connection.

II. The History of Terrestrial Oxygen

The ages of the oldest meteorites is 4.6×10^9 years, while the oldest known terrestrial surface rocks are 3.8×10^9 years. It is not surprising that still more ancient rocks have not been preserved, in view of weathering and the extensive vulcanism that must have been prevalent at that time. Furthermore, the evidence from the moon shows that the very heavy meteoritic bombardment, to which the earth must have been exposed to an even greater degree because of its larger mass, tapered off about 3.9×10^9 years ago. The density of meteoritic infall on the moon was so high that the artifacts of bombardment in still earlier times have apparently been obliterated. We assume that this cessation represented the end of the accretion of the earth from solidified material in the flattened disk surronding the young sun. What was the composition of this material from which our planet was assembled?

Table 1. *Relative Compositions of Solar Atmosphere, Continental Crust of Earth, and Crust Augmented by Oceans and Atmosphere*[a]

Element	Sun	Crust	Crust + oceans + atmosphere
H	4.8×10^{10}	1.4×10^5	1.1×10^6
C	2.2×10^7	1.8×10^3	1.8×10^3
N	4.7×10^6	1.4×10^2	2.0×10^3
O	4.0×10^7	2.9×10^6	3.4×10^6
Si	1.0×10^6	1.0×10^6	1.0×10^6

[a]Data expressed as numbers of atoms, normalized to silicon = 1.0×10^6.

Table 1 is a listing of several principal chemical elements, and their abundances in solar surface material, which we regard as a sample of the original cloud from which the sun and planets originated. There have probably been only very minor changes in the solar surface composition since the sun arrived on the main sequence. The solar wind and solar flare activity carry away small quantities of surface material, but deep convective circulation must erase any selective evaporation effects. Only the surface abundance of lithium, a very minor constituent, has certainly diminished during the sun's lifetime.

It has long been known that the mean densities of the inner planets (Earth = 5.5 g cm^{-3}, Venus = 5.3, Mercury = 5.5) are too high for them to be composed of solar material with its overwhelming abundance of H and He. On the other hand, the giant planets—particularly Jupiter with a mean density of 1.35 gm cm^{-3}—are compatible with a cold, compressed solar composition. It is easy to understand how gaseous H and He could have escaped thermally from the weak gravitational fields of the inner planets, but one cannot account for the huge deficiences of carbon, nitrogen and the noble gases (neon, argon, krypton, xenon) in this way. Table 1 also lists a few abundances in the Earth's crust, to illustrate the differences with respect to the sun. An attractive explanation of this pattern is that the earth was formed by the subsequent accretion of meteoritic material which had solidified in space, and thus was composed largely of minerals and pure metals which were condensible under those conditions. It is then hypothesized that a powerful "solar wind" swept out through the inner planetary system, carrying away the free gases, volatile elements and the finer solid particles, all of which were therefore lost to the embryo inner planets before their accretion. At the outer planets, the combination of their stronger gravitational fields and greater distances from the sun caused this selectivity to be unimportant. The fact that many T Tauri stars, which are believed to be sunlike stars in approximately this stage of their evolution, show spectroscopic evidence of just such "stellar winds" (both steady and sporadic) encourages one to take this hypothesis seriously.

The material of the young Earth, therefore, although derived from gas of solar composition, was subject to selectivities which swept away elements in proportions dependent upon the volatility of the compounds that carried the bulk of that element at that place in the primitive disk nebula. Oxygen was protected to a degree, because although much must have been lost as gaseous CO and H_2O, some was protected by being locked up in refractory mineral solids (silicates, titanates, aluminates, oxides) that constitute the bulk of the meteoritic material which survives today. Not all volatile gases could have been swept away: for example, according to Sill and Wilkening (1978), all terrestrial H_2O could have been preserved in minerals, as water of hydration, while noble gases could have been trapped as clathrates in H_2O ice. It is generally believed that the atmosphere and oceans originated by subsequent outgassing of the outer layers of the Earth over geological time, and that the present composition results from the modification of that original mix by geochemical and photochemical processes, and (relatively recently) by the effects of photosynthesis.

The evolution of the terrestrial atmosphere has long been a subject for speculation, but only since about 1950 has it been a matter for serious scientific investigation. The monograph by Walker (1977) discusses the subject thoroughly, with emphasis on atmospheric chemistry. A computer simulation of atmospheric evolution by Hart (1978) illustrates the interplay of the issues that are believed to be involved, and results in numerical prediction of atmospheric composition and other parameters that, while probably not unique, do seem to fit the known facts.

Briefly, the history of the atmosphere is believed to be as follows: It originated from slow outgassing of the crust and outer mantle, the main constituents being H_2O and CO_2 as in modern volcanic gas, with a minor amount of N_2. The H_2O quickly condensed into the oceans, while the concentration of CO_2 slowly fell as it dissolved in the oceans, reacted with silicates, and deposited out as limestone and other sedimentary carbonate rocks. Thus, after about 2.0 or 2.5×10^9 years, the atmosphere must have been largely nitrogen. Free oxygen was not yet present in significant quantity. It is too active chemically to have been present in the original outgassed atmosphere. The most likely abiotic source of free oxygen is the photodissociation of H_2O by ultraviolet sunlight at the top of the atmosphere, followed by the escape of the hydrogen. The surface concentration of O_2 would be small and rise slowly if only this mechanism were acting. But there is evidence from sedimentary deposits that free oxygen in significant amount first appeared only 2×10^9 years ago, and there is reason to believe that this was due to the appearance of photosynthetic organisms. The rise in the oxygen concentration must have been accompanied by the appearance of an ozone layer in the high atmosphere, and the absorption by O_2 and O_3 of the lethal solar ultraviolet should in turn have led to the proliferation of more complex surface organisms and a higher level of photosynthetic activity. Hart's (1978) calculations suggest that a sharp upturn in the O_2 concentration (to about ⅓ of the present value) occurred as the accumulated result of these effects about 400–500 million years ago. This might be related to the sudden increase in multicellular life forms which took place slightly less than 600 million years ago, near the pre-Cambrian/Cambrian boundary.

The fraction of O_2 in the atmosphere has now risen to 21%, and ignoring the intervention of man, would continue to increase slowly. However, the atmospheric content of free oxygen is negligible compared to that locked up as H_2O in the oceans (about 1000 times as much) or in various minerals of the Earth's crust (about 6000 times as much, in the continental crust alone).

To some degree, it is possible to understand the very different compositions of the atmospheres of Venus and Mars in terms of the same effects that operated here, but as modified by the special circumstances of those planets. Table 2 lists the major constituents and total pressures of the atmospheres of the three terrestrial planets. We can assume that the primeval material from which all three formed, including its complement of trapped gases, was the same.

Venus, of essentially the same mass and radius but nearer the sun than is the Earth, from the beginning had too high a surface temperature for H_2O to form oceans, so the rapid disappearance of CO_2 into carbonate rocks was prevented; Venus' dense, CO_2-rich atmosphere may well contain the total amount of CO_2 outgassed since the beginning. This thick infrared-opaque atmosphere, through the greenhouse effect, is largely responsible for the present high surface temperature of the planet. In the Earth, the oceans took up the halogens (chlorine, bromine, etc.) that are today a significant component of volcanic gases, but in Venus these survive as atmospheric acids. That the original H_2O of Venus' atmosphere has

***Table 2.** Compositions of the Atmospheres of the Terrestrial Planets*[a]

Constituent	Venus[b]	Earth	Mars[c]
N_2	3.4	78	2.7
O_2	0.0069	21	0.13
H_2O	0.14	Variable	0.03 (variable)
CO_2	96.4	0.031	95.3
Ar	0.0019	0.93	1.6
Total pressure (bars)	17.7[b]	1.0	0.0065

[a]Data expressed in percent by volume.

[b]Venus: sampled at approximate height of 24 km by Pioneer Venus probe (Oyama et al, 1979). The total pressure at the surface is about 90 bar.

[c]Mars: the composition of the atmosphere near surface, from the Viking Landers (Owen et al., 1977). The total pressure quoted is for the surface of the Martian reference ellipsoid, not as measured at the Viking landing sites. There is a significant diurnal and seasonal variation (Seiff and Kirk, 1977).

disappeared is clear (Table 2), but the mechanism is not so certainly identified. Probably it was the same as initiated the original O_2 production on the Earth: photodissociation of H_2O by solar ultraviolet light at the top of the atmosphere, followed by escape of the hydrogen, and takeup of the free oxygen by surface material.

Our reconstruction of the history of Mars' atmosphere is much less satisfactory. If Mars originally outgassed the same materials as did the Earth, an arguable point, then much of the H_2O would have frozen out and must now be incorporated in the surface layers. But if a proportional amount of N_2 and CO_2 was also released, the low total pressure, isotope ratios, and the concentrations of noble gases preent indicate that most of it has vanished, the N_2 possibly to space through photochemical processes at the top of the atmospheres, the CO_2 by gradual chemical takeup in the surface. A discussion of the issues and alternatives is given by Anders and Owen (1977).

The principal isotope of oxygen is ^{16}O; in terrestrial ocean water ^{17}O and ^{18}O occur in lesser abundance by factors of 2700 and 490, respectively. However, variations in these ratios occur in various terrestrial samples because the rates of the physical and chemical processes involved are mass dependent. The signature of such fractionation is that it produces approximately twice the change in the $^{18}O/^{16}O$ ratio as in $^{17}O/^{16}O$. Advantage has been taken of this process to infer the rates of ancient fractionation processes from the oxygen isotope ratios in terrestrial materials (Epstein et al., 1977). It has now been recognized that another effect is present in the oxygen of certain inclusions in many very primitive stone meteorites in which the departure from the norm of $^{17}O/^{16}O$ is equal to that of $^{18}O/^{16}O$ (Clayton, 1978). This must be due to the addition of some ^{16}O-rich material to the meteoritic mineral before it solidified. Not only does this show that such solid material in the solar system is a mix of components with different nucleosynthetic histories, but it raises the question where one might find oxygen rich in ^{16}O. The most obvious and least objectionable proposal is that, a few million years before the formation of the planets, our parent interstellar cloud was contaminated by ^{16}O-rich material ejected from one or more nearby supernovas (see Reeves, 1979, for a discussion of this and other isotopic anomalies), and which was not taken up everywhere in the same proportions when solids froze from the gas. The degree by which the total oxygen abundance was thus raised is of course small; a 5% increment of pure ^{16}O is a possibility. The extreme inhomogeneity of the stony meteorites, apparently assembled from materials that tell very different chemical and nucleosynthetic stories, shows that the early years of the inner solar system were a time whose complexities we have begun to appreciate, but not yet to understand.

III. Summary

The oxygen in the atmosphere and oceans, only a small fraction of that imbedded in crustal rocks, is believed to have been synthesized from lighter nuclei in the interiors of massive stars, and then ejected into space when these stars exploded as supernovas. These heavier elements found their way into interstellar clouds, from which subsequent generations of stars were born. The disk around our own sun from which the planets condensed carried in its frozen-out minerals some of this oxygen, plus trapped gases, and from these the Earth formed. The subsequent history of the atmosphere, oceans, and crust reflect the chemical and photochemical processes that operated here, largely abiotically for approximately the first half of the Earth's history, and subsequently as modified by the rise of photosynthesis.

References

Anders, E., and Owen, T. (1977). Mars and Earth: Origin and abundances of volatiles. Science 198:453–465.

Arnett, W. D. (1973). Explosive nucleosynthesis in stars. Annu. Rev. Astron. Astrophys. 11: 73–94.

Arnett, W. D. (1978a). Abundances: An overview of thermonuclear astrophysics for supernova observers. Mem. Soc. Astron. Ital. 49: 431–444.

Arnett, W. D. (1978b). On the bulk yields of nucleosynthesis from massive stars. Astrophys. J. 219:1008–1016.

Audouze, J., and Tinsley, B. M. (1976). Chemical evolution of galaxies. Annu. Rev. Astron. Astrophys. 14:43–79.

Clayton, R. N. (1978). Isotopic anomalies in the early solar system. Annu. Rev. Nucl. Part. Sci. 28:501–522.

Epstein, S., Thompson, P., and Yapp, C. J. (1977). Oxygen and hydrogen isotopic ratios in plant cellulose. Science 198:1209–1215.

Hart, M. H. (1978). The evolution of the atmosphere of the Earth. Icarus 33:23–39.

Mitchell, G. F. Ginsburg, J. L., and Kuntz, P. J. (1978). A steady-state calculation of molecule abundances in interstellar clouds. Astrophys. J. Suppl. 38:39–68.

Owen, T., Biemann, K., Rushneck, D. R., Biller, J. E., Howarth, D. W., and Lafleur, A. L. (1977). The composition of the atmosphere at the surface of Mars. J. Geophys. Res. 82: 4635–4639.

Oyama, V. I., Carle, G. C., Woeller, F., and Pollack, J. B. (1979). Venus lower atmospheric composition: Analysis by gas chromatography. Science 203:802–805.

Reeves, H. (1979). Cosmochronology after Allende. Astrophys. J. 231:229–235.

Reeves, H., and Meyer, J.-P. (1978). Cosmic-ray nucleosynthesis and the infall rate of extragalactic matter in the solar neighborhood. Astrophys. J. 226:613–631.

Seiff, A., and Kirk, D. B. (1977). Structure of the atmosphere of Mars at mid-summer at midlatitudes. J. Geophys. Res. 82:4364–4379.

Sill, G. T., and Wilkening, L. L. (1978). Ice clathrate as a possible source of the atmospheres of the terrestrial planets. Icarus 33: 13–22.

Trimble, V. (1975). The origin and abundances of chemical elements. Rev. Mod. Phys. 47: 877–976.

Walker, J. C. G. (1977). "Evolution of the Atmosphere," New York: Macmillan Publishing Co.

5

Significance of Oxygen on Earth

Daniel L. Gilbert

The earth's crust is composed of four shells, commonly termed spheres. These are the lithosphere, or the solid portion; the hydrosphere, or the liquid portion; the atmosphere, or the gaseous portion; and the biosphere, or the living portion. Oxygen is the most prevalent element in the earth's crust, having an atom abundance of 53.8% (Gilbert, 1964). Oxygen plays a significant role in each of these components. We will briefly discuss the interrelationships of oxygen in these spheres at the present time. Then we will speculate on the past and attempt to forecast the future.

Table 1. *Oxygen Reservoirs in the Earth's Crust*

Earth shell	Emoles O_2	% of total
Lithosphere	371,000[a]	90.5
Hydrosphere	38,500[b]	9.4
Atmosphere	37.7[b]	0.009
Biosphere	0.0812[c]	0.00002
Total	410,000	100

[a]Calculated from percent composition of the constituents as given by Day (1979) and a rock mass of 2.3666×10^7 Eg. The prefix, E, represents the multiple, 10^{18} (Gilbert, 1964). For details, see Table 2.

[b]From Gilbert (1964).

[c]In the biosphere, the ratio of Eg atoms of O to C = 24.9/10.6, which corresponds to a ratio of Emoles of O_2 to C = 1.17 (Gilbert, 1964). Assuming that the carbon mass is 45% of the 1.843 Eg of the biospheric dry matter (Likens et al., 1977), then there are 0.0691 Emoles of carbon in the biosphere. Thus, the Emoles of O_2 = 1.17 times 0.0691 Emoles of carbon = 0.0812 Emoles of O_2.

I. Geochemical Oxygen Cycle

The content and exchange of oxygen in each of the four shells of the earth's crust has been estimated. The lithosphere comprises the bulk of the earth's crust. On a weight basis, it is 95% of the total. The bulk of the atoms, or 84% of the total atoms of the crust, resides in the lithosphere (Gilbert, 1964). Oxygen is the predominant element in the lithosphere; for this reason, Goldschmidt pointed out that the lithosphere can be considered as the oxysphere (Mason, 1966). Thus, 58% of all the atoms in the lithosphere are oxygen atoms (Gilbert, 1964). Table 1 shows that 90% of the oxygen atoms in the crust of the earth reside in the lithosphere. There are about 364,000 Emoles of O_2 in the lithosphere. The symbol E represents the prefix erda, which signifies the value of 10^{18} (Gilbert, 1964). The quantities of oxygen in Table 1 are expressed in the form of molecular oxygen.

Most of the oxygen atoms of the lithosphere are associated with silicon, as can be

Table 2. *Oxygen Reservoirs in the Lithosphere*

Reservoir	Emoles O_2[a]	% of total
SiO_2	237,000	64
Al_2O_3	54,300	14
H_2O	27,500[b]	7
CaO	11,000	3
MgO	10,400	3
Na_2O	7,470	2
Fe_2O_3	6,980	2
FeO	6,390	2
K_2O	4,010	1
TiO_2	3,140	1
Others	2,800[c]	1
Total	371,000	100

[a]Unless otherwise noted, values were calculated using percent composition as given by Day (1979) and using a rock mass of 2.3666×10^7 Eg (Gilbert, 1964).

[b]From Gilbert (1964).

[c]Part of this is from carbonate in the sediments and was taken from Gilbert (1964).

seen in Table 2. Fifteen percent of the oxygen atoms in the lithosphere are with aluminum, and 8% occur as water. There is a significant amount associated with calcium, magnesium, and iron. Only 3% of the oxygen atoms in the lithosphere are combined with the remaining elements. In spite of the fact that the lithosphere contains the bulk of the oxygen in the earth's crust, the lithosphere has the capability of taking up even more oxygen. Thus, the iron in the lithosphere has the potential of taking up an additional 3195 Emoles of O_2 (Table 2).

The hydrosphere is the next biggest reservoir of oxygen in the earth's crust, with about 10% of the total atoms of oxygen. Practically all of this oxygen is in the form of water (Table 3).

Table 3. *Oxygen Reservoirs in the Hydrosphere*

Reservoir	Emoles O_2[a]	% of total
H_2O	38,400	99.8
SO_4^{2-}	84[b]	0.2
Inorganic CO_2	4.15	0.01
H_3BO_3	0.86	0.002
Dissolved O_2	0.39	0.001
Total	38,500	100

[a]From Gilbert (1964), unless otherwise noted.

[b]From Garrels et al. (1976).

About 0.01% of the total oxygen reserves in the earth's crust are in the atmosphere (Table 1). Almost all the oxygen atoms in the atmosphere are in the form of molecular oxygen (Table 4). There are other atmospheric constituents that contain oxygen, but quantitatively they are minor reservoirs. For example, the atmospheric ozone reservoir contains about 10^{-4} Emoles O_2 (Gilbert, 1968). The amount of oxygen in dry air is 20.95 mole % or 23.14 mass %. Assuming 0.66 mole % of water vapor in the atmosphere, the partial pressure of oxygen at sea level is 158 torr or 211 mbar (Gilbert, 1964). A table of conversion factors between various pressure units is given by Gilbert (1981b).

The extreme range of barometric pressures at sea level is 877 to 1084 mbar (McWhirter, 1979). The corresponding oxygen pressures range from 137 to 169 torr (182 to 225 mbar). In 1975, there were 327 ppm (parts per million by volume) of atmospheric carbon dioxide (Bolin, 1977); this corresponds to 0.247 torr. Thus, there are almost 650 more oxygen atoms in atmospheric O_2 than in atmospheric CO_2.

Using the approximate barometric equation (Gilbert, 1964), the partial pressure of a given constituent as a function of altitude can be estimated, if the mole % composition in the atmosphere is constant. The percent composition of the major constituents does remain constant up to 100 km due to vertical mixing in the atmosphere (Walker, 1977).

Table 4. *Oxygen Reservoirs in the Atmosphere*

Reservoir	Emoles O_2	% of total
O_2	37.0[a]	98.25
H_2O	0.6[a]	1.60
CO_2	0.058[b]	0.15
Total	37.7	100

[a]From Gilbert (1964).

[b]Bolin (1977) gives the value of CO_2 in 1975 as 327 ppm, which corresponds to 0.0327 vol %. Since the ratio of the gas constant for air to the gas constant for CO_2 is 0.08206/0.08149, or 1.007, the vol % equals this ratio times the vol %, or 0.0329 vol %. The total moles of air equals 176.6 Emoles (Gilbert, 1964). The moles of CO_2 equals 0.000329×176.6 Emoles, or 0.0581 Emoles (Gilbert, 1964).

For oxygen at a temperature of 0°C, the equation is:

$$P_{O_2} = 158 \times \exp(-h/8) \quad (1)$$

where h is the altitude in km and P_{O_2} is the partial pressure of oxygen in torrs. This equation agrees with the standard atmosphere within 1% up to an altitude of 6 km. However, at higher altitudes, the equation overestimates the pressure by about 5% at 8 km and 10% at 10 km. The temperature decrease at higher altitudes (Gilbert, 1964) is the main cause for this overestimation.

The solubility coefficient for oxygen in sea water is 0.00180 m*M*/torr (Gilbert, 1964), corresponding to 0.284 m*M* oxygen gas dissolved in sea water exposed to air at 10°C. Both decreasing the temperature and lowering the salinity increases the oxygen solubility. Thus, for sea water, the oxygen solubility is 0.212 m*M* at 25°C and 0.319 m*M* at 5°C; whereas the oxygen solubility in distilled water is 0.258 m*M* at 25°C, 0.355 m*M* at 10°C, and 0.402 m*M* at 5°C (Green and Carritt, 1967). The oxygen in the surface water is generally in equilibrium with the atmosphere. However, the oxygen concentration decreases to a minimum as the depth of the water is increased, and then often increases as the depth is further increased (Duedall and Coote, 1972). Convection currents and photosynthetic activity play an important role in the oxygen concentration depth profile. The O_2 exchange between hydrosphere and atmosphere appears to be

Table 5. *Oxygen Reservoirs in the Biosphere*[a]

Component	Emoles O_2 ($\times 10^{-4}$)	% of total
Plant		
Marine		
Upwelling zones	0.004	0.0004
Continental shelf	0.12	0.02
Open ocean[b]	0.44	0.05
Algal beds	0.53	0.07
Estuaries	0.61	0.08
Total	1.70	0.22
Continental		
Forest		
Tropical rain[c]	336	41.5
Other	410	50.7
Total	746	92.2
Other	60	7.5
Total	806	99.7
Total	808	99.9
Animal		
Marine Total	0.438	0.05
Continental		
Forest	0.298	0.04
Other	0.147	0.02
Total	0.445	0.06
Total	0.883	0.11
Total plant and animal	809[d]	100

[a]All values based on assumption that the carbon mass is 45% of the dry mass (Likens et al., 1977), and that the ratio of Emoles of O_2 to C equals 1.17 (Gilbert, 1964). See Table 1 for further details. The plant data are from Whittaker and Likens (1975) and the animal data are from Likens et al. (1977).

[b]The plant communities in the open ocean contain the lowest average plant biomass density, amounting to only 0.13 mole O_2/m^2.

[c]The tropical rain forests contain the highest average plant biomass density, amounting to 2000 moles O_2/m^2.

[d]This corresponds to 0.0691 Emoles of carbon.

very rapid, amounting to 0.18 Emoles O_2/yr (Gilbert, 1964).

An infinitesimal quantity of oxygen in the earth's crust is in the biosphere (Table 1). Yet, oxygen is the second most abundant element in the biosphere, having a percent atom abundance of 24.9%. Hydrogen has an atom abundance of 62.6%. The third and fourth most abundant elements are carbon and nitrogen, which have atom abundances of 10.6% and 1.1%, respectively (Gilbert, 1964).

Table 5 shows that over 99% of the oxygen reservoir in the biosphere (total = 0.0809 Emoles O_2) occurs in land plants. Forest lands account for over 90% of this reservoir, and tropical rain forests contribute more than 40%. The biomass density of the tropical rain forest is 15,000 times that of the plant communities in the open sea (Whittaker and Likens, 1975). Even the extreme desert regions have an average biomass density six to seven times greater than in the open sea plant biomass.

There is an interchange between the oxygen reservoirs in the shells composing the earth's crust. In spite of the small quantity of oxygen in the biosphere, it is intimately related to the atmospheric O_2 exchange. Photosynthesis is the major source of atmospheric oxygen, and can replenish the entire atmospheric O_2 pool in about 2500 years (Table 6 and Fig. 1).

$$CO_2 + H_2O \underset{\text{respiration}}{\overset{\text{photosynthesis}}{\rightleftharpoons}} CH_2O + O_2 \quad (2)$$

There is the reverse process of respiration and decay which deletes the atmospheric O_2

Table 6. *Atmospheric O_2 Exchange*

O_2 exchange	Emoles/Myr[a]	Reservoir turnover (Myr)[b]
Input into O_2 Reservoir		
Photosynthesis		
Ocean	3,000[c]	0.0123
Continent	12,000[c]	0.00308
Total	15,000	0.00247
Photodissociation of water	0.007[d]	5300
Output from O_2 Reservoir		
Respiration and decay		
Plant ocean	1,000[c]	0.0370
Plant continent	7,500[c]	0.00493
Animal (and decay)	6,440[e]	0.00575
Total	14,940[e]	0.00248
Methane oxidation	50[f]	0.74
Rock weathering		
C to CO_2	6[g]	6.2
S to SO_4	3[g]	12.3
FeO to Fe_2O_3	1[g]	37.0
Total	10	3.7
Fossil fuel combustion	350[g]	0.106

[a]Quantities are given in Erdamoles (1 Emole = 10^{18} moles); time is given in megayears or millions of years (Myr).

[b]O_2 reservoir of 37 Emoles divided by the rates in Emoles/Myr.

[c]Calculated from Whittaker and Likens (1975), with the assumption that the carbon mass is 45% of the dry mass (Likens et al., 1977). One Emole of C was taken to be equivalent to 1 Emole of O_2. Net photosynthesis in the ocean equals 2000 Emoles O_2/Myr and on land equals 4500 Emoles O_2/Myr for a total of 6500 Emoles O_2/Myr.

[d]From Walker (1977).

[e]Adjusted so that the total respiration and decay is the gross photosynthetic value of 15,000 Emoles/Myr minus the sum of the methane oxidation rate of 50 Emoles/Myr, plus the weathering rate of 10 Emoles/Myr.

[f]Erhalt and Schmidt (1978) estimate that about 50 Emoles/Myr of CH_4 are produced. It is assumed that this equals the net O_2 consumed.

[g]From Holland (1978).

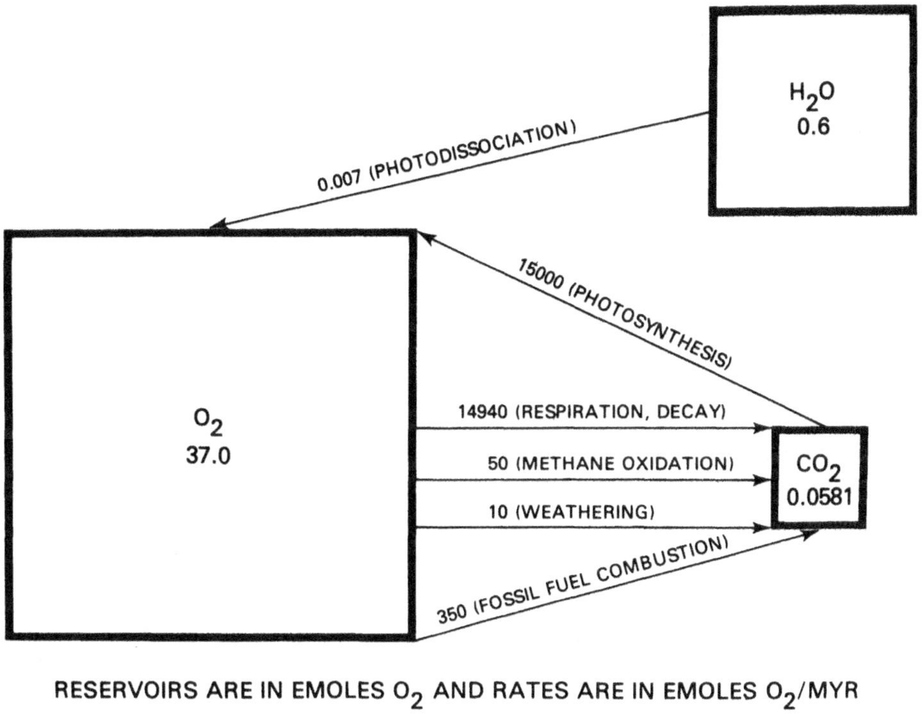

Figure 1. O_2 exchanges in the atmosphere. One Emole = 10^{18} moles (Gilbert, 1964). One Myr = 10^6 years (Cloud, 1976). The rate of O_2 production from photosynthesis equals the sum of the rates of O_2 removal from respiration, decay, methane oxidation, and weathering. This is the O_2 cycle.

reservoir, and which proceeds at almost the same rate of photosynthesis. Photosynthesis liberates more ^{16}O relative to ^{18}O than present in the atmosphere; but this isotopic exchange is counterbalanced by respiratory consumption of more ^{16}O relative to ^{18}O than present in the atmosphere (Dole, 1965).

The net photosynthetic rate equals the gross photosynthetic rate of 15,000 Emoles/Myr minus the plant respiration rate of 8500 Emoles/Myr. The abbreviation, Myr, represents megayear, i.e., 1 Myr equals 10^6 years or one million years (Cloud, 1976).

This net rate of 6500 Emoles/Myr can replenish the 37 Emoles of O_2 in 5700 years. What would happen if this net rate of 6500 Emoles/Myr suddenly ceased? Assuming that the O_2 utilization rate would remain constant, then the O_2 content in the atmosphere would decrease by less than 0.02% in one year.

Four times more photosynthesis occurs on land than in the ocean, in spite of the fact that the land area is less than half of the ocean area. The most productive areas are swamps and marshes, where the net photosynthesis rate equals 110 moles $O_2/(yr\ m^2)$. Photosynthesis in the sea is confined to the upper 200 m, but oxygen is consumed at all depths (Tappan, 1974).

The algal bed areas also have a very high net rate, being 95 moles $O_2/(yr\ m^2)$. One of the most productive areas in the world is in the tropical rain forests, where the net rate is 1400 Emoles O_2/Myr, corresponding to 80 moles $O_2/(yr\ m^2)$. On the other hand, the open ocean is one of the least productive areas, with a net rate of only 5 moles $O_2/(yr\ m^2)$; however, due to its large area, the total net rate of ocean oxygen production amounts to 1500 Emoles O_2/Myr (Whittaker and Likens, 1975).

The global efficiency of the gross photosynthetic production for the visible spectrum is 0.6% (Whittaker and Likens, 1975). Broda (1975) has pointed out that the oxygen in the water of the biosphere is turned over very rapidly by this process. The amount of water consumed in photosynthesis equals the carbon biospheric reservoir of 0.0691 Emoles. The turnover time of the oxygen in this water by the photosynthetic process is only 46 years.

If removal processes ceased, so that photosynthesis would release O_2 from all the inorganic CO_2 in the combined hydrosphere and atmosphere, then the atmospheric oxygen would increase only to 41.2 Emoles, an increase only 11% above its present value of 37.0 Emoles.

Respiration and decay processes not only consume O_2, but also convert organic carbon to CO_2. However, there are small alternate pathways, which accomplish this same conversion to CO_2 and the accompanying consumption of O_2. One pathway is through methane production and subsequent oxidation (Watson et al., 1978). When organic matter is reduced to methane, the net reaction is:

$$CH_2O + H_2O \rightarrow CH_4 + O_2 \qquad (3)$$

Methane is destroyed by $^{\cdot}OH$ radicals, producing $CH_3^{\cdot}$ radicals. These methyl radicals undergo a series of oxidations, resulting in the formation of carbon monoxide. The hydroxyl radicals convert the CO to CO_2 (Walker, 1977; Holland, 1978). The overall reaction is:

$$CH_4 + 2O_2 \rightarrow CO_2 + 2H_2O \qquad (4)$$

However, the methane is just an intermediate in the decay of organic carbon into CO_2, as can be seen by summing up these reactions:

$$CH_2O + O_2 \rightarrow CO_2 + H_2O \qquad (5)$$

The moles of methane produced equal the moles of carbon in the organic matter oxidized by O_2. In other words, the 50 Emoles/Myr of CH_4 produced are equivalent to the removal of 50 Emoles/Myr of O_2 in the atmosphere. This process depletes the atmospheric O_2 in about 750,000 years.

The photosynthetic production of oxygen minus respiration, decay, and methane removal of oxygen is the net biological production of oxygen; this rate is about 10 Emoles/Myr (Table 6). Figure 2 shows the cycle of this slow rate on the oxygen exchange processes. Since oxygen is reactive, it will react with the reduced substances on earth. In particular, the weathering of sedimentary rocks by oxygen oxidizes the exposed carbon, sulfur, iron (Holland, 1978), and manganese (Goldschmidt, 1954). Figure 2 represents these substances as X in their reduced states and as XO_2 in their oxidized states. For the carbon oxidation, water is released from the organic carbon before combining with oxygen. Note that this process of carbon oxidation is chemically the same as respiration; it differs essentially in proceeding at a very slow rate. The total weathering process occurs at a rate of about 10 Emoles/Myr (Holland, 1978). This carbon arises from the sinking and burial of the organic carbon in the marine environment (Garrels et al., 1976; Walker, 1977; Holland, 1978). The slow cycle is completed when the remainder of this organic carbon becomes oxidized to carbon dioxide and the oxidized XO_2 is reduced back to X (Walker, 1977).

The entire oxygen content of the atmosphere is removed by the weathering process in about 4×10^6 years. This is a short time in comparison to the age of the earth. If the earth did not possess an oxygen-producing biosphere to replenish this removal of oxygen from the atmosphere, then little or no free oxygen would be present in the atmosphere. Margulis and Lovelock (1974) estimate that under these conditions, the atmospheric oxygen would be 1 mbar or less. The significance of an appreciable content of molecular oxygen in any planetary atmosphere is evidence for the presence of an oxygen-producing biosphere.

The methane cycle plus the carbon burial-weathering cycle account for about 60 Emoles O_2/Myr; the sum of these cycles

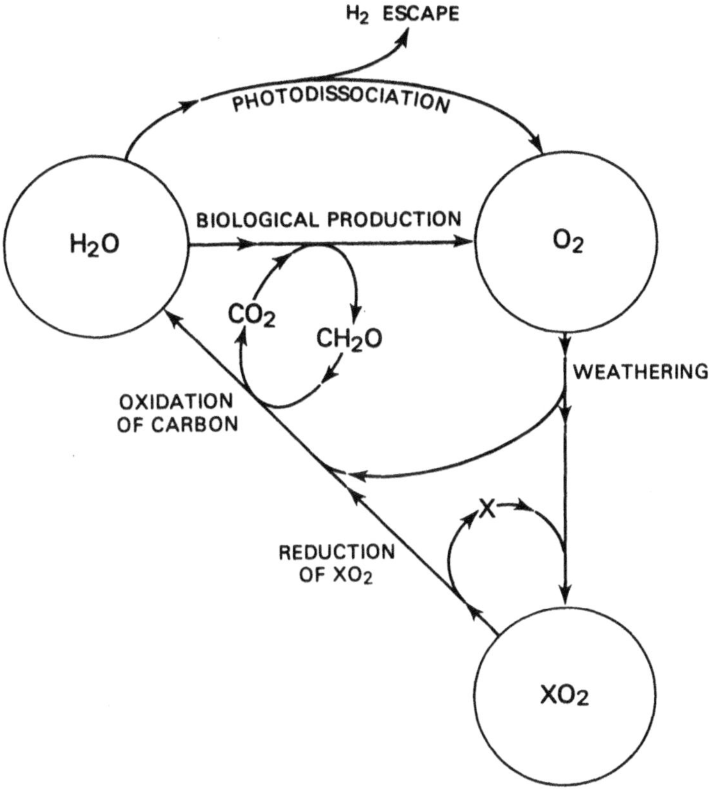

Figure 2. Slow processes in the oxygen cycle. Biological production equals photosynthetic production of O_2 (15,000 Emoles/Myr) minus the sum of the removal rates of O_2 (14,990 Emoles/Myr) from respiration, decay, and methane oxidation. Hence, the net production of O_2 by these fast processes is the biological production (10 Emoles/Myr). The biological production is essentially a splitting of water into its components, oxygen and hydrogen. The carbon acts like a sponge, soaking up the hydrogen to form carbohydrates (CH_2O). The O_2 is returned to H_2O by the carbon which becomes available by sinking in the sea to form sediments, the sedimentary rocks subsequently being oxidized by weathering. The carbon can become dehydrated in this process, and release H_2O. The O_2 can react directly with C to form CO_2. An alternate weathering process is for this carbon oxidation to proceed through an intermediate mechanism. In this alternate scheme, the O_2 oxidizes some chemical (X), such as reduced iron to form an oxide (XO_2) like oxidized iron. Then XO_2 reacts with CH_2O to form X, CO_2, and H_2O. The rate of this removal of O_2 and the formation of H_2O equals 10 Emoles/Myr.

replenish the O_2 in the air every 600,000 years.

The fossil fuel reservoir of carbon is only 0.9 Emoles (Gilbert, 1964). If this reservoir continues to be consumed at its present rate of 350 Emoles/Myr, then this reservoir will be depleted in 2500 years. The quantity of O_2 in the air would have only decreased 2.4%, assuming that the rates of O_2 production and removal remain constant.

Water vapor in the upper atmosphere is photodissociated into hydrogen and oxygen. The hydrogen atoms escape from earth, and the oxygen becomes part of the O_2 pool at the very slow rate of 7 Emoles per 10^9 years. Quantitatively this rate of oxygen production is insignificant in comparison to the oxygen production by photosynthesis.

The biosphere, in spite of its small mass, is important in the oxygen exchange between the atmosphere, hydrosphere, and lithosphere.

II. Ozone

Trace constituents in the atmosphere, such as ozone (Rowland and Molina, 1975; Prinn et al., 1978), atomic oxygen, oxides of nitrogen (Crutzen, 1979), and sulfur (Crutzen et al., 1979; Logan et al., 1979) are all intimately related to ozone. Ozone plays a significant role in the geochemical and geophysical balance of the earth. Ozone is an important shield for solar radiation in the ultraviolet range for wavelengths between 240 to 320 nm and in the visible spectrum for wavelengths up to 710 nm. Most of the ozone is located in the stratosphere between 15 and 30 km, reaching a maximum concentration of about 0.01 μM. If not for the ozone layer acting as a barrier to this lethal radiation, much of the present day biosphere could not exist. Skin cancer is more prevalent when ultraviolet radiation is increased, especially in the 290 to 320 nm range (Prinn et al., 1978).

Recently, much attention has been directed to a possible depletion of the ozone layer in the stratosphere. Ozone is produced by the photodissociation of molecular oxygen into oxygen atoms, which is subsequently followed by atomic oxygen reacting with molecular oxygen (Rowland and Molina, 1975):

$$O_2 \xrightarrow{h\nu} O + O \qquad \textbf{(6)}$$

$$O_2 + O + M \rightarrow O_3 + M \qquad \textbf{(7)}$$

where M is any particle. Both atomic oxygen and ozone contain an odd number of oxygen atoms and it is for this reason, they are designated as odd oxygen species. Odd oxygen is very reactive. Reaction (6) produces odd oxygen, which requires energy, whereas reaction (7) does not (Rowland and Molina, 1975). Below 45 km, ozone makes up almost all the odd oxygen, and at 60 km, ozone just makes up about 50% of the odd oxygen (Prinn et al., 1978).

Ozone can be depleted by a chain catalytic reaction:

$$X + O_3 \rightarrow XO + O_2 \qquad \textbf{(8)}$$

$$XO + O \rightarrow X + O_2 \qquad \textbf{(9)}$$

where X is some reactive species. The net result of reactions (8) and (9) is

$$O_3 + O \rightarrow 2O_2 \qquad \textbf{(10)}$$

Odd oxygen is destroyed by reaction (10). A chlorine atom, Cl, can act as the X species. The photolysis of chlorofluoromethanes, CF_2Cl_2 and $CFCl_3$, results in a chlorine atom being liberated from these chemicals. These man-made chemicals help to deplete the ozone layer (Molina and Rowland, 1974; Rowland and Molina, 1975). They have been used mainly as propellants in aerosol cans, refrigeration units, and plastics (Prinn et al., 1978).

Nitric oxide, NO, can also act as the X species. Then, XO represents nitrogen dioxide, NO_2 (Prinn et al., 1978; Crutzen, 1979). The sources of nitrogen oxides in the stratosphere include solar flares, galactic radiation, organic nitrate molecules such as peroxyacetyl nitrate, nitrogen fertilizers which release nitrous oxide (N_2O) from the soil, nuclear bomb production of nitric oxide, and supersonic aircraft injection of nitric oxide (Crutzen, 1979). Human civilization has to be constantly on the alert not to deplete the protective radiation filtering ozone layer.

If X is represented by the hydrogen atom, H, or by the hydroxyl radical, OH; then XO is represented by OH or by the perhydroxyl radical, HO_2, respectively. The hydroxyl radical is derived from water or methane (Prinn et al., 1978).

In the lower atmosphere or troposphere, the production of ozone occurs via the following chain reaction:

$$HO_2 + NO \rightarrow OH + NO_2 \qquad \textbf{(11)}$$

$$NO_2 \xrightarrow{h\nu} NO + O \qquad \textbf{(12)}$$

The net reaction is

$$HO_2 \rightarrow OH + O \qquad \textbf{(13)}$$

Now odd oxygen is produced in reaction (13) and in order to obtain ozone, reaction (7) is needed. In "clean" air, the source of HO_2 can be derived from carbon monoxide. In "polluted" air, the radical R, which represents unburned hydrocarbons, can be substituted for the hydrogen atom in reaction (11): the RO_2, which is required in reaction (11), can be obtained through oxidation of the R radical (Crutzen, 1979):

$$R + O_2 + M \rightarrow RO_2 + M \qquad \textbf{(14)}$$

The highest recorded concentration of surface ozone is 0.04 μM. This value was from southern California. Peak values of ozone in large cities, such as New York or London, are about 0.01 μM (Goldstein, 1979).

Since CO_2, O_3, and H_2O absorb long wavelength radiation, a decrease in any one of these species would decrease the surface temperature (Crutzen, 1979). Ozone does absorb infrared radiation at a wavelength of about 9.6 μm (Prinn et al., 1978). If there were no ozone layer, then the decrease in the earth's surface temperature would be 1.5°–3°C (Crutzen, 1979).

III. Beginning of Evolution on Earth

To discuss the evolution of our planet, many writers have used organic metaphors which stress the analogies between the earth and the biosphere. Thus, the earth can be compared to a living organism: the atmosphere and hydrosphere function as the circulatory media or blood; rocks in the lithosphere function as the cells; and the biosphere functions as the enzymes.

The ancients in many cultures made the analogy between the universe and animals (Partington, 1970). The Sumerians had a goddess, Min-Tu, Mother Earth (Woolley, 1965). In the Homeric Hymns, XXX (Eliade, 1974), we read, "I will sing of well-founded Earth, mother of all, eldest of all beings."

Alcmaeon (c. 500 B.C.) held the view that man was a microcosmic copy of the macrocosm, or universe (Mason, 1953). In the ancient Indian hymn "Atharva Veda, XII, 1", again we read of "Mother Earth" (Eliade, 1974). Plato expressed the view that the universe can be presented as an animal (Hall, 1975). Aristotle (1953) also wrote about the analogy between animals and the universe, in his *Physica* (Book VIII, line 252b): "Now if this can occur in an animal, why should not the same be true also of the universe as a whole?" The Chinese also used this analogy (Needham, 1969). Harvey (1628) noted that, "The heart [is] the sum of the microcosm."

The microcosm–macrocosm concept has not died. In fact, Lovelock (1972) has introduced the concept of "Gaia," which represents Mother Earth as a living entity. Margulis and Lovelock (1974, 1978) argue that the biota regulate the atmosphere for the optimum functioning of the biota. Thomas (1974) views the earth's atmosphere as "the world's biggest membrane." He writes, "When the earth came alive, it began constructing its own membrane, for the general purpose of editing the sun.... It breathes for us."

We will now discuss the formation of this membrane and its relationship to the biosphere. About 4.5×10^9 years ago the earth was formed from a dust cloud in the solar nebula. To get an idea of how old the universe was then, it must be remembered that the "big bang" or the beginning of the universe was 20×10^9 years ago; the beginning of our galaxy began 17.6×10^9 years ago, and the beginning of our solar system was 5×10^9 to 5.5×10^9 years ago (Day, 1979). In other words, the earth was formed when the universe had attained about 75% of its present age. Thus, our planet is a relative new comer to the universe.

We will not discuss the birth of the earth here (Herbig, 1981). However, we will point out the cosmic abundance of the elements and how these relate to the biosphere, and in particular to oxygen. We note that the cosmos is relatively rich in oxygen, its third most abundant element (Table 7). This is the same order of atom abundance as given above for the biosphere, with the exclusion of the chemically inert helium.

Table 7. *Extraterrestrial Atomic Abundance*

	% atom abundance				
Element	Meteorites[a]	Solar photosphere[a]	Solar corona[a]	Solar[b]	Cosmic ray sources[a]
H	93.37	92.48	92.45	93.93	92.20
He	6.49	7.40	7.40	5.93	6.97
O	0.06	0.06	0.07	0.07	0.29
C	0.04	0.04	0.05	0.04	0.27
N	0.01	0.01	0.01	0.01	0.03
Others	0.03	0.01	0.02	0.02	0.24
Total	100	100	100	100	100

[a]Calculated from Trimble (1975).

[b]Calculated from Pagel (1979).

The percent abundance of the stable isotopes of oxygen on earth are 99.759% ^{16}O, 0.037% ^{17}O, and 0.204% ^{18}O (Weast and Astle, 1979). These isotopic abundances are about the same for the cosmos (Trimble, 1975). The radioactive isotope ^{15}O, with a half-life of only 124 seconds (Weast and Astle, 1979), participates as an intermediate in nucleosynthetic reactions (Rolfs and Trautvetter, 1978; Penzias, 1979).

Urey (1952, 1959) postulated that the earth's primitive atmosphere consisted of hydrogen plus the hydrogenated forms of oxygen, nitrogen, and carbon, i.e., water, ammonia, and methane. This would follow from the predominance of hydrogen, which would combine with the most common elements. Previously, Oparin (1936) had postulated that the origin of life required a reducing atmosphere. Haldane (1954) and Bernal (1949) held similar beliefs. Miller (1953, 1974) in his famous experiments synthesized amino acids from such a mixture of gases, using an electric discharge as an energy source. Many other syntheses of organic molecules have been performed by several investigators using somewhat different gas mixtures and different energy sources (Miller and Orgel, 1974). If these precursor molecules of the biosphere were really synthesized in such a reducing environment, then this would account for the striking correlation between the atom abundance of the cosmos and of the biosphere.

However, a problem arises in this scenario. The velocity of particles increases with a decreasing mass and/or an increasing temperature. If the particle velocity is great enough to overcome the gravitational pull of a planet, then the particle escapes from the planet. Since neon is much less abundant on earth than in the cosmos, this implies that the earth has lost all gas particles with a mass less than that of neon. Neon has an atomic weight greater than that of hydrogen, helium, water, ammonia, and methane. Thus, it would appear that the earth has lost its primitive atmosphere. Almost all of the organic matter synthesized would have been destroyed or lost (Miller and Orgel, 1974).

Hence our present atmosphere is probably secondary in origin, derived from the degassing occurring when radioactive heating released volatiles from the interior. Volcanic gases release interior volatiles and recycled surface volatiles. Both volatiles come into equilibrium with molten material below the crust, which has been assumed to have a constant oxidation state. These gases would, therefore, be expected to comprise the original secondary atmosphere (Walker, 1977). These gases are mainly H_2O and CO_2, but also contain SO_2, CO, and H_2. At 1200°C, they are in equilibrium with an oxygen pressure of about 10^{-8} bar (Holland, 1978).

Chemical evolution preceding the prelife stage of the earth may have actually occurred in outer space (Brooks and Shaw, 1978;

Oparin, 1978). Many organic species exist in space (Buhl, 1974; Oró, 1976; Huntress, 1977; Day, 1979). Since amino acids from meteorites have been detected (Shimoyani et al., 1979), meteorites might have contaminated the earth with organic molecules, the precursors of life. Comet capture (Oró, 1961; Hoyle and Wickramasinghe, 1978) has been suggested as another way in which the earth could have obtained organic compounds. Hoyle and Wickramasinghe (1978) have put forward the possibility that a primitive biosphere may have evolved on comets. Crick and Orgel (1973) have even considered the possibility that living organisms were purposely sent to earth by higher organisms.

The possibility that extraterrestrial life could exist has been presented before modern times (Cohen, 1974). Some worlds were considered to be inhabited by Democritus (fl. 420 B.C.) (Sagan, 1980). Aristotle (1958) wrote that "there certainly ought to be some animal corresponding to the element of fire,... Such a kind of animal must be sought on the moon." During the early part of the seventeenth century, Kepler imagined inhabitants on the moon in his *Somnium* (Sagan, 1980). Newton (Partington, 1961) believed that the part of the air which is necessary for life was derived from comets. At about the same time, Hooke (1665) wrote: "we may perhaps be able to discover *living Creatures* in the Moon, or other Planets." At the end of the seventeenth century, Christiaan Huygens wrote about life on other worlds (Sagan, 1980). Later, Diderot (1776) had a character in one of his plays say "Human polyps in Jupiter or Saturn!" At the beginning of the twentieth century, Percival Lowell, a well known astronomer, believed that Mars was inhabited by intelligent life (Sagan, 1980).

More recently, Shapley (1958) estimated that the probability of a star having in its planetary system a planet with some form of life is 10^{-12} to 10^{-6}. He estimated that there are 10^{20} stars, so that the number of planets with life on them would equal 10^8 to 10^{14}. Pollard (1979) estimated that the probability that a star will have an earthlike planet is 10^{-5} to 10^{-7}. Hart (1979) has pointed out that if a planet is too close to a star, excessive heating will occur due to a runaway greenhouse effect. If it is too distant from a star, runaway glaciation will occur when free O_2 is evolved.

No matter what mechanism could account for the initial metastable compounds on earth, there is no doubt that the macrocosmic abundance of elements has left its imprint on the microcosmic biosphere. This is noted also for the less abundant elements (Gilbert, 1960).

Since hydrogen and helium are so light, they are escaping from the earth. This process of dehydrogenation is oxidizing the earth. The hydrogen lost by this procedure is in both the free and bound forms. Energy sources on earth are solar radiation, ultraviolet radiation, electrical discharges, cosmic rays, radioactivity, volcanoes, shock waves, and solar wind (Miller et al., 1976). These energy sources could be used for the net synthesis of metastable compounds with the simultaneous release of H_2, as depicted in Table 8. The probability of the synthesizing reaction A occurring increases when the H_2 decreases. At high pressures of H_2, there is a high probability of a back reaction occurring

Table 8. *Sequence of Events in the Reducing Atmosphere*

Reaction	Free energy (kcal)
A. Metastable compound synthesis	
$H_2O + CH_4 \rightsquigarrow CH_2O + 2H_2$	32.5
B. Biological energy utilization	
$CH_2O + 2H_2 \rightarrow H_2O + CH_4$	−32.5
C. Biological photoevolution of hydrogen	
$H_2O + CH_4 \rightarrow CH_2O + 2H_2$	32.5

and destroying the metastable compounds, such as $(CH_2O)_n$. Of course, some of these metastable compounds would contain nitrogen, such as amino acids, peptides, and purine bases. Ammonia, hydrogen cyanide, and cyanamide might play a role in the formation of these important building blocks for the biosphere (Oró et al., 1978). The elements phosphorus and sulfur would also play an important role. Thus, adenosinetriphosphate (ATP) and deoxyribonucleic acid (DNA) are important compounds containing just the elements H, C, O, N, and P. ATP is used as the useful energy storehouse and DNA is the genetic material which is necessary for the propagation of the biosphere (Hoffman, 1975).

As the hydrogen decreased, the metastable compounds increased. It seems likely that the concentration of these organic compounds was extremely small (Sillén, 1965). For evolution to occur, there must have been one or more mechanisms for concentrating these produced compounds; these products could be concentrated in lagoons and pools (Bernal, 1949). Bernal also suggests the mechanism of adsorption onto clays and quartz occurring as sand for accomplishing this compound concentration. Molecular order occurs at phase boundaries (Menger, 1972); certainly one of the prequisites for the origin of life.

A primordial oil slick on the water surface by polymerization could have resulted if the methane concentration were a major atmosphere constituent (Lasaga et al., 1971). This slick could also have served as a concentrating mechanism. The beaches, with periodic wetting and drying, would provide an ideal place for adsorption onto crystals, thus concentrating and polymerizing the metastable compounds (Neuman et al., 1970). Mineral surfaces (Otroshchenko and Vasilyeva, 1977) and clays (Paecht-Horowitz, 1978) can exhibit specificity for the metastable compounds as well. In addition, these surfaces possess catalytic properties (Siegel, 1957).

Perhaps a gaseous medium can be a vehicle for a biosphere (Morowitz and Sagan, 1967; Dimmick and Chatigny, 1976). However, present day organisms do not multiply in the air (Rüden et al., 1978).

IV. The First Cells

It is certainly a big step to produce a living organism under such conditions. This step will not be discussed here, simply because our knowledge is lacking. Keosian (1974) has pointed out some of the philosophical problems in postulating this step.

Fatty acids, the precursors of membranes, would be present and set up phase differences with their concentrating effects. Artificial lipid bilayer membranes are susceptible to destruction by oxygen (Huang et al., 1964). Hence, it seems likely the primitive biological membranes could not exist in the presence of oxygen.

There would have evolved chemical pathways in the biosphere (Table 8) for the energy-releasing reaction B. We propose here that the back rate, reaction B, approached the forward synthesis rate, reaction A, but could not exceed the forward rate. If it did exceed the forward rate for a significant period of time, then the metastable compounds would necessarily disappear. Clearly, the back reactions are destructive in nature. The hydrogen producing reaction C is the same as reaction A, except that biological catalysts would be involved for reaction C. In order for the process to continue, perhaps pathways for reaction C were also developing, so that there was a feedback mechanism. These reactions would be occurring in some place where phase separations could concentrate the biochemical pathways. Thus, the evolution involved in reactions B and C would have proceeded from the molecular level of integration, to the molecular system level of integration, to the polymolecular system level of integration, and to the cellular level of integration. An isolated enzyme system would be classified as a molecular system, whereas a biochemical pathway would be classified as a polymolecular system (Schueler, 1960). Keosian (1974) noted that evolution should proceed from the simple to the complex level of integration. Microspheres as forerunners to cells have been extensively studied (Fox and Dose, 1977; Oparin, 1978). Finally, cells or a symbiotic relationship between cells were formed in which both net reactions B and C occurred (Gilbert, 1966; Decker, 1978). For such a biosphere, H_2 serves as an energy store.

However, if the H_2 was not carefully controlled, metastable compounds within the cell itself would be destroyed. In other words, reaction B might tend to occur through uncontrolled reductions. Thus, H_2 would serve not only as an useful biological source of energy but also as a destructive agent for the integrity of the cell. The biosphere would have had to develop antihydrogen or antireductant defenses against this hydrogen toxicity in order to survive. However, since the atmosphere is losing hydrogen into space, the atmosphere is becoming more oxidized. As the atmospheric hydrogen decreased, these antireductant defenses became better qualified to cope with the toxic effects of hydrogen. Figure 3 shows this energy relationship.

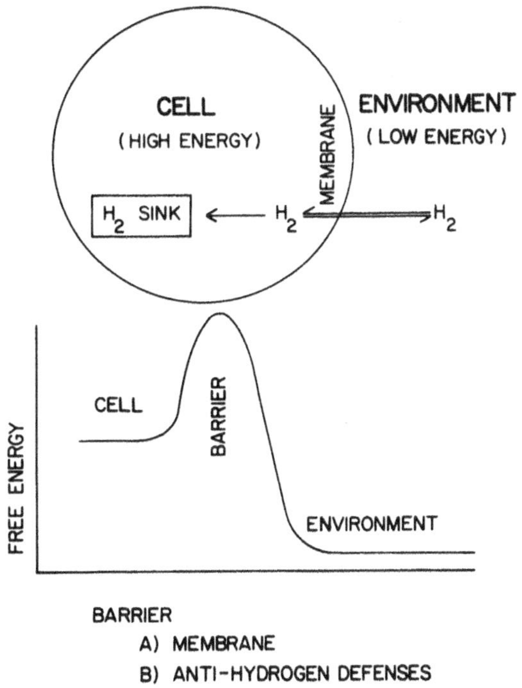

Figure 3. Life-energy profile in reducing atmosphere. This is a simplified diagram to illustrate the energetics of a postulated primitive cell. A membrane barrier prevents the passive diffusion of some substances across the membrane; perhaps the first active transport was for the extrusion of the hydrogen ion out of the cell, thus maintaining an energy gradient. Hydrogen gas easily diffuses across the membrane into the cell, where it is consumed in controlled reductions of the metastable cellular metabolites, energy being released in this process for the cellular requirements. The reductions are controlled by various types of antihydrogen or antireductant defenses.

The cell is like a bag of energy. The membrane permits the net transport across it of other substances by active transport mechanisms, such as the hydrogen ion pump. Thus, the membrane is an energy barrier and serves to maintain the energy difference between the internal milieux and the external milieux. At the same time, H_2 can diffuse across the membrane and reduce the cellular constituents, thus decreasing the internal energy. The energy level within the cell is hence maintained at a higher level than in the external environment by two mechanisms: active transport processes and selectivity of the membrane, and anti-reductant defenses. Indeed, there is some reason to suspect that one of the first active transport systems which evolved was the one involved with the extrusion of H^+ out of the cell (Wilson and Maloney, 1976).

It is interesting to make some calculations on the size of the biosphere and its rate of turnover. There are 0.0691 Emoles of carbon now in the biosphere and the net rate of carbon incorporated is 6500 Emoles/Myr, resulting in a turnover of the biospheric carbon in 10.6 years. As for the increase in the number of organisms, let us consider a bacterium of 1 μm^3. If it reproduced by binary fission, then after 130 generations, the size of the resulting cells would be bigger than the earth. These numbers do not have much meaning except to point out that after a lag period, it need not take a long period of time to reach the present biospheric mass.

Many of the biochemical pathways might still exist today in organisms in spite of the fact that cells from that era have not survived. Methanogenic bacteria have recently been shown to be unique (Woese et al., 1978; Balch et al., 1979). They are strict anaerobes and they utilize energy from the following reaction:

$$CO_2 + 4H_2 \rightarrow 2H_2O + CH_4 \qquad (15)$$

This reaction has a standard free energy of −31.1 kcal (Latimer, 1952). These bacteria contain coenzyme M, the smallest known coenzyme. They also contain coenzyme F_{420}, and both coenzymes are not found elsewhere. No cytochromes or menaquinoines have been found in these bacteria (Balch et al., 1979). These organisms are commonly grown in 0.5 to 0.8 bar of H_2 (Zeikus, 1977), but they have also been able to survive an exposure to 3 bar of H_2 (Balch and Wolfe, 1976). Thus, their antireductant defenses seem very well developed. Perhaps some of these bacteria will be found to be capable of reducing the less oxidized form of carbon, namely as $(CH_2O)_n$. Then we would have a present day example for the existence of reaction B (Table 8) occurring in the biosphere. The methanogenic bacteria are extremely old and have been classified with extreme halophiles and various thermoacidophiles (Woese et al., 1978).

Hydrogen production does occur in the biosphere. There is photohydrogen production as well as a hydrogen production which can occur in the dark. The enzyme hydrogenase is needed for this activity. In some organisms, the complete photosynthetic process is required (Bishop and Jones, 1978). The process is composed of two photosystems, I and II (Mauzerall and Piccioni, 1981). Photosystem I is activated by light to generate a photoreductant. Light activation of photosystem II dissociates water. The cyanobacteria, or blue-green algae (Stanier and Cohen-Bazire, 1977), do not require photosystem II for the photoproduction of hydrogen (Bishop and Jones, 1978). Where X represents the hydrogen donor, the overall reaction is

$$XH_2 \rightarrow X + H_2 \qquad (16)$$

Methanogenic bacteria often have close symbiotic relationships with those bacteria that produce hydrogen. The result is that for some of these interspecies H_2 transfers, the H_2 concentration can be kept at a minimum (Mah et al., 1977). For example, the liberated H_2, produced by the so-called S organism, is taken up by the methanogenic bacteria to produce methane. H_2 inhibits the growth of the S organism: 0.5 bar H_2 completely inhibits its growth (Reddy et al., 1972a). An explanation for this inhibition is that the free energy for the H_2 production is unfavorable unless the H_2 is low (Reddy et al., 1972b). Hence, the S organisms provide the H_2 needed by the methanogenic bacteria, and in return the methanogenic bacteria serve as an antihydrogen defense for the S organism. A similar inhibition by H_2 occurs for *Clostridium cellobioparum* growth (Chung, 1976).

Other examples of H_2 inhibition are the enzyme formation and regulation of fructose in the hydrogen bacterium, *Nocardia opaca* Strain 1b (Probst and Schlegel, 1973); the glucose-6-phosphate dehydrogenase activity of *Hydrogenomonas eutropha* H16 (Bowien et al., 1974); and competitive inhibition of nitrogenase activity (Hwang et al., 1973).

V. Oxygen Production

As the hydrogen dissipated from the earth, the net photodissociation of water became more probable. This resulted in the release of bound hydrogen to outer space, with the release of oxygen. The energy released by this process amounts to 113.4 kcal/mole of oxygen (reaction D in Table 9). Oxygen would then react with many substances including the metastable ones of the functioning biosphere. Note that the magnitude of the energy released by this process of aerobic respiration (reaction E) is practically the same as reaction D, or 114.8 kcal/mole of oxygen. Energetically, the carbon in the system acts merely as a sponge soaking up the H_2:

$$CO_2 + 2H_2 \rightarrow CH_2O + H_2O \qquad (17)$$

Table 9. Sequence of Events in the Oxidizing Atmosphere

Reaction	Free energy (kcal)
D. Oxygen production $2H_2O \rightsquigarrow O_2 + 2H_2$	113.4
E. Biological energy utilization $CH_2O + O_2 \rightarrow H_2O + CO_2$	−114.8
F. Photosynthetic oxygen production $H_2O + CO_2 \rightarrow CH_2O + O_2$	114.8

This reaction has only a free energy of 1.4 kcal/mole, and occurs in the biosphere as photoreduction or as dark carbon dioxide fixation (Bishop and Jones, 1978). Respiration in the biosphere could not proceed any faster than the rate of oxygen release from the net photodissociation of water. Lightning could also possibly supply organic carbon to the biosphere (Chameides et al., 1979).

The next phase in evolution was the reverse step, photosynthesis (reaction F in Table 9). At first, there would be practically no free oxygen in the atmosphere; respiration in the photosynthesizing organisms would utilize almost all the oxygen. Later, when photosynthesis was more prevalent, the liberated oxygen combined with the lithospheric constituents. Finally, there would be an increase of atmospheric oxygen. But this oxygen pollution changed the environment of the earth, and this environmental alteration in turn changed the biosphere.

Let us examine this change. In Fig. 4, we see a diagram of the biosphere in an oxidizing atmosphere. Notice the similarity to Fig. 3, in which the biosphere is exposed to a reducing atmosphere. The only difference is that oxygen replaces hydrogen in the scheme. The membrane is still an energy barrier and maintains an energy differential between the internal and external environments. However, oxygen can permeate through the membrane and oxidize the cell constituents. Antioxidant defenses can inhibit this destructive tendency of oxygen, but can not completely prevent them from occurring. This is the price that the biosphere pays for its existence.

Could we find an alternate available energy source for a biosphere in the predominately hydrogen universe which would equal or even surpass the qualifications of oxygen? To be a good energy source, the source must be abundant. Oxygen is the third most abundant element in the universe, being surpassed only by helium, which is chemically unreactive,

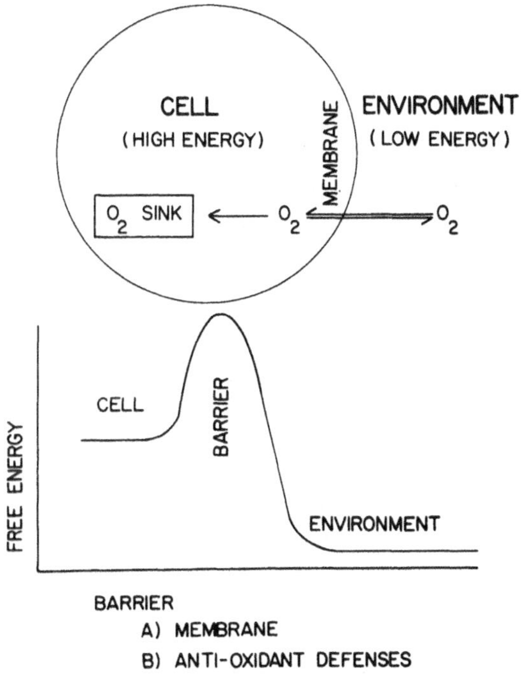

Figure 4. Life-energy profile in oxidizing atmosphere. This is a simplified diagram to illustrate the energetics of an aerobic cell. Note the similarity to Fig. 3. The membrane barrier is more sophisticated. There are more active transport processes than in the cell depicted in Fig. 3. Instead of hydrogen gas being used as an energy source, oxygen gas is used. Cellular metabolites consume oxygen, which releases more energy for the cell than the analagous hydrogen consumption. The oxidations are controlled by various types of antioxidant defenses. However, these defenses are not perfect as shown by oxygen toxicity.

and by hydrogen. The source must be easily available. Unless the temperature is below −183° C, oxygen is a gas. Other gases with their corresponding boiling points are −253° C for hydrogen, −196°C for nitrogen, −192°C for carbon monoxide, −187°C for fluorine, −161°C for methane, −112°C for ozone, −78°C for carbon dioxide, −60°C for hydrogen sulfide, −35°C for chlorine, and −33°C for ammonia (Weast and Astle, 1979). Not only must the energy source have a high potential, but it must also react relatively slowly. If it reacted rapidly, it could not store the energy. There are other species which have a higher potential such as fluorine, ozone, atomic oxygen, and chlorine; but oxygen is sluggish in reacting. Hence, oxygen is the best qualified biological energy source in the universe (Gilbert, 1960).

Hydrogen has been photoproduced in the green algae in which both photosystems I and II participate. Since photosystem II is involved with the dissociation of water producing oxygen, the source of hydrogen arises from water. The net reaction is the dissociation of water into O_2 and H_2 (Pow and Krasna, 1979). Oxygen is known to inhibit hydrogenase, the enzyme which is necessary for the production and/or utilization of H_2 (Bishop and Jones, 1978; Pow and Krasna, 1979). However, this inhibition is very variable (Adams and Hall, 1979). Oxygen also inhibits nitrogenase, the enzyme responsible for nitrogen fixation (Brill, 1975).

The enzyme superoxide dismutase (SOD) removes the superoxide free radical ion (Fridovich, 1981) and thus acts like an antioxidant (Forman and Fisher, 1981). When light intensities and O_2 levels are high, for a period of time, the SOD level may drop drastically, and cyanobacteria can be killed. On the other hand, in the cyanobacteria, oxygen can induce SOD formation (Stanier and Cohen-Bazire, 1977). The SOD present in prokaryotes, i.e., organisms with no nucleus, is either iron SOD or manganese SOD. These two forms are closely related (Fridovich, 1976). SOD has been found in both the obligate anaerobic photosynthetic bacteria (the green bacteria and the purple sulfur bacteria) and in the facultative aerobic photosynthetic bacteria (the purple, nonsulfur bacteria) (Lumsden et al., 1977). The anaerobic bacteria contain the iron SOD (Kanematsu and Asada, 1979), whereas the purple, nonsulfur bacteria contain the manganese SOD (Lumsden et al., 1977).

Perhaps the photosynthetic organisms would be more likely to generate traces of superoxide ion due to exposure to radiation. Generally, anaerobic bacteria contain little or no SOD (Fridovich, 1981). Several aerobic bacteria including a cyanobacterium, as well as *Euglena*, an eukaryote, i.e., a nucleated cell, contain both the iron SOD and the manganese SOD (Kanematsu and Asada, 1979).

As the atmosphere became more oxidizing, the bound hydrogen in ammonia and methane was lost, with the ammonia becoming nitrogen and the methane becoming carbon dioxide. The presence of a chemically inert gas such as nitrogen would inhibit fires; thus, nitrogen acts as an antioxidant on a global scale (Gilbert, 1968). The rate of burning increases as the percentage of oxygen in a gas mixture increases (Hamilton and Sheffield, 1977) and as the total pressure increases (Watson et al., 1978).

The question arises, when did this all happen? For that, we will have to examine the geological record. Since oxygen will tend to oxidize the metal elements in the earth's crust, we would not expect to find free oxygen but oxygen bound to these elements. The banded iron formations (BIFs) contain alternating bands of iron; the iron being in the oxidized and reduced states. The presence of the iron in the oxidized state could be due to the precipitation of the soluble ferrous iron being converted to the insoluble ferric form (Cloud, 1978). Walker (1978) is of the opinion that in order for BIFs to occur, the oxidation required is too large to be accounted for by the release of oxygen from the photodissociation of water. Kasting et al. (1979) also point out that the oxygen supply from water photodissociation cannot be too large. Therefore, some other oxygen source is required, and an obvious source is from photosynthesis. Hence, since the BIF in Greenland is 3.8×10^9 years old, photosynthesis has been an important process for at least 3.8×10^9 years (Cloud, 1976; Walker, 1978). From their carbon isotope measurements of the 3.7×10^9-year-old Isua sediments in West Greenland, Schidlowski et al. (1979) theorize that at that time there was a sizable biosphere capable of producing oxygen.

As mentioned before, the origin of life preceded the process of the liberation of oxygen by the biosphere. Probably, it would have taken a few million years after the earth was formed before life on earth was present. Hence, it would not be unreasonable to suspect that life on earth began 4.0 to 4.2×10^9 years ago. Microstructures from rocks 3.4×10^9 years old have been interpreted to be of biological origin (Knoll and Baarghorn, 1977). Geological evidence shows that algal mats probably were present 3×10^9 years ago (Mason and Von Brunn, 1977). Perhaps, the surface temperature was about 50°C higher than at present (Pollack, 1979).

About 2×10^9 years ago, the BIFs ceased to exist and red beds came into existence (Cloud, 1978). These beds contain hematitite, the reddish-brown ferric oxide pigment, and are formed in a proper oxidizing environment (Van Houten, 1973). Today, these beds are still being formed (Walker, 1967). This red bed formation has been taken as evidence to show that free oxygen in substantial quantitites appeared at this time (Cloud, 1976; Walker, 1977).

During this interval, from 3800 to 2000 million years ago, the iron acted as an oxygen sink (Cloud, 1976). The single cell organisms used the highly efficient respiration, and any of the photosynthetic oxygen produced which leaked out of the cell would either be taken up by other neighboring cells or by the reduced iron. Thus, the reduced iron acted as an external antioxidant to the biosphere. The biosphere necessarily developed antioxidant defenses during this transition from an aerobic to an anaerobic environment, as the biosphere experimented with a more productive energy source. Cloud (1976) believes that the reason it took 1.8×10^9 years from the time that oxygen was liberated by photosynthesis to the time that free oxygen appeared was that this time was necessary for the biosphere to build up the necessary antioxidant defenses. Thus there was an evolutionary pressure for the development of antioxidant defenses (Gilbert, 1960; Sagan, 1961; Gerschman, 1964). This reasoning seems correct, since oxygen is very toxic (Gerschman, 1964).

Antioxidant defenses can be of a varied nature (Gilbert, 1963) and in the biosphere as many of these as possible are used to combat oxygen toxicity. For example, one function of the carotenoids is to protect against oxygen toxicity in the presence of light (Swain, 1974). Thus, the halobacteria, which can use light energy to pump hydrogen ions out of their cells contain carotenoids in their cell membranes (Stoeckenius et al., 1979). Superoxide dismutase acts as an antioxidant, as mentioned before. Other enzymes and chemicals which act as antioxidants are catalase, peroxidase, vitamins C and E (Forman and Fisher, 1981), and cytochrome *P*-450 (Wickramasinghe and Villee, 1975).

The first organisms using light energy to produce metastable compounds should be the anaerobes. The present day photosynthetic bacteria are anaerobes and do not liberate oxygen (Schopf, 1978). For example, the green sulfur bacteria use H_2S as a reductant and produce free sulfur (Broda, 1975):

$$2H_2S + CO_2 \rightarrow CH_2O + H_2O + 2S \qquad \textbf{(18)}$$

The free energy for this reaction is 14.4 kcal (Latimer, 1952). Another bacterial photosynthetic reaction involving sulfur is (Schidlowski, 1979):

$$\tfrac{1}{2}H_2S + CO_2 + H_2O \rightarrow CH_2O + H^+ + \tfrac{1}{2}SO_4^{2-} \qquad \textbf{(19)}$$

This reaction has a standard free energy of 29.4 kcal (Latimer, 1952). These reactions have standard free energies much less than when oxygen is released in producing CH_2O. The energy producing reaction of the sulfate reducing bacteria (Schidlowski, 1979) has a standard free energy of only −10.3 kcal (Latimer, 1952).

$$2CH_2O + SO_4^{2-} \rightarrow 2CO_2 + H_2S + 2OH^- \qquad \textbf{(20)}$$

Schidlowski (1979) gives evidence that photosynthetic sulfur bacteria preceded the sulfate-reducing bacteria and that this biological sulfur cycle emerged between 2.8 to 3.1×10^9 years ago.

Later, the cyanobacteria were developed. These organisms are capable of performing photosynthetic production of oxygen and yet are prokaryotes. Their metabolism is not as well adapted to an oxygen environment as that of the later eukaryotes. Some cyanobacteria have an optimum growth when exposed to an oxygen pressure of 100 mbar. Photosynthesis had evolved from the anaerobic bacteria to the aerobic cyanobacteria (Schopf, 1978).

It might be possible for some primitive biochemical systems to be studied using living microfossils. *Kakabekia umballata* is an unicellular organism found only as a microfossil in the 2×10^9-year-old Gunflint Iron formation in Canada (Barghoorn and Tyler, 1965). Subsequently, a living representative of this genus *Kakabekia barghoorniana*, has been described; it is probably a prokaryote (Siegel, 1977). The living manganese oxidizing bacterium, *Metallogenium*, also shows a striking resemblance to the microfossil, *Eoastrion*. This microfossil is 1.6 to 2×10^9 years old and is probably also a prokaryote (Cloud, 1976).

VI. Development of Eukaryotes and Multicellular Organisms

The eukaryotes probably evolved about 2×10^9 years ago when the oxygen pressure in the atmosphere was significant. From geological evidence, the oldest eukaryotes are only about 1.3×10^9 years old (Cloud, 1976). However, according to Schopf (1975), compelling evidence for the oldest eukaryotes date only from 0.9×10^9 years ago. On the other hand, Durham (1978) believes that the eukaryotes appeared between 1.9 to 2.25×10^9 years ago.

The origin of the eukaryotes has been hotly debated. Margulis (1970) is one of the leading proponents of the endosymbiotic theory. This theory has it that the eukaryotes ingested other smaller organisms, which evolved into such cellular organelles as chloroplasts and mitochondria. The modern eukaryotic cell is generally about ten times bigger than the prokaryotic cell. Photosynthesis occurs in chloroplasts and the reverse process of respiration occurs in mitochondria. The other theory holds that eukaryotes were derived from prokaryotes, which evolved such organelles from themselves (Uzzell and Spolsky, 1974; Raff and Mahler, 1975). A viral invasion has also been suggested (Taylor, 1979).

In support of the endosymbiotic theory, it should be mentioned that many cells do have symbionts living in them (Smith, 1979; Whatley et al., 1979). The organelles also contain their own separate deoxyribonucleic acid (DNA), the genetic material (Margulis, 1970). Protein and nucleic acid sequence data support the endosymbiotic theory (Schwartz and Dayhoff, 1978). The manganese SOD found in bacteria is very similar to the mitochondrial manganese SOD and does not seem to be related to the copper zinc SOD generally found in the cytosol of the eukaryotes (Fridovich, 1976). Therefore, the endosymbiotic theory is probably correct.

The cyanobacteria were probably the precursors of the chloroplasts in the eukaryotic cell. The chloroplasts in plants probably evolved from some prokaryote similar to *Prochloron didemni*. On the other hand, aerobic bacteria were probably the mitochondrial ancestors. *Paracoccus denitrificans*, an aerobic bacterium, has many features which are common with mitochondria. Another DNA containing organelle, the hydrogenosome, is similar to *Clostridium pasteurianum*. The hydrogenosome produces hydrogen gas (Whatley et al., 1979). Mitochondria are present in all eukaryotes (Schopf, 1978).

Peroxisomes are cellular oxidizing particles which contain oxidases and catalase. They lack oxidative phosphorylation capability. The host cell may have had them before the mitochondria were acquired (Broda, 1975).

Thus, the first prokaryotes were anaerobes and later some of them became aerobes. Almost all the eukaryotes are aerobic (Margulis, 1970). It appears that in photosynthesis, photosystem II was a later development than photosystem I (Swain, 1974; Schwartz and Dayhoff, 1978; Towe, 1978). Photosystem II is the system which releases oxygen (Mauzerall and Piccioni, 1981).

Biochemical synthesis of metabolic components is dependent upon the lack or presence of oxygen. The syntheses of squalene and the carotenoids do not require oxygen (Margulis, 1970; Schopf, 1978). Vertebrates obviously have the capability of synthesizing carotenoids (Florkin, 1974). However, the syntheses from squalene of steroids require oxygen. One pathway of oleic acid synthesis from acetyl coenzyme A requires oxygen, whereas another pathway does not. However, unsaturated fatty acid synthesis from oleic acid requires oxygen (Margulis, 1970; Schopf, 1978).

The chloroplasts contain the magnesium containing chlorophyll, the phorbin derivative responsible for photosynthesis. The mitochondria contain the iron-containing cytochromes, the porphin derivatives in the respiratory chain (De Ley and Kersters, 1975). The oxygen acceptor in the respiratory chain is cytochrome oxidase (Chance, 1981). The oxygen carrying hemoglobin (Wittenberg and Wittenberg, 1981) and the antioxidants, catalase and the peroxidases, are other iron-containing or heme-porphin derivatives. Protoporphyrin IX is the common precursor for both porphin and phorbin derivatives. Other tetrapyrroles, the corrin derivatives, such as vitamin B_{12}, are biosynthesized from uroporphyrinogen III, which is also a precursor to protoporphyrin IX. The corrin derivatives occur in anaerobes. The porphins are more oxidized than the corrins and appear in almost all bacteria except the methanogenic bacteria and the anaerobic clostridia. Thus, it seems that the corrin derivatives existed in the early bacteria (DeLey and Kersters, 1975), perhaps 3.8×10^9 years ago. Since carotenoid biosynthesis requires no oxygen, it would seem most probable that this substance was ready to act as an antioxidant even before the environment became oxidizing. Margulis (1970) believes that carotenoids existed before porphyrins.

As the environment became more oxidizing, hydrogen peroxide would be produced. Soluble ferric ion could catalyze peroxide decomposition and thus act as an antioxidant. Perhaps the iron containing catalase was then developed by the cell. Catalase is 10^8 to 10^{10} times more effective than the ferric ion in catalyzing this decomposition (Calvin, 1969). Organisms which contain chlorophyll also contain cytochromes (Margulis, 1970).

The oxygen level in the atmosphere continued to rise as the net photosynthesis increased. The organisms would have special mechanisms to sniff out the sparse oxygen; perhaps these would be cytochrome *c* oxidases which are responsible for the migration of *Euglena* toward oxygen (Miller and Diehn, 1978). As oxygen continued to pollute, ozone was produced in the upper atmosphere. Since ozone absorbs uv radiation (Prinn et al., 1978), it formed a protective layer against this lethal radiation (Berkner and Marshall, 1967; Sagan, 1973; Whitfield, 1976). When the oxygen pressure was increased to 2 torr, then the ozone protective screen for the biosphere was established (Margulis et al., 1976).

Microenvironments existed then as they do now. Thus, there would be environments in which there was little or no oxygen present. When the intracellular oxygen concentration reaches a critical value, anaerobic metabolism changes to aerobic metabolism, which releases more energy for the cell to use. This is the Pasteur effect (Krebs, 1972; Gottlieb,

1981). When the oxygen pressure reaches about 10 torr, then the Pasteur point is reached for many organisms. At this oxygen pressure, the surface waters would be protected against this radiation and the biosphere could expand (Tappan, 1974).

The biosphere became more mobile as more energy became available to it under aerobic conditions. The oxygen level continued to increase. The cell size had increased about ten fold in the transition from prokaryote to eukaryote. However, the bigger the cell, the greater is the problem of transporting substances in and out of the cell. A maximum cell size is reached when the cell can not cope with the diffusion problem (Tappan, 1974; Piiper and Scheid, 1981).

About 700 million years ago, the metazoa, i.e., multicellular animal organisms, evolved. Perhaps multicellular algae were present even earlier (Cloud, 1976). Metazoa require oxygen for their metabolism. The early metazoa still had to cope with the problem of diffusion in the same way as their unicellular ancesters (Tappan, 1974). Multicellular animals and plants had to increase their antioxidant defenses continually in order to survive. Damage to the organisms from oxygen would include events such as enzyme inactivation and lipid peroxidation (Kovachich and Haugaard, 1981).

A circulatory system, with an oxygen carrier such as hemoglobin, was developed 500 to 600 million years ago. Also, a respiratory system was evolved, so that between 440 to 500 million years ago, when the first vertebrate was developed, gills, a closed circulatory system, and hemoglobin were present (Tappan, 1974). Between 400 to 700 million years ago, land plants appeared (Cloud, 1976). At about 400 million years ago, air-breathing insects and vertebrates evolved (Tappan, 1974). Probably, the air-breathing vertebrates arose in fresh water (Graham et al., 1978).

Antioxidant defenses continued to be built up. Glutathione peroxidase is an antioxidant not present in microorganisms but is present in vertebrates (Smith and Shrift, 1979). The chloroplast must necessarily be well equipped with antioxidant defenses, e.g., copper and zinc SOD, vitamins C and E, glutathione, and the carotenoids (Halliwell, 1978).

Going from a water to a terrestrial environment constituted a big step in biological evolution. New mechanisms were evolved for transporting oxygen to the cells of the organisms. Tracheoles, or air tubes going directly to the tissues, were used by insects for this purpose. Lungs were involved in the oxygen transport to a fluid circulatory system for air breathing vertebrates. Air breathers have a distinct advantage compared to water breathers (Lenfant and Johansen, 1972; Tappan, 1974; Piiper and Scheid, 1981). The oxygen in the atmosphere close to sea level is relatively constant, whereas the oxygen in the ocean is variable. The concentration of oxygen in air is 30 times greater than in water. Aquatic animals, in contrast to land animals, are in constant danger of not obtaining a sufficient oxygen supply.

There have been two large scale extinctions during the past 700 million years. One happened about 250 million years ago and the other about 70 million years ago. Tappan (1974) suggests that oxygen depletion may have been at least partly responsible for both extinctions.

About 250 million years ago, a glacial age was ending (Pollack, 1979), which could conceivably cause an increase in the oxygen consumption by cold blooded organisms. If the atmospheric oxygen was not sufficient to maintain an adequate oxygen supply to the tissues under this postulated condition of increased demand, then it is possible that this increased demand could be a contributing factor in the cause of the extinctions which occurred at this time.

In spite of the lack of good evidence (Schopf et al., 1971), it seems reasonable that large oxygen-consuming animals would generally be more susceptible to a decreased oxygen pressure, which in turn, if severe enough, could lead to the possible extinction of these organisms (McAlester, 1970). Tappan (1974) suggests that the high oxygen-consuming dinosaurs became extinct about 70 million years ago due to a postulated lowering of the oxygen partial pressure. On the other hand, Schatz (1957) speculated that an increase in the atmospheric oxygen pressure was responsible for the dinosaur extinction due to oxygen poisoning.

Russell (1979) speculates that the extinctions were due to increased radiation from cosmic events such as a supernova. Depending on how close the cosmic event was to earth, the protective ozone layer could be destroyed (Napier and Clube, 1979).

If the oxygen tension were higher in the past, then antioxidant defenses would have been developed to cope with this increased stress. However, many organisms are adversely affected by a moderate increase in the oxygen tension, which suggests that the oxygen tension was probably never greater than the present atmospheric level (Gilbert, 1968). It seems most likely that the atmospheric oxygen pressure never underwent a significant decrease in its history.

VII. Possible Future Events

How stable is the earth's atmosphere? It appears that the earth is continually losing hydrogen to outer space, but at an extremely slow rate (Walker, 1977). Two hundred years ago, Lavoisier (1862) theorized that hydrogen should be in the upper layer of the earth's atmosphere. Indeed, at an altitude of 1000 km, helium and hydrogen are the predominant atmospheric species (Minzer, 1977). This hydrogen loss will eventually have an effect on the atmosphere. At the present rate of hydrogen loss, it will take 5.5×10^{12} years to remove all the hydrogen in the hydrosphere, and thus dehydrate the earth. However, there is a good chance that the sun will remain relatively stable for only another 5×10^9 years (Herbig, 1981). Catastrophic events, such as nearby supernovae, solar flares, comet collisions, and man-made explosions, can alter this time course.

The earth is undergoing oxidation by losing hydrogen. It is interesting to note that combustion was also explained first by the loss of something, namely phlogiston. Indeed, the postulated decredited dephlogisticated process had much in common with the dehydrogenation process (Gilbert, 1981a). The liberated oxygen from water would not remain free, but would combine with reduced substances, such as iron. The only reason for free oxygen in the atmosphere is the biospheric production of oxygen.

Perhaps, Mars is in such an oxidized state, where all the hydrogen in the water has been liberated to outer space. There is evidence that Mars had liquid water earlier in its history (Pollack, 1979). It does appear that the red color of Mars is indicative of oxidized iron and that the surface is in an extremely oxidized state (Oyama and Berdahl, 1977; Ponnamperuma et al., 1977). The oxidized atmosphere was possibly just carbon dioxide and nitrogen, with no free oxygen. In fact, the atmosphere of Mars is principally 6 mbar of CO_2 (Pollack, 1979). Under the present conditions, it does not seem likely that Mars contains any biosphere. Rather, it seems that there is a possibility that Mars has already gone through both a prelife stage and a life stage, and is now in a postlife stage. Eventually, the thin atmosphere on Mars should be lost.

Biological feedback mechanisms control the oxygen pressure in the atmosphere. If oxygen tends to increase, then photosynthesis is inhibited by the increased pressure of oxygen (Gibbs, 1970). If carbon dioxide tends to increase, then photosynthesis is accelerated (Gilbert, 1964). Hence, if oxygen is increased, then the production of oxygen is decreased; and if carbon dioxide is increased, then the removal of carbon dioxide is increased. When photosynthesis is increased, then the biosphere is increased; which is followed by an increase in respiration; and vice versa. Burning of fossil fuels increases carbon dioxide, which stimulates photosynthesis and increases the oxygen, thus tending to offset any oxygen loss due to combustion. These regulatory mechanisms are for short-term effects. For longer-term effects, the oxygen seems to be stabilized by the opposing influences on the atmospheric oxygen, which are weathering (oxidation) and carbon burial (removal of reduced carbon). Thus, for periods of time up to a few million years, oxygen pressure in the atmosphere will probably not change (Gilbert, 1980). There are enough feedback mechanisms over the short range to maintain the oxygen pressure close to its original value. The atmosphere in general shows a remarkable stability to withstand drastic physical changes (Hunt, 1976).

However, man is burning fossil fuels at an alarming rate. What is this doing to the atmospheric balance? If man burns all the 0.9 Emoles of carbon in the fossil fuel reserves, the oxygen pressure would drop from 158 torr to only 154 torr. However, there would be an increase in the carbon dioxide pressure of about ninefold (Gilbert, 1964). From 1910 to 1970, the volume % of oxygen in the atmosphere has been constant at 20.946 (Machta and Hughes, 1970). On the other hand, the carbon dioxide concentration in the atmosphere has increased by 13% since the beginning of the industrial revolution to 1975 (Bolin, 1977).

About half of the carbon dioxide released in burning fossil fuels remains in the atmosphere; most of the remainder enters the upper layer of the ocean (Broecker et al., 1979). Woodwell et al. (1978) pointed out that clearing of forest land decreases the biomass and releases carbon dioxide into the atmosphere. However, Broecker et al. (1979) calculate that there is no net change in the biosphere mass, probably indicating that reforestation is taking place. At the present rate of carbon dioxide production, the atmospheric carbon dioxide content will double in the next 50 years (Mercer, 1978). Since carbon dioxide prevents the long wavelength radiation from leaving, there is an increase in the surface temperature on the earth. This is the "greenhouse" effect (Mercer, 1978; McLean, 1978).

Civilization has also been polluting the atmosphere with particulate matter which scatters light and should produce cooling. This cooling does not occur however, since the particulates do not remain in the air more than 5 days; as most of the matter is black, they actually absorb heat. Thus, burning of fossil fuels increases the temperature on the earth's surface (Kellogg, 1979). The increase in surface temperature can have very serious effects, such as melting the ice of West Antarctica. If the rate of carbon dioxide continues to increase at the present rate into the atmosphere, then the atmospheric carbon dioxide will double in about 50 years; the result will be a temperature increase of 1.5° to 3°C. The consequence of this temperature increase would be to melt the ice and raise the sea level by 5 m (Mercer, 1978).

Clearly, we have to use less of the fossil fuels as an energy source. We must conserve energy and increase the development of alternate energy sources, such as solar energy, wind energy, water energy, and geothermal energy. Cloud (1978) points out that the energy from the sun received on earth is 26,000 times the present global energy demand.

VIII. Summary

The crust of the earth contains a very large predominance of oxygen. Since oxygen is reactive, it would not remain in the free molecular form very long on a geological time scale. The only reason molecular oxygen is present on earth is that the biosphere rapidly produces it.

When the earth was formed about 4.5×10^9 years ago, the atmosphere was reducing. The universe is predominately hydrogen. Hydrogen is escaping very slowly from the earth, making this planet a pinpoint of oxidation in a reducing universe.

The biosphere was born in a reducing environment and slowly developed the capacity to produce oxygen by photosynthesis. The biosphere has changed the earth's crust and, in turn the earth's crust has changed the biosphere. There is an intimate symbiotic relationship between the biosphere and its environment.

Oxygen is an energy source par excellence. Oxygen gave the biosphere a greater mobility and permitted the appearance of multicellular organisms. However, there is a price for this mobility, and it is that the biosphere has to cope with oxygen toxicity. To combat this toxicity, antioxidants of a varied nature were developed by the biosphere. However, this defense can never be perfect, and so the biosphere still has to suffer from oxygen toxicity. Thus, we see here the dual effect of oxygen, i.e., life-giving energy for aerobic organisms and destructive influence on all forms of life. Is oxygen good or bad? In this context, the question is unanswerable. Oxygen does what it does.

Changes by man in the foreseeable future will not greatly affect the quantity of free oxygen in the atmosphere. However, eventually, the atmosphere will dissipate and the earth will go into its postlife stage. Perhaps, we might find evidence of a previous biosphere in the depths of the lithosphere on Mars!

Acknowledgments

Acknowledgments are given to Dr. Claire Gilbert for her helpful discussions and advice, and to Mrs. Brenda Corbin, librarian of the Naval Observatory Library, Washington, D.C., for her assistance in locating some references.

References

Adams, M. W. W., and Hall, D. O. (1979). Properties of the solubilized membrane-bound hydrogenase from the photosynthetic bacterium *Rhodospirillum rubrum*. Arch. Biochem. Biophys. 195:288–299.

Aristotle (1953). The Works of Aristotle Translated into English. W. D. Ross (Ed.). Vol. 2. Physica. R. P. Hardie and R. E. Gaye (Eds.) New York: Oxford University Press.

Aristotle (1958). The Works of Aristotle Translated into English. J. A. Smith and W. D. Ross (Eds.) Vol. 5. De Generatione Animalium. A. Platt (Ed.) New York: Oxford Univ. Press.

Balch, W. E., and Wolfe, R. S. (1976). New approach to the cultivation of methanogenic bacteria: 2-Mercaptoethanesulfonic acid (HS-CoM)-dependent growth of *Methanobacterium ruminantium* in a pressurized atmosphere. Appl. Environ. Microbiol. 32:781–791.

Balch, W. E., Fox, G. E., Magrum, L. J., Woese, C. R., and Wolfe, R. S. (1979). Methanogens: reevaluation of a unique biological group. Microbiol. Rev. 43:260–296.

Barghoorn, E. S., and Tyler, S. A. (1965). Microorganisms from the Gunflint Chert. Science 147:563–577.

Berkner, L. V., and Marshall, L. C. (1967). The rise of oxygen in the earth's atmosphere with notes on the Martian atmosphere. Adv. Geophys. 12:309–331.

Bernal, J. D. (1949). The physical basis of life. Proc. Phys. Soc. A 62:537–558.

Bishop, N. I., and Jones, L. W. (1978). Alternate fates of the photochemical reducing power generated in photosynthesis: Hydrogen production and nitrogen fixation. Curr. Top. Bioenerg. 8:3–31.

Bolin, B. (1977). Changes of land biota and their importance for the carbon cycle. Science 196:613–615.

Bowien, B., Cook, A. M., and Schlegel, H. G. (1974). Evidence for the *in vivo* regulation of glucose 6-phosphate dehydrogenase activity of *Hydrogenomonas eutropha* H 16 from measurements of the intracellular concentrations of metabolic intermediates. Arch. Microbiol. 97:273–281.

Brill, W. J. (1975). Regulation and genetics of bacterial nitrogen fixation. Annu. Rev. Microbiol. 29:109–129.

Broda, E. (1975). The Evolution of the Bioenergetic Processes. New York: Pergamon Press.

Broecker, W. S., Takahashi, T., Simpson, H. J., and Peng, T.-H. (1979). Fate of fossil fuel carbon dioxide and the global carbon budget. Science 206:409–418.

Brooks, J., and Shaw, G. (1978). A critical assessment of the origin of life. In: Noda, H. (Ed.). Origin of Life. Proceedings of the Second ISSOL Meeting. The Fifth ICOL Meeting. Tokyo: Center for Academic Publications Japan/Japan Scientific Societies Press, pp. 597–606.

Buhl, D. (1974). Galactic clouds of organic molecules. Origins Life 5:29–40.

Calvin, M. (1969). Chemical Evolution. Molecular Evolution Towards the Origin of Living Systems on the Earth and Elsewhere. New York: Oxford University Press.

Chameides, W. L., Walker, J. C. G., and Nagy, A. F. (1979). Possible chemical impact of planetary lightning in the atmospheres of Venus and Mars. Nature 280:821–822.

Chance, B. (1981). The reaction of oxygen with cytochrome oxidase: The role of sequestered intermediates. This volume.

Chung, K-T. (1976). Inhibitory effects of H_2 on growth of *Clostridium cellobioparum*. Appl. Environ. Microbiol. 31:342–348.

Cloud, P. (1976). Beginnings of biospheric evolution and their biogeochemical consequences. Paleobiology 2:351–387.

Cloud, P. (1978). Cosmos, Earth, and Man. A Short History of the Universe. New Haven, Conn.: Yale University Press.

Cohen, S. S. (1974). On the origins of cells: The development of a Copernican revolution. In: Neynman, J. (Ed.). The Heritage of Copernicus: Theories "Pleasing to the Mind." Cambridge, Massachusetts: M.I.T. Press, pp. 207–221.

Crick, F. H. C., and Orgel, L. E. (1973). Directed panspermia. Icarus 19:341–346.

Crutzen, P. J. (1979). The role of NO and NO_2 in the chemistry of the troposphere and stratosphere. Annu. Rev. Earth Planet. Sci. 7:443–472.

Crutzen, P. J., Heidt, L. E., Krasnec, J. P., Pollack, W. H., and Seiler, W. (1979). Biomass burning as a source of atmospheric gases CO, H_2, N_2O, NO, CH_3Cl and COS. Nature 282:253–256.

Day, W. (1979). Genesis on Planet Earth. The Search for Life's Beginning. East Lansing, Michigan: The House of Talos Publications.

Decker, P. (1978). Inverse assimilation: A general principle of evolution of planetary surfaces. In: Noda, H. (Ed.). Origin of Life. Proceedings of the Second ISSOL Meeting. The Fifth ICOL Meeting. Tokyo: Center for Academic Publications Japan/Japan Scientific Societies Press, pp. 631–637.

De Ley, J., and Kersters, K. (1975). Chapter III. Biochemical evolution in bacteria. In: Florkin, M., and Stotz, E. H. (Eds.). Comprehensive Biochemistry. Vol. 29. Part B. Comparative Biochemistry, Molecular Evolution. New York: Elsevier Scientific Pub. Co., pp. 1–77.

Diderot, D. (1776). D'Alembert's Dream. In: Diderot, D. Rameau's Nephew and D'Alembert's Dream (Tancock, L. W., translator). Baltimore: Penguin Books. 1966, pp. 133–233.

Dimmick, R. L., and Chatigny, M. A. (1976). Possibility of growth of airborne microbes in outer planetary atmospheres. In: Ponnamperuma, C. (Ed.). Chemical Evolution of the Giant Planets. New York: Academic Press, pp. 95–106.

Dole, M. (1965). The natural history of oxygen. J. Gen. Physiol. 49 (Part 2):5–27.

Duedall, I. W., and Coote, A. R. (1972). Oxygen distribution in the Pacific Ocean. J. Geophys. Res. 77:2201–2203.

Durham, J. W. (1978). The probable metazoan biota of the Precambrian as indicated by the subsequent record. Annu. Rev. Earth Planet. Sci. 6:21–42.

Eliade, M. (1974). Gods, Goddesses, and Myths of Creation. A Thematic Source Book of the History of Religions. Part 1 of From Primitive to Zen. New York: Harper and Row.

Erhalt, D. H., and Schmidt, U. (1978). Sources and sinks of atmospheric methane. Pure Appl. Geophys. 116:452–464.

Florkin, M. (1974). Chapter I. Concepts of molecular biosemiotics and of molecular evolution. In: Florkin, M., and Stotz, E. H. (Eds.). Comprehensive Biochemistry. Vol. 29. Part A. Comparative Biochemistry, Molecular Evolution. New York: Elsevier Scientific Publishing Co., pp. 1–124.

Forman, H. J., and Fisher, A. B. (1981). Antioxidant defenses. This volume.

Fox, S. W., and Dose, K. (1977). Molecular Evolution and the Origin of Life. Revised Ed. New York: Marcel Dekker.

Fridovich, I. (1976). Oxygen radicals, hydrogen peroxide, and oxygen toxicity. In: Pryor, W. (Ed.). Free Radicals in Biology. Vol. I. New York: Academic Press, pp. 239–277.

Fridovich, I. (1981). Superoxide radical and superoxide dismutases. This volume.

Garrels, R. M., Lerman, A., and Mackenzie, F. T. (1976). Controls of atmospheric O_2 and CO_2: Past, present, and future. Am. Sci. 64:306–315.

Gerschman, R. (1964). Biological effects of oxygen. In: Dickens, F., and Neil, E. (Eds.). Oxygen in the Animal Organism. New York: Pergamon Press, Macmillan Co., pp. 475–494.

Gibbs, M. (1970). The inhibition of photosynthesis by oxygen. Am. Sci. 58:634–640.

Gilbert, D. L. (1960). Speculation on the relationship between organic and atmospheric evolution. Perspect. Biol. Med. 4:58–71.

Gilbert, D. L. (1963). The role of pro-oxidants and antioxidants in oxygen toxicity. Radiat. Res. Suppl. 3:44–53.

Gilbert, D. L. (1964). Cosmic and geophysical aspects of the respiratory gases. In: Fenn,

W. O., and Rahn, H. (Eds.). Handbook of Physiology—Section 3:Respiration. Vol. I. Washington, D.C.: American Physiological Society, pp. 153–176.

Gilbert, D. L. (1966). Antioxidant mechanisms against oxygen toxicity and their importance during the evolution of the biosphere. In: Brown, I. W., Jr., and Cox, B. G. (Eds.). Proceedings of the Third International Conference on Hyperbaric Medicine, Publication 1404, National Research Council. Washington, D.C.: National Academy of Science, pp. 3–14.

Gilbert, D. L. (1968). The interdependence between the biosphere and the atmosphere. Respir. Physiol. 5:68–77.

Gilbert, D. L. (1980). Discussion: What controls atmospheric oxygen? BioSystems 12: 123–124.

Gilbert, D. L. (1981a). Perspective on the history of oxygen and life. This volume.

Gilbert, D. L. (1981b). Oxygen: An overall biological view. This volume.

Goldschmidt, V. M. (1954). Geochemistry. Muir, A. (Ed.) New York: Oxford University Press.

Goldstein, B. D. (1979). The pulmonary and extrapulmonary effects of ozone. In: Fitzsimons, D. W. (Ed.). Oxygen Free Radicals and Tissue Damage. Ciba Foundation Symposium 65 (New Series). New York: Elsevier/North-Holland, pp. 295–319.

Gottlieb, S. F. (1981). Oxygen toxicity in unicellular organisms. This volume.

Graham, J. B., Rosenblatt, R. H., and Gans, C. (1978). Vertebrate air breathing arose in fresh waters and not in the oceans. Evolution 32: 459–463.

Green, E. J., and Carritt, D. E. (1967). New tables for oxygen saturation of seawater. J. Mar. Res. 25:140–147.

Haldane, J. B. S. (1954). The origins of life. New Biol. 16:12–27, Penguin Books. In: Cloud, P. (Ed.). Adventures in Earth History. San Francisco: W.H. Freeman and Co., 1970, pp. 377–384.

Hall, T. S. (1975). History of General Physiology. 600 B.C. to A.D. 1900. Vol. One. From Pre-Socratic Times to the Enlightenment. Chicago: University of Chicago Press.

Halliwell, B. (1978). The chloroplast at work. A review of modern developments in our understanding of chloroplast metabolism. Prog. Biophys. Mol. Biol. 33:1–54.

Hamilton, R. W., Jr., and Sheffield, P. J. (1977). Hyperbaric chamber safety. In: Davis, J. C., and Hunt, T. K. (Eds.). Hyperbaric Oxygen Therapy. Bethesda, Maryland: Undersea Medical Society, pp. 47–59.

Hart, M. H. (1979). Habitable zones about main sequence stars. Icarus 37:351–357.

Harvey, W. (1628). An anatomical disquisition on the motion of the heart-blood in animals. In: Knickerbocker, W. S. (Ed.). Classics of Modern Science (Copernicus to Pasteur). Boston: Beacon Press, 1962, pp. 47–48.

Herbig, G. H. (1981). The origin and astronomical history of terrestrial oxygen. This volume.

Hoffmann, G. W. (1975). The stochastic theory of the origin of the genetic code. Annu. Rev. Phys. Chem. 26:123–144.

Holland, H. D. (1978). The Chemistry of the Atmosphere and Oceans. New York: John Wiley.

Hook, R. (1665). Extracts from Micrographia: or Some Physiological descriptions of Minute bodies made by magnifying Glasses with Observations and Inquires thereupon. Alembic Club Reprints. No. 5. Edinburgh: Oliver and Boyd, Ltd., 1944.

Hoyle, F., and Wickramasinghe, C. (1978). Lifecloud: The Origin of Life in the Universe. New York: Harper and Row.

Huang, C., Wheeldon, L., and Thompson, T. E. (1964). The properties of lipid bilayer membranes separating two aqueous phases: Formation of a membrane of simple composition. J. Mol. Biol. 8:148–160.

Hunt, B. G. (1976). On the death of the atmosphere. J. Geophys. Res. 81:3677–3687.

Huntress, W. T., Jr. (1977). Ion-molecule reactions in the evolution of simple organic molecules in interstellar clouds and planetary atmospheres. Chem. Soc. Rev. 6:295–323.

Hwang, J. C., Chen, C. H., and Burris, R. H. (1973). Inhibition of nitrogenase-catalyzed reductions. Biochim. Biophys. Acta 292: 256–270.

Kanematsu, S., and Asada, K. (1979). Ferric and manganic superoxide dismutases in *Euglena gracilis*. Arch. Biochem. Biophys. 195: 535–545.

Kasting, J. F., Liu, S. C., and Donahue, T. M. (1979). Oxygen levels in the prebiological atmosphere. J. Geophys. Res. 84:3097–3107.

Kellogg, W. W. (1979). Influences of mankind on climate. Annu. Rev. Earth Planet. Sci. 7: 63–92.

Keosian, J. (1974). Life's beginnings—Origin or evolution? Origins Life 5:285–293.

Knoll, A. H., and Barghoorn, E. S. (1977). Archean microfossils showing cell division from the

Swaziland System of South Africa. Science 198:396–398.

Kovachich, G. B., and Haugaard, N. (1981). Biochemical aspects of oxygen toxicity in the metazoa. This volume.

Krebs, H. A. (1972). The Pasteur effect and the relations between respiration and fermentation. Essays Biochem. 8:1–34.

Lasaga, A. C., Holland, H. D., and Dwyer, M. J. (1971). Primordial oil slick. Science 174: 53–55.

Latimer, W. M. (1952). The Oxidation States of the Elements and their Elements in Aqueous Solutions. 2nd Ed. New York: Prentice-Hall.

Lavoisier (1862). Vues générales sur la formation et la constitution de l'atmosphère de la terre. Rec. Mém. Chim. de Lavoisier. Vol. 2, p. 398. In: Oeuvres de Lavoisier. Vol. II. Mémoires de Chimie et de Physique. Paris: Imp. Impériale, pp. 804–811.

Lenfant, C., and Johansen, K. (1972). Gas exchange in gills, skin, and lung breathing. Resp. Physiol. 14:211–218.

Likens, G. E., Bormann, F. H., Pierce, R. S., Eaton, J. S., and Johnson, N. M. (1977). Biogeochemistry of a Forested Ecosystem. New York: Springer-Verlag.

Logan, J., McElroy, M. B., Wofay, S. C., and Prather, M. J., (1979). Oxidation of CS_2 and COS: Sources for atmospheric SO_2. Nature 281:185–188.

Lovelock, J. E. (1972). Gaia as seen through the atmosphere. Atmos. Environ. 6:579–580.

Lumsden, J., Henry, L., and Hall, D. O. (1977). Superoxide dismutase in photosynthetic organisms. In: Michelson, A. M., McCord, J. M., and Fridovich, I. (Eds.). Superoxide and Superoxide Dismutases. New York: Academic Press, pp. 437–450.

Machta, L., and Hughes, E. (1970). Atmospheric oxygen in 1967 to 1970. Science 168: 1582–1584.

Mah, R. A., Ward, D. M., Baresi, L., and Glass, T. L. (1977). Biogenesis of methane. Annu. Rev. Microbiol. 31:309–341.

Margulis, L. (1970). Origin of Eukaryotic Cells. New Haven, Connecticut: Yale University Press.

Margulis, L., and Lovelock, J. E. (1974). Biological modulation of the earth's atmosphere. Icarus 21:471–489.

Margulis, L., and Lovelock, J. E. (1978). The biota as ancient and modern modulator of the earth's atmosphere. Pure Appl. Geophys. 116: 239–243.

Margulis, L., Walker, J. C. G., and Rambler, M. (1976). Reassessment of roles of oxygen and ultraviolet light in Precambrian evolution. Nature 264:620–624.

Mason, B. (1966). Principles of Geochemistry. 3rd Ed. New York: John Wiley and Sons.

Mason, S. F. (1953). Main Currents of Scientific Thought. A History of the Sciences. New York: Henry Schuman.

Mason, T. R., and Von Brunn, V. (1977). 3-Gyr-old stromatolites from South Africa. Nature 266:47–49.

Mauzerall, D. C., and Piccioni, R. G. (1981). Photosynthetic oxygen production. This volume.

McAlester, A. L. (1970). Animal extinctions, oxygen consumption, and atmospheric history. J. Paleontol. 44:405–409.

McLean, D. M. (1978). A terminal Mesozoic "greenhouse": Lessons from the past. Science 201:401–406.

McWhirter, N. (1979). Guinness Book of World Records. 17th Ed. New York: Bantam Books.

Menger, F. M. (1972). Reactivity of organic molecules at phase boundaries. Chem. Soc. Rev. 1: 229–240.

Mercer, J. H. (1978). West Antarctic ice sheet and CO_2 greenhouse effect: a threat of disaster. Nature 271:321–325.

Miller, S., and Diehn, B. (1978). Cytochrome c oxidase as the receptor molecule for chemoaccumulation (chemotaxis) of *Euglena* toward oxygen. Science 200:548–549.

Miller, S. L. (1953). A production of amino acids under possible primitive earth conditions. Science 117:528–529.

Miller, S. L. (1974). The first laboratory synthesis of organic compounds under primitive earth conditions. In: Neynman, J. (Ed.). The Heritage of Copernicus: Theories "Pleasing to the Mind." Cambridge, Massachusetts: M.I.T. Press, pp. 228–242.

Miller, S. L., and Orgel, L. E. (1974). The Origins of Life on the Earth. Englewood Cliffs, New Jersey: Prentice Hall.

Miller, S. L., Urey, H. C., and Oró, J. (1976). Origin of organic compounds on the primitive earth and in meteorites. J. Mol. Evol. 9:59–72.

Minzner, R. A. (1977). The 1976 standard atmosphere and its relationship to earlier standards. Rev. Geophys. Space Phys. 15:375–384.

Molina, M. J., and Rowland, F. S. (1974). Stratospheric sink for chlorofluoromethanes: Chlorine atom-catalysed destruction of ozone. Nature 249:810–812.

Morowitz, H., and Sagan, C. (1967). Life in the clouds of Venus? Nature 215:1259–1260.

Napier, W. M., and Clube, S. V. M. (1979). A theory of terrestrial catastrophism. Nature 282: 455–459.

Needham, J. (1969). Science and Civilization in China. Vol. 2. History of Scientific Thought. New York: Cambridge University Press.

Neuman, M. W., Neuman, W. F., and Lane, K. (1970). On the possible role of crystals in the origins of life. (III) The phosphorylation of adenosine to AMP by apatite. Curr. Mod. Biol. 3:253–259.

Oparin, A. I. (1936). The Origin of Life (Translated by S. Morgulis). New York: Dover, 1953.

Oparin, A. I. (1978). The nature and origin of life. In: Ponnamperuma, C. (Ed.). Comparative Planetology. New York: Academic Press, pp. 1–6.

Oró, J. (1961). Comets and the formation of biochemical compounds on the primitive earth. Nature 190:389–390.

Oró, J. (1976). Prebiological chemistry and the origin of life. A personal account. In: Kornberg, A. (Ed.). Reflections on Biochemistry. Oxford: Pergamon Press, pp. 423–443.

Oró, J., Sherwood, E., Eichberg, J., and Epps, D. (1978). Formation of phospholipids under primitive earth conditions and the role of membranes in prebiological evolution. In: Deamer, D. W. (Ed.). Light Transducing Membrane:. Structure, Function and Evolution. New York: Academic Press, pp. 1–21.

Otroshchenko, V. A., and Vasilyeva, N. V. (1977). The role of mineral surfaces in the origin of life. Origins Life 8:25–31.

Oyama, V. I., and Berdahl, B. J. (1977). The Viking gas exchange experiment results from Chryse and Utopia surface samples. J. Geophys. Res. 82:4669–4676.

Paecht-Horowitz, M. (1978). The influence of various cations on the catalytic properties of clays. J. Mol. Evol. 11:101–107.

Pagel, B. E. J. (1979). Solar abundances. A new table (October 1976). Phys. Chem. Earth 11: 79–80.

Partington, J. R. (1961). A History of Chemistry. Vol 2. New York: St. Martin's Press.

Partington, J. R. (1970). A History of Chemistry. Vol. 1. Part 1: Theoretical Background. New York: St. Martin's Press.

Penzias, A. A. (1979). The origin of the elements. Science 205:549–554.

Piiper, J., and Scheid, P. (1981). Oxygen exchange in the metazoa. This volume.

Pollack, J. B. (1979). Climatic change on the terrestrial planets. Icarus 37:479–553.

Pollard, W. G. (1979). The prevalence of earthlike planets. Am. Sci. 67:653–659.

Ponnamperuma, C., Shimoyama, A., Yamada, M., Hobo, T., and Pal, R. (1977). Possible surface reaction on Mars: Implications for Viking biology results. Science 197: 455–457.

Pow, T., and Krasna, A. I. (1979). Photoproduction of hydrogen from water in hydrogenase-containing algae. Arch. Biochem. Biophys. 194: 413–479.

Prinn, R. G., Alyea, F. N., and Cunnold, D. M. (1978). Photochemistry and dynamics of the ozone layer. Annu. Rev. Earth Planet. Sci. 6: 43–74.

Probst, I., and Schlegel, H. G. (1973). Studies on a gram-positive hydrogen bacterium, *Nocardia opaca* Strain 1b. II. Enzyme formation and regulation under the influence of hydrogen or fructose as growth substrates. Arch. Mikrobiol. 88:319–330.

Raff, R. A., and Mahler, H. R. (1975). The symbiont that never was: An inquiry into the evolutionary origin of the mitochondrion. Symp. Soc. Exp. Biol. 29:41–92.

Reddy, C. A., Bryant, M. P., and Wolin, M. J. (1972a). Characteristics of S organism isolated from *Methanobacillus omelianskii*. J. Bacteriol. 109:539–545.

Reddy, C. A., Bryant, M. P., and Wolin, M. J. (1972b). Ferredoxin- and nicotinamide adenine dinucleotide-dependent H_2 production from ethanol and formate in extracts of S organism isolated from "Methanobacillus omelianskii." J. Bacteriol. 110:126–132.

Rolfs, C., and Trautvetter, H. P. (1978). Experimental nuclear astrophysics. Annu. Rev. Nucl. Part. Sci. 28:115–159.

Rowland, F. S., and Molina, M. J. (1975). Chlorofluoromethanes in the environment. Rev. Geophys. Space Phys. 13:1–35.

Rüden, H., Thofern, E., Fischer, P., and Mihm, U. (1978). Airborne microorganisms: Their occurrence, distribution and dependence on environmental factors—Especially on organic compounds of air-pollution. Pure Appl. Geophys. 116:335–350.

Russell, D. A. (1979). The enigma of the extinction of the dinosaurs. Annu. Rev. Planet Sci. 7: 163–182.

Sagan, C. (1961). On the origin and planetary distribution of life. Radiat. Res. 15: 174–192.

Sagan, C. (1973). Ultraviolet selection pressure on the earliest organisms. J. Theor. Biol. 39: 195–200.

Sagan, C. (1980). Cosmos. New York: Random House.

Schatz, A. (1957). Some biochemical and physiological considerations regarding the extinction of the dinosaurs. Proc. Penn. Acad. Sci. 31: 26–36.

Schidlowski, M. (1979). Antiquity and evolutionary status of bacterial sulfate reduction: Sulfur isotope evidence. Origins Life 9:299–311.

Schidlowski, M., Appel, P. W. U., Eichmann, R., and Junge, C. E. (1979). Carbon isotope geochemistry of the 3.7×10^9-yr-old Isua sediments, West Greenland: implications for the Archaean carbon and oxygen cycles. Geochim. Cosmochim. Acta 43:189–199.

Schopf, J. W. (1975). Precambrian paleobiology: problems and perspectives. Annu. Rev. Earth Planet. Sci. 3:213–249.

Schopf, J. W. (1978). The evolution of the earliest cells. Sci. Am. 239:110–138.

Schopf, T. J. M., Farmanfarmaian, A., and Gooch, J. L. (1971). Oxygen consumption rates and their paleontologic significance. J. Paleont. 45: 247–252.

Schueler, F. W. (1960). Chemobiodynamics and Drug Design. New York: McGraw-Hill.

Schwartz, R. M., and Dayhoff, M. O. (1978). Origins of prokaryotes, eukaryotes, mitochondia, and chloroplasts. Science 199:395–403.

Shapley, H. (1958). Of Stars and Men. The Human Response to an Expanding Universe. New York: Washington Square Press.

Shimoyani, A., Ponnamperuma, C., and Yanoi, K. (1979) Amino acids in the Yamato carbonaceous chondrite from Antarctica. Nature 282: 394–396.

Siegel, B. Z. (1977). *Kakabekia*, a review of its physiological and environmental features and their relation to its possible ancient affinities. In: Ponnamperuma, C. (Ed.). Chemical Evolution of the Early Precambrian. New York: Academic Press, pp. 143–154.

Siegel, S. M. (1957). Catalytic and polymerization-directing properties of mineral surfaces. Proc. Nat. Acad. Sci. U.S.A. 43:811–816.

Sillén, L. G. (1965). Oxidation state of Earth's ocean and atmosphere. I. A model calculation on earlier states. The myth of the "prebiotic soup." Ark. Kemi 24:431–456.

Smith, D. C. (1979). From extracellular to intracellular: The establishment of a symbiosis. Proc. Roy. Soc. B 204:115–130.

Smith, J., and Shrift, A. (1979). Phylogenetic distribution of glutathione peroxidase. Comp. Biochem. Physiol. 63B:39–44.

Stanier, R. Y., and Cohen-Bazire, G. (1977). Phototrophic prokaryotes: The cyanobacteria. Annu. Rev. Microbiol. 31:225–274.

Stoeckenius, W., Lozier, R. H., and Bogomolni, R. A. (1979). Bacteriorhodopsin and the purple membrane of halobacteria. Biochim. Biophys. Acta 505:215–278.

Swain, T. (1974). Chapter II. Biochemical evolution in plants. In: Florkin, M., and Stotz, E. H. (Eds.). Comprehensive Biochemistry Vol. 29. Part A Comparative Biochemistry, Molecular Evolution. New York: Elsevier Scientific Publishing Co., pp. 125–302.

Tappan, H. (1974). Molecular oxygen and evolution. In: Hayaishi, O. (Ed.). Molecular Oxygen in Biology: Topics in Molecular Oxygen Research. New York: American Elsevier Publishing Co., pp. 81–135.

Taylor, F. J. R. (1979). Symbionticism revisited: A discussion of the evolutionary impact of intracellular symbioses. Proc. Roy. Soc. B 204: 267–286.

Thomas, L. (1974). The Lives of a Cell. Notes of a Biology Watcher. New York: Viking Press.

Towe, K. M. (1978). Early Precambrian oxygen: A case against photosynthesis. Nature 274: 657–661.

Trimble, V. (1975). Origin and abundances of the chemical elements. Rev. Mod. Phys. 47: 877–976.

Urey, H. C. (1952). On the early chemical history of the earth and the origin of life. Proc. Nat. Acad. Sci. U.S.A. 38:351–363.

Urey, H. C. (1959). The atmospheres of the planets. Handb. Phys. 52:363–418.

Uzzell, T., and Spolsky, C. (1974). Mitochondria and plastids as endosymbionts: A revival of special creation? Am. Sci. 62:334–343.

Van Houten, F. B. (1973). Origin of red beds. A review—1961–1972. Annu. Rev. Earth Planet. Sci. 1:39–61.

Walker, J. C. G. (1977). Evolution of the Atmosphere. New York: Macmillan.

Walker, J. C. G. (1978). Oxygen and hydrogen in the primitive atmosphere. Pure Appl. Geophys. 116:222–231.

Walker, T. R. (1967). Formation of red beds in modern and ancient deserts. Geol. Soc. Am. Bull. 78:353–368.

Watson, A., Lovelock, J. E., and Margulis, L. (1978). Methanogenesis, fires and the regulation of atmospheric oxygen. BioSystems 10: 293–298.

Weast, R. C., and Astle, M. J. (Eds.) (1979). CRC Handbook of Chemistry and Physics; A Ready-Reference Book of Chemical and Physical Data. 60th Ed. 1979–1980. Boca Raton, Florida: CRC Press, Inc.

Whatley, J. M., John, P., and Whatley, F. R. (1979). From extracellular to intracellular: The establishment of mitochondria and chloroplasts. Proc. Roy. Soc. B 204:165–187.

Whitfield, M. (1976). The evolution of the oceans and the universe. In: Bligh, J., Cloudsley-Thompson, J. L., and MacDonald, A. G. (Eds.). Environmental Physiology of Animals. New York: John Wiley, pp. 30–45.

Whittaker, R. H., and Likens, G. E. (1975). The biosphere and man. In: Lieth, H., and Whittaker, R. H. (Eds.). Primary Productivity of the Biosphere. New York: Springer-Verlag, pp. 305–328.

Wickramasinghe, R. H., and Villee, C. A. (1975). Early role during chemical evolution for cytochrome P450 in oxygen detoxification. Nature 256:509–511.

Wilson, T.H., and Maloney, P.C. (1976). Speculations on the evolution of ion transport mechanisms. Fed. Proc. 35:2174–2179.

Wittenberg, J. B., and Wittenberg, B. A. (1981). Facilitated oxygen diffusion by oxygen carriers. This volume.

Woese, C. R., Magrum, L. J., and Fox, G. E. (1978). Archaebacteria. J. Mol. Evol. 11: 245–252.

Woodwell, G. M., Whittaker, R. H., Reiners, W. A., Likens, G. E., Delwiche, C. C., and Botkin, D. B. (1978). The biota and the world carbon budget. Science 199:141–146.

Woolley, L. (1965). History of Mankind. Cultural and Scientific Development. Vol. I, Part 2. The Beginnings of Civilization. New York: New American Library.

Zeikus, J. G. (1977). The biology of methanogenic bacteria. Bacteriol. Rev. 41:514–541.

6
Photosynthetic Oxygen Production

DAVID C. MAUZERALL AND RICHARD G. PICCIONI

It has been said that we exist on this earth because of an accident of quantum mechanics. It is even more likely that we exist because of two accidents. The first is that oxygen does not consume us instantly as the thermodynamic driving force of free energy insists that it will. A barrier to this reaction, traceable to delicate interelectronic interactions in the oxygen molecular anion, O_2^-, allows us to enjoy our nanoera (a few years out of the universe lifetime of 10^{10} years) in productive endeavors. The second, and far more incredible, fact or accident is that this selfsame oxygen is produced by living material itself. The energy of a dozen quanta from the sun are integrated into a chemical reaction that oxidizes water to oxygen, an almost magical reaction carried out in a crucible which is thermodynamically unstable with respect to reaction with its product! The reason this process, clearly defined two hundred years ago, has remained enigmatic in this period of explosive growth of scientific knowledge is buried in the implications of "mammalian chauvinism," a term applied by Martin Kamen, a great originator, in reference to our preoccupation with the biosphere's energy consumers, to the detriment of our understanding of its energy producers (Kamen, 1963). What little we do know of this second providential accident, photosynthetic oxygen evolution, will be the subject of this chapter.

I. Photosynthesis and Evolution

Before delving into the facts, we will indulge in some speculative thoughts on photosynthesis. Photosynthesis is usually described as the conversion of the energy of solar photons to chemical energy by the oxidation of water to oxygen and the concomitant reduction of carbon dioxide to the equivalent of formaldehyde or sugars. Recombination of the oxygen and reduced carbon or foods supports all the other forms of life on this earth. Figure 1 summarizes the overall energy and material balance between the processes of photosynthesis and aerobic respiration. In the "light reactions" of photosynthesis, the energy derived from the absorption of solar photons drives the oxidation of water to yield reducing equivalents ($H\cdot$) and, as a by-product, molecular oxygen (O_2). While the latter is casually released, the reducing power is carefully preserved in the form of NADPH and used for the reduction of CO_2 to carbohydrate in the "dark reactions" of photosynthesis. For each carbon atom reduced, four electrons are required, hence two water molecules must be oxidized and one oxygen molecule released.

The location of the light reactions is the thylakoid membrane of chloroplasts or cyanobacteria (blue-green algae). The requirement for chlorophyll a in this part of photosynthesis appears to be absolute. The dark reactions, which do not involve chlorophyll, take place in the soluble phase which surrounds the thylakoid. Included among the dark reactions of the carbon fixation pathway, the Calvin–Benson–Bassham cycle (Gregory, 1976), are three steps requiring ATP. The needed ATP is generated in the light reactions of the thylakoid membrane as a by-product of the flow of reducing equivalents. Thus, the light reactions provide not only the reducing power, but also the phosphorylation potential needed in the dark reactions of CO_2 fixation. Some ATP may also be made available for other energy-requiring processes within the plant cell.

In the mitochondria of plants and animals, the transformation of carbon achieved by photosynthesis is reversed (Fig. 1). The energy made available from the conversion of carbohydrate to CO_2 and reducing equivalents, as well as the much larger amount of energy from the recombination of these reducing equivalents with molecular oxygen, is conserved in the form of ATP, and utilized for all energy-requiring reactions in the cell. Normally, a plant will recycle only a small fraction of the reduced carbon from the

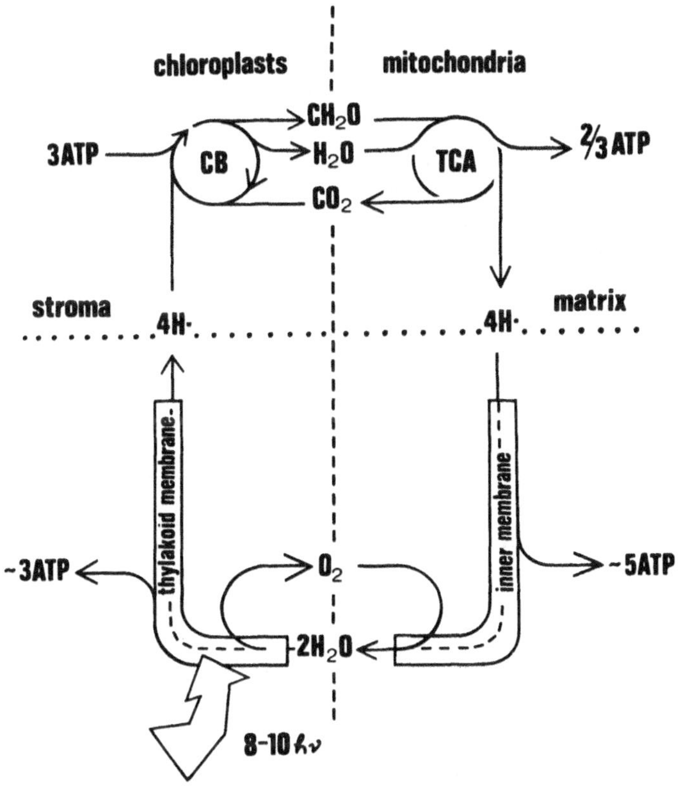

Figure 1. Photosynthesis in relation to respiration. The events depicted may occur within a single photosynthetic cell or organism, but also apply to the net production of oxygen and reduced carbon by plants and its consumption by animals in the biosphere as a whole. CB, Calvin–Benson cycle and gluconeogenic pathway; TCA, tricarboxylic acid cycle and glycolysis. In chloroplasts, reducing equivalents (4H·) are carried by 2NADPH; in mitochondria, by 2NADH. The ATP stoichiometries are for glucose formation and combustion. A precise yield of ATP in photosynthetic and respiratory electron transport has not yet been established.

chloroplast to provide energy in its mitochondria; hence, there is an accumulation of reduced organic material, i.e., the plant grows. This net yield of photosynthesis, experimentally manifested as the evolution of molecular oxygen, is the basis of the accumulation of all the biomass and oxygen on which animal life depends.

One can, however, define photosynthesis in a more general way. Living organisms are organized structures whose stability depends on a continual flow of energy through them (Prigogine and Nicolis, 1971). It is photosynthesis which supplies the gradient of free energy, oxygen/foods, necessary for the continued existence of life. However, in the beginning things differed. In particular, the environment was far more reduced. The present oxidative atmosphere is a product of modern photosynthesis. Thus, as we have argued elsewhere (Mauzerall, 1978a), the gradient of free energy developed from solar photons would involve the oxidation of the then prevalent reduced organic compounds and the emission of hydrogen. Now the irradiation of uroporphyrin, the first porphyrin on the biosynthetic path to chlorophyll, does lead to the oxidation of organic compounds and the formation of reduced porphyrins (Mauzerall, 1976). Similar reactions in the presence of a suitable catalyst leads to the formation of hydrogen (Amouyal et al., (1978). We have now observed the formation of hydrogen on irradiating uroporphyrin, an organic compound and a catalyst (Mercer-Smith and Mauzerall, 1981). It is even possible to observe photoproduction of hydrogen from anaerobic green algae (Gaffron and Rubin, 1942). Thus, the reactions evolutionarily favorable at that time can be observed in the laboratory. Moreover, taking the concept of Granick (1965) that a biosynthetic pathway is a window looking backward in evolution, we assign each intermediate on the pathway a function in its time. Mutations produce variations that are selected if their function improves the survival of the cell and its progeny. By this view the most primitive porphyrin, uroporphyrin, accomplished the most primitive photosynthetic function. At Granick's branch point of the pathway, protoporphyrin incorporates iron leading to heme and eventual mammalian chauvinism, or incorporates magnesium leading to chlorophyll and modern photosynthesis. It is a fact that the incorporation of a filled shell metal such as zinc or magnesium into a porphyrin changes the electron distribution such as to favor the reduction of another molecule by the excited state. That is, the excited state (and also the ground state) is more readily oxidized than the free-base porphyrin, where the negative charge is tied up by bonding to protons in the central N—H bonds. The as yet unfathomed discovery of evolution was the method of combining four of these oxidation equivalents to generate oxygen (and protons) from water. The much larger gradient of free energy available from oxygen and reduced organic compounds allowed the evolution of more complex forms of life. This invention was probably the cause of the explosion of the evolutionary sequence about a half-billion years ago (Mauzerall, 1977).

A second clear characteristic of the biosynthetic pathway of chlorophyll is the systematic modification of the substituents on the porphyrin nucleus. These changes transform the polar, water-soluble uroporphyrin into the waxy lipid-soluble chlorophyll. Several selection forces lead evolution along this particular path. One concerns efficiency. In a watery environment hydrophobic molecules cluster together. This allows the photoreactions to occur more rapidly and effectively than if by random collision in dilute solution. If several pigment molecules cluster correctly, not too close and not too far apart, energy transfer among them allows more efficient photon capture, as in the modern photosynthetic unit. Moreover, the discovery of a pathway to O_2 made very rapid photoreactions mandatory. Excited states are quenched by O_2 at encounter-limited rates. Thus, the primary reaction must occur in < 100 nsec at the present level of O_2. The long life of the triplet state of porphyrins (> 10 msec), which allows efficient reactions in free solution, is no longer useful. Thus, a suitable complex is needed, and in fact the primary charge transfer in the bacterial reaction center complex occurs in a few psec (Clayton, 1978). If this complex is oriented in a membrane, the charge transfer will produce an electrical potential across the membrane. This can be coupled, via the chemiosmotic hypothesis, to the production of ATP. We have observed several of these "forces" in simple model photoreactions with lipid bilayers (Hong and Mauzerall, 1976). Thus, the evolution of photosynthesis and of oxygen have been strongly interdependent.

It is now accepted that the origin of the present enormous pool of atmospheric O_2 is photosynthesis. The UV photodissociation of water could only form O_2 to a level one thousandth of the present atmospheric level (Walker, 1977) because of self-absorption by O_2 and its photoproduct, ozone. The turnover time of this O_2 pool is about 10^4 years. It is interesting to speculate as to what would happen if all photosynthesis on earth were to stop, an event not beyond the realm of the possible given mankind's present activities and continuing infestation of this earth. The amount of reduced carbon at the earth's surface is but 0.2% of the oxygen supply. Thus, even with no further formation of O_2, this vast O_2 pool would only be slightly affected by the oxidation of all available reduced carbon. Moreover if the oxidation of reduced carbon proceeds at its present rate, it would take some 30,000 years to deplete the carbon pool (Broecker, 1970). By contrast, preventing the formation of new foods would cause the disappearance of the higher forms of life within a year or so. Thus although the lower forms of life, bacteria and molds, could survive for some fraction of the time for carbon pool depletion, life at the summit of the biosphere, mankind, is in a far more precarious position. In fact until the adoption of modern agricultural methods, famine was not uncommon. It is only that the basic fact of our dependence on photosynthesis has been hidden by the rapid, and according to Malthus, possibly futile success of energy-intensive agriculture.

Two problems arise from a consideration of the carbon and oxygen pools. The first is the remarkable constancy of the O_2 content of the atmosphere. It has changed less than 0.01% in 60 years (Machta and Hughes, 1970; van Valen, 1971) and possibly not much in 10^8 years. Thus, the formation and utilization of O_2 must be delicately balanced. Yet the respiration of the biosphere is independent of O_2 pressure. Possibly some forms of weathering are proportional to O_2 pressure, and the product, CO_2, certainly determines the rate of photosynthesis of land plants. Still the connection seems tenuous.

The second problem of the carbon/oxygen cycle is the "missing" reduced carbon associated with the immense O_2 pool. For each mole of O_2, a mole of CO_2 must have been reduced (Fig. 1). Possibly it is buried in the lower crust and the turnover goes along with plate tectonics. Walker (1977) has given an excellent discussion of these and related problems.

II. Overview of Modern Photosynthetic Mechanisms

A. The Light Reactions

Before beginning our discussion of the reactions of oxygen formation, a brief summary of current knowledge of the rest of the photosynthetic process is in order. The reader should consult recent reviews for further discussion (Barber, 1976; Trebst and Avron, 1977; Govindjee, 1975; Blankenship and Parson, 1978). As noted above, the primary function of the light reactions of the thylakoid membrane is the reduction of NADP using electrons derived from water. This requires promotion of electrons from a redox potential of 0.81 V positive to 0.32 V negative, relative to the standard hydrogen electrode. Two photoreactions, acting in series, provide the necessary electromotive force. Photosystem II carries out a photoreaction in which water is oxidized and an acceptor at about −0.1 V is reduced. Photosystem I oxidizes the system II acceptor indirectly through a sequence of electron carriers, and reduces its own acceptor at a potential of approximately −0.5 V, providing more than enough potential to reduce NADP. The primary photosystem II acceptor is probably a molecule of plastoquinone. This quinone (called Q) transfers electrons one at a time to a secondary acceptor (Crofts and Wood, 1978) which then transfers pairs of electrons to a pool of approximately 20 plastoquinones, all poised at redox potential similar to that of the primary acceptor.

The part of the photosynthetic electron transport chain located between the two photosystems is in some ways analogous to complex III of the mitochondrial inner membrane (Golbeck et al., 1977). The type *c* cytochromes (cytochrome *f*) (Cramer and Whitmarsh, 1977), type *b* cytochromes, and iron–sulfur proteins cooperate in the transport of electrons from the pool of plastoquinone to plastocyanin, a copper-containing, cyanide-sensitive protein. The precise arrangement of these electron carriers is still very much a matter of debate (Trebst, 1978; Crofts, 1977) but it would appear that a likely sequence, proceeding from photosystem II to photosystem I is that shown in Fig. 2. Cytochrome *f* and the *b*-type cytochromes are probably located off the main pathway of electron flow, cytochrome *f* on a side branch between plastoquinone and plastocyanin, and cytrochrome b_6 at a point of reentry of electrons flowing in a cyclical manner around photosystem I (Arnon, 1977). The experimental basis of this and other schemes is the effect of selective excitation of photosystem I on the redox state of a particular carrier, in conjunction with the use of specific inhibitors of electron transport whose site of action is known (Izawa, 1977), i.e., DCMU (between the primary and secondary electron acceptors of system II), dibromothymoquinone (plastoquinone), and KCN or $HgCl_2$ (plastocyanin). The kinetic and thermodynamic parameters of oxidation and reduction are also useful guides in the placement of a particular carrier in the electron transport scheme, i.e., cytochrome *f* has too slow a rate of turnover to be located on the main path, but its midpoint potential is near that of the system I reaction center (*P*-700). Finally, some information on the relationships of various components is offered by the observed group-

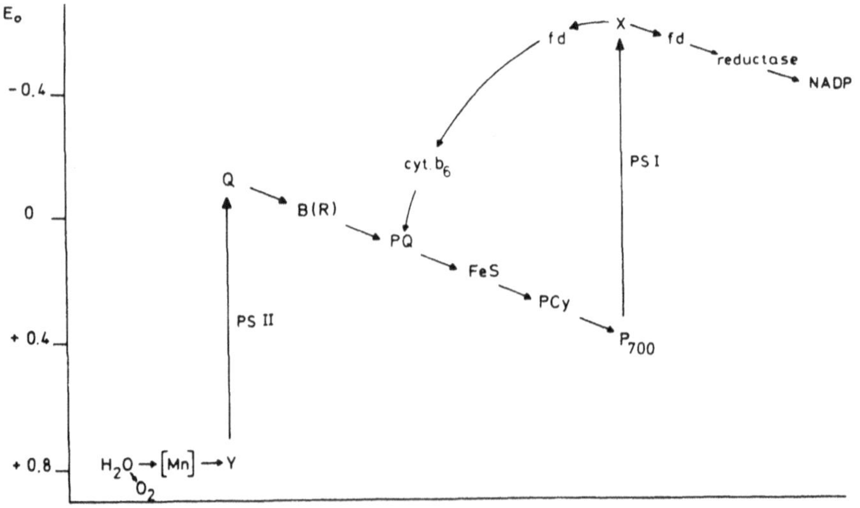

Figure 2. A recently presented scheme for photosynthetic electron transport (Trebst, 1978). The vertical axis indicates the midpoint potential of each component. Abbreviations: [Mn], manganese-containing complex; Y, primary donor of photosystem II; Q, primary acceptor of photosystem II; B(R), secondary acceptor; PQ, plastoquinone pool; FeS, nonheme iron protein; PC_y, plastocyanin; *P*-700, primary donor of photosystem I; X, primary acceptor of photosystem I; fd, ferredoxin; reductase, fd-NADP-oxidoreductase.

ing of carriers into discrete particles isolated by mild detergent treatment of photosynthetic membranes (Wessels, 1977).

At the photosystem I reaction center, a photogenerated chlorophyll cation, designated *P*-700 because of changes in the absorption spectrum associated with its formation, oxidizes plastocyanin. The midpoint potential of *P*-700 is near +0.50 V, while the system I photoelectron is delivered at a potential more negative than −0.55 V. The identity of the system I primary acceptor has been pursued using measurements of both absorption change and electron spin resonance. Recently, workers appear to be reaching agreement upon a non-heme iron protein bound tightly to the thylakoid membrane (Ke, 1978; Malkin, 1977). This species is capable of carrying out a rapid reduction of ferredoxin ($E_0' = -0.42$ V). Yet more recent evidence points to a "more primary" acceptor (Mathis and Conjeau, 1979).

The fate of reduced ferredoxin depends on the metabolic needs of the cell. The alternatives include $NADP^+$ reduction, sulfite reduction, nitrite reduction, and fatty acid desaturation (Arnon, 1977). Ferredoxin can also introduce electrons back into the photosynthetic electron transport chain via cytochrome b_6.

B. The Dark Reactions

In the chloroplast stroma, NADPH is utilized along with ATP for the reduction of 3-phosphoglycerate to glyceraldehyde 3-phosphate. The same reaction sequence employed in mamallian gluconeogenesis then yields fructose 6-phosphate, glucose, and starch. Part of the fructose 6-phosphate and glyceraldehyde 3-phosphate is recycled to form, after ATP-dependent phosphorylation, ribulose 1,5-diphosphate. Carboxylation of this latter compound to generate two molecules of 3-phosphoglycerate is carried out by the key enzyme, ribulose diphosphate carboxylase, a multisubunit enzyme comprising some 50% of the soluble protein of plant leaves (Jensen and Bahr, 1977).

In some species of algae and cyanobacteria capable of nitrogen fixation, NADPH can be oxidized to form molecular hydrogen. The activity of the required hydrogenase is inhibited by molecular oxygen, hence simultaneous hydrogen and oxygen evolution requires a rapid removal of oxygen produced. Direct biophotolysis of water by this mechanism has recently been observed in an algae illuminated in media rapidly flushed with nitrogen (Greenbaum, 1980; Mauzerall and Ley, unpublished).

C. Photorespiration

Under conditions of low intracellular CO_2 tension, molecular oxygen competes efficiently with CO_2 at the active site of RuDP carboxylase, causing the oxidation of ribulose 1,5-diphosphate to phosphoglycolate and CO_2. Phosphoglycolate is hydrolyzed and glycolate transported to the perioxisomes of the plant cell where a flavorprotein oxidase carries out the oxidation of glycolate to glyoxylate. In this fashion, the rate of oxygen consumption of certain plants can increase as much as fivefold under intense illumination, when ribulose 1,5-diphosphate and oxygen are in abundance, and CO_2 is scarce. To prevent the waste of reducing power inherent in this process, called photorespiration, a number of plant species have evolved a mechanism of CO_2 fixation different from that described above. In these so-called C_4 plants, cells in direct contact with the surrounding air fix CO_2 as oxaloacetate from phosphoenolpyruvate. Hydrolysis of a high energy phosphate linkage greatly facilitates this reaction, hence the affinity of these cells for atmospheric CO_2 is very large. Using photosynthetically generated NADPH, the oxaloacetate is reduced to malate and transported to more internally located cells where it is oxidatively decarboxylated to pyruvate and CO_2, NADP serving as an electron acceptor. The released CO_2 is then used in the same manner as in "C_3" plants. The end result is the use of ATP to actively accumulate CO_2. This prevents oxidation of ribulose

1,5-diphosphate, and hence diminishes the rate of photorespiration. Consequently C_4 plants have overall a greater energy efficiency (Hatch, 1976; Zelitch, 1971).

D. Photophosphorylation

In intact thylakoid membranes, photosynthetic electron transport, either from H_2O to NADP, or in the cyclic pathway, brings about the phosphorylation of ADP to form ATP. The bulk, if not all, of the ATP generated by photophosphorylation is utilized in the Calvin cycle (Fig. 1). According to the chemiosmotic theory of Mitchell (see Mitchell, 1976, for a recent review) a necessary intermediate in the transduction of redox to phosphorylation potential is a transmembrane gradient in the chemical potential of hydrogen ion. In the thylakoid membrane, as in the mitochondrial inner membrane and the bacterial plasma membrane, electron flow is coupled to a vectorial movement of protons transverse to the membrane surface. This movement is brought about by an asymmetric arrangement of electron carriers such that oxidation of water and of plastoquinone takes place near the inner surface, and reduction of plastoquinone and NADP near the outer surface of the thylakoid. Protons released during oxidation are thereby released into the inner space of the thylakoid sac, while protons consumed during reduction are taken from the stromal (extrathylakoid) space. Experimental evidence in favor of this scheme includes observation of light-induced transmembrane electric fields and pH gradients (Junge, 1977); the synthesis of ATP under the application of an artificial pH gradient (Jagendorf, 1975) or electric field (Graber et al., 1977); and observed sidedness of the thylakoid membrane with respect to surface accessibility of the components of the electron transport chain (Trebst, 1974; Anderson, 1975).

As is the case for the analogous reactions in mitochondria and bacteria, the mechanism by which the flow of protons down their gradient across the thylakoid membrane generates ATP remains obscure. For a collection of reviews on currently debated alternatives, see Boyer et al. (1977). Recently, Pick and Racker (1979) have succeeded in isolating a membrane-associated complex from thylakoids which, when incorporated into artificial lipid vesicles, catalyzes ATP–P_i exchange. When a pH gradient is artifically imposed across the vesicle, it is possible to observe ATP formation. The complex consists of the components of soluble chloroplast ATPase and, in addition, a small number of additional polypeptides, presumably including the "membrane sector," that portion of the complex which provides the transmembrane proton channel. With the isolation of this complex and similar preparations from bacteria (Kagawa, 1978) a study of the molecular mechanism of phosphorylation can begin in earnest.

III. The Formation of Oxygen

While a good deal of uncertainity clouds our understanding of almost all aspects of photosynthetic electron transport and photophosphorylation, our ignorance regarding the mechanism of oxygen evolution is exceptionally deep. The central question remains unanswered: How are four electrons removed from two molecules of water so as to release protons and molecular oxygen? One reason for our ignorance is the failure of biochemists to isolate from the rest of the photosynthetic machinery those proteins, specialized pigments, and other components required specifically for the evolution of oxygen. Nonetheless a great deal of information has been derived from kinetic studies of intact systems, and a picture of at least some of the molecular species involved begins to emerge. An excellent review of our current knowledge of oxygen evolution has been presented by Radmer and Cheniae (1977).

A. Thermodynamics

The reaction of carbon dioxide and water to form oxygen and a sugar requires about 120 kcal of Gibbs free energy for each mole of carbon fixed. Since each photon of any wavelength absorbed by photosynthetic pigments is rapidly transformed to the lowest excited singlet state of chlorophyll (700 nm or 1.8 eV per molecule) it corresponds to an excitation energy of 40 kcal/mole. Thus, a minimum of three photons per carbon atom would be required to drive this reaction to equilibrium.

It must be clear that the chemical energy is a Gibbs free energy and the photon supplied energy of the excited state is an internal or Helmholtz energy. The difference is the entropy of the process. For the *overall* process the entropy is a small part of the total reaction energy, and so is neglected. Unfortunately this negligance is often carried over to *partial* reactions, where it may be seriously in error. This reflects the fact that thermodynamics does *not* determine mechanism. Bolton (1978) has calculated the minimum wavelength of light necessary to drive dissociation of water to oxygen and hydrogen with one (390 nm), two (615 nm), or four (890 nm) photons by making simple assumptions. They are that one electron is transferred for each photon absorbed, and only about one half of the energy is stored in a stable form. The water-splitting reaction is energetically similar to that carried out by photosystem II. The question of the ultimate thermodynamic efficiency of converting solar photons to chemical energy has been discussed critically by Knox (1979), Ross et al. (1976), and Parson (1978). A simple approach is to assume that each photoact can raise one electron to a certain level of energy. Given two photoacts in series to reach past the thermodynamic level, eight photons are required to remove 4 electrons from H_2O to form one molecule of O_2 and fix one carbon atom. With the exception of an interlude whose pernicious influence has yet to be fully documented, the measurements of the quantum requirement (1/quantum yield) for photosynthesis under optimal conditions have given an average value of 10 ± 2. Radmer and Kok (1977) give a good review of this subject. The very careful experiments of Emerson and Lewis (1943) gave a value of 11 between 560 and 680 nm. This is the best justification of the present scheme for photosynthesis discussed above.

Since the primary photoreactions of photosynthesis have a quantum yield of $95 \pm 5\%$, the observed overall quantum requirement for oxygen evolution indicates a 20–30% "loss" of photoelectrons to energy-requiring processes other than water splitting. These excitations may be dissipated in the generation of ATP via a cyclic electron-flow pathway, in which there is no net oxidation or reduction of substrate (Arnon, 1977). The ATP so formed may be needed to supplement that produced by noncyclic electron flow to fix CO_2 in the dark reactions of the Calvin cycle (Fig. 1).

The energy efficiency of photosynthesis is less than the quantum efficiency. This is an inevitable consequence of using solar black body radiation to excite a single-level photosystem. These systems include silicon photovoltaic cells. The energy efficiency of photosynthesis with 680 nm photons is about 30%, and plants have actually been grown with green light at an efficiency of 20% of the absorbed radiation (Radmer and Kok, 1977). The energy efficiency of solar radiation utilization under field conditions is only about 1%. The crocodile tears of some engineers over this "waste" should be watered by the realization that the photosynthetic process has yet to be equaled by man. One must note that the plant grows and duplicates itself while making excess O_2 and food, a property notably absent from man-made devices. A plant is thus a powerful entropy machine. This is characteristic of living systems. They work at constant temperature and are not thermal machines. Living systems adapt to the environment and not vice versa.

Along with these thermodynamic considerations one must keep in mind that there exists no evidence for reversibility in the O_2 formation reaction. Neither experiments at high pressures (Vidaver, 1969) nor with mass spectrometer (Gerster et al., 1972) have given results interpretable as reversible O_2 reactions. The former show a reversible inhibition of O_2 formation and a lowered efficiency of energy transfer (Schreiber and Vidaver, 1973). The latter provide evidence that the source of O atom is water, not CO_2.

These experiments are valuable and should be extended as much as possible; they show that the O_2 formation reaction is rather irreversible and an extra several tenths of an eV (5–10 kcal/mole) above the thermodynamic limit is required to drive this reaction forward. It is interesting that whereas the electrochemical H^+–H_2 reaction is reversible on platinum, the O_2 formation on this catalyst requires 0.4 V of overpotential beyond the thermodynamic 1.23 V (Hoare, 1968). The only true Nernstian oxygen electrode is the calcium-doped zirconium oxide ceramic operating at 800°C (Greenbaum and Mauzerall, 1976).

B. Oxygen Comes from Water

The early prejudice that the source of the O atoms in O_2 was CO_2 was disproven by the experiments of Ruben et al. (1941), which showed that the O^{18} label of photosynthetic O_2 was similar to that of water and not of the carbon dioxide. Van Niel (1935) had proposed the photolysis of water based on the analogy between plant photosynthesis ($H_2O \rightarrow O_2$) and bacterial photosynthesis ($H_2S \rightarrow S$). This analogy is an example of genius: a wrong hypothesis leading to immensely fruitful results. The rather rapid exchange of O in CO_2 and H_2O, catalyzed by carbonic anhydrase, has always limited the quantitation of these difficult experiments. They have been developed by Gerster and co-workers (1972) to the point where the lag in the rise of the isotopic content of O_2 can be ascribed to an intermediate pool between the H_2O^{18} added and the O_2 evolution.

C. Requirements

Chlorophyll. Aside from the obvious requirement of pigments which can absorb solar radiation, there appears to be an absolute requirement for chlorophyll *a* in O_2 producing systems. The concept of the photosynthetic unit (PSU) is that about 300 pigment molecules act as an antenna to collect solar photons and funnel the excitation to a photochemical reaction center. There are at least two kinds of reaction centers in green plants (see above) and their interconnection on the antenna level is a subject of active research (Butler, 1977). The antenna or accessory pigments need only absorb solar photons and efficiently transfer energy among themselves and to chlorophyll *a*. This relatively weak selection pressure has allowed evolution to experiment colorfully with these pigments. Dozens are known and they account for the reds (phycoblins) and the browns (fucoxanthins, pteridinins) of seaweed. With the single exception of carotenoids, all these pigments are derivatives of porphyrins. Nature knows a good thing when it has it. These accessory pigments tend to transfer energy to the oxygen system (II) and the "bulk" of chlorophyll *a*'s transfer to the reducing system (I). Also, no single organism hogs the whole solar spectrum with a variety of pigments. Nature prefers diversity of form and color over monolithic organization. It must be said, though, that when photons are scarce, antenna pigment, and/or photosynthetic unit density is increased until the organism is optically black, thus negating the effect of color (Ramus et al., 1976).

By contrast the requirement of efficient energy conversion in the reaction center necessitates strict control over molecular energy levels and distance and orientation of electron donors and acceptors (Mauzerall, 1978c); hence, the requirement for chlorophyll *a*. If the final formation of O_2 were a photochemical act tied to chlorophyll *a* (see below) this requirement would be absolute.

Manganese. The difficult and detailed experiments of Cheniae and Martin (1973) have established, by both negative and positive approaches, the requirement for manganese. Not only do manganese-deficient cells have lowered O_2 capacity, but such cells can be reactivated by adding manganese. The reactivation is complex and requires two separate photoevents separated by a dark period of 10 msec, with the intermediate having a lifetime of about 1 sec. An attractive hypothesis is that Mn^{2+} must be photooxidized thru Mn^{3+} (unstable) to Mn^{4+} to be active. A good review of this and related work is provided by Diner and Joliot (1977). Their argument that only Mn^{2+} is involved because coaddition of a reducing agent allows activation in the dark is vitiated by the hideous complexity of Mn induced auto-oxidation with O_2. The formation of H_2O_2 and OH· allows the higher valence states of Mn to be easily formed.

Heating chloroplasts in the presence of hydoxylamine or Tris removes Mn from O_2-forming systems. It is estimated that 4–6 Mn may be bound to the O_2 site with 2–4 Mn being critical for O_2 production. There have been sporadic reports of Mn containing cofactors required for O_2 formation, (Fredricks and Jagendorf, 1971; Tel-Or and Avron, 1975), but a recent claim has been made for the isolation of a definite and replaceable Mn protein (Winget, 1979). There is also a counter claim that a similar protein does not have Mn but contains heme (Nakatani et al., 1981). The pot is beginning to boil!

Protein components. A detailed understanding of the mechanism of photosynthetic oxygen evolution will only come about with the isolation of well-defined assemblies containing the minimum number of proteins, pigments, and other components required to carry out light-driven water oxidation. The search for this "holy grail" of photosynthesis has led primarily to the use of mild detergents which, while capable of effecting some dissociation of the photosynthetic apparatus, preserve the essential structural integrity required for oxygen evolution. So far, success along these lines has been limited to the separation of photosystems I and II into two, still quite complex fractions (Wessels, 1977). Although it is unlikely that "system II" preparations of this kind actually contain the minimum number of components required for oxygen evolution, attempts at further purification have led to irreversible loss of activity.

Recently, the emphasis has shifted to the isolation and identification of chlorophyll–protein complexes, relatively simple aggregates of chlorophyll, carotenoid and protein which retain integrity during electrophoretic analysis (Boardman et al., 1978). In general, these entities lack photochemical activity that is of physiological significance, but rather represent elementary components of light-harvesting antennae or reaction centers. The function of particular chlorophyll–protein (CP) complexes in vivo must be determined indirectly by spectral characteristics, or by studies of mutants in which a defect in the synthesis of a particular polypeptide component is correlated with loss of a particular photosynthetic function. An alternative technique of some utility is the production of antibodies directed against specific CP polypeptides. Inhibitory effects of such antibodies on intact photosynthetic preparations are then examined to attempt to localize the function of the antigen (Schmidt et al., 1978).

A number of laboratories have achieved an extensive resolution of the photosynthetic machinery into its component polypeptides and pigment–protein complexes while generating a minimum of free chlorophyll (Wessels and Borchert, 1978; Delepelaire and Chua, 1979). The combined results of these workers suggest strongly that *in vivo*, all chlorophyll is specifically associated with protein. A particularly promising study was carried out on five distinct CP complexes obtained by lithium dodecyl sulfate digestion of thylakoid membranes of the green alga *Chlamydomonas reinhardtii* (Delepelaire and Chua, 1979). Of the five complexes, two (CP III and IV) are of special interest because they are depleted in mutants specifically deficient in oxygen evolution activity (Bennoun and Chua, 1975). In addition, each of CP III and CP IV contain, per 50 and 47 kilodalton apoprotein, respectively, only 4–5 molecules of chlorophyll *a*, and 1 molecule of β-carotene per molecule of protein. Each displays a shoulder in the absorption spectrum near 680 nm, reminiscent of the system II primary donor proposed by Witt (1979). The use of monospecific antibodies established the nonidentity of the apoproteins of CP III, IV, and other complexes, as well as their similarity to analogous complexes in other plant species. By using the apoprotein immunological "signature" as a basis for identifying CP complexes, a great deal of ambiguity surrounding the study of these components is avoided.

Very recently, some progress has been made toward isolation of a photosystem II particle with both photochemical function and well-defined composition. By digitonin extraction of spinach thylakoids, followed by ion-exchange chromatography, sucrose-gradient centrifugation, and isoelectric focusing, Satoh and Butler (1978) have isolated a chlorophyll *a*-containing fraction active in the light-dependent oxidation of diphenyl carbazide, a reaction closely associated with system II photochemistry. On a chlorophyll basis, the activity of the fraction was 3–4 times that of intact membranes. If the thylakoids are extensively washed with low ionic strength EDTA buffer before digitonin digestion all but two polypeptides (mol. wt. 43,000 and 27,000) are excluded from the "system II reaction center" fraction without altering the isoelectric point, chlorophyll *a*/*b* ratio, or absorption and emission spectra (Satoh, 1979). The EDTA prewash does, however, render the resulting purified preparation less active and more labile; evidently, prewashing removes some component(s) important to photochemical function. The 43 kilodalton polypeptide of Satoh and Butler's digitonin particle may correspond to the apoprotein of CP III or IV discussed above. Diner and Wollman (1980) have isolated a particle with very high system II activity but no O_2 formation. It has an antenna size of only 40 chlorophyll.

Calcium. The activity of cell-free preparations from cyanobacteria is specifically dependent on a rather high (10^{-2} *M*) concentration of calcium (Piccioni and Mauzerall, 1978a). These preparations are also of interest because they are resistant to the pervasive inhibitor of O_2 evolution, DCMU, and specifically require a high potential ($E'_0 > 0.45$ V) Hill oxidant (Piccioni and Mauzerall, (1978b). The calcium does not seem to be required to ensure a proton or ion gradient or to activate the Hill oxident. The calcium requirement needs vigorous cell breakage (French press) to appear and such treatment releases the endogenous cellular calcium. Thus, the level of calcium may be a control of oxygen-forming activity. Both calcium and mangensium cause a twofold decrease of the turnover time in these preparations.

Anions. The stimulation of Hill activity of broken chloroplasts by chloride ion has been known for some time, but its function remains obscure (Diner and Joliot, 1977). The effect is not specific, being observed with bromide ions and less well with nitrate and perchlorate ions. A requirement for bicarbonate ion has also been extensively studied (Govindjee and van Rensen, 1978). Possibly ion pairing with the high valence state of Mn is involved.

D. Kinetics

O_2 yield oscillations. Although it was known from the time of James Franck (Allen and Franck, 1955) that thoroughly dark-adapted algae did not produce O_2 from a single short flash of light, it remained for the elegant experiments of Joliot and Kok to show spectacular oscillations in yield of O_2 in a sequence of such flashes separated by about $\frac{1}{3}$ sec. An example is given in Fig. 3. Joliot and Kok (1975) have reviewed their valuable work on this remarkable phenomena. The first flash never produces O_2; the second a small, variable amount; the third a maximum about twice the steady state value; and succeeding yields show a damped oscillation of period four. The requirements are dark-adapted (15–30 min) healthy cells, a sufficiently sensitive O_2 detector, and saturating single-turnover flashes. To avoid exciting the units more than once, the flash must be shorter than the turnover time (~200 μsec). The detector used by Kok and Joliot is the large area bare platinum polarograph introduced by Haxo and Blinks (1950). Joliot refined the detector, including the use of modulated light and phase sensitive detection of the O_2 current (Joliot, 1966) to a point of great precision and versatility. Much of our best knowledge concerning O_2 production comes from this instrument and its gifted users.

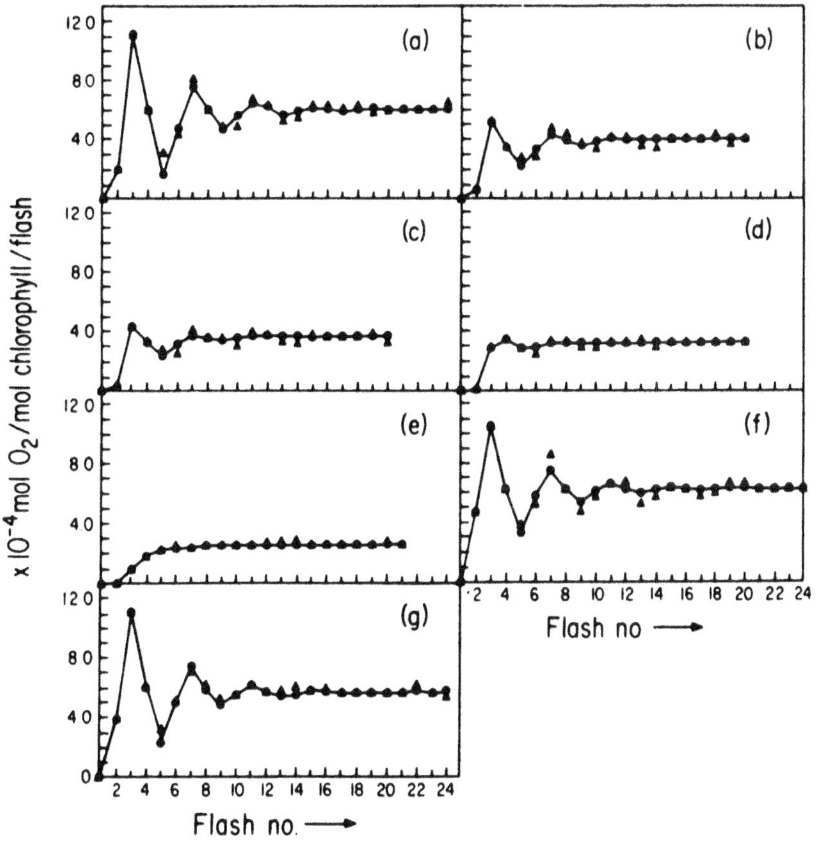

Figure 3. Oxygen yield per flash of *Chlorella* in the presence of oxidants and reductants. The ordinate is the absolute yield of O_2 per mole of chlorophyll per flash. The concentration of benzoquinone (BQ) equals 1.0 m*M* for all the experiments reported here. Additional concentrations are as follows: (a) BQ the only added redox component; (b) HQ (hydroquinone) = 0.5 m*M*; (c) HQ = 1.0 m*M*; (d) HQ = 1.5 m*M*; (e) HQ = 2.0 m*M*; (f) Fe $(CN)_6^{3-}$ = 3.0 m*M*; (g) $IrCl_6^{2-}$ = 3.0 m*M*. ▲, experimental data; ●, theoretical points. *Chlorella* were suspended in 20 m*M* potassium phosphate buffer at pH 7. All O_2 measurements performed at 10 ppm O_2 background concentration. Flash rate = 1 per 10 sec (Greenbaum, 1977).

The oscillations in O_2 yield are explained by Kok's clock model (Kok et al., 1970): Four oxidizing equivalents are successively accumulated at a single reaction center as it advances from a reduced state, S_0, through a series of four progressively more oxidized states, S_1, S_2, S_3, and S_4. In a concerted fashion, the oxidizing power of S_4 was postulated to cause the splitting of two water molecules to yield molecular oxygen and in doing so return to S_0. The observed oscillations in oxygen yield flash reflect, according to this view, the more or less synchronized progression of the ensemble of photosynthetic units through the "S states" to S_4. This synchrony is established experimentally by the dark adaptation period during which all units come to rest in a redox "ground state." Damping of the oxygen yield oscillations, i.e., randomization of units among the S states, is explained by "misses" and "double hits." The former occur when centers do not advance even if a photon is absorbed, and the latter occur when units advance two steps in a single flash. These effects are rather peculiar in that (1) misses occur even when the light flash is operationally saturating; (2) the propensity toward double hits and misses appears randomly among the photosynthetic units in a sample, and migrates from one center to another in the time interval between flashes, not during a flash; and (3) neither depends very much on flash intensity. Since, after dark adaptation, the maximum yield of oxygen occurs on the third flash, it follows that the most stable state in the dark is S_1, not S_0; the data are best fit by a ratio of S_1 to S_0 populations equal to 3:1. The rate of misses consistent with the data is 10% for chloroplasts and 20% for *Chlorella*—hardly negligible. Since the complete model has been shown to contain no less than 28 independent parameters (Lavorel, 1975), a respectable fit to the observed flash yield is not only possible, but rather easy.

The lifetimes of the S states have been estimated by varying the time interval following a selected flash in the flash sequence, and observing the effect upon the oxygen yield from subsequent flashes. In all such experiments, the analysis is greatly complicated by the inevitable mixture of states occupied by the ensemble of photosynthetic units in the sample. It is clear from the available data that S_0 and S_1 are stable under normal conditions. State S_2 lives about a minute and state S_3 about 5 sec (*Chlorella*) or a minute (chloroplasts). State S_4 is presumed to form O_2 very rapidly. Anaerobic (reducing) conditions and certain compounds increase the rate of loss of states S_2 and S_3, but the results are complex. For example, in the presence of CCCP the O_2 yield decreases with increasing flash separation but the "miss" parameter does not change. The redox properties of the S states are of great interest. Since, at physiological pH, the redox potential of H_2O/O_2 couple is +0.81 V, and that of the primary electron acceptor is near −0.1 V, thermodynamics simply requires that a total of at least $4 \times .91 = 3.6$ eV be provided per molecule of O_2 produced. How this total is partitioned among the four photon-requiring steps depends on the particularities of the mechanism. If the four steps are not equi-energetic, the more energetic steps must have redox potentials greater than 0.91 V, thus requiring a very efficient conversion of the energy of the lowest singlet excited state of chlorophyll (1.8 eV above the ground state) to chemical energy. Bouges-Boucquet (1973) found that the ratio of the yields of the third to fourth flashes was increased by ferricyanide and decreased by ascorbate, and suggested that S_0/S_1 has a rather low potential, 0.2 V. Kok et al. (1975) countered that S_0 was oxidized to S_1 by O_2, and reduced to S_0 by a reductant, the ratio S_1 to S_0 being determined by a steady state. Greenbaum and Mauzerall (1976) weakened this agrument by showing that not only S_1 but even more dramatically S_2, determined by the O_2 yield on the second flash, increased as a function of redox level under highly anaerobic ($<$ 10 ppm O_2) conditions. These experiments were possible only with the use of a novel oxygen-detection system. Rather than being reduced at a platinum cathode, oxygen released by algae during a flash of light is carried by a flow of inert gas to a very sensitive Nernstian O_2 electrode. The latter consists of sample and reference compartments separated by a calcium-doped zirconium oxide ceramic maintained at 800°C, a temperature at which the ceramic is freely permable to O^{2-} ions. A difference in the partial pressures of O_2 in the sample and reference gases is detectable as a Nernst potential. Because of complete separation of the detector and the algae, there is no interference between the experimental redox environment of the photosynthetic preparation and the oxygen measurement. The bare platinum electrode is very restrictive in this respect. At a substantial loss in sensitivity, the platinum electrode can be made more versatile by interposing a very thin, O_2-permeable membrane (Teflon or silicone) between the photosynthetic preparation and the electrode surface.

An attempt can be made to resolve the apparent contradictions that direct redox titrations cast upon the four-step clock mechanism by postulating "double hitting" within the flash duration. However, the "double hits" cannot have their traditional meaning, since our experiments were carried out with a laser flash of 0.3 μsec halfwidth and none of the "tail" which plagues electronic flashes. Moreover, our study of the turnover time with short flashes showed no rise in the microsecond time range required by the "double hit" concept.

Although an attempt can be made to save the picture by various ad hoc modifications, e.g., postulating a "new" acceptor with the required properties, Greenbaum (1977) has proposed a more general approach. From the simple assumption that the yield of O_2 on the $(k+4)$th flash, Y_{k+4}, is proportional to that on the kth flash, Y_k, and the fact that the yield reaches an asymptotic limit Y_{ss}, for the large number of flashes (the Emerson–Arnold yield) one arrives at the fourth order finite difference equation:

$$Y_{k+4} = C_1 Y_k + Y_{ss}/(1 - C_1)$$

Where C_1 is a constant, essentially the memory of the kth flash yield at the $(k+4)$th flash. The equation can be solved by standard methods:

$$Y_k = [4\sqrt{C_1}]^k\ [K + A\ \cos(\tfrac{1}{2}k\pi + B)] + Y_{ss}$$

where K, A, and B are adjustable constants. The memory constant C_1 determines the damping of the oscillations (cosine term). The data are fit by adjusting C_1 for the best fit after calculating A, B, and K by yields of flashes 2, 3, and 4 with an assumed C_1. Examples are given in Fig. 3. It is interesting that as the benzoquinone to hydroquinone ratio is lowered the oscillations damp out and the fitted C_1 becomes progressively smaller.

The class of equations $Y_{k+2} = C'_1 Y_k + C_2$ can also give rise to oscillating solutions of periodicity four although the memory is only two steps. This occurs if C'_1 is negative. A possible interpretation is negative feedback by some reductant on the oxygen evolving apparatus. This model is of interest because of the many observations of cycle-of-two phenomena.

Although this phenomenological model is somewhat abstract, it allows fitting of widely varying data with a single set of comparable parameters without introducing ad hoc assumptions. The systematic variation of the parameters (e.g., C_1 with quinone–hydroquinone ratio) then allows interpretation of the parameters in more molecular terms.

Proton NMR relaxation time oscillations. Wydrzynski et al. (1978) have shown that the rate of water proton spin–lattice relaxation depends on the amount of manganese in suspensions of chloroplast membranes. Addition of reductants increased and of oxidants decreased tne relaxation rate. The differing redox states of manganese, through their differing electron spin relaxation rates and reversible binding of water, are known to cause differing proton relaxation rates. These authors have further shown that the rate of water proton spin–spin relaxation is an oscillating function of the number of flashes given to the chloroplast suspension just before the magnetic measuring pulses. The oscillation of the rate constant has a period of four, and like the O_2 yields, peaked on the 3rd, 7th, and 11th flash. A model based on four Mn's of varying valence and two bound water molecules with differing numbers of protons in a four step cycle partially accounts for the data (Govindjee et al., 1977). These experiments appear to give very direct information on changes at the manganese site of O_2 formation. Unfortunately other experimenters (Robinson et al., 1981) argue that the proton relaxation does not monitor Mn function. The original experiments cannot be reproduced in detail (Wydrzynski et al., 1981).

Proton release. Since the oxidation of water releases four protons along with O_2, proton formation at specific points in the S cycle is of great interest. Fowler and Kok (1976) first detected the uptake and release of protons from single flashes in uncoupled chloroplasts using a large but fast glass electrode. The proton yields oscillate with a period of four, but determining the quantitative output at each flash is much more difficult. Fowler's (1977) most recent claim is that 2 protons are emmitted at $S_3 \rightarrow S_4$, $\frac{3}{4}$ at $S_0 \rightarrow S_1$, and $1\frac{1}{4}$ at $S_2 \rightarrow S_3$. Saphon and Crofts (1977) also claim protons are emitted at all transitions except $S_1 \rightarrow S_2$. The consensus is, therefore, that the interaction of water with oxidized system II donors is not restricted to the final, O_2-releasing step, as envisioned initially (Kok et al., 1970). Rather, water molecules are oxidized in a series of steps. Velthuys (1979) has suggested that the oxidized donor in the $S_1 \rightarrow S_2$ reaction remains cationic and correlated this with its apparent higher potential and reactivity with nucleophiles such as amines.

Other indicators. Many other indicators of the activity of the O_2-generating system have been studied. The most common are changes in fluorescence yield and in delayed luninescence. The former are based on the Kautsky effect (Kautsky and Hirsch, 1931), wherein the quantum yield of fluorescence increases on saturating the system with light, and the latter is one of several provocative discoveries of Arnold (Strehler and Arnold, 1951). Oscillation of these changes in luminescence yield with light flash sequences have been observed along with a vast number of other effects. Unfortunately the measurements are nowadays too easy to make and thus maybe too much has been measured. The Byzantine complexities of this subfield are beyond this review and we can only refer the reader to recent reviews (Lavorel, 1975; Papageorgiou, 1975; Mauzerall, 1978b).

The extensive and somewhat more informative measurements on changes in absorption are, for better or for worse, not directly applicable to O_2 production. They mainly concern electron transport and have been well summarized by Junge (1977) and Witt (1979).

E. Theory

As is usually the case, lack of specific knowledge has stimulated the imagination of workers in this field. The arrival of firm, and thus restrictive, data on O_2 formation allows us to ignore the woollier contributions. This firm data can unfortunately be readily summarized: (1) the O atoms come from water, with possible involvement of a small pool of intermediates between O_2 and free water; (2) two to four Mn atoms are critically involved and valence changes occur; (3) at least two and, more likely, four states of the system are cycled through in making one molecule of O_2. Point (3) requires some elaboration. If there are n photodriven steps, there are a minimum of $2n$ states since the actual excited state of the system must be differentiated from the product state. Without this dark energy loss, or trapping step, the cycle could not move in a single direction. This directed motion is a negative entropy flow and requires care and feeding. If there are z intermediates (electron carriers, proton transfers) then there could be $2nz$ states differentiable in principle. At present only the tip of the icebergian apparatus is visible.

The Kok group (1970) added a corollary to this list: The O_2 centers are independent and do not exchange charge. Were charge to freely equilibrate, no oscillations would be possible. This point is further supported by observation of the same flash yield pattern when the O_2 activity is depleted severalfold by addition of DCMU, UV radiation, or Mn depletion. However, it seems to us quite possible that a weak coupling of two O_2 centers could allow the variation seen in the "misses," and "double hits," and "forbidden" Y_2.

The Kok group has kept their S state model at the phenomenological level. Renger (1978) has attempted to identify the S states with serially oxidized water, the intermediates being stabilized by binding to manganese, thus smearing the valence states of each. He has found it necessary to add an oxidation step independent of water or protonation [the symbol (+) indicates a photogenerated oxidizing equivalent]:

$$M \overset{h\nu(+)}{\rightarrow} M^+ \qquad S_0 \rightarrow S_1 \tag{1}$$

$$\text{Mn H}_2\text{O Mn H}_2\text{O} \overset{h\nu(+)}{\rightarrow} \text{Mn H}_2\text{O HOMn} + H^+ \qquad S_1 \rightarrow S_2 \tag{2}$$

$$\text{Mn H}_2\text{O HOMn} \overset{h\nu(+)}{\rightarrow} \text{Mn}\cdots\overset{\text{H}}{\text{O}}\text{—}\underset{\text{H}}{\text{O}}\cdots\text{Mn} + H^+ \qquad S_2 \rightarrow S_3 \tag{3}$$

$$
\left.
\begin{array}{l}
M^{+} + Mn\cdots \overset{H}{O}-\underset{H}{O}\cdots Mn \xrightarrow{h\nu\,(+)} \\
\quad Mn\cdots O{=}O\cdots Mn + 2H^{+} + M \\
\qquad\qquad\qquad\qquad S_3 \rightarrow S_4 \\
\\
2H_2O + Mn\ O_2\ Mn \ \rightarrow \\
\quad Mn\ H_2O\ Mn\ H_2O + O_2 \\
\qquad\qquad\qquad\qquad S_4 \rightarrow S_0
\end{array}
\right\} \quad (4)
$$

By means of this assumption Renger is able to explain the +0.4 V acceptor (M) often associated with the O_2-producing system. However, additional hypotheses about "intrinsic" and "extrinsic" proton release are necessary to fit the proton release pattern. If reactions (1) and (2) were interchanged, the fit would be good without gates to proton movement. The compound M could be one or more of the other Mn atoms known to be present, and its oxidation could be dependent on the redox state of the H_2O–Mn complex.

The redox state of the initial $(MnH_2O)_2$ complex is not specified in this model. From the reactivation experiments (Cheniae and Martin, 1973) we would suggest either two Mn^{3+} or Mn^{4+} and Mn^{2+}. The redox potentials of Mn, as for Fe and other transition elements, depend so strongly on their environment ($\Delta\Delta E > 1$ V) that no simple predictions are possible. It has been known for a long time, and we have verified with our specific detection method for O_2 (Greenbaum and Mauzerall, unpublished) that the higher valence states of Mn (4+ to 7+) readily form O_2 from H_2O, particularly in acid. Thus, the rather large number (16–4096) of combinations for 4 ± 2 Mn atoms, each of four different valence states will have to be narrowed by observation. By valence we refer to formal charge on the Mn, it being understood that the covalent binding to the O atoms (and others?) spreads out the charge, leading to the critical formation of the first O—O bond. The rest is relatively easy.

A more chemical mechanism has been proposed which allows a variable number of quanta to drive the cycle (Mauzerall and Chivvis, 1973). The key intermediate is the "dioxolium" ion, named by anology to the known dithiolium ion containing sulfur in place of oxygen. The reaction is based on the β-dicarbonyl structure of chlorophyll, and so would couple photochemistry and redox reactions. The function of the Mn would lie in holding the O atoms to aid formation of the critical O—O bond (Fig. 4). The reoxidation of the β-hydroxy alcohol to the β-dicarbonyl is a low potential reaction, and could go in the dark if the cellular redox reagents (e.g., NADP) are sufficiently oxidized. As these become more reduced, i.e., at high light intensity, it may require light to drive these reactions forward, i.e., via oxidation of the Mn. This cycle thus fits the observation that the first and/or second state of the system has a somewhat lower redox potential and that the redox condition of the photosynthetic apparatus affects the shape of the O_2 flash yield curve. The scheme is consistent with the observed proton spin relaxation times in that the Mn is further oxidized only on two reactions during a cycle. The proton release also follows the observed sequence, in particular having no release on the second step ($S_1 \rightarrow S_2$).

It is likely that the Mn at the time of the formation of O_2 from the dioxolium intermediate is in a high spin state ($S \geq 1$). This would allow formation of O_2 in its ground triplet state via spin exchange reactions. The close coupling of the Mn and the photoactive chlorophyll also requires that the ESR of the Chl^{+} be affected by the high spin states of Mn. This agrees with the difficulty in observing such a signal, and with the recent observation of spin polarized ESR signals of very short lifetime (Warden et al., 1976).

The proposed mechanism clearly predicts a lag in O^{18} labeling of the O_2 from the recently added labeled H_2O. The system *must* cycle once to incorporate H_2O into the carbonyl groups, assuming their exchange is slow under neutral condidtions. This lag can be inferred from recent data on such experiments (Gerster et al., 1972).

This proposed mechanism couples the oxidation and photoreactions rather closely. This is based on the absolute requirement of

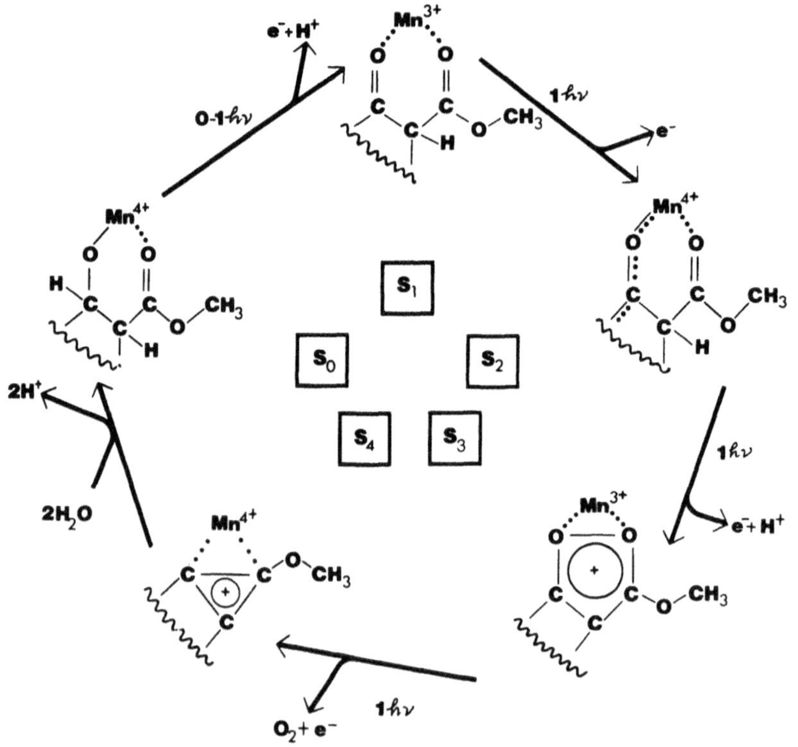

Figure 4. Cyclic process forming O_2 via the dioxolium ion. Only the β-keto ester part of the chlorophyll molecule is shown, the remainder is represented by the wavy line. The correspondence to the "S" states are shown in the squares. The oxidation of the β-hydroxy ester ($S_0 \rightarrow S_1$) can occur slowly in the dark or rapidly in the light and involves concomitant reduction of the Mn and emmission of a proton. The Mn is reoxidized in $S_1 \rightarrow S_2$. The next step ($S_2 \rightarrow S_3$) forms the dioxolium ion, with concomitant reduction of Mn and proton emission. Another excitation, $S_3 \rightarrow S_4$, reoxidizes the Mn and emits O_2 in the triplet ground state. Hydrolysis of the cyclopropenium ion, $S_4 \rightarrow S_0$, reforms the β-hydroxy ester emitting two protons. In both the dioxolium and cyclopropenium ions, the charge can be delocalized into the remainder of the chlorophyll aromatic system.

chlorophyll *a* to form O_2 and on our observations that we have never observed O_2 formation on mixing photoactive preparations with strong oxidants in the dark (Greenbaum, Diner, and Mauzerall, unpublished). The sensitivity of our oxygen detectors, luminometer (Burr and Mauzerall, 1968), or ZnO electrode (Greenbaum and Mauzerall, 1976) was sufficient to detect ~10% yield of O_2 per PSU. There are any number of claims in the literature that (photo)oxidation of chlorophyll yields O_2, but in our hands we were able to arrive at O_2 yields $<10^{-4}$ (Mauzerall and Chivvis, 1973; Greenbaum and Mauzerall, 1974), i.e., into the impurity levels of even highly purified chlorophyll. Other claims for the photoproduction of O_2 using phenoxazine, quinones, or chlorophyll-silver chloride were also easily disproved. However, old reports of O_2 formation in low yield from silver chloride or zinc oxide with UV light could be easily verified. Krasnovsky and Brin (1978) have done much work on O_2 formation with these and other semiconductor systems. Recently there has been a great surge of interest in the "photoassisted" electrolysis of water (Hardee and Bard, 1977).

Finally, the "dioxolium" model has been written with one Mn and one chlorophyll for simplicity, but clearly several may be involved in the actual reaction. One can appreciate the complex subtlety of the photosynthetic system by looking more critically at the usual view that four photoreactions forming four positive charges or "holes," are "added" to make O_2. What is the "adding?"

While the experiments of Joliot and others have demonstrated that separate photosystem II centers act independently to accumulate the four oxidizing equivalents (holes) needed to form O_2 from water, detailed models of the mechanisms of oxygen evolution must confront an analogous issue within each photosystem II center. Stated in an extreme form: Are the four holes accumulated within a single redox component or are they distributed over four independent sites? The former model, which implies strong interaction between accumulated charges, is consistent with some data, as discussed above (Bouges-Bocquet, 1973; Greenbaum, 1977); however, the maximum energy available from a single red photon sets a firm upper limit to the redox potential difference between the system II acceptor and the system II center in its most oxidized state. If the system II center is allowed to "float," forming holes at ever-increasing potentials, then, assuming constant efficiency of photon-energy utilization, the electrons will be given off at ever-increasing potentials also. Although 0.1 to 0.2 V of potential difference can be bridged by thermal energy, with a concomitant decrease in reaction rate, a difference of, say, 0.4 V would require an acceptor with a higher redox potential than required for the first electron.

One possibility which would accommodate these energetic difficulties within this "centralized accumulation" model is that the acceptor of the last electron or last two electrons is not the conventional "primary" system II acceptor ($E_m = -0.1$ V), but rather a species with considerably higher midpoint potential. In order to preserve electron flow to photosystem I, the potential of the postulated acceptor would have to be less than that of *P*-700 ($E_m = 0.5$ V). Electrons from this acceptor would not enter the plastoquinone pools. Experimental evidence in favor of alternative routes of electron flow from photosystem II has recently appeared (Piccioni and Mauzerall, 1978b; Huzisige et al., 1979).

If the four holes required for oxygen evolution appear on four independent redox components (distributed accumulation), then there is no need for the redox potential to rise as charge is accumulated, and therefore no need for more than one acceptor. Now, however, it is the dark chemistry which is complicated; the connections are in parallel and must be combined to produce useful output. The four holes must interact, e.g., by chemical disproportionation, to localize sufficient total energy on adjacent oxygen atoms to form O_2. This appears to be the traditional interpretation of the Kok clock scheme. However, if the chemistry occurred only at the last step forming O_2, the S states would be identical, which they are not. Thus the actual mechanism must lie between these two extremes of series and parallel input. Our model and the recent data on different redox levels for acceptors imply a mixed accumulation scheme.

IV. Conclusion

Over 200 years have passed since Priestley observed bubbles of O_2 on plant leaves under water. We can now discuss the process in some technical detail, but the mechanism of the photosynthetic formation of O_2 still eludes us. It is somewhat embarrassing in this age of megascience that this question, so basic to all life on earth, has been repeatedly bypassed. One can only hope the weakening of mammalian chauvinism, the assignment of relevant priorities, the application of powerful modern technologies, and the innovative research of motivated workers will soon remove this blot of ignorance from the book of Science.

This article is dedicated to those few but brilliant scientists whose work led to our present knowledge of the photosynthetic formation of oxygen: Hans Gaffron, Ralph Emerson, William Arnold, Martin Kamen, Bessel Kok, and Pierre Joliot.

Acknowledgment

This review was prepared with help form the National Science Foundation, grant PCM 77-09102 and NIH fellowship #5F32 GM06866-02 (Richard Piccioni).

References

Allen, F. L., and Franck, J. (1955). Photosynthetic evolution of oxygen by flashes of light. Arch. Biochem. Biophys. 58:124–143.

Amouyal, E., Keller, P., and Kagan, H. (1978). Hydrogen production by visible light irradiation of aqueous solution of tris(2,2-bipyridine)ruthenium (2+). Nouv. J. Chim 2:547–549.

Anderson, J. M. (1975). The molecular organization of chloroplast thylakoids. Biochim. Biophys. Acta 416:191–235.

Arnon, D. I. (1977). Photosynthesis 1950–1975: Changing concepts and perspectives. In A. Trebst and M. Avron (Eds.), Encyclopedia of Plant Physiology, Vol. 5. Berlin-Heidelberg-New York: Springer-Verlag.

Barber, J. (Ed.) (1976). Topics in Photosynthesis, Vol. 1:The Intact Chloroplast. Amsterdam: Elsevier Press.

Bennoun, P., and Chua, N.-H. (1975). Thylakoid membrane polypeptides of *Chlamydomonas reinhardtii*: Wild-type and mutant strains deficient in photosystem II reaction center. Proc. Nat. Acad. Sci. U.S.A. 72:2175–2179.

Blankenship, R. E., and Parson, W. W. (1978). The photochemical electron transfer reactions of photosynthetic bacteria and plants. Annu. Rev. Biochem. 47:635–653.

Boardman, N. K., Anderson, J. M., and Goodchild, D. J. (1978). Chlorophyll-protein complexes and structure of mature and developing chloroplasts. Curr. Top. Bioenerg. 7:36–109.

Bolton, J. R. (1978). Solar fuels. Science 202: 705–711.

Bouges-Bocquet, B. (1973). Limiting steps in photosystem II and water decomposition in *Chlorella* and spinach chloroplasts. Biochim. Biophys. Acta 292:772–785.

Boyer, P. D., Chance, B., Ernster, L., Mitchell, P., Racker, E., and Slater, E. C. (1977). Oxidative phosphorylation and photophosphorylation. Annu. Rev. Biochem. 46: 955–1026.

Broecker, W. S. (1970). Man's oxygen reserves. Science 168:1537–1538.

Burr, A., and Mauzerall, D. (1968). The oxygen luminometer. An apparatus to determine small amounts of oxygen, and application to photosynthesis. Biochim. Biophys. Acta 153:614–624.

Butler, W. L. (1977). Chlorophyll fluorescence: A probe for electron transfer and energy transfer. In A. Trebst and M. Avron (Eds.), Encyclopedia of Plant Physiology, Vol. 5. Berlin-Heidelberg-New York: Springer-Verlag.

Cheniae, G. M., and Martin, I. F. (1973). Absence of oxygen-evolving capacity in dark-grown *Chlorella*: The photoactivation of oxygen evolving centers. Photochem. Photobiol. 17:441–459.

Clayton, R. K. (1978). Physiochemical mechanisms in reactions centers of photosynthetic bacteria. In R. K. Clayton and W. R. Sistrom (Eds.), The Photosynthetic Bacteria. New York: Plenum Press.

Cramer, W., and Whitmarsh, J. (1977). Photosynthetic cytochromes. Annu. Rev. Plant Physiol. 28:133–172.

Crofts, A. R. (1977). Organization of electron transport. In D. O. Hall, J. Coombs, and T. W. Goodwin (Eds.), Proceedings of the Fourth International Congress on Photosynthesis. London: The Biochemical Society.

Crofts, A. R., and Wood, P. M. (1978). Photosynthetic electron transport chains of plants and bacteria and their role as proton pumps. Curr. Top. Bioenerg. 7:175–244.

Delepelaire, P., and Chua, N.-H. (1979). Lithium dodecyl sulfate/polyacrylamide gel electrophoresis of thylakoid membranes at 4°C: Characterization of two additional chloroplyll a-protein complexes. Proc. Nat. Acad. Sci. U.S.A. 76:111–115.

Diner, B. A., and Joliot, P. (1977). Oxygen evolution and manganese in photosynthesis I. In A. Trebst and M. Avron (Eds.), Encyclopedia of Plant Physiology, Vol. 5. Berlin-Heidelberg-New York: Springer-Verlag.

Diner, B. A., and Wollman, F.-A. (1980). Isolation of highly active photosystem II particles from a mutant of *Chlamydomonas reinhardtii*. Eur. J. Biochem. 110:521–526.

Emerson, R., and Lewis, C. M. (1943). The dependence of the quantum yield of *Chlorella* photosynthesis on wave length of light. Am. J. Bot. 30:165–178.

Fowler, C. F. (1977). Proton evolution from photosystem II: Stoichemistry and mechanistic considerations. Biochim. Biophys. Acta 462: 414–421.

Fowler, C. F., and Kok, B. (1976). Determination of H^+/e^- ratios in chloroplasts with flashing light. Biochim. Biophys. Acta 423:510–524.

Fredricks, W. W., and Jagendorf, A. T. (1964). A soluble component of the Hill reaction in *Anacystis nidulans* Arch. Biochem. Biophys. 104:39–49.

Gaffron, H., and Rubin, J. (1942). Fermentative and photochemical production of hydrogen in algae. J. Gen. Physiol. 26:219–240.

Gerster, R., Dupuy, J., and Guerin de Montgareuil, R. (1972). Isotopic exchange, photosynthesis and oxygen^{18}. In G. Forti, M. Avron, and A. Melandri (Eds.). Second International Congress on Photosynthesis. The Hague: Junk Publishers.

Golbeck, J. H., Lein, S., and San Pietro, A. (1977). Electron transport in chloroplasts. In A. Trebst and M. Avron (Eds.), Encyclopedia of Plant Physiology, Vol. 5. Berlin-Heidelberg-New York: Springer-Verlag.

Govindjee (Ed.) (1975). Bioenergetics of Photosynthesis. New York: Academic Press.

Govinjee and van Rensen, J. J. S. (1978). Bicarbonate effects on the electron flow in isolated broken chloroplasts. Biochim. Biophys. Acta 505:183–213.

Govinjee, Wzdrzynski, T., and Marks, S. B. (1977). The role of manganese in the oxygen evolving mechanisms of photosynthesis. In L. Packer, C. G. Papageorgiou, and A. Trebst (Eds.), Bioenergetics of Membranes. Amsterdam: Elsevier Press.

Graber, P., and Witt, H. T. (1977). Conformational change of the chloroplast ATPase induced by a transmembrane electric field and its correlation to phosphorylation. Biochim. Biophys. Acta 461:426–440.

Granick, S. (1965). Evolution of heme and chlorophyll. In E. Bryson and H. J. Vogel (Eds.), Evolving Hemes and Proteins. New York: Academic Press.

Greenbaum, E. (1977). Photosynthetic oxygen evolution under varying redox conditions: New experimental and theoretical results. Photochem. Photobiol. 25:293–298.

Greenbaum, E. (1980). Simultaneous photoproduction of hydrogen and oxygen by photosynthesis. Biotech. and Bioeng. Symposium no. 20: pp. 1–13.

Greenbaum, E., and Mauzerall, D. (1974). A search for evidence of the dioxolium ion mechanism of oxygen production in photosynthesis. 2nd Annual Meeting of American Society for Photobiology (Abstract).

Greenbaum, E., and Mauzerall, D. (1976). Oxygen yield per flash of *Chlorella* coupled to chemical oxidants under anaerobic conditions. Photochem. Photobiol. 23:369–372.

Gregory, R. R. F. (1976). Biochemistry of Photosynthesis. New York: John Wiley & Sons, Ltd.

Hardee, K. L., and Bard, A. J. (1977). Semiconductor electrodes. X. Photochemical behavior of several polycrystalline metal oxide electrodes in aqueous solutions. J. Electrochem. Soc. 124:215–224.

Hatch, M. D. (1976). Photosynthesis: The path of carbon. In J. Bonner and J. E. Varner (Eds.), Plant Biochemistry, 3rd ed.. New York: Academic Press.

Haxo, F. T., and Blinks, L. R. (1950). Photosynthetic action specta of marine algae. J. Gen. Physiol. 33:389–422.

Hoare, J. P. (1968). The Electrochemistry of Oxygen. New York: Interscience Publishers.

Hong, F. T., and Mauzerall, D. (1976). Tunable voltage clamp method: Application to photoelectric effects in pigmented bilayer lipid membranes. J. Electrochem. Soc. 123:1317–1324.

Huzisige, H., Doi, M., and Natuga, T. (1979). Relationships among the three light-induced absorbance changes related to the reaction-center of photosystem II. Plant Cell Physiol. 20:935–946.

Izawa, S. (1977). Inhibitors of electron transport. In A. Trebst and M. Avron (Eds.), Encyclopedia of Plant Physiology, Vol. 5. Berlin-Heidelberg-New York: Springer-Verlag.

Jagendorf, A. T. (1975). Mechanism of photophosphorylation. In Govindjee (Ed.), Bioenergetics of Photosynthesis. New York: Academic Press.

Jensen, R. G., and Bahr, J. T. (1977). Ribulose 1,5-biphosphate carboxylase-oxygenase. Annu. Rev. Plant Physiol. 28:370–400.

Joliot, P. (1966). Oxygen evolution in algae illuminated by modulated light. Brookhaven Symp. Biol. 19:418–433.

Joliot, P., and Kok, B. (1975). Oxygen evolution in photosynthesis. In Govindjee (Ed.), Bioenergetics of Photosynthesis. New York: Academic Press.

Junge, W. (1977). Membrane potentials in photosynthesis. Annu. Rev. Plant. Physiol. 28: 503–536.

Kagawa, Y. (1978). Reconstitution of the energy transformer, gate, and channel subunit reassembly, crystalline ATPase, and ATP synthesis. Biochim. Biophys. Acta 505:45–93.

Kamen, M. P. (1963). Primary Processes in Photosynthesis. New York. Academic Press.

Kautsky, H., and Hirsch, A. (1931). Neue versuche zur kohlensaure-assimilation. Naturwissenschaften 19:964.

Ke, B. (1978). The primary electron acceptors in green-plant photosystem I and photosynthetic bacteria. Curr. Top. Bioenerg. 7:76–138.

Knox, R. S. (1979). Conversion of light into free energy. In H. Gerischer and J. J. Katz (Eds.), Light-Induced Charge Separation in Biology and Chemistry (Dahlem Kongerenzen). Berlin: Verlag-Chemie.

Kok, B., Forbush, B., and McGloin, M. (1970). Cooperation of charges in photosynthetic oxygen evolution. I. A linear four-step mechanism. Photochem. Photobiol. 11:457–475.

Kok, B., Radmer, R., and Fowler, F. (1975). Electron transport in photosystem II. In M. Avron (Ed.), Proceedings of the Third International Congress on Photosynthesis, Vol. 1. Amsterdam: Elsevier Press.

Krasnovsky, A. A., and Brin, G. P., (1978). Photosensitization by titanium dioxide and zinc oxide: Oxygen and hydrogen evolution. In H. Metzner (Ed.), Photosynthetic Oxygen Evolution. New York: Academic Press.

Lavorel, J. (1975). Luminescence. In Govindjee (Ed.), Bioenergetics of Photosynthesis. New York: Academic Press.

Machta, L., and Hughes, E. (1970). Atmospheric oxygen in 1967 to 1970. Science 168: 1582–1584.

Malkin, R. (1977). Primary electron acceptors. In A. Trebst and M. Avron (Eds.), Encyclopedia of Plant Physiology, Vol. 5. Berlin-Heidelberg-New York: Springer-Verlag.

Mathis, P., and Conjeau, H. (1979). Rapid reduction of P-700 photooxidized by a flash at low temperature in spinach chloroplasts. Photochem. Photobiol. 29:833–837.

Mauzerall, D. (1976). Chlorophyll and photosynthesis. Phil. Trans. R. Soc. London B 273: 287–294.

Mauzerall, D. (1977). Porphyrins, chlorophyll and photosynthesis. In A. Trebst and M. Avron (Eds.), Encyclopedia of Plant Physiology, Vol. 5. Berlin-Heidelberg-New York: Springer Verlag.

Mauzerall, D. (1978a). Photoredox reactions of porphyrins and the origins of photosynthesis. In E. van Tamlen (Ed.), Bioorganic Chemistry, Vol. 4. New York: Academic Press.

Mauzerall, D. (1978b). Multiple excitation and the yield of chlorophyll a fluorescence in photosynthetic systems. Photochem. Photobiol. 28: 991–998.

Mauzerall, D. (1978c). Electron transfer photoreactions of porphyrins. In D. Dolphin (Ed.), The Porphyrins, Part C, Vol. 5. New York: Academic Press.

Mauzerall, D., and Chivvis, A. (1973). A novel cyclical approach to the oxygen producing mechanism of photosynthesis. J. Theor. Biol. 42:387–395.

Mercer-Smith, J., and Mauzerall, D. (1981). Molecular hydrogen production by uroporphyrin and coporoporphyrin: A model for the origin of photosynthetic function. Photochem. Photobiol., in press.

Mitchell, P. (1976). Vectorial chemistry and the molecular mechanics of chemiosmotic coupling: Power transmission by proticity. Biochem. Soc. Trans. 4:399–430.

Nakatani, H. Y., Barber, J. and Evans, M. C. W. (1981). Cholate-extracted protein from thylakoids. 5th International Congress on Photosynthesis. Abstract, p. 409.

Papagergiou, G. (1975). Chlorophyll fluorescence: An intrinsic probe of photosynthesis. In Govindjee (Ed.), Bioenergetics of Photosynthesis. New York: Academic Press.

Parson, W. W. (1978). Thermodynamics of the primary reactions of photosynthesis. Photochem. Photobiol. 28:389–393.

Piccioni, R., and Mauzerall, D. (1978a). Calcium and photosynthetic oxygen evolution in cyanobacteria. Biochim. Biophys. Acta 504: 384–397.

Piccioni, R., and Mauzerall, D. (1978b). A high-potential redox component located within cyanobacterial photosystem II. Biochim. Biophys. Acta 504:389–405.

Pick, U., and Racker, E. (1979). Purification and reconstitution of the N,N-dicyclohexylcarbodimide-sensitive ATPase complex from spinach chloroplasts. J. Biol. Chem. 254:2793–2799.

Prigogine, I., and Nicolis, G. (1971). Biological order, structure and instabilities. Q. Rev. Biophys. 4:107–148.

Radmer, R., and Cheniae, G. (1977). Mechanisms of oxygen evolution. In J. Barber (Ed.), Primary Processes of Photosynthesis. Amsterdam: Elsevier Press.

Radmer, R. J., and Kok, B. (1977). Light conversion efficiency in photosynthesis. In A. Trebst and M. Avron (Eds.), Encyclopedia of Plant Physiology, Vol. 5. Berlin-Heidelberg-New York: Springer-Verlag.

Ramus, J., Beale, S. I., and Mauzerall, D. (1976).

Correlation of changes in pigment content with photosynthetic capacity of seaweeds as a function of water depth. Mar. Biol. 37:231–238.

Renger, G. (1978). Theoretical studies about the functional and structural organization of the photosynthetic oxygen evolution. In H. Metzner (Ed.), Photosynthetic Oxygen Evolution. New York: Academic Press.

Robinson, H. H., Scharp, R. R., and Yocum, C. F. (1981). Topology of manganese release upon inactivation of the oxygen evolving complex. 5th International Congress on Photosynthesis. Abstract, p. 478.

Ross, R. T., Anderson, R. J., and Hsio, T.-L. (1976). Stochastic modeling of light energy conversion in photosynthesis. Photochem. Photobiol. 24:267–278.

Ruben, S., Randall, M., Kamen, M. P., and Hyde, L. L. (1941). Heavy oxygen (O^{18}) as a tracer in the study of photosynthesis. J. Am. Chem. Soc. 63:877–879.

Saphon, S., and Crofts, A. R. (1977). Protolytic reactions in photosystem II: A new model for the release of protons accompanying the photooxidation of water. Z. Naturforsch. 32e: 617–626.

Satoh, K. (1979). Polypeptide composition of the purified photosystem II pigment-protein complex from spinach. Biochim. Biophys. Acta 546: 84–92.

Satoh, K., and Butler, W. L. (1978). Low temperature spectral properties of subchloroplast fractions purified from spinach. Plant Physiol. 61: 373–379.

Schmidt, G. H., Menke, W., Radnuz, A., and Koenig, F. (1978). Polypeptides of the thylakoid membrane and their functional characterization. Z. Naturforsch. 33c:723–730.

Schreiber, U., and Vidaver, W. (1973). Photosynthetic energy transfer reversibility inhibited by hydrostatic pressure. Photochem. Photobiol. 18:205–208.

Strehler, B. L., and Arnold, W. (1951). Light production by green plants. J. Gen. Physiol. 34:809–820.

Tel-Or, E., and Avron, M. (1975). Isolation and characterization of a factor which restores the Hill reaction from *Phormidium luridum*. Proceedings of the Third International Congress on Photosynthesis. Amsterdam: Elsevier Press.

Trebst, A. (1974). Energy conservation in photosynthetic electron transport of chloroplasts. Annu. Rev. Plant. Physiol. 24:423–458.

Trebst, A. (1978). Organization of the photosynthetic electron transport system of chloroplasts in the thylakoid membrane. In G. Schäfer and M. Klingenberg (Eds.), Energy Conservation in Biological Membranes. Berlin-Heidelberg-New York: Springer-Verlag.

Trebst, A., and Avron, M. (Eds.) (1977). Encyclopedia of Plant Physiology, Vol. 5: Photosynthesis I. Berlin-Heidelberg-New York: Springer-Verlag.

Van Niel, C. B. (1935). Photosynthesis of bacteria. Cold Spring Harbor Symp. Q. Biol. 3: 138–150.

van Valen, L. (1971). The history and stability of atmospheric oxygen. Science 171:439–443.

Velthuys, B. R. (1979). Abstract: page 113, TAM-B7. Seventh Annual Meeting of American Society for Photobiology.

Vidaver, W. (1969). Hydrostatic pressure effects on photosynthesis. Int. Rev. Ges. Hydrobiol. 54:697–747.

Walker, J. C. G. (1977). Evolution of the Atmosphere. New York: Macmillan & Co.

Warden, J. T., Blankenship, R. E., and Sauer, K. (1976). A flash photolysis ESR study of photosystem II. Signal II v.f. the physiological donor to P-680. Biochim. Biophys. Acta 423: 462–478.

Wessels, J. S. C. (1977). Fragmentation. In A. Trebst and M. Avron (Eds.), Encyclopedia of Plant Physiology, Vol. 5. Berlin-Heidelberg-New York: Springer-Verlag.

Wessels, J. S. C., and Borchert, M. T. (1978). Polypeptide profiles of chlorophyll-protein complexes and thylakoid membranes of spinach chloroplasts. Biochim. Biophys. Acta 503: 78–93.

Winget, G. T. (1979). Oxygen evolution: A reconstituted system. Presented at Seventh Annual meeting of the American Society for Photobiology. Pacific Grove, Calif. (Abstract, page 114).

Witt, H. T. (1979). Energy conversion in the functional membrane of photosynthesis. Analysis by light pulse and electrical pulse methods. Biochim. Biophys. Acta 505:355–427.

Wydrzynski, T., Govindjee, and Gutowsky, H. S. (1981). Proton relaxation by chloroplast membranes: Effects of EDTA. 5th International Congress on Photosynthesis. Abstract, p. 631.

Wydrzynski, T. J., Marks, S. B., Schmidt, P. G., Govindjee, and Gutowski, H. S. (1978). Nuclear magnetic relaxation by the manganese in aqueous suspensions of chloroplasts. Biochemistry 17:2155–2162.

Zelitch, I. (1971). Photosynthesis, Photorespiration and Plant Productivity. New York: Academic Press.

7

Oxygen Toxicity in Unicellular Organisms

SHELDON F. GOTTLIEB

The world is full of mysteries. Life is one.
J.B.S. Haldane

I. Introduction

Interest in the toxic effects of oxygen on unicellular organisms derives from intrinsic heuristic aspects of the subject and its wider relationship to the biological regulation of growth and metabolism and to the biological mechanisms of oxygen toxicity as well as from the practical uses to which such knowledge can be put.

The scientific interest in oxygen toxicity has its origins in understanding the benefits and challenges that the evolution of oxygen afforded living organisms and the resulting evolutionary responses of organisms to the unique environmental challenges brought about by the advent of an oxidizing environment. The evolutionary significance of oxygen was discussed by Gilbert (66–73) and more recently by Fridovich (62,63) and represents the underlying theme of this book.

From a practical view, repeated attempts have been made to exploit clinically Pasteur's observation that some organisms can exist without air and that others are strictly confined to an anaerobic environment, lest their growth be inhibited or they be killed by the exposure to molecular oxygen. The cidal effects of oxygen led to the therapeutic use of oxygen and/or hydrogen peroxide (98,101, 196). Hyperbaric oxygenation as a therapeutic modality (34,80) has been a recent major stimulus to the study of the physiological, biochemical, and pharmacological effects of oxygen on microbes, especially pathogens (82,83).

Previous recent reviews (62,80,82,83,179, 198) discussed in detail many of the theoretical and practical experimental and clinical problems related to oxygen effects on microorganisms. This review will summarize some aspects of previous work and supplement it with later information as well as data not previously reviewed.

A. Nomenclature

What's in a name?

Shakespeare

The phrase *increased oxygen tensions* (IOT) will be used throughout this chapter as opposed to the terms and abbreviations commonly used in the hyperbaric field. A detailed discussion of nomenclature has appeared (82). The response of unicellular organisms represents a continuum of effects; those effects occurring under strict anaerobic conditions to those occurring under a variety of artificial environments including hyperbaric environments of 100% oxygen. IOT represents a phrase more descriptive of what occurs naturally, as well as artificially, and avoids the possibility of considering events occurring in hyperbaric oxygen environments as phenomena discontinuous with phenomena occurring at oxygen tensions equal to or less than 1 atmosphere absolute (ATA). IOT shares the same deficiency as the other terms, i.e., impreciseness (82): biological effects vary markedly with oxygen tension and exposure time (80,83,87). Thus, as should be done for other commonly used terms, it is necessary to include, at least paranthetically, the specific oxygen pressures and durations of exposure used in each experiment (82). Ideally, one should know the oxygen tension profile of each growth tube or reaction vessel throughout the experiment. Technology is available for continuously monitoring P_{O_2} (128,129), as well as maintaining organisms in steady-state growth conditions in media having a constant composition using a continuous flow apparatus. However, the costs and, in many instances, the increased complexity of experimentation that would be associated with these techniques, especially for hyperbaric environments, make them somewhat formidable and almost prohibitive.

II. Factors Affecting Responses to Increased Oxygen Tensions

Let (truth) and falsehood grapple: who ever knew truth put to the worse, in a free and open encounter?

Milton

Organismal responses to IOT are affected by a variety of factors, i.e., the partial pressure of oxygen, CO_2, hydrogen ion concentration, and nutritional status of the medium, temperature of incubation, total atmospheric pressure, as well as the unique physiological, biochemical, and genetic attributes of an organism, and its nutritional state.

A. Oxygen Tensions

As previously discussed (82,83), oxygen is a drug and should be viewed pharmacologically. The implication is that one must not only be concerned with the effects of various concentrations (pressures or tensions) of oxygen and factors that influence these effects, but also with factors that influence its uptake and distribution in vitro and in vivo.

Oxygen is relatively insoluble, about 60–70 μg/ml H_2O/ATA O_2 at 0°C (10,20). From considerations of Henry's law, it can be deduced that the volume of O_2 physically dissolved in a given medium will be directly proportional to its partial pressure.

As with the other atmospheric gases, O_2 solubility decreases with increasing temperature; at 25°C, the solubility of O_2 is 39 μg/ml H_2O/ATA O_2 and, at 37°C, it is about 34 μg/ml H_2O/ATA O_2 (10). This solubility is further reduced by the presence of salts.

Thus, depending on the size of the inoculum and intrinsic rate of respiration of the organism, which is related to age as well as the immediate previous growth history of the organism, the oxygen in the medium may be consumed extremely rapidly. Unless precautions are taken to maintain a specific constant P_{O_2} in the growth medium, as can be done with a chemostat (179), the organism is constantly growing in an environment with a diminishing O_2 tension. Thus, O_2-dependent physiological and biochemical reactions should be constantly undergoing change (19,58). Similarly the oxidation-reduction potential of the medium and the cells should be changing as the oxygen tensions decrease, and different end products of metabolism may accumulate (8,19,88,149).

The control of oxygen tensions in growth media is of extreme industrial importance (38,107,143,188). In industrial fermentations (IF) it is essential to optimize growth conditions in order to increase the efficiency of converting substrate to cell mass or specific product. In contrast to IF, waste treatment involves the conversion of substrates to CO_2 with a minimum increase in cell mass. In both of these processes, knowledge of microbial physiology as affected by P_{O_2} is of marked importance.

B. Oxidation-Reduction Potential

It is well known that bacteria develop reducing environments during growth, the lowest potentials being generated during the logarithmic phase of growth. The reasons for the development of reducing conditions is not readily understood. The redox potential of any bacterial culture is a resultant of many complex interactions, among them the degree of oxygenation of the medium and its composition. The medium composition, altered by microbial metabolism, may include metabolic products and cellular constituents. Thus, the redox potential may not reflect any intrinsic characteristic of a culture. Although it is not really known exactly what the redox potential signifies, Wimpenny and Necklen (200) ask, "...is E_h merely a useful number that changes drastically as aeration increases..., or is it a parameter that has some deeper significance in cell regulatory mechanisms?"

Wimpenny and Necklen (200) have shown that above a certain critical level of oxygen, the E_h rises sharply as P_{O_2} increases and makes a plateau at +300 mV. Under conditions of IOT, it is not possible to study the effects of E_h independent of the P_{O_2}.

C. Partial Pressure of Oxygen

The use of chemostats, with their ability to accurately control environmental P_{O_2}, markedly enhanced the study of the effects of oxygen as a regulator of microbial growth, especially in studying the transition from anaerobiosis to aerobiosis (87,198,200). Unfortunately, in the study of IOT, such precise control of P_{O_2} has not generally been employed. Most studies concerning IOT depend on decreasing the diffusional barriers between the growing culture and the ambient P_{O_2} (80,82,83). Since IOT, especially O_2 tensions greater than 2–3 ATA, tends to inhibit growth and/or enhance the bacteriostatic or bactericidal actions of antimicrobial agents, it would be helpful, from a theoretical as well as practical point of view, to know at what P_{O_2} the static, cidal, or drug interaction phenomena occur. Relatively precise P_{O_2} measurements are being made in living tissue with respect to wound healing and hyperbaric therapy (165).

Knowledge of the precise P_{O_2} (and its concomitant, duration of exposure) which adversely affects microbial growth, metabolism, and/or toxin production, coupled with knowledge of the P_{O_2}-duration of exposure relationship which produces tissue O_2 toxicity, could lead to a more rational approach to hyperbaric therapy.

D. Respiration

To study respiration one can use either nongrowing cells or growing cells. Until the introduction of the chemostat, emphasis was placed on measuring the oxygen consumption of nongrowing cells. Since the advent of the chemostat, it has been possible to study respiration in growing cultures. The effects of oxygen tensions on respiration are varied and complex. Primarily, efforts were and are directed at studying low oxygen tensions from a P_{O_2} of 150 mm Hg (air saturated medium) to complete anaerobiosis. Relatively few studies have been done using IOT, $P_{O_2} \geq 760$ mm Hg.

The available evidence indicates that over a relatively wide range of oxygen tensions, microbial oxygen consumption may be independent of the P_{O_2} (51,87,110), irrespective of whether the cells are resting or growing. Above a certain P_{O_2}, respiration is inhibited reversibly (9,22,50,51,82,139,146,159, 183). Below a certain critical P_{O_2}, microbial respiration of growing cultures as a function of P_{O_2} may become more complex. Harrison and Pirt (88) found that the respiratory response of *Klebsiella aerogenes* to P_{O_2} constituted a complex relationship. Above 10 mm Hg O_2, the respiratory rate was independent of P_{O_2}; at or below 10 mm Hg, the respiratory rate oscillated in a very complex relationship to the available P_{O_2}; at a P_{O_2} of

0.5 mm Hg, respiration increased to a value of at least 20% greater than that obtained in the presence of a $P_{O_2} > 10$ mm Hg. Below a P_{O_2} of 0.2 mm Hg, respiration decreased. The mechanism of this complex respiratory behavior to diminished oxygen tensions seems to be related to feedback control mechanisms probably involving NAD(P)H (87).

Whether the oscillating response to low P_{O_2} is unique to *K. aerogenes* or is more universal is not known. Rate of respiration of *Azotobacter* species increases with an increase in P_{O_2} in the range of 0.1 to 0.3 ATA, at which point a maximum respiratory rate is reached; as the P_{O_2} increases beyond 0.3 ATA, the respiratory rate decreases at a linear rate (50,51,139,146,159,183). There is a report of *Azotobacter* respiration increasing linearly up to 0.8 ATA O_2 (59). The oxygen-induced decrease in respiration is reversible (50).

In contrast to *Azotobacter*, the respiratory rate of *Bacillus subtilis, Saccharomyces cerevisiae*, and *Rhizobium trifolii* remained constant over the range of 0.05 to 1.0 ATA O_2. However, in none of these studies was the P_{O_2} as low as that used by Harrison and Pirt (88) when they observed the oscillatory respiratory rate in *K. aerogenes*. Harrison (87) discussed in detail the regulation of respiration in growing bacteria.

Longmuir (125) showed that the rate of respiration of eight species of bacteria is related to the oxygen concentration by the Michaelis-Menten equation. The K_m's of the intact organisms were determined and found to vary according to the size of the cell. Although this relationship seemed to hold for bacteria, and could be expressed by a general equation, it did not hold for yeast, thus indicating the limited usefulness of the equation. Differences were found between the K_m's of intact cells and cell-free preparations of the larger organisms. The apparent disparity was explained by an increased diffusional barrier to oxygen in the larger organisms; there is no appreciable effect of diffusion of the K_m's of smaller organisms. Button and Garver (36) reported higher values than did Longmuir and, in addition, they did not find a correlation between the size of *Torulopsis utilis* and the K_m for growth. The apparent discrepancy may be explained by the differences in techniques used; Longmuir may have been studying the rate of O_2 absorption by an O_2-starved cell, whereas Button and Garver may have been measuring an oxygen-regulated rate-limiting reaction.

The respiratory chain is susceptible to changes in P_{O_2}. *E. coli* growing in the presence of excess O_2 has a respiratory chain that terminates with cytochrome *O* oxidase; in the presence of reduced P_{O_2} there is, in addition to cytochrome *O* oxidase, a synthesis of cytochrone *d* oxidase (152). This latter oxidase has a greater affinity for O_2 than cytochrome *O* oxidase. Despite this change in the concentration of the terminal oxidases, there does not appear to be any change in the efficiency of respiratory energy conservation (90,152). This conclusion is at variance with that of Meyer and Jones (138). The conflict will be readily resolved when P_{O_2} values are determined for a variety of organisms grown in oxygen and oxygen-limited environments. Experiments with mutants lacking one or the other of the oxidases should also be considered.

In line with the observation that oxygen is a potent regulator of microbial metabolism (198), Bai et al. (6) found that oxygen tensions alter the nature of the aldolase produced by *Mycobacterium tuberculosis*; in the presence of what is presumed to be a low P_{O_2}, i.e., stationary surface culture (P_{O_2} not measured), only class II aldolase, the enzyme functional in glycolysis, is found; in fermenter grown cells, high P_{O_2} (P_{O_2} not stated), the ratio of class I to class II aldolase is 90:10. Class I aldolase is associated with gluconeogenesis.

Propionobacteria are considered to be anaerobic, yet, after anaerobic growth, four species contained cytochromes *b*, *a* or a_1, a_2, and a carbon monoxide-binding pigment. None of these four species grew aerobically on surface culture; they did grow aerobically in broth. Two of the organisms, *P. shermanii* and *P. freudenreichii*, grew slower aero bically than anaerobically and had a markedly diminished synthesis of cytochrome pigments (49); *P. rubrum* and *P.*

pentosaceum grew faster aerobically than anaerobically; oxygen inhibition of cytochrome synthesis was less marked than in the other two propionobacter species. The failure to grow on surface culture may be explained by the direct exposure to a higher P_{O_2} than that attained in broth with a resultant inhibition of oxidative phosphorylation.

The role of oxygen in bacterial heme synthesis (108) as well as its effect on the type and amount of cytochromes produced under a variety of conditions is under active investigation (157,168) as well as the physiological role of the specific cytochromes (193). Sapshead and Wimpenny (157) showed that molecular O_2, and not nitrate, has the greatest regulatory effect on the type and concentration of cytochromes in *Micrococcus denitrificans*.

Broman et al. (17) recently reported that O_2 and nitrate have the greatest regulatory effect on the type and concentration of cytochromes in *Micrococcus denitrificans*.

Broman et al. (17) recently reported that O_2 and nitrate tend to repress the deiminase route and the arginase pathway of arginine metabolism in anaerobic cultures of *Bacillus licheniformis*. Control of these alternate pathways of arginine metabolism is more complex than is intimated by the above observations. The mechanism of the repression and derepression is unknown; Broman et al. (17) speculate that it is not a direct effect of O_2 or nitrate.

Oxygen is known to induce the synthesis of mitochondria in yeast (192) and is involved in the conversion of nonrespiring protomitochondria into respiratory mitochondria (83).

E. Pasteur Effect

The role of oxygen in regulating the metabolism of microbes has been extensively studied since Pasteur's observation that facultative anaerobes, transferred from anaerobic to aerobic environment, demonstrate sparing effect, thereby making more cell yield, and a greater production of CO_2 per molecule of glucose consumed, while the simultaneous rate of CO_2 production and glucose decreased.

The Pasteur effect may be defined as a conservation of substrate under conditions in which metabolic energy is abundantly available.

A comparison of the anerobic and aerobic energetics reveals that the greatest energy conservation afforded by aerobic metabolism, in the form of ATP, results in a substrate sparing effect, thereby making more carbon available for anabolic reactions and conversion to biomass. A variety of theories have been formulated concerning the mechanism of the Pasteur effect. These include the effect of oxygen on altering the rate of substrate uptake (123), on catabolite repression (144), on ADP/ATP ratios and their effects on specific enzymes (79,87), on availability of specific cytochromes (122), on altering the affinity of enzymes for oxygen (195) on the pyridine nucleotides (87), and on the nature of the specific enzyme (180). With respect to the latter point, it has been shown that phosphofructokinase (PRK) obtained from an anaerobic culture of *E. coli* was ATP-sensitive, whereas the same enzyme obtained from aerobically grown cells was ATP-insensitive (180). Thus, the loss of allosteric control, of at least PFK, appears to be involved in the conversion from anaerobiosis to aerobiosis. Whether the loss or gain of allosteric properties of other enzymes during the anaerobic–aerobic transition remains to be investigated.

The anaerobic–aerobic transition is also dependent upon the composition of the medium. In the presence of media low in K^+, *Saccharomyces cerevisiae* show unusual behavior in the presence of oxygen (189). At 0.4 ATA O_2, growth is inhibited and the cells are susceptible to hypotonic shock, i.e., they are unable to maintain ionic gradients. Pyruvate kinase and alcohol dehydrogenase activities behaved differently to low K^+; the concentration of the former was reduced whereas the latter was increased. Normal yeast cells can be obtained by increasing the K^+ concentration in the medium to 3 m*M*.

F. Carbon Dioxide

The evidence indicating that presence or absence of CO_2 from the atmospheric environment may alter organismal responsiveness to IOT has been reviewed (82). The powerful effect CO_2 lack can have, may be seen from the cidal effects of 1 ATA O_2 on *E. coli* in the absence of CO_2 and the absence of cidal effects in the presence of CO_2. The rate of killing follows first-order kinetics and decreases with decreasing temperature or the removal of organic nutrients (25,27). It is now known that the organic nutrients provide for O_2 metabolism and the generation of toxic oxygen radicals. It was pointed out by Gottlieb (82) that differences in results among investigators in the past may have been due, in some situations, to the omission of CO_2 from the experimental atmosphere.

Bornside et al. (16) have shown that the bactericidal effects of 3.0 ATA O_2 on surface cultures of a variety of gram-negative, aerobic enteric bacteria were enhanced by the absence of CO_2. Repaske et al. (151) discuss the potent control CO_2 exerts on the lag period and on the growth of *Streptococcus sanguis*.

The mechanism whereby CO_2 deficiency exerts its effect is unknown. CO_2 is known to alter permeability (80) and is required for fatty acid synthesis—important compounds in the synthesis of cell envelopes—as well as TCA cycle intermediates, which, in turn, are required for the synthesis of metabolic intermediates. It has been shown that *E. coli* and *Serratia marcescens* demonstrate an increase in osmotic fragility when exposed suddenly to a hypotonic medium (43). Although there was no overt structural change, the organisms were unable to multiply.

Cairney (37) omitted CO_2 from his experiment and found *Candida albicans* to be sensitive to O_2 inhibition of growth when exposed to 3 ATA O_2 for as little as seven 90-min periods; the time distribution was three 90-min exposures, 8 hr apart, in the first 24 hr, and two 90-min exposures, 12 hr apart on each of the next 2 succeeding days. These results are at variance with those of Robb (153) who reports *C. albicans* to be oxystabile to 10 ATA O_2 for 7 days. She, too, omitted CO_2 from the experimental atmosphere; she claimed CO_2 deficiency did not affect the results. Robb (153) as well as Stuart et al. (174) are also at variance with Bornside (12). The latter showed that CO_2 deprivation only enhanced the microbicidal activity of 3 ATA O_2 for only 2 of 21 yeasts; in 4 yeasts, the presence of CO_2 enhanced the cidal effects of IOT, the cidal response of the remaining 15 yeasts to 3 ATA O_2 were not affected by the presence or absence of CO_2.

G. Other Factors

To date there is still no systematic study of the effect of pH on susceptibility to oxygen toxicity. It has been noted that Staphylococci are more acid tolerant in an aerobic than anaerobic environment (7). The effect of pressure has been reviewed (82). The role of nutrition will be discussed under mechanisms of oxygen toxicity.

III. Kinetics

Reports abound in the literature concerning the relationship of oxygen supply to rates of bacterial growth (1,39,54,110,136,137). In contrast, most reports concerning the effects of IOT on bacterial growth are concerned with the presence or absence of inhibition and whether the growth inhibition is bacteriostatic or bactericidal. Generally, bacteriostatic effects are reported if the organism, after exposure to a given O_2 pressure-duration regimen, with or without transfer to fresh medium, grows. Consideration is usually not given to the fact that even under those conditions a variable number of organisms may have been killed. Viable cell counts are not routinely performed. Cidal effects are defined by the failure of growth in fresh media following exposure to a given O_2-duration regimen.

Few reports are concerned with the kinetics of inhibition. Hill and Osterhout (94) reported an increase in the rate of inactivation of *Clostridium perfringens* inoculated on trypticase soy agar when the P_{O_2} increased from 2 to 5 ATA. As expected, the divergence became more marked with increasing duration of exposure and was especially marked by 6 hr. At 2 hr there was sufficient variability such that a clear generalization could not be made.

With aerobic organisms, there is a wide variability in the rate of killing of organisms depending on the relative susceptibility of the organism to IOT. Irvin and Smith (106) found that the mean viable count of *Pseudomonas pyocyanea* began to decrease after 24 hr of exposure to 3 ATA O_2 for organisms that were exposed while in the log phase of growth, and decreased after 72 hr of exposure to 3.0 ATA O_2 of organisms that were in the lag phase of growth. Bornside (13) reported that exposure of *P. aeruginosa* to 3.0 ATA O_2 (no CO_2) in tripticase soy broth markedly inhibited growth by extending the lag. When the organisms were removed from the IOT, they grew at rates indistinguishable from unexposed controls. Kaye (113) exposed *E. coli, Aerobacter aerogenes, P. aeruginosa, Salmonella typhimurium, Proteus mirabilis, Diplocococcus pneumoniae, Staphylococcus aureus,* and *Streptococcus pyogenes* to 2 ATA O_2 and found that all of the organisms grew; the rate of growth and total amount of growth tended to be decreased by the IOT. Brown (22) reported that the growth of membrane cultures of *E. coli* on a minimal salts–acetate medium was completely inhibited by $P_{O_2} \geq$ 1.7 ATA. As the P_{O_2} exceeded 1.7 ATA, the time for the onset of inhibition to become manifest decreased; at 4.2 ATA O_2, inhibition began to occur at about 1.5 hr. Although he did not report on the reversibility of the O_2-induced inhibition, one may surmise that the O_2 effect was bacteriostatic, since there was no decrease in cell number up to 6 hr of exposure and since a decrease in respiration was rapidly reversed following removal of the organism from the IOT. The 2 ATA O_2 inhibition of growth of *E. coli* in trypticase soy broth reported by Marquis et al. (132) appeared to be due to a slight increase in the lag phase and a possible slight decrease in the log phase; the final growth titer was unaffected. The O_2 effect on the log phase growth was markedly potentiated by the addition of 20.4 ATA helium or 6.8 ATA N_2O. As P_{O_2} increases, the toxic effect of O_2 on *E. coli* increased. *S. aureus* was more sensitive to O_2 than *E. coli*, while *S. faecalis* was least sensitive.

Hendricks (91), studying the kinetics of growth inhibition of *Corynebacterium diphtheriae* and *S. typhosa* in the presence and absence of sulfisoxazole or trimethoprim at 2.87 ATA O_2, found that O_2 exerted a static effect on *S. typhosa* even in the presence of the drugs. The growth kinetics of *C. diphtheriae* showed that there is growth of at least one generation followed by a marked bactericidal effect, as evidenced by an exponential rate of decline in cell numbers. The addition of 100 or 500 μg% sodium sulfisoxazole did not significantly alter the kinetics of the bactericidal effect. Mader et al. (130) did not find any effect in the rate of growth of *S. aureus* following a single 2-hr exposure to 2 ATA O_2 either in the presence or absence of 1.0 μg/ml cephalothin.

There are relatively few kinetic studies performed on the effect of IOT on number of organisms in wounds. Irvin et al. (105,106) found that standard wounds in guinea pigs infected with either *S. aureus* or *P. pyocyanea* and exposed continuously to 2 ATA O_2 showed a statistically significant decrease in number of organisms up to 72 hr as compared to the air controls. Near the end of the experiment, when wounds were exposed to air, there was no significant difference between test and control wounds.

There is a need for doing intensive kinetic studies in vitro on a variety of organisms and a variety of media, as well as in the presence or absence of antimicrobial agents. Corresponding kinetic studies are also required for a variety of in vivo models. For in vivo models, attention has to be paid to the method of administering the oxygen; surface application, as used by Irvin et al. (105,106), is useful for surface wounds because of the minimal risk of oxygen toxicity to the host while applying virtually full P_{O_2} to the organism. However, it is a partially diffusion-limited technique. Oxygen delivery via the cardio-pulmonary system has the advantage of maximal distribution of oxygen to a wound or phlegmon, but suffers from the disadvantage of exposing the host to oxygen toxicity. Detailed discussions of these subjects have appeared (82,83).

IV. Mechanisms of Oxygen Toxicity

The grandeur of the universe is always commensurate with the grandeur of the soul that surveys it.

Heine

Locating the site(s) of oxygen action as well as the mechanism of oxygen toxicity has been one of the prime goals in the hyperbaric field. Approaches to this investigation have been varied and include the effects on cell permeability, respiration and energetics, protein synthesis, metabolic pathways including individual enzymes, and nutrition, as well as studying the end products of oxygen metabolism.

In the past decade, significant advances were made in understanding the mechanism of oxygen action, especially as it relates to the metabolism of oxygen.

Molecular oxygen is not exceedingly reactive in biological systems, due to the nature of its electron structure. However, the reduction of molecular oxygen in biological systems yields toxic intermediates. It is these intermediates that are involved in phenomena such as lipid peroxidation, enzyme inactivation and denaturation of proteins, and structural changes in nucleic acids (47,64). The toxic intermediates of oxygen metabolism are the result of the univalent reduction of oxygen.

In the course of evolution, many organisms evolved means of coping with the ever present danger of oxygen toxicity. Some organisms simply avoid contact with molecular oxygen by residing in anaerobic environments. Some organisms are aerotolerant and do not use molecular oxygen as an oxidant. However, most aerobic organisms, by virtue of their metabolic requirement of oxygen as the final electron acceptor for substrate oxidation and energy extraction, resort to other means of defense against oxygen toxicity. One line of defense involves the binding of molecular oxygen to carriers such as hemoglobin, myoglobin, and cytochrome *P*-450 (126, 160,201). A second line of defense involved the evolution of multivalent pathways for the reduction of oxygen so that toxic intermediates do not occur. The multivalent reduction of oxygen in biological systems involves the use of transition cations such as Fe(II) and Cu(II) (61), and includes enzymes such as cytochrome *c* oxidase, ascorbic acid oxidase, and tyrosinase (61). A third line of defense involves the evolution of quenching mechanisms, i.e., the removal of toxic oxygen species by substances that can react with toxic compounds and which will inhibit the propagation of oxidizing reactions. A fourth line of defense involves the evolution of mechanisms designed to repair cellular damage induced by toxic intermediates of univalent O_2 reduction (100,102,177,178). A fifth mechanism involves the evolution of scavenging mechanisms such as those provided by catalases, peroxidases, and especially by superoxide dismutases (61,64). For detailed reading on the nature of the toxic intermediates and cell defense mechanisms, see Forman and Fisher (59a) and Fridovich (64a).

Although the knowledge of the toxic intermediates of oxygen metabolism along with the knowledge of the various scavenging agents deepened the understanding of the molecular mechanisms of oxygen toxicity, there still remain questions as to what enzyme(s) or structures are being attacked and inactivated or destroyed, and how these systems relate to growth inhibition and death. Further questions are concerned with whether or not there is a single or multiple site of O_2 action that can explain its toxic behavior.

It had been pointed out that an organism's responsitivity to oxygen in part is dependent on the growth medium and that nutrients may mitigate the toxic effects of oxygen (82).

Gottlieb (81,82) suggested that nutritional backtracking should help determine sites of oxygen action. Using an organism that utilized 1-amino-2-propanol as the sole source of carbon, nitrogen, and energy, he found that amino acids decreased oxygen inhibition of growth of an *Achromobacter* species. Gottlieb and Pakman (84) showed that various sugars or yeast extract on brain heart infusion permitted growth of aerobic vibrios under oxygen tensions that are normally inhibitory. Additional data bearing on this subject have been reviewed (82,83). Boehme et al. (11) exposed log phase *E. coli* E-26 to 4.2 ATA O_2 for 1 hr in a medium supplemented with 20 amino acids. By singly deleting the amino acids, they were able to show that the absence of valine, tyrosine, isoleucine, tryptophan, leucine, phenylalanine, cysteine, methionine, asparagine, or threonine caused a statistically significant loss of protection as compared to the full 20 amino acids. Using intermediates in the biosynthesis of these 10 amino acids they found that oxygen may block the acetohydroxy acid synthetase, acetoxy acid reductoisomerase or dihydroxy acid dehydratase enzymes, whereas the enzymes involved in the synthesis of leucine from α-ketoisovalerate were unaffected. Valine was one of the most critical of the amino acids required for protection. Since homocysteine substituted for methionine, this indicated that the transmethylase was relatively oxygen insensitive and that the O_2-induced inhibition occurred prior to homocysteine synthesis. In the aromatic amino acids, it appeared that O_2 inhibited the conversion of shikimate to chorisomate. Except for those vitamins dependent upon amino acid biosynthesis (nicotinic acid, folic acid, pantothenic acid), vitamins did not protect against oxygen toxicity. Indeed, Brown (24) found that the addition of sodium lipoate did not protect *E. coli* against O_2 toxicity despite the fact that evidence existed that O_2 inhibits enzyme reactions involving lipoate as a coenzyme (181,182). They did find that lipoate, in the presence of IOT, was somewhat toxic. Thus, they confirmed the observation of Gottlieb (81,82) that reducing substances may be toxic even in the presence of oxygen. The observation that the thiol compounds, cysteine and lipoic acid, do not reverse O_2 inhibition of growth, and may even be toxic under IOT conditions, is somewhat suprising in view of the role of glutathione (57,92,154) and sulfhydryl groups in cell functioning, and the fact that increased hydrostatic pressure could cause oxidation of sulfhydryl groups (158). It should be noted that there are *E. coli* mutants that can grow in the absence of glutathione (5). The mechanism whereby SH compounds inhibit growth is not known; however, the observation of the Kosowers (117) that oxidized glutathione inhibits protein synthesis in vitro, could be extrapolated to a similar occurrence in vivo.

It should be noted that the oxygen-dependent inactivation of gramicidin S synthetase could be partially protected against by the addition of the reducing agent dithiothreitol (65).

Using a strain of *S. typhosa* which had an absolute growth requirement for L-leucine and thiamine even at 1 ATA air, Gottlieb and Solosky (unpublished results) found that the vitamins biotin and nicacinamide were among several nutrients that prevented inhibition of growth at 1.87 ATA O_2. Of the amino acids, asparagine, proline, serine, tyrosine, valine, methionine, and tryptophan tended to afford protection (72); there is a marked similarity to the amino acids Boehme et al. (11) found to be O_2 protective.

The data of Gottlieb and Solosky (unpublished results) and of Hendricks (91) cited above were not as sharp as those of Boehme. The reason for the greater variability in the data of the former investigators can be gleaned from the differences in technique. Boehme et al. exposed the organisms to 4.2 ATA O_2 for 1-hr exposure and absorbance was measured. Thus, they tended to measure the more oxygen-sensitive pathways, whereas the other investigators exposed their cultures to 1.87 ATA O_2 continuously for 24 hr. Thus, during the 24-hr exposure a greater number of variables could be involved, especially inhibition of more resistant pathways, as well as the emergence of more resistant strains.

Following their nutritional leads, Brown and Yein (28) found the enzyme dihydroxy-acid dehydratase was reduced 78% in extracts of *E. coli* exposed to 4.2 ATA O_2 for as little as 10 min. Other enzymes involved in the synthesis of branch chain amino acids, acetolactate synthetase and dihydroxyisovalerate dehydrogenase (isomerizing), were either unaffected or showed a negligible inhibition. Further studies (29) revealed that phosphoribosylpyrophosphate (PRPP) synthetase was exceedingly oxygen sensitive; more than 90% of the enzyme activity was reversibly inactivated by a 10-min exposure to 4.2 ATA. Quinolinate synthetase activity was reduced 92% after 1-hr exposure to 4.2 ATA O_2 in a medium containing protective amino acids. The significance of the amino acids can be seen from the fact that only a 25% inactivation occurred in their absence. Apparently the stimulation of metabolism by the amino acids in the presence of IOT permitted the production of toxic oxygen radicals. Along these same lines PRPP synthetase was not inhibited when nonmetabolizing cells were exposed to hyperoxia.

Additional studies as to the site and mechanism of oxygen toxicity include the interactions of oxygen and antimicrobial agents, specifically the sulfonamides and the sulfonamide potentiator trimethoprim (83). These studies suggest that PABA-folic acid pathways may be oxylabile.

The inhibition of amino acid synthesis could give rise to a stringent response (40,41,45). Stringency is the inhibition of RNA synthesis when the cells are deprived of a required amino acid and leads to the inhibition of many anabolic processes. In amino acid deficiency, the concentration of the regulatory nucleotides, guanosine 5′-diphosphate, 3′-disphospate (ppGpp), and guanosine 5′-triphosphate, 3′-diphosphate (pppGpp) is increased. If energy is limiting, only the concentration of the diphosphate increases. The apparent signal for the stringent response is an uncharged tRNA (deficiency of aminoacylation of tRNA) in the acceptor site of the ribosome.

Brown et al. (30) reported that 4.2 ATA O_2 inhibition of growth of *E. coli* was associated with inhibition of protein synthesis as well as RNA and DNA synthesis, and was most likely due to a stringent response.

In view of the fact that growth inhibition could be due to a variety of factors including inhibition of transport mechanisms, respiration and energy transduction mechanisms, lipid and membrane synthesis, etc. (82,83), it becomes mandatory to not just learn that oxygen adversely affects these reactions, but also to learn their temporal relationships (80,82,83).

Brown's group (21,23,26,32,35,134,172, 173) in an interesting series of experiments has shown that various cellular functions were reversibly inhibited by IOT, but they were not inhibited prior to the onset of growth inhibition. Brown's observation that 4.2 ATA O_2 does not significantly interfere with lipid synthesis or lipid structure (26) was essentially confirmed by Harley et al. (86). These investigators (86) showed that the sensitivity of an unsaturated fatty acid auxotroph of *E. coli* to 20 ATA O_2 increased as the degree of unsaturation of the fatty acid addition to the medium increased. However, the kinetics of survival, although approximating first-order, required incubation times from 6 to 96 hr. Considering the extraordinarily high P_{O_2}, it seems unlikely that O_2 effects on the lipid contributed to the mechanism of oxygen toxicity as described by Brown. However, Harley et al. exposed resting cells suspended in a minimal salt medium in the absence of an energy source and fatty acid supplement. These are conditions which tend not to favor the formation of toxic O_2 radicals and which tend to alter the nature of the stringent effect.

Although Stees and Brown (173) claim that O_2 inhibition of growth of *E. coli* is not due to a generalized oxidation of SH groups, they have not ruled out the possible oxidation of SH groups of one or more essential enzymes or structural proteins.

One can visulize the sequence of events that occurs in the inhibition of growth of *E coli* exposed to 4.2 ATA O_2 as follows: a rapid (within 0.05 of a generation) interference in the biosynthesis of branch chained and aromatic amino acids which leads to the stoppage of protein, niacin, and thiamine synthesis and induction of the stringent response (31).

It should be stressed that most of the above studies were done using *E. coli*. It becomes imperative to learn if the same mechanism holds true for other prokaryotes as well as eukaryotes and for unicellular as well as multicellular organisms. It is interesting that *S. typhosa* appears to have similar amino acid requirements as does *E. coli* as O_2 protectants, while also showing differences. Since the *S. typhosa* studies were done at 1.87 ATA O_2, as compared to the 4.2 ATA O_2 for *E. coli*, and since there may be differences between the two organisms in their response to IOT, the question arises, does the same sequence of events occur at a lower P_{O_2}? Perhaps not (30,54,63).

Questions also arise as to what is the sequence of events leading to growth inhibition when nutrients are not limiting: deficient media represent highly artificial conditions. A systematic study of the role of ions in or on these systems has not been done. One wonders what role such a system may have in the anaerobic-aerobic transformation.

Vaughn and Wilbur (189) showed that growth inhibition of *Schizosaccharomyces pombe* by 2.0 ATA O_2 was reversible and not related to oxygen interference with respiration. It is interesting, in light of Brown's work, that growth rate and protein synthesis was reduced 44% by 1 ATA O_2. However, they also noted marked reductions in the transport of glycine, leucine, and uracil. The kinetics indicated a noncompetitive inhibition. They interpreted the growth inhibition to be due to an O_2 interference with transport mechanisms.

V. Some Miscellaneous Effects of Oxygen

In addition to the events centering around the anaerobic-aerobic transformation and the possible different mechanisms involved in O_2 inhibition of growth, there are additional effects of oxygen.

One interesting observation is the unbalanced growth reported for *Streptococcus sanguis* (54) in a 0.95 ATA O_2 + 0.05 ATA CO_2 environment. Anaerobically grown cells continued to show cell division, despite the fact that there was no simultaneous increase in turbidity; i.e., cell division apparently lags mass increase during log growth but catches up during the stationary phase. In contrast, O_2 grown cells showed increased turbidity even though total cell number and the number of colony forming units remained constant. Direct observation coupled with these data indicate that O_2 grown cells are larger than anaerobic cells at the end of the log phase and in the stationary phase of growth, so that cell division did not keep pace with increase in cell mass. Continued exposure to oxygen for 6 hr resulted in an exponential decrease in viability without a decrease in cell mass or cell number. The O_2 grown cells became pleomorphic.

RNA synthesis continued for 2 hr, after which it abruptly ceased, while protein synthesis continued. Thus, this response is different from the stringent response reported by Brown (30). Eisenberg also noted that the O_2 cells became pleomorphic. It was found that the factors responsible for pleomorphism were generated during the first 2 hr of growth, in the early to mid log phase (54).

The above events occurred at 0.95 ATA O_2. What sequence of events would have occurred if the organism had been exposed to a higher P_{O_2}?

Another interesting effect of oxygen is its role in endospore formation. During sporulation there is a loss of the capacity for de novo nucleotide synthesis. The inactivation of one of the key enzymes, aspartate transcarbamylase, involves the generation of metabolic energy (48,190). In contrast, the inactivation of glutamine phosphoribosylpyrophosphate amido-transferase (ATase), the first enzyme of de novo purine synthesis, has a specific requirement for molecular O_2, and does not appear to be dependent on the generation of metabolic energy (184). It is tentatively thought that the ATase exists in an equilibrium between an aggregated and a disaggregated state. Adenosine 5′-monophosphate tends to stabilize the enzyme whereas guanosine 5′-monophosphate destabilizes it. Thus, the relative stability of ATase may depend on an AMP/GMP ratio, and O_2 may be exerting its effect by the oxidation of one or more groups on the enzyme that are essential for its activity (185). The inactivation of gramicidin S synthetase in *Bacillus brevis* is also oxygen dependent and independent of metabolism (65).

Oxygen is a well-known inhibitor of nitrate reductase, although the mechanism is still not known (148,166,199). MacGregor and Bishop (127) claim, along with Pichinoty (148), that molecular O_2 is the repressor and that cytochrome *b* does not function as the sensor of oxygen levels (127).

In face of the evidence of O_2 repression of nitrate reductase, Jetschmann et al. (109) report the activation of nitrate reductase proenzyme by molecular oxygen and ferricyanide, and the inhibition of the O_2 activation by carbon monoxide.

Biological nitrogen fixation, a subject of immense scientific and practical importance, is directly affected by molecular oxygen (1,52,53,111,115,150,164,169,202). Nitrogenase, the enzyme complex which reduces molecular nitrogen to ammonia is either sensitive to oxygen or can be resolved into protein components, one of which is very oxylabile. Which nitrogenase is oxylabile depends on its source (150). All nitrogenases require ATP and Mg^{2+} for activity. Yates (202) found that nucleotide triphosphates rendered oxygen-tolerant nitrogenase extracts from *Azotobacter chroococcum* oxygen-sensitive. This effect was reversed by Mg^{2+}. Cellular respiration may protect nitrogenase from O_2 inhibition (111,115,164, 169). Apparently ATP induces a reversible conformational change in the Ac2 component of AS1 which renders it oxylabile.

VI. Anaerobiosis and Mechanisms of Resistance to Oxygen Toxicity

Recently, the problems related to anaerobes have taken on new significance; this is especially true in medical microbiology (4, 76–78,163,197). Marked improvement in techniques has permitted the isolation and study of a larger number of anaerobic species from a wide variety of environments (39,133, 171,175,187). Anaerobes are ubiquitous. No longer are anaerobes just found in "traditional" anaerobic environments, e.g., the large intestine or deep in body orifices. As Rosebury (156) discussed, they are also isolated from aerobic environments in sites such as the human skin, where they outnumber aerobic organisms 10:1, and in the mucous membranes of the mouth, where the anaerobic bacteria outnumber the aerobes 30:1.

It is not surprising that one finds, as noted earlier, a continuous spectrum of microbial sensitivity to increasing oxygen tensions. Based on growth on surface cultures, Loesche (124) identifies at least three subgroups of anaerobes: (a) strict anaerobes, i.e., those that grow in an atmosphere containing less than 0.005 ATA O_2 on a PPLO supplemented medium; (b) moderate anaerobes, i.e., organisms that grow in an atmosphere containing more than 0.005 ATA O_2 but less than 0.1 ATA O_2 and show maximal growth in a range up to 0.03 ATA O_2; and (c) microaerophiles, i.e., organisms which require a low P_{O_2} for optimal growth (P_{O_2} greater than 0.003 ATA but less than 0.2 ATA) but which are unable to grow in air. Tally et al. (176) report that anaerobic bacteria isolated from infections are almost always oxygen tolerant.

Although the nature of the growth medium may influence the growth rates of the organisms (161) and their sensitivity to O_2 (82), there are also factors intrinsic to the organisms that determine their ultimate sensitivity to oxygen. A detailed discussion of these factors is beyond the realm of this chapter. Factors that determine relative anaerobiosis are closely lined to the factors related to the mechanism of resistance to oxygen toxicity (59a). Additional information may be gleaned from Morris (140,141), McCord et al. (135), and Gregory et al. (85).

VII. The Use of Increased Oxygen Tensions as an Antimicrobial

He who saves a single life, it is as though he has saved the entire world.

Talmud

Much of the information that would be included in this section has recently been reviewed (44,82,83,89), including the technical and ethical constraints associated with the in vivo studies.

It is appropriate to reiterate some factors that affect in vivo P_{O_2} and the relationship to using IOT as an antimicrobial alone, or in combination with other antimicrobials. For a detailed treatment of these subjects, it is suggested the reader consult references from which some of the ensuing material has been obtained (44,80,82,83).

The goal of hyperbaric therapy is to arrive at a partial pressure of oxygen within the tissue which damages, rather than aids the microbe, without significant negative effects on the host.

In microbes, as in mammals, oxygen toxicity is time–pressure-dependent. Dose–response versus time is critical. This time factor has important clinical implications. In therapeutics one looks for an exploitable difference in specificity and sensitivity to drugs between host and parasite. The greater the specificity of the drug for the parasite, the safer the drug is for clinical use. When a drug has more general toxicological properties, differences in concentration are important, since the concentration required to inhibit microbes is one or more orders of magitude less than the LD_{50} for humans. With oxygen it is different. As a drug, oxygen acts indiscriminately. In mammals the P_{O_2} required to raise tissue P_{O_2} to a level which will adversely affect growth and/or metabolism (including toxin production) of invading microbes is in the same range as that which gives rise to pulmonary and CNS manifestations of oxygen toxicity in humans. The exploitable difference in sensitivity in hyperbaric therapy is time. The goal is exposure to a given P_{O_2} long enough to affect microbial physiology adversely, and to give the host's defense mechanisms an opportunity to prevail, but not long enough to cause toxicity in the host. Another goal is to raise P_{O_2} sufficiently to bolster body defense mechanisms with or without simultaneously altering the physiology of a microbe.

Anatomical, physiological, and biochemical considerations determine cellular P_{O_2} in mammals. Perhaps the most practical method of elevating cellular P_{O_2} significantly is by increasing the P_{O_2} of the arterial blood ($P_{a_{O_2}}$). The $P_{a_{O_2}}$ is primarily determined by the P_{O_2} in the alveoli ($P_{A_{O_2}}$) and the ventilation/perfusion relationship of the entire lung. Assuming a normal lung, the $P_{A_{O_2}}$, and thereby the P_{O_2} of the blood in contact with the functioning alveoli, can be elevated only by altering the composition of the inspired air. Compared to air breathing, the inhalation of O_2 at 1 ATA results in an increase of the P_{O_2} of the inspired air ($P_{I_{O_2}}$) and alveolar air by factors of 4.75 and 6.6, respectively. If 100% oxygen is inhaled at 3 ATA, there is a 14.8-fold increase in the $P_{I_{O_2}}$, and a 21.5-fold elevation of the $P_{A_{O_2}}$ as compared to breathing air at 1 ATA. The theoretical consequences of breathing 100% O_2 at 3 ATA are that there is enough O_2 physically dissolved in the plasma capable of meeting the normal O_2 requirements of most tissues, and that this volume of O_2 is being delivered at a high pressure (theoretically 2193 mm Hg compared to 100 mm Hg breathing air at 1 ATA), thereby increasing the rate of diffusion of O_2 to the tissues. That O_2 is delivered via the circulatory system implies that the area for diffusion is maximal and O_2 is readily accessible to all parts of the body.

It is difficult to extrapolate from the in vitro to the in vivo situation. In vitro, depending on the technique used, the organisms can be exposed to the full P_{O_2} (82,83). In vivo, depending on the nature of the wound, location of the wound, and degree of inflammation, the diffusion of oxygen will be interfered with and the organism may never be exposed to the full P_{O_2}. As a result, there is the possibility that the IOT may enhance bacterial growth.

Attempts have been made to simulate the in vivo interaction in vitro by use of intermittent exposures (82,83). However, no systematic study has been done to show whether or not there is a correlation between in vivo results and intermittent in vitro exposures, especially with aerobic organisms.

A. Anaerobic Infections

Clostridial gas gangrene. The initial success in hyperbaric oxygen therapy of infectious diseases was obtained in the treatment of anaerobic infections, especially clostridial infections. The in vitro and in vivo interactions of oxygen and clostridia have been reviewed (82,83,89,142). As concerns in vitro studies, Hill and Osterhout (94) showed that 3 ATA O_2 is bactericidal to a variety of clostridial species in the log phase of growth, whereas spores were not affected. However, within a 2-hr exposure approximately 20–80% of the organisms, depending on the species, still survived. Whole blood tended to protect against the bactericidal effects of O_2; protection probably resulting from the presence of catalase.

It is difficult for a laboratory scientist to comment on clinical areas, especially since many factors enter into clinical studies that do not enter into animal experimentation. However, there are two controversies as concerns the use of hyperbaric oxygenation as a therapeutic modality that should be mentioned. The first controversy concerns the relative beneficial effects of hyperbaric oxygen therapy as opposed to radical surgery in combination with massive antibiotic administration; the second controversy involves the time IOT should be given, i.e., before or after debridement. Heimbach et al. (89) discussed these controversies in relation to the data they obtained on their patients. Admittedly it is difficult to compare mortality figures from different institutions; however, one cannot help but be impressed with the reduction in mortality and the extensive saving of tissue that occurs when early hyperbaric therapy is used as compared to what occurs when radical surgery and antibiotic therapy are the prime mode of therapy. The data of other clinical investigators (82,99, 103,119) support the observations of Heimbach et al. (89) that hyperbaric oxygenation be initiated prior to other forms of therapy other than perhaps those which are required for diagnosis, or extremity fasciotomy, etc.

Gottlieb (82) discussed the available experimental and clinical data bearing on the problems concerning the time of administering IOT, and concluded that the data support the view that surgery for removal of necrotic tissue be performed after IOT when the patient is nontoxemic and improved (34,89, 155). As had been pointed out, IOT not only facilitates demarcation of the necrotic tissue, but also obviates the need to perform radical surgery through healthy tissue (34,89).

It is instructive to understand the reasons behind these controversies. In the first controversy—hyperbaric oxygen therapy vs. massive antibiotic therapy and radical surgery—the opponents of IOT (2) argue that the beneficial effects have not been proved. Although they cite comparative mortality figures to substantiate their case [which the analysis of Heimbach et al. (89) shows to be questionable], it is interesting to note that these clinicians (2) admitted to having no firsthand clinical experience with hyperbaric therapy. In contrast, virtually all clinicians using hyperbaric therapy have experience with treating gas gangrene without it. The detractors of IOT (2) also raise the spector of pulmonary oxygen toxicity. Actually the extensive worldwide experience using hyperbaric therapy with a negligible incidence of O_2 toxicity tends to negate this argument.

In cases where investigators (33) have used IOT and concluded that IOT did not necessarily influence the progression of myonecrosis or change the prognosis for recovery, it is instructive to note that no details were supplied as to when IOT was given, the pressures used, the number of exposures, the frequency of exposure, how O_2 was administered, etc., even though a strong appeal had been made (82,83) and is reiterated in this chapter (vide supra) that such information accompany all discussions concerning IOT. It should be remembered that gas gangrene in humans will vary depending, inter alia, on the site of primary injury, the type of injury, the degree of injury, the state of the host defense mechanisms (nutrition, metabolic diseases, etc.) and organismal virulence.

The second controversy—over whether IOT should be administered before or after surgery—arose, in part, as a result of different therapeutic regimens being used: 2 ATA O_2 vs. 3 ATA O_2 (82). This a controversy to which a definitive resolution may never be forthcoming. It is a controversy not concerned with the efficacy of hyperbaric therapy but rather when it should be used. Those clinicians who use hyperbaric therapy are convinced that it is an important complement to other forms of therapy. Alvis (3) points out that, "we know no one these days who advocates hyperbaric oxygenation alone for gas gangrene."

Some conclusions concerning the beneficial effects of hyperbaric oxygenation were derived from studies in which O_2 was used subsequent to surgery; however, no cases were presented in which O_2 was administered prior to surgery (60).

In at least one case where doubt is cast on the efficacy of hyperbaric therapy (97), there are no details present as to how much pressure was used or how administered. The first exposure of 1 hr is one half of what is recommended.

There is a marked difference in therapeutic benefit that can be derived from the extra atmosphere of pressure at 3 ATA: diffusion of O_2 is enhanced and a greater decrease in volume and diameter of gas pockets occurs at 3 ATA than would occur at 2 ATA, thereby resulting in a smaller diffusion barrier to O_2. Clinicial and animal experimentation support the need for the higher pressure (82,83).

Part of the difficulty in assessing the time for hyperbaric therapy arises from problems in execution and interpretation of animal data and their extrapolation to humans. For example, Hitchcock et al. treat their gangrene patients at 2.5 to 3 ATA, yet their animal studies (46) are performed at 2.0 ATA. From their data they conclude that "the combination of antibiotics, surgery and OHP was the most effective form of treatment of experimental gas gangrene in guinea pigs and rabbits. . . ." It would appear that the inability to resolve the question of oxygen before or after surgery involves clinical decisions concerning the severity of the case and how early it was detected and brought for treatment. Moral constraints and the lack of sufficient numbers of human patients make statistically significant, definitive clinical studies virtually impossible to obtain.

The question of the optimal time for the application of hyperbaric therapy may not be resolved by animal studies since available animal models have some marked limitations (94,95). These include differences in the status of host defense mechanisms, site of infection, the manner of producing the infection, organismal virulence, numbers of organisms injected, and the specific O_2 pressure-duration exposure regimens, including the frequency of exposure and the time of onset of treatment following the introduction of the virulent organisms.

In assessing the efficacy of hyperbaric oxygenation in treating gas gangrene, it is important tht initial and follow-up bacteriological studies be made. As was recently pointed out (2,42,194), it is difficult to diagnose clostridial gas gangrene, since a variety of other bacteria may produce infections with gas, and since the presence of gas may be due to pathology unrelated to the infection. Because of the ubiquitious nature of clostridia, it is difficult to associate directly their presence to the active infection. The efficacy of any form of therapy, including hyperbaric oxygenation, is dependent not only on the effects of O_2 on the organism but on enhancing host defense mechanisms. Thus, as recently emphasized by Ledingham and Tehrani (121), patients' overall defense mechanisms must also be considered.

Clostridium tetani. C. tetani is anaerobic and IOT both inhibit tetanus toxin production (60) and is bactericidal (112). Yet, despite this, hyperbaric oxygenation is without any significant beneficial effect on the clinical course of tetanus (82,83). In fact tetanus intoxication was found to enhance oxygen toxicity in mice (112). Such results suggest that hyperbaric oxygenation should be considered as being contraindicated in tetanus therapy.

Non-spore-forming anaerobes. Non-spore-forming anaerobes (NSA) may be found in all types of infections; no tissue or organ appears to be immune. In fact, NSA are more commonly found in infections than are clostridia (163). Given that NSA are associated with urinary tract infections (162), and that women with recurrent urinary tract infections have large numbers of pathogenic bacteria in the vagina (55,170), one cannot help but wonder about the efficacy of IOT in treating these NSA associated infections. There are no animal or human studies which report on the effect of IOT on the long term changes in the vaginal flora, or on whether there is a resultant change in the pattern of recurrent urinary tract infections.

Similarly, no long-term studies on the effects of various increased oxygen tensions on the flora of the mouth or respiratory system have been undertaken. Paegle et al. exposed rats to 1 ATA O_2 for 14 days and noted a change in the flora of the upper respiratory tract; the flora shifted from one which is predominantly gram-positive to one which is mainly gram-negative (145). Specifically, *Streptococcus moniliformis* disappeared from the upper trachea, whereas the α-hemolytic streptococci and *Staphylococcus albus* decreased. The predominant gram-negative organisms found were species of *Pseudomonas* and *Proteus*. However, a careful reading of the data should raise the question as to whether these changes truly reflect a direct action of O_2; perhaps they result from a change in the host defense mechanisms due to the undoubted presence of pulmonary oxygen toxicity. Perhaps the changes reflect an interaction between possible altered host defense mechanisms, the organisms, and the IOT. There were no studies done to indicate the duration of the effect and the rate of return to the normal flora.

In contrast to the paucity of genitourinary tract and oral and respiratory tract studies, more information is available concerning the fecal flora (82,83). The work of Bornside et al. (14,15) and Gillmore and Gordon (74) indicates that IOT will alter the fecal flora. The extent and nature of the change is dependent on the P_{O_2}-duration relationship (15).

Paegle and Tewari (145) did not observe any significant change in the fecal flora of their 1 ATA O_2-exposed rats. This is similar to the findings of Gillmore and Gordon (74) in their 0.6–0.77 ATA exposed mice. However, it is interesting to note that in an 100% O_2 environment at 0.2 ATA total pressure, Gillmore and Gordon did find changes. Suprisingly, these investigators did not find alterations in the anaerobic flora. However, these studies failed to include the length of time the fecal flora remained altered and the rate of return, if at all, to the normal.

Further studies are required on the quantitative effects of IOT on individual organisms isolated from feces (191). Also, further in vivo studies such as those of Hill (93) are required; she showed that 2 ATA O_2 (17 daily treatments each of 3-hr duration) was efficacious in treating experimental mixed *Bacteroides–Fusobacterium* liver abscesses in mice.

An interesting aspect of hyperbaric environments is the temporary increase in susceptibility of mice to endotoxic challenge. Apparently under the quite high pressures, 20 and 35 ATA normoxic helium environments, gram-negative rods and gram-positive cocci escape the large intestine and can be found in the liver and peritoneal cavities. The mechanism whereby the organisms escape the colon under extreme hyperbaric conditions (75) is unknown.

B. Aerobic Microorganisms

The relative success of hyperbaric oxygenation in treating various anaerobic infections naturally led to questioning whether it would be as effective in treating infectious diseases associated with aerobic organisms (80,82, 83,118). A careful review of the literature revealed that there are beneficial effects of IOT on altering the course of some infections either in animal models or in humans (82,83).

Questions arise as to what is the mechanism of the beneficial effects: Are these direct effects of IOT on the organism? Is the sole effect of IOT to bolster host defense mechanisms? Or does IOT both exert a direct deleterious effect on the invading parasite as well as enhance host defense mechanisms? Yes, in certain "aerobic" diseases, IOT may well be exerting a direct toxic effect on the organism (82,83). In other diseases, e.g., osteomyelitis (82,83,131), it would appear that the P_{O_2} is of insufficient magnitude and, according to in vitro data, administered for an insufficient duration to adversely affect the organism. The conclusion is that IOT is strengthening host defense mechanisms (82, 83,89,102,104).

The nature of the host defense mechanisms that are being enhanced is still open to question; however, improvement in tissue oxygenation and possible enhancement of microbial action of leukocytes is probably involved (102,104,131,147).

To facilitate the antimicrobial action of O_2, investigations had been instituted to study the interaction of IOT with antimicrobial agents. This work has been recently reviewed in depth (82,83). As alluded to earlier, the interactions of IOT with antimicrobials may also provide insight into the mechanisms of O_2 toxicity (82,83). Such studies suggest that PABA-folic acid pathways are oxylabile.

However, there is a marked paucity of in vivo studies on the combined effects of IOT and antimicrobials. Recently, Keck and Gottlieb demonstrated a synergistic action between 2.87 ATA O_2 and sulfisoxazole on the in vitro and in vivo (goldfish) growth of *Vibrio anguillarum* (114).

Such data suggest that IOT may be used in conjunction with sulfonamides in treating various infectious diseases, especially in cases where the host defense enhancing effect of IOT may be useful, i.e., wound infections as well as venereal diseases (80,82,83,120).

The in vivo demonstration of synergism follows successfully several years of study on the in vitro interactions of IOT and sulfonamides on bacterial growth (82,83). While studying the factors involved in synergism, at least synergism involving the PABA-folic acid system, it was learned that organismal responsiveness to the sulfonamide is more important in determining the degree of synergism than is organismal responsiveness to IOT (83).

VIII. Conclusions

To produce a mighty book you choose a mighty theme.

Melville

Throughout this review I have attempted to provide insight as to where there is need for further work. The area of sites and mechanisms of oxygen toxicity is giving proof to Gottlieb's prediction (82) that "experimentation on the biochemical effects of gaseous environments should add depth to knowledge of intermediary metabolism with all its attendant ramifications." Further, comparative physiological studies of representative organisms of the various genera are required to ascertain the universality of the findings of Brown et al. The knowledge gained from such metabolic studies may also provide new insights into the development of new antimicrobial agents, some of which may act synergistically with IOT. Studying the factors that favor synergism between known antimicrobials (83) and O_2 should provide valuable insights into metabolic pathways as well as providing insights into designing new antimicrobials.

The appearance of antibiotic-resistant organisms has raised serious sociomedical concerns (18,56,116,167,186) and has increased the need for developing new or improved techniques for combating infectious diseases, i.e., vaccines, antibiotics, or agents that increase host defense mechanisms. One new approach is to use IOT in conjunction with antimicrobials. New model systems for infectious diseases have be be developed. The use of the armadillo for studying Hansen's disease (83) is a good example of just such a model. Another example is the use of the goldfish as an in vivo model for studying the interaction of IOT and antimicrobial agents (114). In addition to indicators in this article as to where research energies might be profitably placed, there are additional recommendations of a more applied clinical nature that have been discussed elsewhere (82,83).

Acknowledgments

The writing of this manuscript was supported in part by a grant from the Hearst Foundation.

References

1. Ackrell, B. A. C., and Jones, C. W. (1971). The respiratory system of *Azotobacter vinelandii*. 2. Oxygen effects. Eur. J. Biochem. 20:29–35.
2. Altemeier, W. A., and Fullen, W. D. (1971). Prevention and treatment of gas gangrene. J. Am. Med. Assoc. 217:806–813.
3. Alvis, H. J. (1971). Hyperbaric oxygen therapy of gas gangrene. J. Am. Med. Assoc. 218: 445.
4. Anderson, C. B., Marr, J. J., and Balinger, W. F. (1976). Anaerobic infection in surgery: Clinical review. Surgery 79:313–324.
5. Apontoweil, P., and Berends, W. (1975). Isolation and initial characterization of glutathione-deficient mutants of *Escherichia coli* K12. Biochim. Biophys. Acta 399:10–22.
6. Bai, N. J., Pai, M. R., Murthy, P. S., and Venkitasubramanian, T. A. (1974). Effect of oxygen tension on the aldolases of *Mycobacterium tuberculosis* H_{37} Rv. FEBS Lett. 45:68–70.
7. Barber, L. E., and Deibel, R. H. (1972). Effect of pH and oxygen tension on staphylococcal growth and enterotoxin formation in fermented sausage. Appl. Microbiol. 24:891–898.
8. Bartholomew, W. H., Karow, E. O., Sfat, M. R., and Wilhelml, R. H. (1950). Oxygen transfer and agitation in submerged fermentations. Ind. Eng. Chem. 42:1801–1809, 1810–1815.
9. Bartley, W., and Broomhead, V. M. (1972). The effect of oxygen concentration on the growth and metabolism of *Saccharomyces cerevisiae* grown with excess of potassium or in potassium-deficient media. Biochem. J. 130:251–258.
10. Battino, R. (1966). The solubility of gases in liquids. Chem. Rev. 66:395–463.
11. Boehme, D. E., Vincent, K., and Brown, O. R. (1976). Oxygen toxicity: Inhibition of amino acid biosynthesis. Nature 262:418–420.
12. Bornside, G. H. (1978). Quantitative cidal activity of hyperbaric oxygen for opportunistic yeast pathogens. Aviat. Space Environ. Med. 49:1212–1214.
13. Bornside, G. H. (1967). Exposure of *Pseudomonas aeruginosa* to hyperbaric oxygen: Inhibited growth and enhanced activity of polymyxin B. Proc. Soc. Exp. Biol. Med. 125:1152–1156.
14. Bornside, G., Cherry, G. W., and Myers, M. B. (1973). Intracolonic oxygen tension and in vivo bactericidal effect of hyperbaric oxygen on rat colonic flora. Aerospace Med. 44:1282–1286.
15. Bornside, G. H., Donovan, W. E., and Myers, M. B. (1976). Intracolonic tensions of oxygen and carbon dioxide in germfree conventional, and gnotobiotic rats. Proc. Soc. Expt. Biol. Med. 151:437–441.

16. Bornside, G. H., Pakman, L. M., and Ordonez, A. A., Jr. (1975). Inhibition of pathogenic enteric bacteria by hyperbaric oxygen: Antibacterial activity in the absence of carbon dioxide. Antimicrob. Agents Chemother. 7:682–687.
17. Broman, K., Lauwers, N., Stalen, V., and Wisme, J. M. (1978). Oxygen and nitrate in utilization by *Bacillus lichenformis* of the arginase and arginine deiminase routes of arginine catabolism and other factors affecting their syntheses. J. Bacteriol. 135:920–927.
18. Brooks, G. F., Gotschlich, E. C., Holmes, K. K., Sawyer, W. D., and Ysung, F. E. (1978). Immunobiology of *Neisseria gonorrhoeae*. Washington, D.C.: American Society for Microbiology, pp. 422.
19. Brown, C. M., and Johnson, B. (1971). Influence of oxygen tension on the physiology of *Saccharomyces cerevisiae* in continuous culture. Antonie van Leeuwenhoek J. Microbiol. Serol. 37:477–487.
20. Brown, D. E. (1970). Aeration in submerged culture of microorganisms. In: Methods in Microbiology, Vol. 2. Eds. J. R. Norris and D. W. Ribbons. New York: Academic Press, pp. 125–174.
21. Brown, O. R. (1971). Resistance to oxidative phosphorylation in *Escherichia coli* to hyperoxia. Bioenergetics 2:217–220.
22. Brown, O. R. (1972). Reversible inhibition of respiration of *Escherichia coli* by hyperoxia. Microbios 5:7–16.
23. Brown, O. R. (1972). Correlations between sensitivities to radiation and to hyperoxia in microorganisms. Radiat. Res. 50:309–318.
24. Brown, O. R. (1975). Failure of lipoic acid to protect against acute cellular oxygen toxicity in *Escherichia coli*. Microbios 14:205–217.
25. Brown, O. R., and Howitt, H. F. (1969). Growth inhibition and death of *Escherichia coli* from CO_2 deprivation. Microbios 3:241–246.
26. Brown, O. R., Howitt, H. F., Stees, J. L., and Platner, W. S. (1971). Effects of hyperoxia on composition and rate of synthesis of fatty acids in *Escherichia coli*. J. Lipid Res. 12:692–698.
27. Brown, O. R., Stees, J. L., Mills, D. F., Davis, R., and Major, S. (1969). Killing kinetics of *Escherichia coli* in a carbon dioxide deficient, pure oxygen environment. Microbios 3:267–272.
28. Brown, O. R., and Yein, F. (1978). Dihydroxyacid dehydratase: The site of hyperbaric oxygen poisoning in branch-chain amino acid biosynthesis. Biochem. Biophys. Res. Commun. 85:1219–1224.
29. Brown, O. R., and Yein, F. (1979). Sensitivity to and site of oxygen poisoning in *Escherichia coli*. Am. J. Clin. Nutr. 32:267.
30. Brown, O. R., Yein, F., and Boehme, D. (1977). Role of the "stringent response" in O_2 inhibition of protein, RNA, and DNA synthesis. Undersea Biomed. Res. 4:A18.
31. Brown, O. R., Yein, F., and Boehme, D. (1978). Bacterial sites of oxygen toxicity potentially common to red cells and erythropoiesis. In: The Red Cell. Ed. G. L. Brewer. New York: Alan R. Liss, Inc., pp. 701–714.
32. Brown, O. R., Yein, F., Mathis, R., and Vincent, K. (1977). Oxygen toxicity: Comparative sensitivities of membrane transport, bioenergetics and synthesis in *Escherichia coli*. Microbios 18:7–25.
33. Brown, P. W., and Kinman, P. B. (1974). Gas gangrene in a metropolitan community. J. Bone Jt. Surg. 56-A:1445–1454.
34. Brummelkamp, W. H. (1965). Considerations of hyperbaric oxygen therapy at three atmospheres absolute for clostridial infections type Welchii. Ann. N.Y. Acad. Sci. 117:688–699.
35. Brunker, R. L., and Brown, O. R. (1971). Effects of hyperoxia on oxidized and reduced NAD and NADP concentrations in *Escherichia coli*. Microbios 4:193–203.
36. Button, D. K., and Garver, J. C. (1966). Continuous culture of *Torulopsis utilis*: a kinetic study of oxygen limited growth. J. Gen. Microbiol. 45:195–204.
37. Cairney, W. J. (1978). Effect of hyperbaric oxygen on certain growth features of *Candida albicans*. Aviat. Space Environ. Med. 49:956–958.
38. Carpenter, D. F., and Silverman, G. J. (1974). Staphylococcal enterotoxin b and nuclease production under controlled dissolved oxygen conditions. Appl. Microbiol. 28:628–637.
39. Casciato, D. A., Stewart, P. R., and Rosenblatt, J. E. (1975). Growth curves of anaerobic bacteria in solid media. Appl. Microbiol. 29:610–614.
40. Cashel, M. (1969). The control of ribonucleic acid synthesis in *Escherichia coli* IV. Relevance of unusual phosphorylated compounds from amino acid-starved stringent

strain. J. Biol. Chem. 244:3133–3141.

41. Cashel, M., and Gullant, J. (1969). Two compounds implicated in the function of the RC gene of *E. coli*. Nature 221:838–841.
42. Chambers, C. H., Bond, G. F., and Morris, J. H. (1974). Synergistic necrotizing myositis complicating vascular injury. J. Trauma 14:980–984.
43. Choe, B. K., and Bertani, G. (1968). Osmotic fragility of bacteria deprived of carbon dioxide. J. Gen. Microbiol. 54:59–66.
44. Davis, J. C., and Hunt, T. K., Eds. (1977). Hyperbaric Oxygen Therapy. Bethesda, Maryland: Undersea Medical Society, 348 pp.
45. DeBoer, H. A., Raué, H. A., Ab, G., and Gruber, M. (1971). Role of the ribosome in stringent control of bacterial RNA synthesis. Biochim. Biophys. Acta 246:157–160.
46. Demello, F. J., Hitchcock, C. R., and Haglin, J. J. (1974). Evaluation of hyperbaric oxygen, antibiotics, and surgery in experimental gas gangrene. In: Fifth International Hyperbaric Congress Proceedings. Eds. W. G. Trapp, E. W. Bannister, J. A. Davison, and P. A. Trapp. Vancouver Simon Fraser University, 534–561.
47. Demopoulos, H. B. (1973). The basis of free radical pathology. Fed. Proc. 32:1859–1861.
48. Deutscher, M. P., and Kornberg, A. Biochemical studies on bacterial sporulation and germination. VIII. Patterns of enzyme development during growth and sporulation of *Bacillus subtilis*. J. Biol. Chem. 243:4653–4660.
49. DeVries, W., Van Wijck-Kapteijn, and Stouthamer, A. H. (1972). Influence of oxygen on growth, cytochrome synthesis and fermentation pattern in propionic acid bacteria. J. Gen. Microbiol. 71:515–524.
50. Dilworth, M. J. (1962). Oxygen inhibition in *Azotobacter vinelandii* pyruvate oxidation. Biochim. Biophys. Acta 56:127–138.
51. Dilworth, M. J., and Parker, C. A. (1961). Oxygen inhibition of respiration in *Azotobacter*. Nature 191:520–521.
52. Döbereiner, J., Day, J. M., and Dart, P. J. (1972). Nitrogenase activity and oxygen sensitivity of the *Paspalum notatum-Azotobacter paspali* association. J. Gen. Microbiol. 71:103–116.
53. Drozd, J., and Postgate, J. R. (1970). Effects of oxygen on acetylene reduction, cytochrome content and respiratory activity of *Azotobacter chroococcum*. J. Gen. Microbiol. 63:63–73.
54. Eisenberg, R. J. (1973). Introduction of unbalanced growth and death of *Streptococcus sanguis* by oxygen. J. Bacteriol. 116: 183–191.
55. Elkins, I. B., and Cox. C. E. (1974). Perineal, vaginal and urethral bacteriology of young women. I. Incidence of gram-negative colonization. J. Urol. 111:88–92.
56. Eskridge, N. K. (1978). Are antibiotics endangered resources? Bioscience 28:249–252.
57. Fahey, R. C., Brody, S., and Mikolajczyk, S. D. (1975). Changes in the glutathione thioldisulfide status of *Neurospora crassa* conidia during germination and aging. J. Bacteriol. 121:141–151.
58. Farago, D. A., and Gibbins, L. N. (1975). Variation in the activity levels of selected enzymes of *Erwinia anuylovora* 595 in response to changes in dissolved oxygen tension and growth rate of D-glucose-limited-chemostat cultures. Can. J. Microbiol. 21:343–352.
59. Fife, J. M. (1943). The effect of different oxygen concentrations on the rate of respiration of *Azotobacter* in relation to the energy involved in nitrogen fixation and assimilation. J. Agr. Res. 66:421–440.

59a. Forman, H. J., and Fisher, A. B. (1981). Anti-oxidant defenses. This volume.

60. Fredette, V. J. (1964). Action de l'oxygene hyperbare sur le bacille tetanique et sur la toxine tetanique. Rev. Can. Biol. 23:241–246.
61. Fridovich, J. (1974). Superoxide dismutases. Annu. Rev. Biochem. 44:147–159.
62. Fridovich, I. (1974). Superoxide and evolution. Horizons Biochem. Biophys. 1:1–37.
63. Fridovich, I. (1975). Oxygen: Boon and bane. Am. Sci. 63:54–59.
64. Fridovich, I. (1977). Oxygen is toxic! Bioscience 27:462–465.

64a. Fridovich, I. (1981). Superoxide radical and superoxide dismutases. This volume.

65. Friebel, T. E., and Demain, A. L. (1977). Oxygen-dependent inactivation of gramicidin S synthetase in *bacillus brevis*. J. Bacteriol. 130:1010–1016.
66. Gilbert, D. L. (1960). Speculation on the relationship between organic and atmospheric evolution. Perspect. Biol. Med. 4:58–71.

67. Gilbert, D. L. (1963). The role of prooxidants and antioxidants in oxygen toxicity. Radiat. Res. Suppl. 3:44–53.
68. Gilbert, D. L. (1964). Cosmic and geophysical aspects of the respiratory gases. In: Handbook of Physiology, Section 3: Respiration, Vol. I. Eds. W. O. Fenn and H. Rahn. Washington, D.C.: American Physiological Society, pp. 153–176.
69. Gilbert, D. L. (1964). Atmosphere and evolution. In: Oxygen in the Animal Organism. Eds. F. Dickens and E. Neil. New York: Pergamon Press, Macmillan, pp. 641–655.
70. Gilbert, D. L. Atmosphere and oxygen. (1965). Physiologist 8:9–34.
71. Gilbert, D. L. (1966). Antioxidant mechanisms against oxygen toxicity and their importance during the evolution of the biosphere. In: Proceedings of the Third International Conference on Hyperbaric Medicine, Publ. 1404. Eds. I. W. Brown, Jr., and B. G. Cox. Washington, D.C.: National Academy of Sciences, National Research Council, pp. 3–14.
72. Gilbert, D. L. (1968). The interdependence between the biosphere and the atmosphere. Resp. Physiol. 5:68–77.
73. Gilbert, D. L. (1972). Oxygen and life. Anesthesiology 37:100–111.
74. Gillmore, J. D., and Gordon, F. B. (1975). Effect of exposure to hyperoxic, hypobaric, and hyperbaric environments on concentrations of selected aerobic and anaerobic fecal flora of mice. Appl. Microbiol. 29:358–367.
75. Gillmore, J. D., and Walker, R. I. (1977). Evidence of bacteremia and endotoxemia in mice undergoing hyperbaric stress. Undersea Biomed. Res. 4:67–73.
76. Gorbach, S. L., and Bartlett, J. G. (1974). Anaerobic infections. N. Engl. J. Med. 290: 1177–1184.
77. Gorbach, S. L., and Bartlett, J. G. (1974). Anaerobic infections. N. Engl. J. Med. 290: 1237–1245.
78. Gorbach, S. L., and Bartlett, J. G. (1974). Anaerobic infections. N. Engl. J. Med. 290: 1289–1294.
79. Gosalvez, M., Garcia-Suarez, S., and Lopez-Alarcon, L. (1978). Metabolic control of glycolysis in normal and tumor permeabilized cells. Cancer Res. 38:142–148.
80. Gottlieb, S. F. (1965). Hyperbaric oxygenation. Adv. Clin. Chem. 8:69–139.
81. Gottlieb, S. F. (1966). Bacterial nutritional approach to mechanism of oxygen toxicity. J. Bacteriol. 92:1021–1027.
82. Gottlieb, S. F. (1971). Effect of hyperbaric oxygen on microorganisms. Annu. Rev. Microbiol. 25:111–152.
83. Gottlieb, S. F. (1977). Oxygen under pressure and microorganisms. In: Hyperbaric Oxygen Therapy. Eds. J. C. Davis and T. K. Hunt. Bethesda, Maryland: Undersea Medical Society, pp. 79–99.
84. Gottlieb, S. F., and Pakman, L. M. (1968). Effect of high oxygen tensions on the growth of selected, aerobic, gram-negative bacteria. J. Bacteriol. 95:1003–1010.
85. Gregory, E. M., Moore, W. E. C., and Holdeman, L. V. (1978). Superoxide dismutase in anaerobes: Survey. Appl. Environ. Microbiol. 35:988–991.
86. Harley, J. B., Santangelo, G. M., Rasmussen, H., and Goldfine, H. (1978). Dependence of *Escherichia coli* hyperbaric oxygen toxicity on the lipid acyl chain composition. J. Bacteriol. 134:808–820.
87. Harrison, D. E. F. (1977). The regulation of respiration rate in growing bacteria. Adv. Microbiol. Physiol. 14:243–313.
88. Harrison, D. E. F., and Pirt, S. J. (1967). The influence of dissolved oxygen concentration on the respiration and glucose metabolism of *Klebsiella aerogenes* during growth. J. Gen. Microbiol. 46:193–211.
89. Heimbach, R. D., Boerema, I., Brummelkamp, W. H., and Wolfe, W. G. (1977). Current therapy of gas gangrene. In: Hyperbaric Oxygen Therapy. Eds. J. C. Davis and T. K. Hunt. Bethesda, Maryland: Undersea Medical Society, pp. 153–176.
90. Hempfling, W. P., and Mainzer, S. E. (1975). Effects of varying the carbon source limiting growth on yield and maintenance characteristics of *Escherichia coli* in continuous culture. J. Bacteriol. 123:1070–1087.
91. Hendricks, R. D. (1977). Effects of increased partial pressures of oxygen on the growth of bacteria. Thesis. Purdue University, West Lafayette, Indiana.
92. Hibberd, K. A., Berget, P. B., Warner, H. R., and Fuchs, J. A. (1978). Role of glutathione in reversing the deleterious effects of a thiol-oxidizing agent in *Escherichia coli*. J. Bacteriol. 133:1150–1155.
93. Hill, G. B. (1976). Hyperbaric oxygen exposures for intrahepatic abscesses produced in mice by nonsporeforming anaero-

bic bacteria. Antimicrob. Agents Chemother. 9:312–317.
94. Hill, G., and Osterhout, S. (1972). Experimental effects of hyperbaric oxygen on selected clostridial species. I. In vitro studies. J. Infect. Dis. 125:17–25.
95. Hill, G. B., and Osterhout, S. (1972). Experimental effects of hyperbaric oxygen on selected clostridial species. II. In vivo studies in mice. J. Infect. Dis. 125:26–35.
96. Hill, G. B., and Osterhout, S. (1973). Exposure to hyperbaric oxygen not beneficial for murine tetanus. J. Infect. Dis. 128: 238–242.
97. Himal, H. S., McLean, A. P. H., and Duff, J. H. (1974). Gas gangrene of the scrotum and perineum. J. Surg. Gynecol. Obstet. 139:176–178.
98. Hinton, D. (1947). A method for the arrest of spreading gas gangrene by oxygen injection. Am. J. Surg. 73:228–232.
99. Hitchcock, C. R. F., Demello, J., and Haglin, J. J. (1975). Gangrene infection, new approach to an old infection. Surg. Clin. N. Am. 55:1403–1410.
100. Hoekstra, W. G. (1975). Biochemical function of selenium and its relation to vitamin E. Fed. Proc. 34:2081–2089.
101. Hoge, S. F. (1932). Oxygen in gas infection. J. Arkansas Med. Soc. 28:4–9.
102. Hohn, D. C. (1977). Oxygen and leukocyte microbial killing. In: Hyperbaric Oxygen Therapy. Eds. J. C. Davis and T. K. Hunt. Bethesda, Maryland: Undersea Medical Society, pp. 101–110.
103. Holland, J. A., Hill, G. B., Wolfe, W. G., Osterhout, S., Saltzman, H. A., and Brown, J. W., Jr. (1975). Experimental and clinical experience with hyperbaric oxygen in the treatment of clostridial myonecrosis. Surgery 77:75–85.
104. Hunt, T. K., Niinikoski, J., Zederfeldt, B. H., and Silver, I. A. (1977). Oxygen in wound healing enhancement: Cellular effects of oxygen. In: Hyperbaric Oxygen Therapy. Eds. J. C. Davis and T. K. Hunt. Bethesda, Maryland: Undersea Medical Society, pp. 111–122.
105. Irvin, T. T., Norman, J. N., Suwanagul, A., and Smith, G. (1966). Hyperbaric oxygen in the treatment of infections by aerobic microorganisms. Lancet I:392–394.
106. Irvin, T. T., and Smith, G., (1968). Treatment of bacterial infections with hyperbaric oxygen. Surgery 63:363–376.
107. Ishizaki, A., Shibai, H , Hirose, Y., and Shiro, T. (1973). Estimation of aeration and agitation in respect to oxygen supply and ventilation. Agr. Biol. Chem. 37:107–113.
108. Jacobs, N. J., Jacobs, J. M., and Morgan, H. E., Jr. (1972). Comparative effect of oxygen and nitrate on protoporphyrin and heme synthesis from Δ-amino levulinic acid in bacterial cultures. J. Bacteriol. 112:1444–1445.
109. Jetschmann, K., Solomonson, L. P., and Vennesland, B. (1972). Activation of nitrate reductase by oxidation. Biochim. Biophys. Acta 275:276–278.
110. Johnson, M. J. (1967). Areobic microbial growth at low oxygen concentrations. J. Bacteriol. 94:101–108.
111. Jones, C. W., Brice, J. M., Wright, V., and Ackrell, B. A. C. (1973). Respiratory production of nitrogenase in *Azotobacter vinelandii*. FEBS Lett. 29;77–81.
112. Kaye, D. (1964). Effect of hyperbaric oxygen on bacteria in vitro and in vivo. Clin. Res. 12:454.
113. Kaye, D. (1967). Effect of hyperbaric oxygen on aerobic bacteria in vitro and in vivo. Proc. Soc. Expt. Biol. Med. 124:1090–1093.
114. Keck, P. E., and Gottlieb, S. F. (1980). Interaction of increased pressures of oxygen and sulfonamides on the in vitro and in vivo growth of pathogenic bacteria. Undersea Biomed. Res. 7:95–106.
115. Klucas, R. (1977). Nitrogen fixation by *Klebsiella* grown in the presence of oxygen. Can. J. Microbiol. 18:1845–1850, 1972.
116. Koppes, G. M., Ellenbogen, C., and Gebbhart, R. J. (1977). Group Y meningococcal disease in United States Air Force recruits. Am. J. Med. 62:661–666.
117. Kosower, N. S., and Kosower, E. M. (1974). Effects of GSSG on protein synthesis. In: Glutathione. Proc. of the Gen. Soc. Biol. Chem. Eds. L. Floke, H. Ch. Benohr, H. Sies, H. D. Walker, and A. Wendel. New York: Academic Press, pp. 276–286.
118. Kozinin, P. J., Lynfield, J., and Seelig, M. S. (1974). successful treatment of systemic candidiasis following cardiac surgery. Am. J. Dis. Child. 128:106–108.
119. Lambertşen, C. J. (1972). Oxygen in the therapy of gas gangrene. J. trauma 12:825–827.
120. Lassus, A., and Renkonen, O. V. (1979).

Short-term treatment of gonorrhoea with intramuscular and oral forms of trimethoprim-sulphamethoxazole. Br. J. Vener. Dis. 55:24–25.

121. Ledingham, J. McA., and Tehrani, M. A. (1975). Diagnosis, clinical course and treatment of acute dermal gangrene. Br. J. Surg. 62:364–372.

122. Linton, J. D., Harrison, D. E., and Bull, A. E. (1977). Molar growth yields, respiration and cytochrome profiles of *Benekea natriegens* when grown under carbon limitation in a chemostat. Arch. Microbiol. 115: 135–142.

123. Lobo, Z., and Miatra, P. K. (1977). Physiological role of glucose-phosphorylating enzymes in *Saccharmoyces cerevisiae*. Arch. Biochem. Biophys. 182:639–645.

124. Loesche, W. J. (1969). Oxygen sensitivity of various anaerobic bacteria. Appl. Microbiol. 18:723–727.

125. Longmuir, I. S. (1954). Respiration rate of bacteria as a function of oxygen concentration. Biochem. J. 57:81–87.

126. Longmuir, I. S., Gottlieb, S. F., Pashko, L., and Martin, P. (1978). In vivo rate of induction of cytochrome P-450 synthesis in hyperoxia. Fed. Proc. 37:430.

127. MacGregor, C. H., and Bishop, C. W. (1977). Do cytochromes function as oxygen sensors in the regulation of nitrate reductase biosynthesis? J. Bacteriol. 131:372–373.

128. MacLennan, D. G. (1970). Principles of automatic measurement and control of fermentation growth parameters. In: Methods of Microbiology, Vol. 2. Eds. J. R. Norris and D. W. Ribbons. New York: Academic Press, pp. 1–21.

129. MacLennan, D. G., and Pirt, S. J. (1966). Automatic control of dissolved oxygen concentration in stirred microbial cultures. J. Gen. Microbiol. 45:289–302.

130. Mader, J. T., Guckian, J. C., Glass, D. L., and Reinarz, J. A. (1978). Therapy with hyperbaric oxygen for experimental osteomyelitis due to *Staphylococcus aureus* in rabbits. J. Infec. Dis. 138:312–318.

131. Mainous, E. G. (1977). Hyperbaric oxygen in maxillofacial osteomyelitis, osteoradionecrosis, and osteogenesis enhancement. In: Hyperbaric Oxygen Therapy. Eds. J. C. Davis and T. K. Hunt. Bethesda, Maryland: Undersea Medical Society, pp. 191–216.

132. Marquis, R. E., Thom, S. F., and Crookshank, C. A. (1978). Interactions of helium, oxygen, and nitrous oxide affecting bacterial growth. Undersea Biomed. Res. 5:189–198.

133. Martin, W. J. (1974). Isolation and identification of anaerobic bacteria in the clinical laboratory. Mayo Clinic Proc. 49:300–308.

134. Mathis, R. R., and Brown, O. R. (1976). ATP concentration in *Escherichia coli* during oxygen toxicity. Biochim. Biophys. Acta 440:723–732.

135. McCord, J., Keele, B. B., Jr., and Fridovich, J. (1971). An enzyme-based theory of obligate anaerobiosis: The physiological function of superoxide dismutase. Proc. Nat. Acad. Sci. U.S.A. 68:1024–1027.

136. McDaniel, L. E., Bailey, E. G., and Zimmerli, A. (1965). Effect of oxygen-supply rates on growth of *Escherichia coli*. I. Studies in umbaffled and baffled shake flasks. Appl. Microbiol. 13:109–114.

137. McDaniel, L. E., and Bailey, E. G., and Zimmerli, A. (1965). Effects of oxygen-supply rates on growth of *Escherichia coli* II. Comparison of results in shake flasks, and 50-liter fermentor. Appl. Microbiol. 13: 115–119.

138. Meyer, D. J., and Jones, C. W. (1973). Reactivity with oxygen of bacterial cytochrome oxidases a_1, aa_3, and o. FEBS Lett. 33:101–105.

139. Meyerhof, O., Burk, O., and Iwasaki, K. (1928). Über die Fixation des Luftstickstoffs durch *Azotobacter*. Z. Phys. Chem. (A) 139:117–142.

140. Morris, J. G. (1975). The physiology of obligate anaerobiosis. Adv. Microbiol. Physiol. 12:169–246.

141. Morris, J. G. (1976). Oxygen and the obligate anaerobe. J. Appl. Bacteriol. 40:229–244.

142. Morris, J. G., and O'Brien, R. W. (1971). Oxygen and clostridia: A review. In: Spore Research. Eds. A. N. Barker, G. W. Gould, and J. Wolf. London: Academic Press, pp. 1–37.

143. Nagodawithana, T. W., Castellano, C., and Steinkraus, K. (1974). Effect of dissolved oxygen, temperature, initial cell count and sugar concentration on the viability of *Saccharomyses cerevisiae* in rapid fermentations. Appl. Microbiol. 28:383–391.

144. Okinaka, R. T., and Dobrogosz, W. J. (1967). Catabolite repression and the pasteur effect in *Escherichia coli*. Arch. Biochem. Biophys. 120:451–453.

145. Paegle, R. D., Tewari, R. P., Bernhard, W. N., and Peters, E. (1976). Microbial flora of the larynx, trachea, and large intestine of the rat after long-term inhalation of 100 percent oxygen. Anesthesiology 44:287–290.
146. Parker, C. A., and Scutt, P. B. (1960). The effect of oxygen on nitrogen fixation by *Azotobacter*. Biochim. Biophys. Acta 38: 230–238.
147. Perrins, D. J. D., and Davis, J. C. (1977). Enhancement of healing in soft tissue wounds. In: Hyperbaric Oxygen Therapy. Eds. J. C. Davis and T. K. Hunt. Bethesda, Maryland: Undersea Medical Society, pp. 229–248.
148. Pichinoty, F. (1965). L'effect oxygene et la biosynthese des enzymes d'oxydoreduction bacteriens. C.N.R.S. Symp. 124:507–522.
149. Pirt, S. J. (1957). The oxygen requirement of growing cultures of an *Aerobacter* species determined by the continuous culture technique. J. Gen. Microbiol. 16:54–75.
150. Postgate, J. (1971). Relevant aspects of the physiological chemistry of nitrogen fixation. Symp. Soc. Gen. Microbiol. 21:287–307.
151. Repaske, R., Repaske, A. C., and Mayer, R. D. (1974). Carbon dioxide control of lag period and growth of *Streptococcus sanguis*. J. Bacteriol. 117:652–659.
152. Rice, C. W., and Hempfling, W. P. (1978). Oxygen-limited continuous culture and respiratory energy conservation in *Escherichia coli*. J. Bacteriol. 134:115–124.
153. Robb, S. M. (1966). Reactions of fungi to exposure to 10 atmospheres pressure of oxygen. J. Gen. Microbiol. 45:17–29.
154. Roberts, R. B., Abelson, P. H., Cowie, D. B., Bolton, E. T., and Britten, R. J. (1955). Studies of biosynthesis in *Escherichia coli*. Carnegie Inst. Washington Publ. 607:318.
155. Roding, B., Groenveld, H. A., and Boerema, I. (1972). Ten years of experience in the treatment of gas gangrene with hyperbaric oxygen. Surg. Gynecol. Obstet. 134:579–585.
156. Rosebury, R. (1962). Microorganisms Indigenous to Man. New York: McGraw-Hill.
157. Sapshead, L. M., and Wimpenny, J. W. T. (1972). The influence of oxygen and nitrate on the formation of the cytochrome pigments of the aerobic and anaerobic respiratory chain of *Micrococcus denitrificans*. Biochim. Biophys. Acta 267:388–397.
158. Schmid, G., Lüdemann, H. O., and Jaenicke, R. (1978). Oxidation of sulfhydryl groups in lactate dehydrogenase under high hydrostatic pressure. Eur. J. Biochem. 86: 219–224.
159. Schmidt-Lorenz, W., and Rippel-Baldes, A. (1957). Wirkung des Sauerstoff-Partialdracks auf Wachstum und Stickstoffbindung von *Azotobacter chroococcum beiji*. Arch. Mikrobiol. 28:45–68.
160. Scholander, P. F. (1960). Oxygen transport through hemoglobin solutions. Science 131: 585–590.
161. Schwartz, A. C. (1973). Anaerobiosis and oxygen consumption of some strains of *Propionibacterium* and a modified method for comparing the oxygen sensitivity of various anaerobes. Z. Microbiol. 13:681–691.
162. Segura, J. W., Kelalis, P. P., Martin, W. J., and Smith, L. H. (1972). Anaerobic bacteria in the urinary tract. Mayo Clin. Proc. 47:30–33.
163. Sencer, D. J. and Staff of Laboratory Division and Epidemiology Program. (1971). Emerging diseases of man and animals. Annu. Rev. Microbiol. 25:465–486.
164. Shah, V. K., Pate, J. L., and Brill, W. (1973). Protection of nitrogenase in *Azotobacter vinelandii*. J. Bacteriol. 115: 15–17.
165. Silver, I. Oxygen tension in the clinical situation. This volume.
166. Skowe, M. K., and DeMoss, J. (1968). Location and regulation of synthesis of nitrate reductase in *Escherichia coli*. J. Bacteriol. 95:1305–1313.
167. Smilak, J. D. (1974). Group-Y meningococcal disease. Ann. Intern. Med. 81:740–745.
168. Sperry, J. F., and Wilkins, T. O. (1976). Cytochrome spectrum of an obligate anaerobe, *Eubacterium lentum*. J. Bacteriol. 125:905–909.
169. St. John, R. T., Shah, V. K., and Brill, W. J. (1974). Regulation of nitrogenase synthesis by oxygen in *Klebsiella pneumoniae* J. Bacteriol. 119:266–264.
170. Stanley, T. A. (1973). The role of introital enterobacteria in recurrent urinary infections. J. Urol. 109:467–472.
171. Starr, S. E., Thompson, F. S., Dowell, V. R., Jr., and Balow, A. (1973). Micromethod system for identification of anaerobic bacteria. Appl. Microbiol. 25:713–717.

172. Stees, J. L., and Brown, O. R. (1973). Stability of yeast fatty acid synthetase component enzymes to irreversible inactivation by hyperbaric oxygen. Microbiology 8:247–256.
173. Stees, J. L., and Brown, O. R. (1973). Susceptibilities of intracellular and surface sulphydryl groups of *Escherichia coli* to oxidation by hyperoxia. Microbios 7:257–266.
174. Stuart, B., Gerschman, R., and Stannard, J. N. (1962). Effect of high oxygen tension on potassium retentivity and colony formation bakers' yeast. J. Gen. Physiol. 45:1019–1030.
175. Sutter, V. L., Kwok, Y. Y., and Finegold, S. M. (1972). Standardized antimicrobial disc susceptibility testing of anaerobic bacteria. I. Susceptibiity of *Bacterioides fragilis* to tetracycline. Appl. Microbiol. 23:268–275.
176. Tally, F. P., Stewart, P. R., Sutter, V. L., and Rosenblatt, J. E. (1975). Oxygen tolerance of clinical anaerobic bacteria. J. Clin. Microbiol. 1:161–164.
177. Tappel, A. (1963). Vitamin E as the biological lipid antioxidant. Vitamins Hormones 20:493–510.
178. Tappel, A. L., Forstrom, J. W., Zakowitz, J. J., Lyons, D. E., and Hawkes, W. G. (1978). The catalytic site of rat liver glutathione peroxidase as selenocysteine; and selenocysteine in rat liver. Fed. Proc. 37: 706.
179. Tempest, D. W. (1970). The continuous cultivation of microorganisms: 1. Theory of the chemostat. In: Methods in Microbiology, Vol. 2. Eds. J. R. Norris and D. W. Ribbons. New York: Academic Press, pp. 259–276.
180. Thomas, A. D., Doelle, H. W., Westwood, A. W., and Gordon, G. L. (1972). Effect of oxygen on several enzymes involved in the aerobic and anaerobic utilization of glucose in *Escherichia coli*. J. Bacteriol. 112:1099–1105.
181. Thomas, J., Jr., and Neptune, E. M., Jr. (1963). Chemical mechanisms in oxygen toxicity. In: Proceedings of the Second Symposium on Underwater Physiology. Eds. C. J. Lambertsen and L. J. Greenbaum. Washington, D.C.: National Research Council, pp. 139–151.
182. Thomas, J., Jr., Neptune, E. M., Jr. and Sudduth, H. C. (1963). Toxic effects of oxygen at high pressure on the metabolism of D-glucose by dispersions of rat brain. Biochem. J. 88:31–45.
183. Tschapek, M., and Giambiagi, N. (1955). Nitrogen fixation of Azotobacter in soil—its inhibition by oxygen. Arch. Mikrobiol. 21:376–390.
184. Turnbaugh, C. L., Jr., and Switzer, R. L. (1975). Oxygen-dependent inactivation of glutamine phosphoribosylpyrophosphate amidotransferase in stationary-phase cultures of *Bacillus subtilis*. J. Bacteriol. 121: 108–114.
185. Turnbaugh, C. L., Jr., and Switzer, R. L. (1975). Oxygen-dependent inactivation of glutamine phosphoribosylpyrophosphate amidotransferase in vitro: Model for in vivo inactivation. J. Bacteriol. 121:115–120.
186. U.S. Department of Health, Education and Welfare/Public Health Service, Center for Disease Control. (1978). Follow-up on multiple-antibiotic resistant pneumonococci-South Africa. Morbidity Mortality Weekly Rep. 27:1.
187. Van Der Weil-Korstanje, J. A. A., and Winkler, K. C. (1970). Medium for differential count of the anaerobic flora in human feces. Appl. microbiol 20:168–169.
188. VanHemert, P. (1974). Application of measurement and control of dissolved oxygen pressure to vaccine production. Biotechnol. Bioeng. Symp. 4:741–753.
189. Vaughn, G. L., and Wilbur, K. M. (1970). Oxygen toxicity in a fission yeast. Fed. Proc. 29:845.
190. Waindle, L. M. and Switzer, R. L. (1973). Inactivation of aspartic transcarbamylase in sporulating *Bacillus subtilis*: Demonstration of a requirement for metabolic energy. J. Bacteriol. 114:517–527.
191. Walden, W. C., and Hentges, D., Jr. (1975). Differential effects of oxygen and oxidation-reduction potential on the multiplication of three species of anaerobic intestinal bacteria. Appl. Microbiol. 30:781–785.
192. Wallace, P. G., and Linnane, A. W. (1964). Oxygen-induced synthesis of yeast mitochondria. Nature 201:1191–1194.
193. Webster, D. A., and Orii, Y. (1978). Physiological role of oxygenated cytochrome O: Observations on whole-cell suspensions of *Vitreoscilla*. J. Bacteriol. 135:62–67.

194. Weinstein, L., and Barza, M. A. (1973). Gas gangrene. N. Eng. J. Med. 289:1129–1131.
195. White, D. C. (1962). Factors affecting the affinity for oxygen of cytochrome oxidases in *Hemophilus parainfluenzae*. J. Biol. Chem. 11:3757–3761.
196. White, J. H. (1928). Report of case of gas gangrene treated with oxygen injection and peroxide. J. Oklahoma State Med. Assoc. 21:59–60.
197. Wilson, W. R., Martin, W. J., Wikowske, C. J., and Washington, J. A., II. (1972). Anaerobic bacteremia. Mayo Clin. Proc. 47:639–646.
198. Wimpenny, J. W. T. (1969). Oxygen and carbon dioxide as regulators of microbial growth and metabolism. Symp. Soc. Gen. Microbiol. 19:161–197.
199. Wimpenny, J. W. T., and Cole, J. A. (1967). The regulation of metabolism in facultative anaerobes. III. The effect of nitrate. Biochim. Biophys. Acta 148:233–242.
200. Wimpenny, J. W. T., and Necklen, D. K. (1971). The redox environment and microbial physiology. I. The transition from anaerobiosis to aerobiosis in continuous culture of facultative anaerobes. Biochim. Biophys. Acta 253:352–359.
201. Wittenberg, J. B. (1959). Oxygen transport–new function proposed for myoglobin. Biol. Bull. 117:402–403.
202. Yates, M. G. (1972). The effect of ATP upon the oxygen sensitivity of nitrogenase from *Azotobacter chroococcum*. Eur. J. Biochem. 29:386–392.

8

Oxygen Exchange in the Metazoa

JOHANNES PIIPER AND PETER SCHEID

In unicellular organisms and in the smallest metazoa, diffusion constitutes the sole mechanism by which oxygen (O_2) is transported from the environment into the cells. In all higher metazoa, however, O_2 supply by diffusion would be insufficient, and specialized gas exchange organs and transport systems have been developed to secure the O_2 demand of the animal, particularly in situations of elevated demand.

This brief review intends to survey the mechanisms by which oxygen is transported from the environment into the body of animals. We have tried to recognize the general principles underlying the diversity of the O_2 transport mechanisms. Working of these principles in O_2 transport is mainly considered in vertebrates, for which sufficient morphological and physiological information is available to allow a quantitative analysis of gas exchange on the basis of functional models. Lack of pertinent information precludes similar analysis of invertebrate gas exchange. The analysis is extended to include the other respiratory gas, carbon dioxide (CO_2), exchange of which is tightly coupled with that of O_2.

Many important aspects of gas exchange will not be considered, e.g., mechanisms and energetics of ventilation and circulation; their adjustment to the varying internal requirement and varying external O_2 availability; mechanisms of O_2 exchange in tissues; chemistry of O_2 combination in blood; biochemistry of oxidative metabolism. The interested reader is referred to a number of recent monographs on various aspects of comparative physiology of respiration (Hughes, 1963; Irving, 1964; Negus, 1965; Steen, 1971; Jones, 1972; Mill, 1972; Prosser, 1973; Dejours, 1975; Schmidt-Nielsen, 1979; Tenney, 1979).

I. Transport of Oxygen by Diffusion

A. Diffusion Inside the Body

Consider a uniform, flat barrier, of area, F, and thickness, l, which separates compartments 1 and 2, with gas partial pressures P_1 and P_2 (Fig. 1). The diffusional flux of the gas under study through F, $\dot{M}$, in steady state is given by Fick's law of diffusion

$$\dot{M} = K \frac{F}{l}(P_1 - P_2) \quad (1)$$

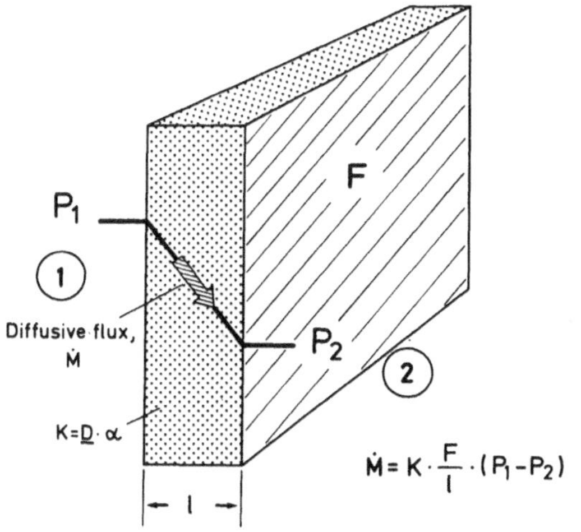

Figure 1. Simple model for diffusive flux at steady state.

in which $K = D \times \alpha$ contitutes Krogh's diffusion constant (D, diffusion coefficient of the gas in the barrier; α, solubility of the gas in the barrier). This equation results from integration of the general, time-dependent diffusion equation for the conditions specified. It is applicable to O_2 as to any other gas.

Equation (1) may be used to describe O_2 flux through a barrier (e.g., skin, tissue barrier in exchange organs), which itself does not consume O_2. However, for animals that rely entirely on diffusion for their O_2 demand, the oxygen consumption of the tissue through which diffusion occurs must be considered. In steady state, the more general diffusion equation applies

$$\dot{M}/F = K\frac{dP}{dx} \tag{2}$$

which states that the diffusive flux in x-direction, $\dot{M}/F$ (substance flow per unit area), at any site in the system, x, is proportional to the partial pressure gradient, dP/dx, at this site.

Equation (2) may be solved for geometrically simple organs, with homogeneously distributed O_2 consumption, to calculate the partial pressure profile of O_2 within the body. This is schematically shown in Fig. 2 for an unlimited flat sheet of half-thickness R (Fig. 2A) for an endless cylinder of radius R (Fig. 2B) and for a sphere of radius R (Fig. 2C).

Of particular interest is the O_2 partial pressure drop, ΔP, from the outer body surface to the center of the body, which may be obtained by integrating Eq. (2) (see Fig. 2). Evidently, there is an upper limit to the dimension R, R_{max}, which is reached when P_{O_2} in the center drops to zero. Integration of Eq. (2) yields for this upper limit R_{max}.

$$R_{max} = \sqrt{2n\frac{K}{a}P_0} \tag{3}$$

in which the parameter a is the specific O_2 consumption (= O_2 consumption per unit tissue volume). The partial pressure of O_2 at the body surface for which P_{O_2} just reaches zero in the center, P_0, is usually referred to as critical P_{O_2}.

The shape parameter, n, in Eq. (3) attains the following values: endless flat sheet, $n = 1$; endless cylinder, $n = 2$; sphere, $n = 3$.

Equation (3) allows to calculate the upper size limit for an animal that is supplied with O_2 exclusively by diffusion. A "typical" value for specific O_2 consumption for a small aquatic animal is: $a = 9 \times 10^{-5}$mmol cm^{-3} min^{-1} (turbellarian Crenobia alpina, 10 mg body weight, at 14.5°C; after Altmann and Dittmer, 1971). For Krogh's diffusion constant for O_2 in tissue, half the value in water may be used (cf. Altman and Dittmer, 1971): $K = 9 \times 10^{-7}$ mmol min^{-1} cm^{-1} atm^{-1}. These values result in the following upper size limits when P_0 is 0.2 atm: infinite sheet, $R_0 = 0.6$ mm; infinite cylinder, $R_0 = 0.9$ mm; sphere, $R_0 = 1.1$ mm.

The specific O_2 consumption, a, decreases in larger animals with increasing body weight (W) according to the allometric relationship (cf. Kleiber, 1961):

$$\log a = m + (b - 1) \log W \tag{4}$$

in which $b \approx 0.75$ (m = constant). Nonetheless, diffusion can only satisfy the O_2 demand of organisms for which the smallest body thickness does not exceed a few millimeters.

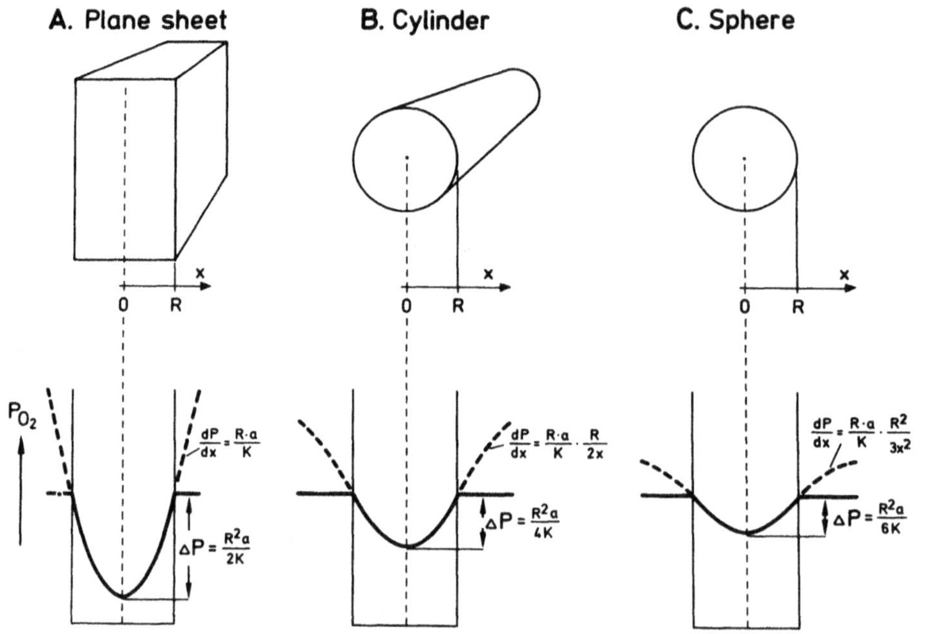

Figure 2. Partial pressure profiles of O_2 in models with O_2 consumption. The P_{O_2} profiles inside the body are parabolic, ΔP representing the total drop from outer surface to center. ΔP is related to geometry (R), to specific O_2 consumption, a, and to Krogh's diffusion constant for O_2 in the tissue, K. In the environment surrounding the body, P_{O_2} is constant if this medium is stirred, but increases if the medium is still.

The calculated results are valid only for the models and values assumed. In reality, shapes are complex, the integument of the animal may attain low K_{O_2} values, and the O_2 consumption may be unevenly distributed.

Convection contributes to some degree to O_2 transport in all animals including those lacking a circulatory system, since movement of the body wall may result in movement of interstitial fluid, and plasma streaming may exist inside cells. These mechanisms would increase the critical body size to some extent.

B. Diffusion in the Environmental Medium

It was assumed in these estimates that a surface P_{O_2}, in this case $P_0 = 0.2$ atm, is constant throughout the medium. This would, however, imply that the surrounding medium is well mixed. If, on the other hand, the organism dwells in stagnant water, then the O_2 has to diffuse through the medium toward the organism giving rise to diffusional gradients within the medium.

Equation (2) yields the gradients of P_{O_2}, dP/dx, as a function of distance, x, from the body surface, $x = R$:

$$\frac{dP}{dx} = \frac{Ra}{K}\,\frac{1}{n}\left(\frac{R}{x}\right)^{n-1} \tag{5}$$

where the symbols are those explained above.

The P_{O_2} profiles that result from Eq. (5) for the three geometrical bodies are indicated in Fig. 2 as dashed lines, but are more clearly visualized in Fig. 3, in which the P_{O_2}, relative to the critical P_{O_2} of the plane sheet, is plotted against the distance, in units of R, for the plane sheet (A), the cylinder (B), and the sphere (C). (K has been assumed to be twice the value for tissue; see above.) To the left of the abscissa value $x/R = 1$ (which represents the body surface) are the (parabolic) P_{O_2}

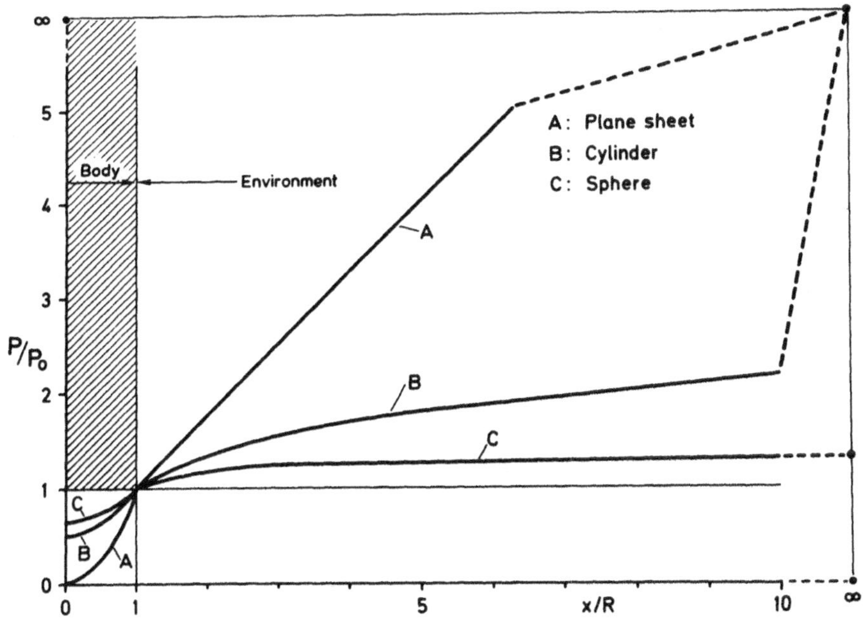

Figure 3. Plot of P_{O_2} profiles (in units of the critical P_{O_2} of the plane sheet, P_o) against distance from center of the body (in units of R). The profiles are given for the three geometrical models of Fig. 2, and for the body interior ($0 \leq x/R \leq 1$) and the environment ($x/R < 1$). Values of P/P_o attained for $x/R \rightarrow \infty$ are indicated.

profiles inside the body; to the right, those in the surrounding medium. For the plane sheet, P_{O_2} rises linearily, while the increase for the other two is less. This is because the density of the O_2 flux for the latter shapes decreases with distance. Whereas both for the plane sheet and the cylinder, P_{O_2} increases without limit, there is a finite value of P_{O_2} for the sphere. This means that only a spherical body can resort to supply of O_2 by diffusion in stagnant water. However, whereas the upper size limit R_{max} in well-stirred medium is about 1.1 mm (see example given above) it drops to below 0.8 mm in absolutely stagnant water.

It follows that the resistance to diffusion residing in the water shell surrounding the animal is an important factor limiting the availability of O_2. In most cases O_2 requirements cannot be met in still water due to diffusive resistance of water. As a matter of fact, any aquatic organism requiring O_2 must make use of convectional transport in the environmental water. Such convective mixing is brought about by movement of cilia or flagellae or by muscular activity moving the body relative to the environment; mixing can also be achieved by convective currents produced by extraneous agents in the aqueous environment.

Since K_{O_2} for air is about 200,000 times higher than for water, convective mixing of the environment is a much less stringent requirement for O_2 uptake in air-breathing animals. Nonetheless, considerable O_2 gradients may occur in gas-filled respiratory organs (e.g., lungs of vertebrates) and may persist despite ventilatory mixing. This is due to the relatively small cross-sectional area of the proximal, conducting airways in relation to the distal end where gas exchange occurs (cross-sectional area of human trachea, 2.5 cm^2; of terminal bronchioles, about 180 cm^2; alveolar surface area about 1,000,000 cm^2). The conducting airways thus display a bottleneck for O_2 flux, which decreases sharply toward the surface through which gas exchange occurs between medium and organism (cf. Section X).

II. Gas Exchange Organs: Structural Features

To secure O_2 supply for all body cells in larger organisms, specialized gas exchange organs have developed. In general, these organs exhibit an enlarged surface area of a very thin tissue barrier that constitutes the surface in this part of the body. Figure 4 shows some general types of gas exchange organs. In Fig. 4A the situation is shown where O_2 flux from the body surface to the cell is by diffusion alone (dashed arrow). This is the case considered in the previous section.

Specialized gas exchange organs are shown in Fig. 4C and E, where an increase in the surface area available for O_2 uptake is by outfolding (evagination, Ç), termed gills, or by infolding (invagination, E), leading to lungs or tracheal systems. In the types A, C, and E, O_2 flux is entirely by diffusion, within the medium, through the medium/body interface and within the organism. In all larger organisms special convective transport mechanisms have evolved, constituting the circulatory system, in which blood or hemolymph acts as a carrier for O_2 from the gas exchange surface to the O_2-consuming cell. This is shown in the models B, D, and F. This convective system shortens considerably the distance that has to be overcome by diffusion and thus removes the size constraint imposed by diffusion.

In type D the gill sticks into the medium, thus partly removing the problem imposed by unstirred medium. This exposure does have a risk, it renders the thin gas exchange epithelium rather vulnerable to mechanical injury. In this respect, the invaginated lung is much better protected. However, the long distance between the environment and the gas exchange epithelium makes convection of the medium imperative.

Not all systems of Fig. 4 are realized. In particular, the gill type without circulatory system (C) would create even larger diffusion distances than in A and is thus not feasible. On the other hand, not all respiratory organs can clearly be categorized into the types of Fig. 4. However, in most cases there exists the serial arrangement for O_2 transport by (a) medium convection (ventilation; to some degree diffusion through medium); (b) diffusion through tissue barrier from medium to body fluid; (c) convection by body fluid. The limiting role of the serial steps may vary greatly. An extreme situation is realized in insects where O_2 is transported, by convection and diffusion, within the air-filled tracheal system whose last branches, the tracheoles, reach, or even penetrate, individual body cells. The circulatory system in these animals plays no role for transport of respiratory gases (respiratory pigments are usually lacking).

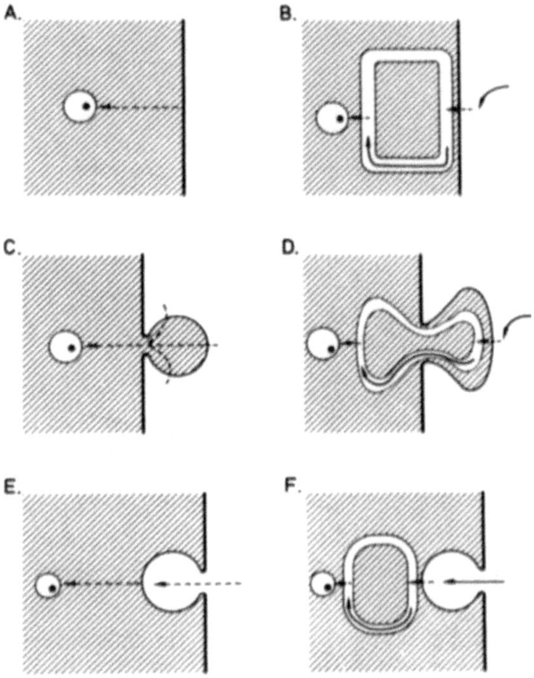

Figure 4. Schematic representation of gas exchange organs without (left) and with (right) circulatory system.

Description of the diversity in the animal kingdom of gas exchange organs and of convective principles both for medium and blood is beyond the scope of this review (see references listed in the Introduction). Only a few evolutionary lines will be highlighted. Some more details of respiratory organs in vertebrates will be considered in Section VI.

A clear tendency of the external gas exchange function to be restricted to certain specialized areas, the respiratory organs, is evident with growing size and complexity of the animals. In higher aquatic animals the gills are usually recessed into gill chambers, mainly for mechanical protection. In aquatic invertebrates, the highest degree of complexity has been reached by decapod crustaceans (unidirectionally ventilated gill chambers under the carapace) and in cephalopod molluscs (gills in the ventilated mantle cavity). The book gills of merostomes (xiphosures) are apparently homologous with the book lungs (for air breathing) of arachnids.

Gills are mostly for water breathing and lungs for air breathing. But in some fishes the gills are stiffened so that they can be used for breathing air, and the holothurians (echinoderms) possess water lungs as invaginations of the posterior intestinal tract. Primordial "lungs" in the form of air-containing cavities with more or less folded internal surface are present in air-breathing terrestrial crustaceans and in air-breathing fishes.

In vertebrates the gas exchange organs (gills or lungs) are circumscript, ventilated, and perfused. In amphibians, however, there exists a phylogenetically interesting group displaying perseverance of the transition from aquatic to terrestrial life, and thus from water to air breathing. In these animals the entire skin is important for gas exchange. In some amphibians, lungs are reduced or are even absent and all gas exchange occurs through the skin.

In many lower animal groups an ample water current onto increased internal body surfaces is provided by the action of flagellae (sponges) or cilia (tunicates and many other groups). The water stream provides both minute food particles ("filter feeding") and oxygen. The perforated branchial basket for filter feeding is at the origin of the gill apparatus of vertebrates as evidenced by comparative anatomy (protochordates, ascidians, amphioxus). A change of the function of the gill basket has taken place during the evolution, the primary purpose of food filtering being replaced by the growing importance of extraction of O_2, only the latter function remaining after the ancestral fishes developed jaws to become predators (cf. Romer, 1970).

For internal O_2 transport, circulatory systems have developed consisting of tubes for channeling flow and of a single heart or multiple contractile vessels for maintenance of this flow. Generally open and closed circulatory systems are discerned (but the distinction is not unambiguous). In an open circulatory system, only major vessels are developed and the liquid (hemolymph) is in open connection with the intercellular space of tissues (arthropods, most molluscs, tunicates). In closed circulatory systems, the circulatory body fluid, blood, is separated from the intercellular space by formation of lacunae or capillaries with a distinct wall (nemertean worms, annelids, cephalopod molluscs, amphioxus, vertebrates).

III. Gas Exchange Organs: Functional Organization

The fully developed gas exchange organ, to be considered in the remaining parts of the review, comprises the specialized gas exchange surface and its structural arrangement with respect to medium flow and blood flow. A current of the medium (air or water) passes over the gas exchange area where gas exchange takes place between the external and the internal medium (body fluid or blood) which also is in continuous flow (Fig. 5).

For a quantitative analysis of gas transfer, the following quantities are required (for dimensions and units, see Piiper et al., 1971).

1. *Transfer rate*, $\dot{M}$, amount of gas exchange per unit time, e.g., O_2 uptake rate (mmol min^{-1}).
2. *Concentration*, C, of a gas species (e.g., O_2) in medium or blood (mmol $liter^{-1}$).
3. *Partial pressure*, P, of a gas species (torr = mm Hg).
4. *Capacitance coefficient*, β, defined as increment in concentration per increment in partial pressure, dC/dP (mmol $liter^{-1}$ $torr^{-1}$); if in liquids no chemical binding is involved, β equals the physical solubility coefficient, α.

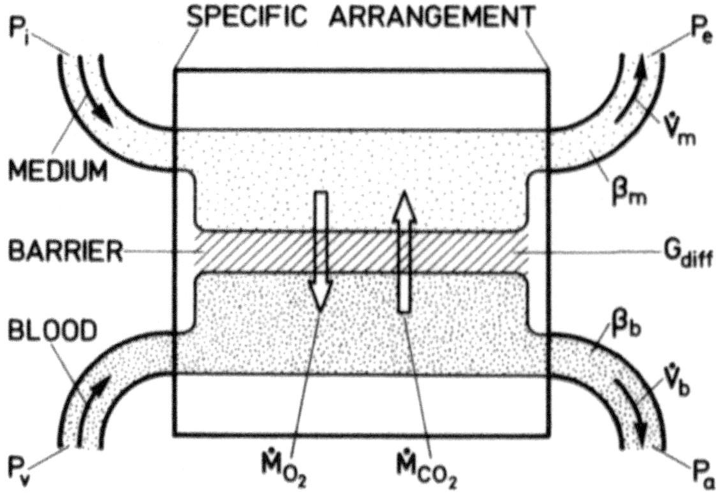

Figure 5. General model for analysis of external gas exchange.

5. *Flow rate*, V, of external medium or blood ($1 \cdot min^{-1}$).
6. *Diffusive conductance*, G_{diff}, of the tissue barrier (see below) ($mmol\ min^{-1}\ torr^{-1}$).

Throughout this review, the following subscripts will be used: m, medium; b, blood; g, gas; w, water; i, inspired; e, expired; a, arterial; v, venous.

For a simple mathematical treatment of the various gas exchange organs the following assumptions will be made, validity of which will be discussed in Section X.

1. The gas exchange system is in steady state, meaning constancy in time of all variables. In particular, ventilation and blood flow are assumed constant, i.e., nontidal and nonpulsatile.
2. The β_m and β_b values are assumed constant, independent of P in the partial pressure ranges considered.

The mass balance yields the following relationships for the external medium

$$\dot{M} = \dot{V}_m\beta_m(P_i - P_e) \qquad \textbf{(6)}$$

and for the internal medium

$$\dot{M} = \dot{V}_b\beta_b(P_a - P_v) \qquad \textbf{(7)}$$

It is useful to define the conductance, G, of a gas species as the uptake rate of this species per unit partial pressure difference, $G = \dot{M}/(P_1 - P_2)$. This allows us to define the component conductances for ventilation (= medium convection) and perfusion (= blood convection):

$$G_{vent} = \dot{V}_m\beta_m \qquad \textbf{(8)}$$

$$G_{perf} = \dot{V}_b\beta_b \qquad \textbf{(9)}$$

These relationships demonstrate the importance of the capacitance coefficient, β, for the convective conductances.

For O_2 in water, β_m is generally equal to the physical solubility of O_2 in water. In body fluids containing a chemical O_2 carrier (e.g., hemoglobin), β_b is elevated above the physical solubility and is equal to the slope of the O_2 dissociation curve.

The diffusive conductance of the tissue barrier, G_{diff}, constitutes the third component conductance. If the tissue barrier consists of a plane sheet, of area F and thickness l, then G_{diff} could be expressed in accordance with Fick's diffusion equation [Eq. (1)] as

$$G_{diff} = K\frac{F}{l} \qquad \textbf{(10)}$$

and this definition would correspond to Eqs.

(8) and (9) for G_{vent} and G_{perf}. However, in real systems, neither the geometry (F and l) nor the physical properties of the tissue barrier (K) are known or measurable with sufficient accuracy. Hence, G_{diff} is usually obtained from the definition of the conductance as mass transfer rate through the membrane per unit partial pressure difference

$$G_{diff} = \int^{area} dG_{diff} = \int^{area} \frac{d\dot{M}}{P_m - P_b} \quad (11)$$

The diffusive conductance of lungs is mostly designated as the diffusing capacity (D) or transfer factor.

Aside from these conductances, a factor of considerable importance is the specific arrangement of the medium and blood flow in the organ. In fact, for given component conductances, G_{vent}, G_{perf}, and G_{diff}, the overall conductance, G_{tot},

$$G_{tot} = \frac{\dot{M}}{P_i - P_v} \quad (12)$$

is generally different for different types of gas exchange system. This aspect will be discussed in Sections VIII and IX.

The following discussion is not restricted to O_2 but includes the other important respiratory gas, carbon dioxide (CO_2), which is metabolically produced as result of substrate oxidation. Consideration of CO_2 is useful, as functional adaptations of gas exchange parameters constitute in most cases a compromise between the demands for O_2 supply and CO_2 excretion.

IV. External Medium: Air Versus Water

The properties of primary importance in determining the carriage of O_2 and CO_2 by the external medium are the respective β values. For the air or gas phase it follows from the ideal gas law,

$$PV = MRT \quad (13)$$

(P, pressure or partial pressure; V, volume; M, amount of substance; R, gas constant; T, absolute temperature), that the capacitance coefficient

$$\beta = \frac{M}{VP} = \frac{1}{RT} \quad (14)$$

is equal for all gas species (insofar as the gas mixture can be considered an "ideal" gas).

For O_2 in water, the β is in every respect equal to physical solubility, its value depending upon temperature and concentration of solutes (salinity).

For CO_2 in water that contains bicarbonate (and/or other buffer substances), the buffering reactions must be taken into account; these lead to an effective increase of β_{CO_2} through formation (or disintegration) of bicarbonate with the changing P_{CO_2}. The underlying chemical reactions in sea water involve formation of bicarbonate (HCO_3^-) in the presence of the bicarbonate/carbonate and boric acid/borate buffer systems:

$$CO_2 + CO_3^{2-} + H_2O \rightleftharpoons 2HCO_3^-$$

$$CO_2 + H_2BO_3^- + H_2O \rightleftharpoons HCO_3^- + H_3BO_3$$

Due to bicarbonate formation with increasing P_{CO_2} (or bicarbonate disintegration with decreasing P_{CO_2}) the "effective CO_2 dissociation curve" of sea water has a steeper slope than would result from the physical solubility alone.

Differences in β values between O_2 and CO_2 and between respective values in water and air (Table 1) lead to some important consequences.

1. Because β_{O_2} is much less for water than for air, water-breathing animals must ventilate much more than air breathers in order to achieve the same O_2 uptake [cf. Eq. (6)].
2. Because in water β_{CO_2} exceeds β_{O_2}, this high water ventilation forcibly means low

Table 1. *Comparison of Physiochemical Properties of Respired Gases in Water and Air*

Property	Air/water[a] CO_2	Air/water[a] O_2	CO_2/O_2 Air	CO_2/O_2 Water
Capacitance coefficient, β	1.05	30	1.0	29
Diffusion coefficient, D	8,000	7,000	0.8	0.7
Krogh's diffusion constant, K	8,000	200,000	0.8	20

[a]Approximate values. For air, $K = D_g\beta_g$; for water, $K = D_w\alpha_w$.

$(P_e - P_i)_{CO_2}$ differences [Eq. [6]). For air, $\beta_{CO_2} = \beta_{O_2}$, and thus with air ventilation $(P_e - P_i)_{CO_2}$ is similar to $(P_i - P_e)_{O_2}$ (the ratio being equal in steady state to the metabolic respiratory quotient, RQ).

These considerations may be used to predict blood and body P_{O_2} and P_{CO_2} values in various water and air breathers provided the process that limits gas exchange is known, i.e., convection of medium or blood or diffusion through tissue barrier. The much higher value of β_{CO_2} compared with β_{CO_2} in water results in a much higher ventilatory conductance for CO_2 than for O_2 $(G_{vent})_{CO_2} > (G_{vent})_{O_2}$. If convection in water limits gas exchange, as is the case for fishes with predominant gas exchange in gills, this discrepancy in G_{vent} leads to small P_{CO_2} values in body and blood. As shown in Fig. 6, the highest possible P_{CO_2} value, that may be attained when O_2 extraction from water is complete, is only about 5 torr. Arterial P_{CO_2} in water-breathing elasmobranchs and teleosts is usually about 2 torr (cf. Fig. 14).

In skin-breathing animals, in which diffusion in tissue limits gas exchange, CO_2 release is favored over O_2 uptake because the diffusion constant of Krogh, is much higher for CO_2 than for O_2 (Table 1) and, therefore, $(G_{diff})_{CO_2} > (G_{diff})_{O_2}$ [Eq. (10)]. This is true for both air and water as media bordering on the skin so long as these media do not limit gas exchange appreciably. Low blood P_{CO_2} values have indeed been found in the exclusively skin-breathing salamander, *Desmognatus fuscus*, in air (arterial $P_{CO_2} = 5$ torr, cf. Fig. 14).

Thus, the important difference in respect of blood P_{CO_2} is not between water breathing and air breathing but rather between transport in water (or tissue) and transport in air as the process limiting gas exchange. In the lungs of mammals and birds, gas transfer is via a tissue barrier into gas spaces (alveoli, parabronchi), and in this transfer CO_2 is favored compared to O_2. However, the limiting process in these lungs is convection by ventilating gas and, therefore, G_{vent} is the decisive parameter for overall gas exchange. Because $\beta_{O_2} = \beta_{CO_2}$ for gas, $(G_{vent})_{CO_2} = (G_{vent})_{O_2}$, O_2 and CO_2 face resistances to gas exchange that are of the same order of magnitude. Thus, for a given O_2 partial pressure gradient between body and environment, much larger body P_{CO_2} values are encountered in lung breathers. Similar considerations hold for gas exchange in bird eggs and in insects in which diffusion in air is limiting, because diffusion coefficients for O_2 and CO_2 in air are similar.

In most amphibians and many air-breathing fishes gas exchange takes place simultaneously via lungs (or lung-like air-breathing organs) and via extrapulmonary organs, i.e., gills and/or skin. With such dual or bimodal (trimodal) breathing, O_2 uptake and CO_2 output are unequally allotted to these alleys, the RQ being lower (than mean metabolic RQ) for the lungs and higher for the extrapulmonary pathways. Since the O_2/CO_2 ratio for the capacitance coefficient β is much higher for air than for water (and tissue), transfer of O_2 is favored in lungs, resulting in pulmonary RQ being much lower than extrapulmonary RQ. The lines in the O_2–CO_2 diagram of Fig. 6 for the overall gas exchange in dual breathers are intermediate between the lines for lung breathers and those for water (or skin) breathers, the exact value for body P_{CO_2} depending on the overall sig-

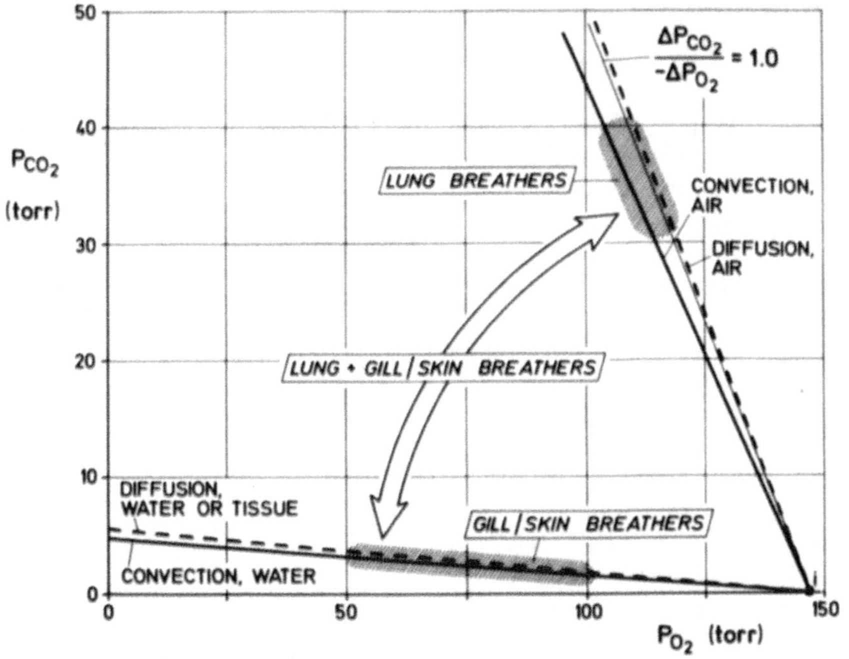

Figure 6. The CO_2–O_2 diagram for transport of respiratory gases by convection (of air or water) or diffusion (in air, water, or tissue), *RQ* being 0.9. The identity line, $\Delta P_{CO_2}:\Delta P_{CO_2} = 1.0$, is attained for convective transport in air at $RQ = 1.0$. Ranges for partial pressures of CO_2 and O_2 observed in arterialized blood of air breathers and water breathers are indicated by shaded areas. Dual (bimodal/trimodal) breathers are located in the intermediate range indicated by the double arrow.

nificance of pulmonary versus extrapulmonary gas exchange.

On the basis of the Henderson-Hasselbalch equation

$$\mathrm{pH} = \mathrm{p}K' + \log \frac{[\mathrm{HCO_3^-}]}{\alpha_{CO_2} P_{CO_2}} \qquad \textbf{(15)}$$

(pK' = first apparent dissociation constant of carbonic acid; α_{CO_2} = physical solubility of CO_2) blood pH is expected to vary inversely with P_{CO_2}. However, at constant temperature, body fluid pH has been shown to be rather independent of the respiratory mode and medium. This is evidently achieved by adjustments of the bicarbonate concentration, which is lower in gill and skin breathers as compared to air breathers.

For the ideal models presented below, β is the only significant property of the medium with respect to gas transfer. In real gas exchange organs, however, a number of other properties are important.

1. *Diffusion properties*, characterized by the diffusion coefficient, D, or by Krogh's diffusion constant, K ($= D\alpha$) determine the development of partial pressure gradients inside the medium (interlamellar water in fish gills; surrounding air or water in skin breathing; "stratification" in mammalian lungs, see Section X).
2. *Viscosity*, η, is a major property determining the mechanical resistance to respiratory medium flow, both with air and water breathing.
3. *Density*, ρ, determines the inertia of the medium and is, therefore, of importance in the respiratory flow whose rate varies with time, e.g., the respiratory cycle.

These factors all tend to make water breathing more costly, i.e., requiring more energy per volume of medium respired, than air breathing, because K is much smaller and ρ and η much higher in water than in air (water to air ratio for ρ is 815, and for η 63).

V. Internal Medium: Carriage of Oxygen by Blood

The convection transport of O_2 by blood (hemolymph) flow is decisively enhanced by specific O_2 carriers, which are proteins with heavy metal ions bound to the protein either directly or via a prosthetic group. The structure (amino acid sequence) of the protein influences the affinity of the carrier for O_2. Table 2 gives an (incomplete) overview of the O_2 carriers, or respiratory pigments, and their occurrence in the blood and hemolymph of animal groups (for details see Jones, 1972; Prosser, 1973). The reversible association/dissociation of O_2 to the pigment is represented by O_2 dissociation curves which are usually visualized as plots of the partial saturation of the O_2 carrier, S_{O_2}, against O_2 partial pressure P_{O_2}.

For consideration of convective O_2 transport by circulation, however, a different O_2 dissociation curve appears to be more relevant, viz., the plot of O_2 concentration, C_{O_2}, against P_{O_2}. The decisive property of O_2 carriage for convective transport is the slope of the C_{O_2}–P_{O_2} relationship, which represents the capacitance coefficient $\beta_b = \Delta C_{O_2}/\Delta P_{O_2}$ (see Section III).

The β_b value is determined, and may be modified, by a number of factors, that are partly relevant as adapting mechanisms (Fig. 7).

Concentration of the carrier defines the maximum value of chemically bound O_2, the O_2 capacity (Fig. 7A). The total β_b (for any given β_b) is the sum of a chemical component, $(\beta_b)_{chem}$, that is proportional to the O_2 capacity, and a physical component, $(\beta_b)_{phys}$, representing the physical solubility. $(\beta_b)_{phys}$, is, in first approximation, independent of the O_2 capacity, its value in higher animals being much smaller than $(\beta_b)_{chem}$ in the physiological range of P_{O_2}.

Cooperativity (Hill's index n). In most cases respiratory pigments are compound protein molecules that are composed of several subunits (4 in vertebrate hemoglobin) each of which is capable of binding one O_2 molecule. Association of O_2 to the subunits, which takes place in successive steps, changes the affinity of the remaining subunits. In most cases, this change is an enhancement, the subunits thus displaying positive cooperativity. The equilibrium of the hemoglobin–oxygen reaction can empirically be described by the Hill equation

$$[HbO_2] = K[Hb][O_2]^n \qquad \textbf{(16)}$$

with Hill's constant $n \geq 1$. [Hb] and $[HbO_2]$ are concentrations of deoxygenated and oxygenated hemoglobin, respectively. The constant K is a measure of the O_2 affinity.

Substituting O_2 saturation (S_{O_2}) into Eq. (16), as the ratio $[HbO_2]/([Hb]+[HbO_2])$, and introducing the half saturation P_{O_2}, P_{50},

Table 2. *Oxygen Carriers in Blood (Hemolymph of Animals)*

Oxygen carrier	Metal	Prosthetic group	Contained in	Animal groups
Hemoglobin	Fe	Protoporphyrin	Cells (or dissolved)	All vertebrates; some holothurians, annelids, crustaceans, and others (protozoans; root nodules of leguminous plants)
Chlorocruorin	Fe	A porphyrin	Dissolved	Some polychaetes
Hemerythrin	Fe	None	Cells	Some polychaetes, sipunculids, priapulids, brachiopods
Hemocyanin	Cu	None	Dissolved	Mollusks, xiphosures, many crustaceans, arachnids

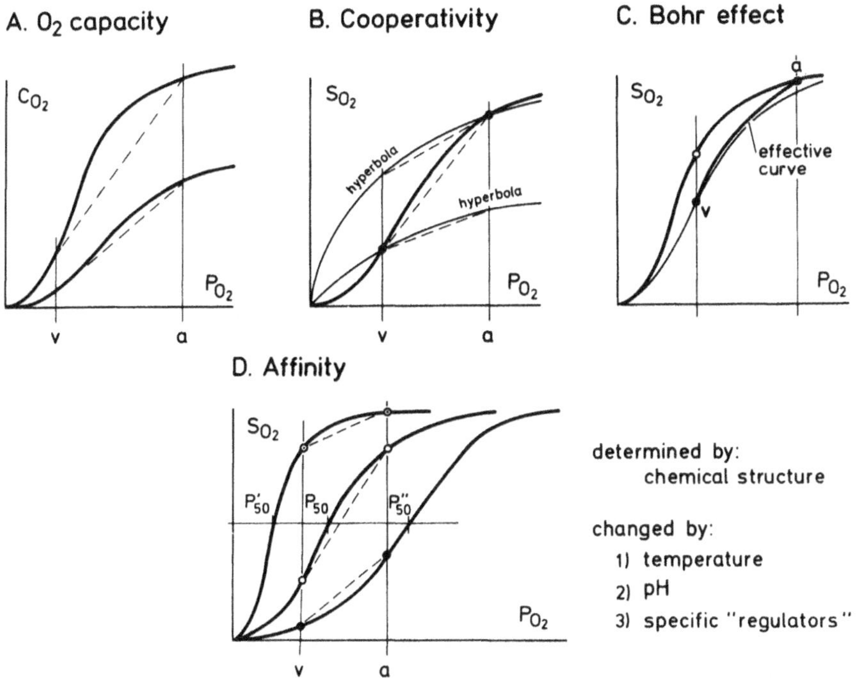

Figure 7. Influences on the shape and location of the O_2 dissociation curve and thereby on β_b for O_2.

as a measure for the constant K, one obtains the Hill equation

$$\log\left(\frac{S_{O_2}}{1 - S_{O_2}}\right) = n \log\left(\frac{P_{O_2}}{P_{50}}\right) \quad \textbf{(17)}$$

In many cases n is nearly constant in the S_{O_2} range from about 0.2 to 0.95. In mammalian blood it is typically about 2.5 to 3 ($n = 2.7$ for human blood). $n = 1$ yields a hyperbolic, $n > 1$ a sigmoid O_2 dissociation curve (Fig. 7B). It is evident from Eq. (17) that the slope of the O_2 dissociation curve (dS_{O_2}/dP_{O_2}) increases with n. The theoretical upper limit for n is the number of subunits in the polymeric respiratory pigment (Antonini and Brunori, 1971).

O_2 affinity. The O_2 affinity, expressed by the P_{50}, determines the P_{O_2} range in which β reaches its maximum (Fig. 7D). It is determined by the chemical structure of the hemoglobin and may be modified by (a) temperature (increase in temperature increases P_{50}, the dissociation of O_2 hemoglobin being exothermic); (b) pH (increase in pH lowers P_{50}); (c) specific regulatory substances that combine reversibly with hemoglobin: CO_2; 2,3-diphosphoglycerate (DPG); inositol pentaphosphate (IPP); adenosine triphosphate (ATP); guanosine triphosphate (GTP).

For the usually occurring sigmoid-shaped O_2 dissociation curves (see below) β is maximum in a range somewhat lower than P_{50}, and for a given P_a and P_v the concentration difference $C_a - C_v$ is highest for $P_a > P_{50} > P_v$. In man (and other mammals), in which these relationships are best known, P_{O_2} in mixed venous blood at rest ($\approx$ 40 torr) is definitely higher than P_{50} ($\approx$ 27 torr), but in heavy exercise mixed venous P_{O_2}, and particularly muscle venous P_{O_2}, drop clearly below P_{50}. Apparently the P_{50} value allows optimum conditions for convective circulatory O_2 transport during heavy exercise rather than during rest. Moreover, the relatively low P_{50} enables the organism to maintain efficient circulatory O_2 transport in hypoxia (respiratory disease, high altitude, etc.).

There has been much controversial

debate on the extent and the significance of the changes of P_{50} during adjustment to high altitude hypoxia. It must be kept in mind that an increase of P_{50} will enhance the delivery of O_2 to tissue by enlarging the P_{O_2} for blood–tissue diffusion, but would at the same time be unfavorable for diffusion-limited O_2 uptake in lungs.

Bohr effect. Increasing acidity decreases the O_2 affinity of hemoglobin (and other respiratory pigments), as evidenced by increasing P_{50}. The Bohr factor, B, as a quantitative measure of the Bohr effect, may be expressed as the change in the logarithm of P_{O_2} with changing pH at constant O_2 saturation

$$B = (\partial \log P_{O_2} / \partial \mathrm{pH})_{S_{O_2}} \qquad \textbf{(18)}$$

B depends on S_{O_2} and is most usually given for half saturation

$$B_{50} = d \log P_{50} / d\mathrm{pH} \qquad \textbf{(19)}$$

For mammalian blood, B_{50} is in the range between -0.4 and -0.8.

The physiologically most important regulator of blood acidity is carbonic acid or CO_2. Changing CO_2 changes pH and thus O_2 affinity. This gives rise to the physiological Bohr effect, by which venous blood, with relatively high P_{CO_2}, displays a lower affinity than arterial P_{CO_2} (Fig. 7C), whereby the β_b of the "physiological" dissociation curve is enhanced above those for constant P_{CO_2}.

Recent experiments have shown that CO_2 exerts a specific effect on the O_2 affinity, whereby a CO_2 Bohr factor may be discerned from a fixed acid Bohr factor, the latter being smaller than the former (Bauer, 1974).

Level of P_{O_2}. Due to the sigmoid shape of O_2 dissociation curves of blood (for $n > 1$), β is lowest at very high P_{O_2} where the pigment is fully saturated with O_2, and β therefore is equal to the physical solubility of O_2. With decreasing P_{O_2}, β increases, reaching its maximum typically somewhat below P_{50}, and decreasing again thereafter.

This dependence on P_{O_2} entails that β, and therefore the perfusive O_2 conductance ($\dot{V}_b\beta$), increases in hypoxia (see below).

O_2combining pigments occur in lower vertebrates without distinctly developed circulation and may also occur in nonmobile cells (myoglobin in vertebrate muscles). Two other functions are attributed to O_2 carriers.

O_2 storage and release. In many cases a high O_2 demand is present only during a short time. This peak demand can be satisfied by release of O_2 stored in the body more easily than by increase of O_2 uptake. Clearly the capacitance coefficient β_b is a pertinent measure of the O_2 storage or release.

Facilitated diffusion of O_2. The facilitated diffusion of O_2 through hemoglobin solution, discovered by Scholander (1960), may be important in providing O_2 to exercising muscle and in may other instances.

These functions of respiratory pigments may be phylogenetically more primitive than the convective transport function which could develop only after development of a circulatory system.

VI. Respiratory Organs of Vertebrates and Their Models

In the following sections the general principles outlined above will be applied to the analysis of external respiration in vertebrates. Omission of invertebrates from a systematic analysis, dictated by the lack of systematic functional understanding of their gas exchange organs, appears to be no great drawback as most of the general principles of external gas exchange can be exemplified in vertebrates.

It will be attempted in this section to simplify the apparently complicated anatomy of gas exchange organs in order to derive a functional model. This model must be simple enough to warrant mathematical treatment applicable to a wide range of physiological situations; yet it must be faithful enough to reveal the specific functional properties of each system. Mathematical analysis is considered in Section VIII; its application to some experimental data is performed in Section IX; and the significance of model simplifications for the analysis is discussed in Section X.

The functional properties of external gas exchange organs in vertebrates can be described by three inherently different models (Piiper and Scheid, 1972, 1975, 1977): the *countercurrent system* for *fish gills*; the *crosscurrent system* for *bird lungs*; the *ventilated pool system* for *mammalian lungs*. A fourth model, the *infinite pool system* for the *amphibian skin* will also be considered, but formally it represents a particular, degenerate case of the other systems.

A. Fish Gills: The Countercurrent System

The fish gill apparatus (Hughes and Shelton, 1962; Randall, 1970; Johansen, 1971; Hughes and Morgan, 1973) comprises a number of gill arches each carrying a double row of gill filaments (Fig. 8). Attached to both sides of the filaments are the secondary lamellae which are spaced densely enough, both along the filaments and between neighboring filaments, to form a fine sieve which must be passed by the respiratory water flow on its way from the mouth to the outside. Venous blood, driven by the heart and through the ventral aorta, flows into the afferent branchial arteries, located in the gill arch, and into the afferent arteries in the gill filaments to enter the blood lacunae in the secondary lamellae. It is in the secondary lamellae where blood and water attain a close contact, separated only by a thin epithelium through which diffusional exchange of respiratory gases (O_2 and CO_2) takes place. Arterialized blood flows then via efferent vessels in the filaments and gill arches and via the dorsal aorta into the arterial system to supply the body tissues.

The anatomical arrangement described forms the basis of a countercurrent system for gas exchange between two flows, the respiratory water flow and the blood flow (cf. Fig. 8), both separated by a tissue barrier that must be passed by diffusion.

B. Bird Lungs: The Crosscurrent System

The respiratory tract of birds comprises two functionally different structures, the air sacs and the parabronchial lung (King, 1966; Duncker, 1971; Scheid, 1979), both con-

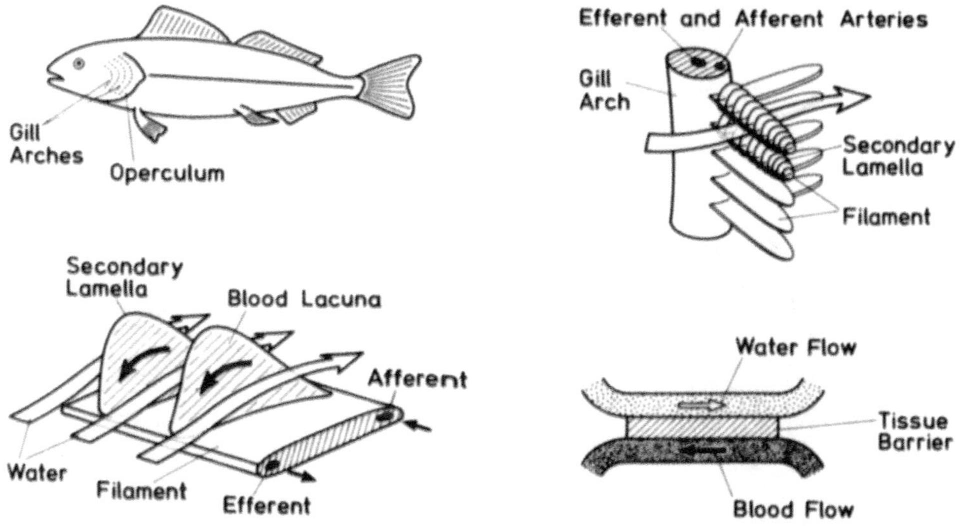

Figure 8. Schema of the gill apparatus in teleosts to show the relationship between the anatomical structure and the simplified model used for analysis of gas exchange.

nected through bronchial tubes to the trachea (Fig. 9). The air sacs are expanded and compressed with the chest wall by the action of respiratory muscles and so act like bellows to provide respiratory air flow through the parabronchial lung, in the same direction both during inspiration and expiration (Fig. 9). The parabronchial lung, the gas exchange organ in the strict sense, represents a parallel assembly of a few hundred long narrow tubes, the parabronchi. All along the length of each parabronchus depart fine (some micrometers wide) air capillaries which form a meshwork that entwines equally fine blood capillaries on their way from arterioles, at the periphery of the periparabronchial tissue, to venules, located near the parabronchial lumen.

Air in the air capillaries and blood in the blood capillaries attain gas exchange contact through a thin tissue barrier. Transport of respiratory gases between the parabronchial lumen and the air capillaries occurs mainly by diffusion; however, the diffusional resistance is close to negligible, at least during resting conditions (Scheid, 1978a). Hence, the air flow encounters on its route through the parabronchial lumen in serial order a great number of blood capillaries for gas exchange. The system may thus be addressed as serial multicapillary or as crosscurrent, since respiratory air and blood flows cross each other in the gas exchange zone. This arrangement is present in all bird species yet investigated (Duncker, 1971).

C. Mammalian Lungs: The Ventilated Pool System

The airways of mammals (Weibel, 1963, 1973) constitute a highly branching system of conducting bronchial passageways that eventually lead, in the respiratory zone, into a great number of alveoli (Fig. 10). The alveoli are surrounded by a network of blood capillaries, originating from the pulmonary artery,

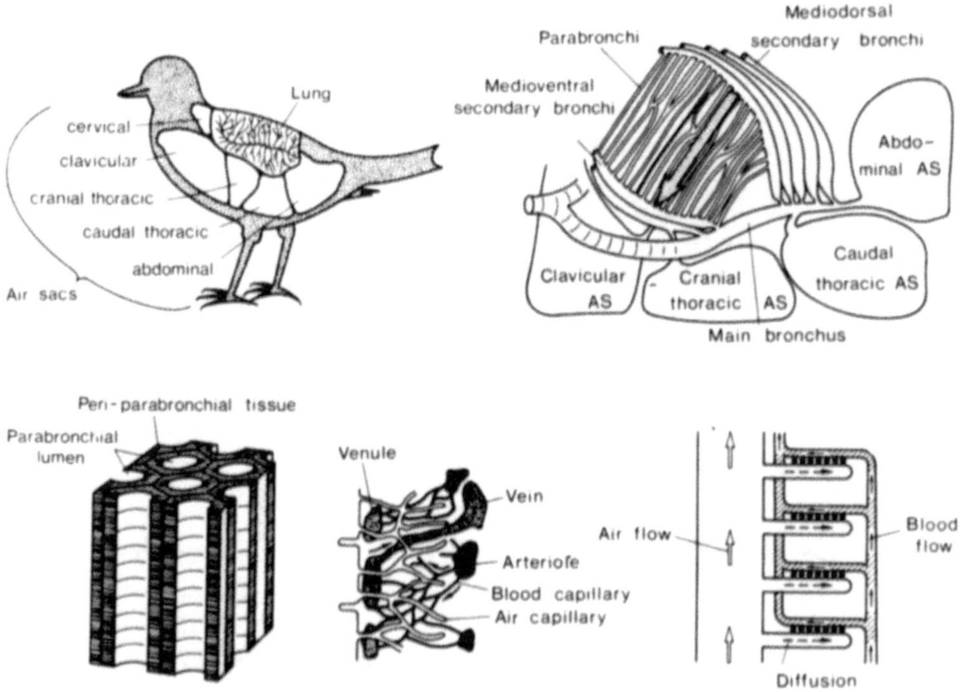

Figure 9. Schematic diagram showing the air sac lung apparatus in a bird and how a model for gas exchange is deduced from the arrangement of air capillaries and blood capillaries in the periparabronchial tissue. Four paired air sacs usually occur in birds: the cervical (not shown), cranial thoracic, caudal thoracic and abdominal air sacs, and the unpaired interclavicular air sac.

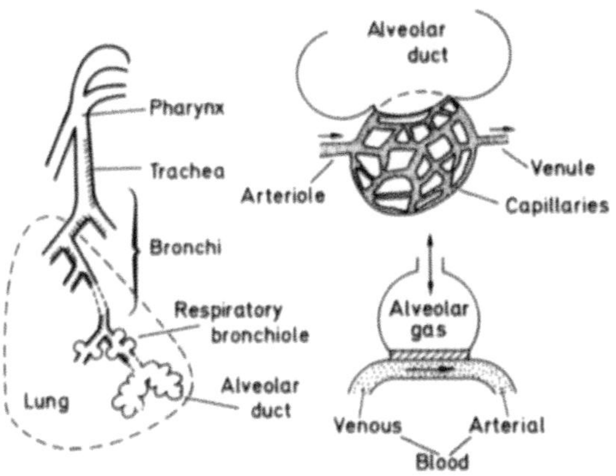

Figure 10. Schema of the alveolar lung to show the relationship between structure and model.

thus interposed between the right and left heart. Hence, virtually the entire cardiac output perfuses the alveolar capillaries to achieve gas exchange contact with the alveolar gas, there being only a thin tissue barrier to separate air and blood. The alveolar air is renewed by the alveolar tidal air, entering and leaving the alveolar space in the respiratory cycle through the efforts of the respiratory muscles.

Alveolar tidal volume constitutes a small fraction of the alveolar residual air volume, thus rendering alveolar concentrations of respiratory gases virtually constant in time. Blood flow in alveolar capillaries is essentially constant despite the cyclic activity of the heart. Hence, the model for alveolar gas exchange (Fig. 10) is a ventilated pool of alveolar gas in contact through a tissue barrier with a constant flow of blood.

D. Amphibian Skin: The Infinite Pool System

Skin breathing is significant for all extant amphibians, but constitutes the only mode of respiratory gas exchange in those aquatic and terrestrial salamanders that lack both lungs and gills (Czopek, 1961, 1965). As the gas exchange epithelium is identical with the skin, a compromise must be reached for its thickness to keep the diffusive resistance at its minimum yet provide enough mechanical protection for the underlying tissues. Gas exchange occurs through the skin with the blood perfusing a dense subepithelial capillary network, supplied in part with arterial blood from the arterial system, in part with venous blood from a branch of the pulmonary arch. Oxygenated cutaneous blood flows into the venous system. This arrangement is in contrast to that of pulmonary gas exchange systems, which provide for at least a partial separation of oxygenated from venous blood.

Despite these differences in the blood circulatory system, the gas exchange system exhibits similarities with the mammalian alveolar lung. In fact, the amphibian skin may be viewed as a degeneration of the ventilated pool system for infinitely high alveolar ventilation, when alveolar gas concentrations are identical with those of the ambient air. Hence, the infinite pool system is proposed as a model for gas exchange across the amphibian skin.

VII. Models for Gas Exchange: Qualitative Description

The idealized models resulting from the structural elements of the external gas exchange organs in vertebrates are shown in Fig. 11. The three basic systems, countercurrent in fish gills (Model I), crosscurrent in avian parabronchi (Model II), and ventilated pool in mammalian alveoli (Model III), are complemented by the degenerated system of the infinite pool for amphibian skin (Model IV). Models I to III conform to the general gas exchange system of Fig. 5 in that a medium flow (water or air) is exposed to a blood flow, both flows being separated by a thin tissue barrier. The essential differences between the systems reside in the relative structural arrangements of the medium to blood flow. In Model IV, there is functionally an infinite medium flow, whereby the structural arrangement of blood flow relative to the medium becomes immaterial. Hence, this model results as the limiting case of infinite medium flow from all the others.

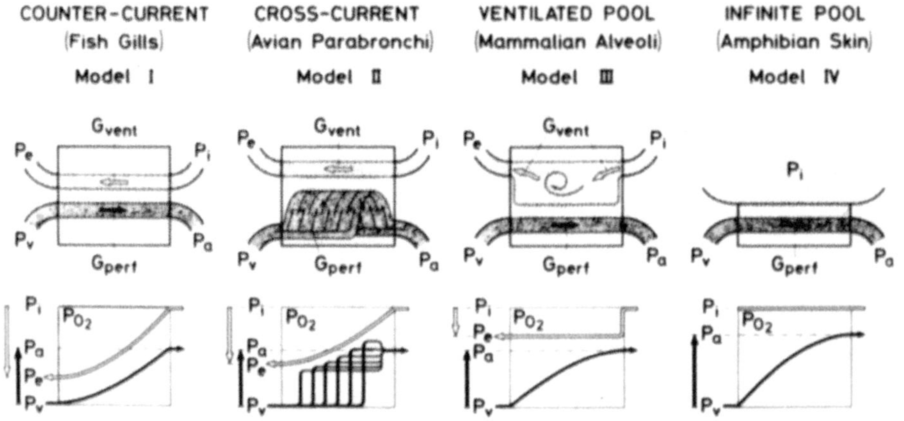

Figure 11. Schema of the four models used or analysis of gas exchange in vertebrates. Bottom, typical partial pressure profile for O_2 in medium and blood.

Figure 11 displays typical partial pressure profiles for O_2 (P_{O_2}) along the gas exchange surface, both in the medium (light lines) and blood (dark lines). These profiles result from O_2 exchange across each element of the tissue barrier, the O_2 transfer rate being proportional to the driving pressure head, that is to the P_{O_2} difference across the element between the medium and blood. In this section, the P_{O_2} profiles will be described qualitatively and a more rigorous quantitative analysis will be presented in the next section.

In the countercurrent system, medium with the highest P_{O_2}, at the inflow end, is in contact with arterial blood, while outflowing medium contacts venous blood with its low P_{O_2}. Hence, arterial P_{O_2} may approach inflowing P_{O_2} and may thus rise above outflowing P_{O_2}. The arrows to the left of the profiles show that the range of medium P_{O_2} may overlap that of blood P_{O_2}. The extent of this overlap depends mainly on the ventilatory and perfusive conductances, G_{vent} and G_{perf}, and on the diffusive conductance, G_{diff}, of the tissue barrier.

Whereas in the countercurrent system each blood capillary contacts the medium flow all along the exchange surface, a single capillary in the crosscurrent system achieves gas exchange contact with parabronchial air only at a small fraction along the parabronchial length. As a consequence, the air progressively loses O_2 (and gains CO_2) on its way through the parabronchus as it exchanges respiratory gases with the blood. The degree of arterialization in the capillaries will thus diminish from the gas inflow to the outflow end. Blood from all capillaries mixes to form (systemic) arterial blood, the arterial P_{O_2} thus reflecting an average of end-capillary P_{O_2} of all capillaries. This average may be above the P_{O_2} in end-parabronchial gas whereby an overlap of P_{O_2} ranges in gas and blood results in this system as in the countercurrent (arrows to the left of the P_{O_2} profiles in Fig. 11).

In the ventilated pool system, a blood capillary contacts a medium pool which is well stirred and is thus homogeneous in respect of partial pressures. Hence, blood P_{O_2} rises nearly exponentially from the venous (inflow) to the end-capillary (arterial) value. The highest blood P_{O_2} possible in this system equals the alveolar value, an overlap being impossible with this model.

The partial pressure profile in the infinite pool system is virtually identical to that of Model III except for the lacking P_{O_2} drop from the ambient medium to the exchange surface.

The partial pressure profiles, and particularly the P_{O_2} values at the medium and blood inflow and outflow ends, may be used to compare the efficacy for gas transfer of the various models. For given O_2 uptake rate and conductances, G_{vent} and G_{perf}, that model is considered to have the highest efficacy which achieves the highest arterial (or venous) P_{O_2} for a given medium P_{O_2} (P_i). Hence, gas exchange efficacy is directly linked with the extent to which the P_{O_2} ranges in blood and medium overlap. The order of efficacy is thus countercurrent > crosscurrent > ventilated pool.

VIII. Models for Gas Exchange: Quantitative Analysis

A. Differential Equations

Three elementary processes govern gas exchange in the general system in Fig. 5, convective transport in both medium and blood and diffusive transport across the tissue barrier. Each process is describable by a differential equation, the peculiarities of the structural arrangement in each system (Fig. 11) being recognized by the individual boundary conditions.

1. At any site within the system, the differential transfer rate from medium to blood, $d\dot{M}$, by diffusion across the differential element of the tissue barrier, of diffusive conductance, dG_{diff}, is proportional to the medium-blood partial pressure difference, $(P_m - P_b)$:

$$d\dot{M} = dG_{diff}(P_m - P_b) \tag{20}$$

2. This transfer rate causes a drop of partial pressure in the medium flowing past the differential element of the exchange element:

$$d\dot{M} = G_{vent} dP_m \tag{21}$$

3. Similarly, $d\dot{M}$ results in a differential rise of partial pressure in blood flowing past the same element of the tissue barrier

$$d\dot{M} = G_{perf} dP_b \tag{22}$$

These differential equations can be integrated analytically, using standard procedures, considering the boundary conditions provided by the structural arrangements of the systems and assuming, most appropriately, the partial pressure of the inflowing medium, P_i, and the total gas exchange rate, $\dot{M}$. The mathematical derivation is facilitated by some assumptions, in particular (1) independence of G_{perf} and G_{vent} on partial pressures, and (2) steady-state conditions, i.e., constancy of all variables in time (see above).

B. Results of Calculation

Integration of Eqs. (20)–(22) yields partial pressures of medium and blood at any site within the system and hence profiles like those exemplified in Fig. 11. Of particular interest are their values at the inflow and outflow ends in medium (P_i and P_e) and in blood (P_a and P_v), both from a theoretical point of view, as they allow a comparison of the various models in respect of their gas exchange efficacy (see below), and from the practical point of view, as they allow estimation of functional parameters from experimental results in a given species (see Section IX).

Gas exchange efficacy. Particularly useful are the relative partial pressure differences (Piiper and Scheid, 1975), Δp, which are dimensionless variables that depend on the (dimensionless) ratios of conductances, G (see below). Three such Δp terms may be defined, attributable to the three elementary processes, ventilation, perfusion, and medium/blood transfer:

$$\Delta p_{vent} = \frac{P_i - P_e}{P_i - P_v} \tag{23}$$

$$\Delta p_{perf} = \frac{P_a - P_v}{P_i - P_v} \tag{24}$$

$$\Delta p_{tr} = \frac{P_e - P_a}{P_i - P_v} \tag{25}$$

The first two terms are easily shown to

relate directly to the total gas exchange conductance of Eq. (12):

$$\Delta p_{\text{vent}} = G_{\text{tot}}/G_{\text{vent}} \quad \textbf{(26)}$$

$$\Delta p_{\text{perf}} = G_{\text{tot}}/G_{\text{perf}} \quad \textbf{(27)}$$

They constitute the fraction of the total gas exchange resistance (inverse of conductance) attributable to ventilation and perfusion, respectively.

As the three elementary processes, ventilation, perfusion, diffusion, can in none of the systems be regarded as being arranged in series, in respect of gas flow, the third term, Δp_{tr}, which is defined as the complement of the other two to unity

$$\Delta p_{\text{tr}} = 1 - (\Delta p_{\text{vent}} + \Delta p_{\text{perf}}) \quad \textbf{(28)}$$

depends on the diffusive medium/blood transfer characteristics and also on the structural peculiarities of the system. It can achieve negative values in some models and is a parameter particularly related to the inherent efficacy of a system.

Table 3 shows the three Δp values in the four models considered. They are presented as functions of the two ratios $G_{\text{vent}}/G_{\text{perf}}$ and $G_{\text{diff}}/G_{\text{perf}}$. The dependence of Δp on these two ratios has been discussed earlier (Piiper and Scheid, 1975). We shall restrict further analysis to a $G_{\text{vent}}/G_{\text{perf}} = 1$, as most experimental values obtained in fishes, birds, and mammals (see Section IX) are not far from this value.

In Fig. 12 is plotted the total gas exchange conductance, G_{tot}, as a function of G_{diff} for the special case of $G_{\text{vent}} = G_{\text{perf}} = 1$. As can be seen from Eqs. (26) and (27), $G_{\text{tot}} = \Delta p_{\text{vent}} = \Delta p_{\text{perf}}$ in this case. Three lines in Fig. 12 relate to the three major systems, Models I to III of Fig. 11. For any value of G_{diff}, G_{tot} attains the largest value in the countercurrent, the smallest in the ventilated pool system. Since $G_{\text{tot}} = \dot{M}/(P_{\text{i}} - P_{\text{v}})$, for any set of input partial pressures, P_{i} and P_{v}, and any set of elemental conductances, G_{vent}, G_{perf}, G_{diff}, the countercurrent system is able to transfer more gas than the crosscurrent, and this system more than the ventilated pool system. Hence, G_{tot} constitutes a measure for the gas exchange efficacy that results from the particular structural arrangement of the system.

The differences in G_{tot} between the systems are most marked for large values of G_{diff} but blur increasingly as G_{diff} is diminished and diffusive transfer becomes the limiting step for gas exchange.

In general, the gas exchange efficacy diminishes for all models of Fig. 11 when $G_{\text{vent}}/G_{\text{perf}}$ is above or below unity (Piiper and Scheid, 1975, 1977), and the differences between the systems become less pronounced.

Limitation index. Whereas the parameter Δp appears particularly useful in the discus-

Table 3. *Formulas for the Relative Partial Pressure Differences*[a]

Partial pressure difference	Countercurrent	Crosscurrent	Ventilated pool	Infinite pool
Δp_{vent}	$\dfrac{1 - \exp(-Z)}{X - \exp(-Z)}$	$1 - \exp(-Z')$	$\dfrac{1 - \exp(-Y)}{X + 1 - \exp(-Y)}$	0
Δp_{perf}	$\dfrac{X[1 - \exp(-Z)]}{X - \exp(-Z)}$	$X[1 - \exp(-Z')]$	$\dfrac{X[1 - \exp(-Y)]}{X + 1 - \exp(-Y)}$	$1 - \exp(-Y)$
Δp_{tr}	$\dfrac{X \exp(-Z) - 1}{X - \exp(-Z)}$	$\exp(-Z') - X[1 - \exp(-Z')]$	$\dfrac{X \exp(-Y)}{X + 1 - \exp(-Y)}$	$\exp(-Y)$

[a] $X = G_{\text{vent}}/G_{\text{perf}}$; $Y = G_{\text{diff}}/G_{\text{perf}}$; $Z = Y(1 - 1/X)$; $Z' = \frac{1}{X}[1 - \exp(-Y)]$.

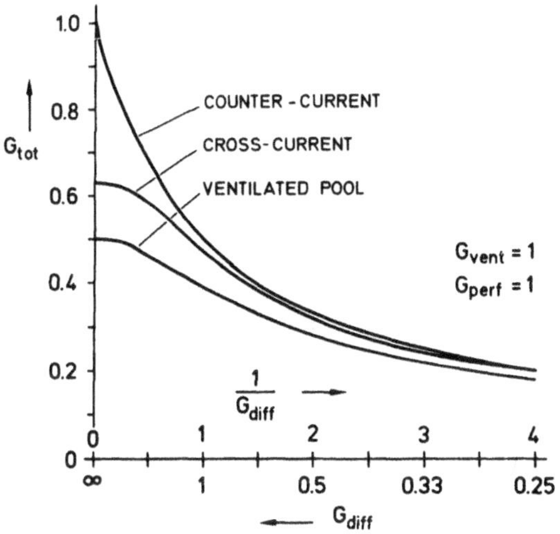

Figure 12. Total gas exchange conductance, G_{tot}, is plotted against the diffusive resistance, $1/G_{diff}$, for the three major models of Fig. 11. The convective conductances, G_{vent} and G_{perf}, are unity.

sion of the gas exchange efficacy, the limitation index, L, is suited for assessment of the significance of the elementary processes, ventilation, perfusion and, diffusion, in limiting the gas exchange rate. The limitation index for the elementary process x, L_x, may be defined as

$$L_x = 1 - \frac{\dot{M}(\text{limited by x})}{\dot{M}(\text{not limited by x})} \quad (29)$$

$$L_x = 1 - \frac{G_{tot}(\text{limited by x})}{G_{tot}(\text{not limited by x})} \quad (30)$$

$\dot{M}$(limited by x) constitutes the actual gas uptake rate with finite values of all elementary conductances, while $\dot{M}$(not limited by x) constitutes the uptake rate calculated with infinite conductance for the process x. Expressions for the L values of the four models of Fig. 11 have been tabulated by Piiper and Scheid (1975).

Consideration of L is particularly illustrative for Model IV, the infinite pool system of the amphibian skin, as here only two elementary processes, diffusion and perfusion, are involved. Using Eqs. (27) and (30), the two limitations may thus be defined:

$$L_{diff} = 1 - \frac{G_{perf}\,\Delta p_{perf}}{\lim\limits_{G_{diff} \to \infty} (G_{perf}\Delta p_{perf})} \quad (31)$$

$$L_{perf} = 1 - \frac{G_{perf}\,\Delta p_{perf}}{\lim\limits_{G_{perf} \to \infty} (G_{perf}\Delta p_{perf})} \quad (32)$$

Using the expressions of Table 3, the L values can be expressed as simple functions of G_{diff}/G_{perf} which are plotted in Fig. 13 on a semilogarithmic scale. For very low values of the ratio G_{diff}/G_{perf}, $L_{perf} = 0$ and L_{diff} approaches unity, while the reverse is true for high values of the ratio. For G_{diff}/G_{perf} between 0.1 and 3 there exists a combined limitation of gas exchange by both processes.

The value of the G_{diff}/G_{perf} ratio depends on the two conductances, which in turn are given by anatomical parameters (area and thickness of diffusion barrier), by physiological parameters (e.g., rate of blood perfusion of exchange system), and by the physical properties of the gas under study (effective solubility in blood, β; Krogh's diffusion constant in the tissue barrier). For O_2 uptake across

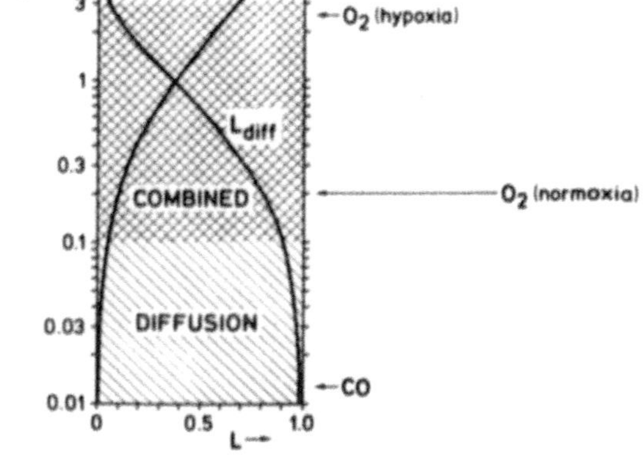

Figure 13. Diffusion and perfusion limitation in amphibian skin and in dog lungs for various gases.

salamander skin (Piiper et al., 1976) there exists a predominant diffusion limitation (Fig. 13).

The analysis may directly be applied to study relative limitation of gas uptake by diffusion and perfusion from alveolar air into capillary blood in mammalian lungs. Figure 13 shows that the limitation of O_2 uptake in dog lungs depends on the alveolar O_2 level itself. This is due to the curvilinear nature of the O_2 dissociation curve, being steep (high β and hence large value of G_{perf}) in the hypoxic, and flat (low β and hence small value of G_{perf}) in the hyperoxic range (see Section V). Carbon monoxide, CO, displays a very large β value, and this results in a virtually exclusive limitation by diffusion. This is the main reason why diffusing capacity in alveolar lungs is frequently determined by CO. On the other hand, all inert gases, like N_2, are virtually exclusively limited by perfusion, and may thus be utilized for bloodless measurement of cardiac output in animals and man (cf. Piiper and Scheid, 1980).

IX. Application of Model Analysis in Animal Experiments

The formal analysis presented in the preceding section can be applied in several ways to experimentally observed data. Usually, it is utilized to determine in quantitative terms a model parameter that is otherwise difficult to obtain, viz., the diffusive conductance, G_{diff} (diffusing capacity, D, or transfer factor; see above). G_{diff} can only be obtained with some accuracy when there exists a significant diffusion limitation (L_{diff}). If O_2 is used as the gas for analysis, hypoxia is applied experimentally, mainly to create conditions with high L_{diff}.

Figure 14 illustrates experimental values of partial pressures for O_2 and CO_2 obtained in resting animals, chosen to constitute representative species for each type of gas exchange organ:

1. Larger spotted dogfish (*Scyliorhinus stellaris*), an elasmobranch fish, for the countercurrent model of fish gills (Piiper and Baumgarten-Schumann, 1968; Baumgarten-Schumann and Piiper, 1968).
2. Domestic hen (*Gallus domesticus*) for the crosscurrent system of avian parabronchial lungs (Scheid and Piiper, 1970).
3. Dog (*Canis familiaris*) for the ventilated pool system of mammalian alveoli (Haab et al., 1964).
4. Common dusky salamander (*Desmognathus fuscus*), a lungless and gill-less salamander, for the infinite pool system of amphibian skin (Piiper et al., 1976).

The following features appear to illustrate differences in the gas exchange organs:

1. For air breathers (bird, mammal) the ranges of P in the medium are nearly identical for O_2 and CO_2. The reason is that G_{vent} is identical for both gases, and $\dot{M}_{O_2}$ nearly equal to $\dot{M}_{CO_2}$ [see eq. (8) in Section III]. For fish gills, the P range in the medium is large for O_2 but small for CO_2, due to the much larger G_{vent} for CO_2 than for O_2 (higher solubility in water of CO_2 than of O_2). The O_2 demand causes fish to hyperventilate in respect of CO_2.
2. In blood, the P ranges are generally smaller for CO_2 than for O_2. The reason is the higher effective blood solubility, β_b, for CO_2 than for O_2.
3. Only for the bird is there an overlap, both for O_2 and CO_2, in blood and medium partial pressure ranges. In mammalian lungs, this overlap cannot occur, $P_a = P_e$ in the limiting case. Although the countercurrent system would allow this overlap to occur, it is not found in fish in normoxia, indicating that the high intrinsic efficacy is not utilized in full extent at rest.

The partial pressures of Fig. 14 allow, with known gas exchange rates, $\dot{M}$ (cf. Piiper and Scheid, 1975), to calculate the component conductances, G_{vent}, G_{perf}, and G_{diff}, and limitation indices, L_{vent}, L_{perf}, and L_{diff} (Table 4).

The following observations can be made:

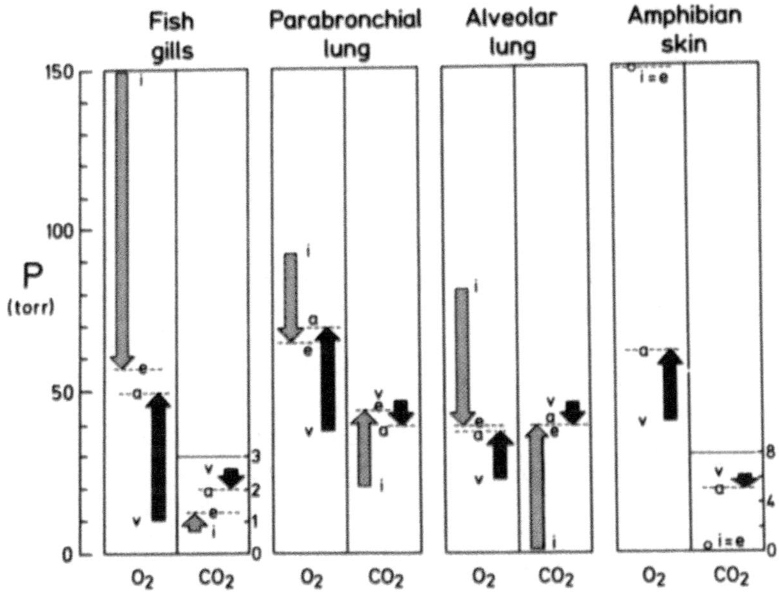

Figure 14. Representation of experimental data on representative species for the four models of Fig. 11. Values in medium (light arrow) extending from inspired (i) to expired (e) partial pressure; in blood (dark arrows) extending from (mixed) venous (v) to arterial (a) partial pressures. Data for both O_2 and CO_2 are indicated. Note the low P_{CO_2} values for fish gills and amphibian skin (expanded P_{CO_2} scales for these two only!). An overlap in medium and blood partial pressure ranges is observed only in the bird.

1. For CO_2 in air breathers, the L_{diff} is generally indistinguishable from 0. The equilibration of P_{CO_2} between medium and blood is virtually complete, meaning equality between P_e and P_a in the alveolar lung and maximal overlap in the parabronchial lung. In these cases, G_{diff} approaches infinity.
2. In fish gills and salamander skin, G_{diff} is finite for CO_2 as for O_2. The G_{diff} value for CO_2 is some 20 times larger than for O_2, largely due to the higher solubility in tissue for the former gas. On the other hand, the ratio G_{diff}/G_{perf} is similar for both gases.
3. For the air-breathing animals, L_{vent} is

Table 4. *Experimental Values for Analysis*

	Countercurrent[a] (fish gills)		Crosscurrent[b] (parabronchial lungs)		Ventilated pool[c] (alveolar lungs)		Infinite pool[d] (amphibian skin)	
	CO_2	O_2	CO_2	O_2	CO_2	O_2	CO_2	O_2
G_{vent}	0.11	0.00084	0.040	0.040	0.16	0.16	∞	∞
G_{perf}	0.10	0.0020	0.11	0.034	0.88	0.48	0.17	0.009
G_{diff}	0.045	0.0010	∞	0.063	∞	1.44	0.05	0.002
L_{vent}	0.13	0.33	0.67	0.29	0.85	0.74	0	0
L_{perf}	0.15	0.09	0.06	0.35	0.15	0.18	0.14	0.10
L_{diff}	0.69	0.36	0	0.11	0	0.01	0.74	0.80

[a]Larger spotted dogfish (*Scyliorhinus stellaris*), an elasmobranch fish.

[b]Domestic hen (*Gallus domesticus*).

[c]Dog (*Canis familiaris*).

[d]Common dusky salamander (*Desmognathus fuscus*), a lungless and gill-less salamander.

larger for CO_2 than for O_2, indicating that for CO_2 ventilation constitutes the elementary process that limits gas exchange most. Hence, increase in ventilation results in a steeper increase of CO_2 output than of O_2 uptake creating a (transitory) increase in the gas exchange ratio, R. The same mechanism is responsible for the local variation of R in lungs with ventilation/perfusion inhomogeneity.

X. Real Systems and Model Analysis

A number of idealizing assumptions have been made to arrive at the models of Fig. 11. These simplifications enabled us to develop a relatively simple analytical treatment of the gas exchange performance. However, it must be discussed to which extent the results of the idealized models apply to the real systems as well. A full discussion of these problems has recently been given for the alveolar lung (Piiper and Scheid, 1980) and for the avian parabronchial lung (Scheid, 1979).

A. Anatomical Arrangement

Anatomical complexity of the real system may render it difficult to sample the medium or blood entering or leaving the organ.

Medium dead space. When mammals inspire, first dead space gas, corresponding to end-expired gas of the preceding breath, and then fresh gas enters the alveoli. This dead space is easily accounted for by subtracting dead space volume from each tidal volume to calculate alveolar (effective) ventilation. A similar approach may be used in birds, while the unidirectional flow in fish gills results in absence of dead space in this system.

Blood capillaries and gas exchange organ. Blood leaving the gas exchange organ may become substantially modified by admixture of venous blood before it can be sampled as arterial blood from systemic arterial vessels. The amount of this venous admixture is small in mammals and birds in which pulmonary and systemic circulations are arranged in series (Piiper and Scheid, 1980; Burger et al., 1979). Similarly in exclusively water-breathing fish, where the gas exchange organ and the metabolizing tissues are arranged in series, only a small venous admixture is to be expected. However, in virtually all other vertebrates there exist very complicated anatomical arrangements which result in extensive functional "venous admixture." In some cases, the composition of blood entering and leaving the gas exchange organ is to be calculated with a number of assumptions from measurements in blood accessible to a puncturing needle.

B. Inhomogeneities

Parallel inhomogeneities. It was assumed in the model analysis that all units within the gas exchange organ are functionally identical, meaning that the ratios of G_{vent}/G_{perf} and G_{diff}/G_{perf} of all units are the same. It is well known that regional inhomogeneities exist in lungs, and their influences in mammalian lungs have been extensively investigated (cf. Piiper and Scheid, 1980). In general, they reduce the gas exchange efficacy of the system. Although not investigated in such detail, functional inhomogeneities are known to exist in other vertebrates as well (Hughes, 1973).

Temporal variations. In the model analysis, the partial pressures of inflowing medium and blood, the conductances (and the gas exchange rates) have been assumed to be constant in time. Tidal ventilation and the cyclic activity of the heart result, however, in some temporal variability of these parameters. In alveolar lungs, tidal ventilation renders alveolar partial pressures changing in time, but the significance for gas exchange is small, and this is true even for birds despite their small ratio of parabronchial tidal volume to parabronchial residual air (Scheid et al., 1977; Scheid, 1978b). Despite the cyclic action of the water-promoting buccal and opercular pumps in fish, water flow through the gill sieve appears to be nearly constant in time (Hughes and Morgan, 1973).

C. Additional Transfer Resistances

Diffusion in medium and blood. In the model analysis, three elementary processes have been assumed to determine the gas exchange, and each has been assumed to contribute to the total resistance to gas exchange: convective resistances of medium and blood, $1/G_{vent}$ and $1/G_{perf}$, and the diffusive resistance, $1/G_{diff}$. This analysis neglects diffusive resistances in medium and blood.

In mammals, the finite diffusivity of O_2 (and of other gases) in alveolar air imposes a "stratified resistance," the significance of which for alveolar gas exchange has not yet been fully evaluated (cf. Scheid and Piiper, 1980).

Whereas gas transport through the parabronchial lumen in bird lungs is by convection, respired gases have to diffuse through the air capillaries to reach the blood–gas interface. Theoretical analysis suggests that this additional diffusion resistance is minimal under most circumstances (Scheid, 1978a).

Much more severe is the diffusion resistance offered to respiratory gases by the water flowing through the gill sieve. This is because the diffusion constant of O_2 and CO_2 in water is several orders of magnitude smaller than in air. This effect incorporates a severe complication into the model analysis of fish gills. Estimations on the basis of simplified models show that the diffusion resistance offered by interlamellar water constitutes 25 to 50% of the total transfer resistance in teleost and elasmobranch fish (Scheid and Piiper, 1971, 1976).

Diffusion inside the blood is usually expected to present no significant limitation due to the smallness of capillaries and to the convective mixing provided by the flowing blood.

Kinetics of chemical reactions. For model analysis, the equilibrium dissociation curves for O_2 and CO_2 in blood have been utilized, implying that all chemical reactions are fast on the scale of contact time in the gas exchange organ. Despite intensive in vivo experiments on the kinetics of O_2 with hemoglobin in several mammalian species (Forster, 1964; cf. Piiper and Scheid, 1980), no definite statement as to its limiting role can be given. For CO_2 excretion, the rate-limiting step appears to be the dehydration of H_2CO_3 which is catalyzed by carbonic anhydrase present in red cells. According to our present knowledge, chemical reaction rates and cell–plasma exchange do not significantly limit gas exchange in alveolar lungs (cf. Piiper and Scheid, 1980). It is, however, not yet determined whether the lower reaction rates expected at the lower body temperature of poikilotherms impose additional transfer resistances.

D. Blood Chemistry

Shape of dissociation curves. It has been assumed in the model analysis that β_b (for O_2 and CO_2) is constant, independent of partial pressure. However, particularly the O_2 dissociation curve is alinear, displaying an S shape. There are several ways to cope with the shape of the O_2 dissociation curve. One is to approximate this curve by simple analytical functions, which allow the differential equations [Eqs. (20) to (22)] to be analytically solved. Another is to use numerical procedures, like the Bohr integration method (cf. Forster, 1964; Piiper and Baumgarten-Schumann, 1968), for their solution. The disadvantage of these techniques is that the solution usually cannot be given for a wide range in underlying parameters.

Another method, used in the experiments of Table 4, is to analyze data obtained in hypoxia and hypercapnia where the dissociation curves are nearly linear. It must be realized, however, that in birds, where blood partial pressure potentially span the entire range from P_i to P_v, hypoxia must be very deep (Burger et al., 1979).

Bohr and Haldane effects. These effects provide links between O_2 and CO_2 binding of blood and result in a dependence of $(\beta_b)_{O_2}$ not only on O_2 but also on CO_2, and vice versa for $(\beta_b)_{CO_2}$. For alveolar lungs, these effects appear to constitute no major problem (Hlastala, 1973). However, in the crosscurrent system of avian lungs, the Haldane effect may exert a peculiar action whereby end-expired P_{CO_2} may exceed even mixed venous P_{CO_2}, which cannot easily be accounted for by the crosscurrent system without the Haldane effect (Meyer et al., 1976).

XI. Summary

Diffusion inside the body of multicellular organisms cannot provide enough oxygen for the metabolic demands of the tissues. Special systems had to be developed to transport oxygen and carbon dioxide between the environment and the body tissues. This paper reviews the structural features of gas exchange organs in vertebrates and provides the basis for a quantitative comparison of these systems. There are two important parameters to determine the gas exchange rates in idealized models:

1. Convective and diffusive conductances, comprising physical and chemical properties of the respired medium and blood, e.g., solubilities and binding characteristics for O_2 and CO_2
2. The structural arrangement of medium and blood flow in the gas exchange apparatus

Important differences may exist between the idealized model and the real system, particularly due to the existence of functional inhomogeneities which impair the gas exchange efficacy of a given system.

References

Altman, P. L., and Dittmer, D. S. (eds.) (1971). Respiration and Circulation. Bethesda, Maryland: FASEB.

Antonini, E., and Brunori, M. (1971). Hemoglobin and Myoglobin in Their Reactions with Ligands. Amsterdam: North-Holland Publishing Company.

Bauer, C. (1974). On the respiratory function of hemoglobin. Rev. Physiol. Biochem. Pharmacol. 70:1–31.

Baumgarten-Schumann, D., and Piiper, J. (1968). Gas exchange in the gills of resting unanesthetized dogfish (*Scyliorhinus stellaris*). Respir. Physiol. 5:317–325.

Burger, R. E., Meyer, M., Graf, W., and Scheid, P. (1979). Gas exchange in the parabronchial lung of birds: Experiments in unidirectionally ventilated ducks. Respir. Physiol. 36:19–37.

Czopek, J. (1961). Vascularization of respiratory surfaces of some Plethodontidae. Zool. Pol. 11:131–148.

Czopek, J. (1965). Quantitative studies on the morphology of the respiratory surface in amphibians. Acta Anat. 62:296–323.

Dejours, P. (1975). Principles of Comparative Respiratory Physiology. Amsterdam: North-Holland Publishing Company.

Duncker, H. -R. (1971). The lung air sac system of birds. Ergebn. Anat. Entwicklungsgesch. 45:Heft 6.

Forster, R. E. (1964). Diffusion of gases. In: Handbook of Physiology, Section 3: Respiration, Vol. I. Eds. W. O. Fenn and H. Rahn. Washington, D.C.: American Physiological Society, pp. 839–872.

Haab, P., Duc, G., Stucki, R., and Piiper, J. (1964). Les échanges gazeux en hypoxie et la capacité de diffusion pour l'oxygène chez le chien narcotisé. Helv. Physiol. Acta 22:203–227.

Hlastala, M. P. (1973). Significance of the Bohr and Haldane effects in the pulmonary capillary. Respir. Physiol. 17:81–92.

Hughes, G. M. (1963). Comparative Physiology of Vertebrate Respiration. London: Heinemann.

Hughes, G. M. (1973). Comparative vertebrate ventilation and inhomogeneity. In: Comparative Physiology: Locomotion, Respiration, Transport and Blood. Eds. L. Bolis, K. Schmidt-Nielsen, and S. H. P. Maddrell. Amsterdam:

North-Holland Publishing Company, pp. 187–220.

Hughes, G. M., and Morgan, M. (1973). The structure of fish gills in relation to their respiratory function. Biol. Rev. 48:419–475.

Hughes, G. M., and Shelton, G. (1962). Respiratory mechanisms and their nervous control in fish. Adv. Comp. Physiol. Biochem. 1:275–364.

Irving, L. (1964). Comparative anatomy and physiology of gas transport mechanisms. In: Handbook of Physiology, Section 3: Respiration, Vol. 1. Eds. W. O. Fenn and H. Rahn. Washington, D.C.: American Physiological Society, pp. 177–212.

Johansen, K. (1971). Comparative physiology: Gas exchange and circulation in fishes. Annu. Rev. Physiol. 33:569–612.

Jones, J. D. (1972). Comparative Physiology of Respiration. London: Edward Arnold.

King, A. S. (1966). Structural and functional aspects of the avian lungs and air sacs. Int. Rev. Gen. Exp. Zool. 2:171–267.

Kleiber, M. (1961). The Fire of Life. An Introduction to Animal Energetics. New York: Wiley.

Meyer, M., Worth, H., and Scheid, P. (1976). Gas-blood CO_2 equilibration in parabronchial lungs of birds. J. Appl. Physiol. 41:302–309.

Mill, P. J. (1972). Respiration in the Invertebrates. London: Macmillan, St. Martin's Press.

Negus, V. (1965). The Biology of Respiration. London: Livingstone.

Piiper, J., and Baumgarten-Schumann, D. (1968). Effectiveness of O_2 and CO_2 exchange in the gills of the dogfish (*Scyliorhinus stellaris*). Respir. Physiol. 5:338–349.

Piiper, J., and Scheid, P. (1972). Maximum gas transfer efficacy of models for fish gills, avian lungs and mammalian lungs. Respir. Physiol. 14:115–124.

Piiper, J., and Scheid, P. (1975). Gas transport efficacy of gills, lungs and skin: Theory and experimental data. Respir. Physiol. 23:209–221.

Piiper, J., and Scheid, P. (1977). Comparative physiology of respiration: Functional analysis of gas exchange organs in vertebrates. In: International Review of Physiology, Respiration Physiology II, Vol. 14. Ed. J. G. Widdicombe. Baltimore: University Park Press, pp. 219–253.

Piiper, J., and Scheid, P. (1980). Blood-gas equilibration in lungs. In: Pulmonary Gas Exchange. Volume I. Ventilation, Blood Flow, and Diffusion. Ed. J. B. West. New York: Academic Press, pp. 131–171.

Piiper, J., Dejours, P., Haab, P., and Rahn, H. (1971). Concepts and basic quantities in gas exchange physiology. Respir. Physiol. 13:292–304.

Piiper, J., Gatz, R. N., and Crawford, E. C., Jr. (1976). Gas transport characteristics in an exclusively skin-breathing salamander, *Desmognathus fuscus* (Plethodontidae). In: Respiration of Amphibious Vertebrates. Ed. G. M. Hughes. New York: Academic Press, pp. 339–356.

Prosser, C. L. (1973). Oxygen: Respiration and metabolism. In: Comparative Animal Physiology, 3rd ed. Ed. C. L. Prosser. Philadelphia: Saunders, pp. 165–211.

Randall, D. J. (1970). Gas exchange in fish. In: Fish Physiology, Vol. IV. Eds. W. S. Hoar and D. J. Randall. New York: Academic Press, pp. 253–292.

Romer, A. S. (1970). The Vertebrate Body. 4th ed. Philadelphia: Saunders.

Scheid, P. (1978a). Analysis of gas exchange between air capillaries and blood capillaries in avian lungs. Respir. Physiol. 32:27–49.

Scheid, P. (1978b). Estimation of effective parabronchial gas volume during intermittent ventilatory flow: Theory and application in the duck. Respir. Physiol. 32:1–14.

Scheid, P. (1979). Mechanisms of gas exchange in bird lungs. Rev. Physiol. Biochem. Pharmacol. 86:137–186.

Scheid, P., and Piiper, J. (1970). Analysis of gas exchange in the avian lung: Theory and experiments in the domestic fowl. Respir. Physiol. 9:246–262.

Scheid, P., and Piiper, J. (1971). Theoretical analysis of respiratory gas equilibration in water passing through fish gills. Respir. Physiol. 13: 305–318.

Scheid, P., and Piiper, J. (1976). Quantitative functional analysis of branchial gas transfer: Theory and application to *Scyliorhinus stellaris* (Elasmobranchii). In: Respiration of Amphibious Vertebrates. Ed. G. M. Hughes. New York: Academic Press, pp. 17–38.

Scheid, P., and Piiper, J. (1980). Intrapulmonary gas mixing and stratification. In: Pulmonary Gas Exchange. Volume I. Ventilation, Blood Flow, and Diffusion. Ed. J. B. West. New York: Academic Press, pp. 87–130.

Scheid, P., Worth, H., Holle, J. P., and Meyer, M. (1977). Effects of oscillating and intermittent ventilatory flow on efficacy of pulmonary O_2 transfer in the duck. Respir. Physiol. 31: 251–258.

Schmidt-Nielsen, K. (1979). Animal Physiology: Adaptation and Environment, 2nd ed. London: Cambridge University Press.

Scholander, P. F. (1960). Oxygen transport through hemoglobin solutions. Science 131: 585–590.

Steen, J. B. (1971). Comparative Physiology of Respiratory Mechanisms. New York: Academic Press.

Tenney, S. M. (1979). A synopsis of breathing mechanisms. In: Evolution of Respiratory Processes: A Comparative Approach. Eds. S. C. Wood and C. Lenfant. New York: Marcel Dekker, pp. 51–106.

Weibel, E. R. (1963). Morphometry of the Human Lung. Berlin: Springer-Verlag.

Weibel, E. R. (1973). Morphological basis of alveolar-capillary gas exchange. Physiol. Rev. 53:419–495.

9

Facilitated Oxygen Diffusion by Oxygen Carriers

JONATHAN B. WITTENBERG AND BEATRICE A. WITTENBERG

Diffusion is a slow process. Oxygen has only a very limited solubility in water. Consequently, the rate of diffusion of oxygen into respiring cells limits the size of cells and limits the rate at which they can do sustained work. In those vertebrate muscles which are dedicated to sustained activity, the red muscles and red fibers in muscles of mixed fiber type, every muscle cell is in contact with at least one and as many as ten capillaries at its periphery. The problem of oxygen delivery to the tissue is reduced to a question of oxygen movement through the cytoplasm of each cell. Populations of separated individual cells can be prepared from the heart and liver of adult animals. Oxygen is supplied to these cells from a homogeneous surrounding medium whose oxygen pressure can be controlled experimentally. In this essay we focus attention on these two favorable preparations and on the legume root nodule and consider only noninvasive probes of intracellular oxygen pressure. We address the questions: How does oxygen traverse the cytoplasm of cells? What are the oxygen pressures within cells? Do intracellular hemoglobins enhance the flow of oxygen through the cytoplasm?

Intracellular hemoglobins are found in the red fibers of vertebrate skeletal muscle, in vertebrate hearts, in some vertebrate smooth muscles, in muscles, nerves, and other tissues of many invertebrates, and within the plant cells which are host to nitrogen-fixing symbiotic bacteria of legume root nodules. The tissue hemoglobin of vertebrate muscle and hearts is called myoglobin; that of legume root nodules is called leghemoglobin (legume hemoglobin). In addition, myohemerythrin, a non-heme, oxygen-binding iron protein is abundant in the muscles of a very few sipunculid worms (53). Myoglobin is generally found in those muscles requiring slow repetitive or sustained activity of considerable force. In such muscles the concentration of myoglobin increases in response to increased demand for oxygen; smaller increases are seen in response to decreased oxygen supply, for instance, at high altitude. The distribution and occurrence of myoglobin suggest that it may be of use to the muscle; we seek experimental proof. Myoglobin (68,111), the facilitated diffusion of oxygen by myoglobin (54,111) and the physiological role of leghemoglobin (7,115) have been reviewed.

I. Steady-State Partial Deoxygenation of Myoglobin in Active Tissue

The discussion of oxygen pressure within cells begins with Millikan (67) whose experiments are at once the first and the best. In these experiments Millikan used the degree of saturation of myoglobin with oxygen as an instantaneous indicator of local oxygen pressure within the cells. The concept of using cell constituents as reporter groups was not new; it had originated with Keilin, who noted the reduction of intracellular cytochrome, and has been widely used since. The elegant instrumentation constructed by Millikan and others in Cambridge at that era was new. Tissue hemoglobin remains the unique, noninvasive reporter of intracellular oxygen pressure. Millikan's experiment stands alone in that the tissue was in its normal, blood-perfused state.

Millikan constructed a photoelectric device by means of which the degree of oxygen saturation of the intracellular hemoglobin of the cat's soleus muscle could be measured instantaneously and recorded continuously. Measurements could be made of the muscle in situ without interfering with either the nerve supply or the blood supply to the tissue. Blood hemoglobin contributed little to the absorption spectrum of the muscle, perhaps because the red blood cells present a much smaller cross section for the absorption of light than the muscle cells; the presence of blood hemoglobin in the muscle capillaries caused an error of less than 10% in the readings. When the muscle was at rest with the blood flowing, myoglobin was largely or completely oxygenated. When the muscle was stimulated to contract, the demand for oxygen rose immediately and desaturation of myoglobin started within the time resolution of the apparatus and reached a maximum value within 1 sec from the onset of contraction.

It is apparent from this experiment that myoglobin acts as a short-time oxygen store and supplies a goodly part of the oxygen consumed at the onset of contraction. Millikan (68) and Hill (41) calculated that the oxygen bound to myoglobin was just sufficient to supply the heart for the duration of systole when flow in the cardiac capillaries may be slowed by constriction of the vessels. Later Åstrand et al. (9) deduced that myoglobin-bound oxygen stores were called on during short spells (say 10–30 sec) of heavy muscular work in man.

There is no doubt that myoglobin serves as a long-term oxygen store in diving mammals and birds. The muscles of these animals often contain manyfold the concentration of myoglobin required to support muscular work and appear almost black by virtue of their contained myoglobin.

If, in Millikan's experiment, stimulation is continued so that the muscle enters a sustained tetanic contraction, the myoglobin rapidly reaches a steady level of desaturation with oxygen which is maintained for the duration of the tetanus, and returns rapidly to full saturation when the muscle relaxes. The rate of return is independent of the duration of tetanus, showing that no oxygen debt was incurred and that the muscle had functioned in a steady state. Blood flow through the soleus is not greatly affected by maximal tetanic contraction. It is this steady state, in which metabolic demand is neatly balanced by the flow of oxygen from the capillaries, that interests us. It must be the normal state of the heart which operates continuously and can contract no debt. It is probably the predominant working state of postural muscle and of muscles of locomotion (111).

Millikan's finding that intracellular hemoglobin functions in states of partial oxygenation appears to be quite general. At least it has been found so in each of the admittedly few cases in which a serious search has been made. An interesting example is the intracellular hemoglobin of the giant neurons of the mollusc, *Aplysia*, which Chalazonitis (24) finds is normally about 25% saturated. Imposed changes in intracellular oxygen pressure and myoglobin saturation change the membrane potential and frequency and rhythmicity of spontaneous firing of the neuron (24,25). Three additional instances discussed below are myoglobin in pigeon breast muscle fibers studied by Wittenberg et al. (106), leghemoglobin in the soybean root nodule (4), and myoglobin in the saline perfused heart with an abundant oxygen supply (35). Myoglobin unsaturation in blood-perfused skeletal muscle and heart has been calculated from partition of carbon monoxide between blood and tissue (27–29).

The partial oxygen saturation of intracellular hemoglobin in respiring cells exposed to oxygen pressures which at equilibrium would saturate the hemoglobin implies gradients of oxygen pressure within the cell. Mammalian myoglobin, at body temperature, is half saturated with oxygen near 3 torr, 4.5 μM dissolved oxygen, whether in situ in heart tissue (94), in isolated heart muscle cells (B. A. Wittenberg, unpublished results), or when isolated as a pure protein. The intracellular myoglobin near the capillary, in which the oxygen pressure does not ordinarily fall below 20 torr (111), must be largely saturated. Somewhere within the cell it must be largely desaturated, with a steep gradient of myoglobin saturation and oxygen pressure from the cell periphery.

II. Myoglobin in Tissue Oxygen Delivery

Forty years after Millikan's experiment, Wittenberg et al. (106) obtained the first direct proof that myoglobin actually is of use to a muscle. They used bundles of essentially undamaged red muscle fibers teased from pigeon breast muscle, and measured oxygen consumption as a function of ambient oxygen pressure. The essential condition of this experiment was that all measurements were made in steady states of oxygen supply and consumption. In the steady state the storage function of myoglobin makes no contribution to oxygen supply; oxygen transport can be studied. The steady state oxygen uptake of their preparations decreased monotonically from a maximal value near 150 torr, where oxygen uptake is independent of oxygen pressure, to zero at zero oxygen pressure, Fig. 1A, indicating that respiration was diffusion limited. Four reagents, nitrite, hydroxylamine, phenylhydrazine, and hydroxyethylhydrazine, were found which abolish the oxygen-binding function of myoglobin without demonstrable effect on the oxygen uptake of isolated mitochondria or of mitochondria in situ at saturating ambient oxygen pressure. The effect of each of these reagents was to decrease the steady state oxygen uptake of muscle within that range where oxygen consumption is oxygen limited, Fig. 1B. Near the middle of this range abolishing myoglobin function halves the rate of oxygen uptake; the original rate can be restored by doubling the oxygen pressure. The effect is reversed as tissue enzymes slowly restore the myoglobin to its native ferrous, oxygenated state. As a control, the pattern of inhibition by myoglobin-reactive reagents was shown to be entirely different from that found with the known inhibitors of mitochondrial oxygen uptake, cyanide and antimycin A (Fig. 1C). These latter inhibit oxygen uptake at all oxygen pressures up to the highest examined.

Observations with a microscope fitted with a microspectroscope ocular showed that the myoglobin of the fiber bundles was extensively deoxygenated under the conditions of the experiment.

The simplest interpretation of these findings is that myoglobin acts to enhance the inflow of oxygen into red muscle fibers. The presence of myoglobin doubles the oxygen uptake that the tissue can maintain. Oxygen uptake is tantamount to ATP generation in muscle, and doubling the capacity of the muscle to sustain work is a large contribution indeed.

As early as 1922 Douglas and Haldane (34) and Haldane and Priestley (39) had come to view diffusion of oxygen through the tissue cells, particularly the muscle cell, as the central limitation which the circulatory system was designed to overcome. Recently, Weibel (102) has amplified this point of view, taking his evidence from the new science of quantitative electron microscopy, and from

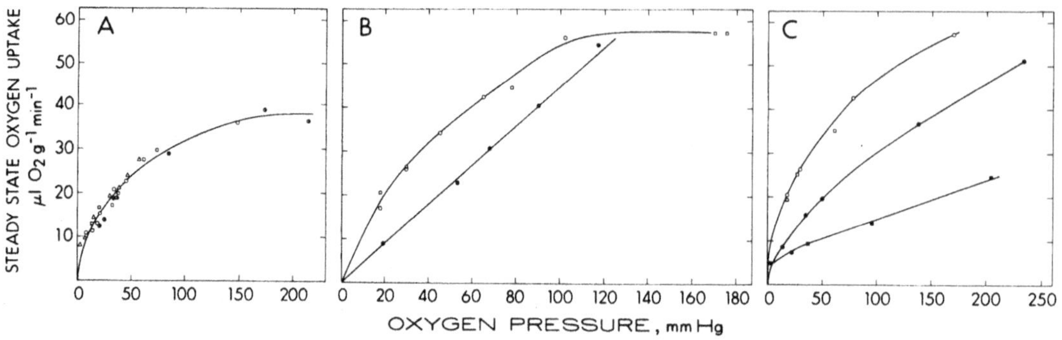

Figure 1. Steady-state oxygen uptake of pigeon breast muscle fiber bundles as a function of steady-state ambient oxygen pressure. ***A.*** Oxygen uptake of four untreated preparations. ***B.*** Effect of hydroxylamine. *Open symbols*, untreated fiber bundles. *Closed circles*, the same fiber bundles in the presence of hydroxylamine. ***C.*** Effect of cyanide and antimycin A. *Open symbols*, untreated fiber bundles. *Closed circles*, fiber bundles in the presence of cyanide at a concentration sufficient to partially inhibit mitochondrial respiration. *Closed squares*, fiber bundles in the presence of antimycin A. From Wittenberg et al. (106).

the biochemical and physiological studies of trained muscles reviewed by Holloszy (42).

Endurance training in man brings about a near doubling of maximal cardiac output which appears to be proportional to the increase in whole body maximal oxygen consumption. The muscles respond to training by augmenting their capacity for aerobic metabolism, often without much hypertrophy of the contractile material, and this is paralleled by increased volume density of the mitochondrial inner membrane (site of the respiratory electron transport chain and of cytochrome oxidase). The marked adaptation of the ability of the heart to deliver blood and of the tissues to consume oxygen contrasts strongly with the apparently small changes in the local vascularization of the trained muscles. Although the average number of capillaries per muscle fiber increases to meet the increased average fiber cross section, the capillary density of the muscle remains unchanged at 600 capillaries/mm^2. This means that the diffusion distance, that is one half the distance between capillaries, remains unchanged at 20 μm. In contrast the myoglobin content of trained muscles almost doubles (42) as did the volume density of mitochondrial inner membrane. This recalls the early experiments of Lawrie (62) who found that the myoglobin content of muscles parallels their content of cytochrome oxidase. Therefore, increased work output is met by cellular adaptations, increased mitochondrial inner membrane density and increased myoglobin content, and by increased cardiac output to bring more oxygen to the muscle.

III. Leghemoglobin-Mediated Oxygen Delivery in Plants

We turn now to nitrogen fixation in the legume root nodule, a system adapted to work at very low oxygen pressure. In this circumstance essentially all of the useful oxygen uptake is mediated by intracellular hemoglobin. By useful oxygen uptake we mean oxygen used for ATP production through oxidative phosphorylation. The bulk of the ATP formed is consumed in the fixation of atmospheric nitrogen by the enzyme complex nitrogenase, and, accordingly, the rate of nitrogenase action provides a continuing measure of the rate of ATP generation. Bergersen et al. (15) devised an ingenious way to monitor nitrogenase activity in the presence of the inhibitor, carbon monoxide, and introduced the idea of measuring simultaneously two rates, the rate of ATP generation and the rate of oxygen utilization. Following this lead, the complex system operating in the root nodule was unraveled very quickly (7,8,11,13,14,115,118). The first finding was that blockade of leghemoglobin function with carbon monoxide nearly abolishes ATP generation without any large effect on the rate of oxygen uptake.

Nitrogen-fixing nodules are formed on the roots of legumes in response to invasion by bacteria of the genus *Rhizobium* which proliferate within the cells of the developing nodule. *Rhizobia*, modified for symbiotic life, are called bacteroids. These occupy about one third of the plant cell volume and are responsible for most of the vigorous oxygen uptake of the nodule. Mitochondria are sparse. Fixation of atmospheric nitrogen into ammonia is accomplished by an enzyme complex, nitrogenase, which occurs wholly within the bacteroids. Nitrogenase is destroyed by traces of oxygen, but at the same time depends for its activity on a supply of ATP formed by bacteroidal oxidative phosphorylation. The nodule is surrounded by a cortex which limits the oxygen influx available to the central tissue. Within the central tissue a system of air passages (12) makes contact with each individual cell. In fact, the density and distribution of these air passages, perhaps coincidentally, is remarkably like the density and distribution of capillaries in cardiac muscle (113). The role of leghemoglobin, accordingly, is to assure the flow of oxygen within each individual cell. In the nodule, as in muscle, the problem of oxygen delivery to the tissues reduces to a question of oxygen movement through the cytoplasm of each cell.

Leghemoglobin in the root nodule, like myoglobin in a muscle, operates in a state of partial saturation with oxygen; in fact it is about 20% oxygenated in the functioning nodule (4). The oxygen affinity of leghemoglobin is extraordinarily great; $P_{1/2}$ for soybean leghemoglobin *a* is 0.04 torr, equivalent to 70 n*M* dissolved oxygen at 20°C. The high affinity is a consequence of very fast, almost diffusion limited combination with oxygen; the oxygen dissociation rate is similar to that of other hemoglobins (44,117). Twenty percent saturation corresponds to a volume-averaged intracellular oxygen pressure of about 0.006 torr, or 10 n*M* dissolved oxygen in the cytoplasm. The concentration of leghemoglobin in the nodule cell, 1.0–1.5 m*M*, is about 100,000 times the concentration of dissolved oxygen. It follows that an overwhelming proportion of oxygen transport to the bacteroids must be leghemoglobin mediated.

Rhizobia, when they become bacteroids, lose the conventional terminal oxidases, cytochrome *c* oxidase and cytochrome *o*, charateristic of macroaerobic life and acquire a new set of spectroscopically demonstrable hemeproteins (2,3). At the same time they acquire two functionally distinct terminal oxidase systems (5,7,115). One, which we call the *effective* bacteroid terminal oxidase, is leghemoglobin dependent. An hemeprotein *P*-450 is a component of this terminal oxidase system (8). Through the action of this system leghemoglobin-delivered oxygen supports an efficient formation of ATP through oxidative phosphorylation in the bacteroids. The *effective* oxidase functions only at very low oxygen pressure, with an optimum near 40 n*M* dissolved oxygen, coincident with half-saturation of leghemoglobin with oxygen. An *ineffective* bacteroid terminal oxidase has a much lower affinity for oxygen, accepts free dissolved oxygen and shows little activity below 1 μM oxygen. ATP is not accumulated by operation of the ineffective oxidase.

Having dissected the system, we can attempt to reconstruct the conditions of oxygen supply to the bacteroids of the intact nodule. We envision two extreme states:

During active nitrogen fixation, the bacteroids consume oxygen solely or predominantly through their leghemoglobin-dependent, *effective* terminal oxidase system. Essentially all of the oxygen flux through the cytoplasm is leghemoglobin mediated. The concentration of free oxygen dissolved in the cytoplasm is less than 1/10,000 of the concentration of leghemoglobin-bound oxygen. The concentration of oxygen at the interface between the cytoplasm and the bacteroid surface is unknown, but clearly is pegged or buffered at some value near the $P_{1/2}$ of leghemoglobin by facilitated diffusion of oxygen in the region immediately adjacent to the bacteroid surface. We foresee that oxygen consumed at the surface will be replenished by oxygen drawn from the bulk solution, carried pick-a-back aboard leghemoglobin molecules undergoing random translational diffusion. To define the gradient of oxygen pressure through the cytoplasm we would have to know the oxygen pressure in the air passages. This is unknown, but assuredly small, and we surmise that the gradient is relatively flat.

A very different situation prevails in nodules in which the oxygen-combining function of leghemoglobin has been blocked by reaction with carbon monoxide (7,15,115). Oxygen now traverses the cytoplasm only as free dissolved oxygen, and in this new steady state the gradient of oxygen pressure from the air passages to the bacteroid must have become sufficiently steep to support this new oxygen flux. The oxygen pressure at the low end of the gradient, adjacent to the bacteroid surface has risen to more than 1000 n*M*. We deduce this because the bacteroids are now accepting oxygen through their *ineffective* terminal oxidase system and this system scarcely operates below this pressure. Nitrogen fixation effectively stops for lack of ATP. The oxygen pressure in the air passages at the high end of the gradient, although unknown, must have risen sufficiently to provide the steeper gradient now required to drive oxygen toward the bacteroids.

Interestingly, the nodule oxygen uptake scarcely differs in these two extreme states (15,85). This fact, which at first appeared inexplicable, finds a simple explanation when one realizes that nodule oxygen consumption is set by diffusion across the nodule cortex (97,115); the central tissue of the nodule, in order to maintain its low oxygen pressure, must consume all the oxygen it receives.

Within the cells of the central tissue, allocation of the available oxygen is under biological control. Nitrogen fixation in the nodule is normally oxygen limited and increases immediately in response to increased oxygen pressure outside the nodule. If rhizosphere oxygen pressure is reduced nitrogen fixation drops immediately. However, if the new rhizosphere pressure is maintained for several hours, the system adapts and returns to its initial rate of nitrogen fixation (33). We do not know what has occurred but surmise that adaptation results from return to an optimal internal oxygen pressure, possibly by adjustment of the relative rates of the *effective* and *ineffective* bacteroid terminal oxidases. This implies cellular control over the internal oxygen pressure gradient.

In conclusion, oxygen can move through the cytoplasm of the nodule cell in two radically different modes. In one essentially all of the oxygen is leghemoglobin carried; in the other none. An outside observer sees only an unimpressive change in oxygen uptake.

IV. Myoglobin-Facilitated Oxygen Diffusion

About twenty years ago the phenomenon of myoglobin-facilitated oxygen diffusion was discovered by Wittenberg (109) and independently by Scholander (80). Wittenberg (110,111) proved the molecular mechanism and Wyman (121), in a simultaneous publication, presented the classic statement of the basic equations governing the phenomnenon. Hemmingsen (40) contributed important measurements of the oxygen flux made using isotopic oxygen. A number of workers have presented solutions to the basic equations. Here we will follow that developed by Murray and colleagues.

When a thin layer or slab of hemoglobin or myoglobin solution is interposed between two chambers, in one of which the oxygen pressure is in the neighborhood of 20 torr or greater, and in the other of which it is kept as nearly as possible at 0, the flux of oxygen across the slab exceeds that corresponding to simple diffusion by an amount which is independent of the oxygen pressure on the high pressure side, Fig. 2. The excess is facilitation. It is lacking in the case of nitrogen and is abolished by conversion of the protein to the ferric state which does not

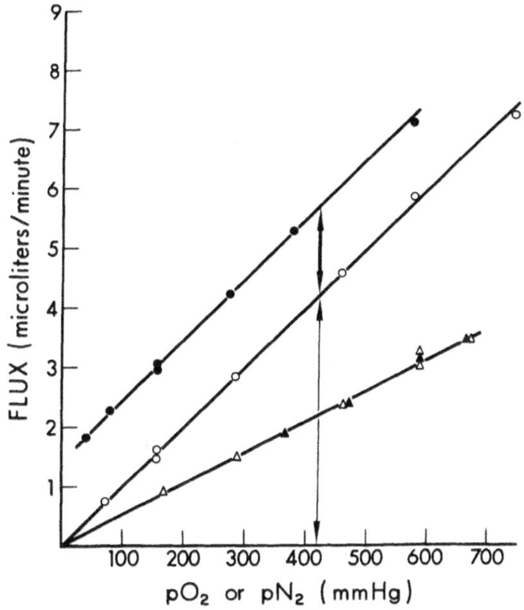

Figure 2. Steady-state diffusion of oxygen (*circles*) and nitrogen (*triangles*) through a solution of oxyhemoglobin (*solid symbols*) compared with diffusion through a solution of ferric hemoglobin (*open symbols*). *Light arrow*, diffusive component of oxygen flux. *Heavy arrow*, hemoglobin-facilitated component. From Wittenberg (110).

combine with oxygen (Fig. 2), or by the action of carbon monoxide which preempts the oxygen-binding sites of the myoglobin. Facilitation is not a special property of the heme group. Any oxygen-binding protein of suitably small molecular size will facilitate oxygen diffusion.

It was at once clear that the phenomenon was in some way the result of the diffusion of the protein molecules, which are constantly taking up and giving off oxygen in a region of an oxygen concentration gradient. Immobile myoglobin molecules can make no contribution to the translational movement of oxygen. The thermodynamic activity of released oxygen molecules is precisely the same as that of oxygen molecules which have not yet been bound, and at steady state the oxygen which is in solution is not affected by the presence of immobile oxygen-binding sites. The experiment has been done; immobilized hemoglobin in fact did not facilitate oxygen diffusion (32). It was originally proposed by Scholander that rotational diffusion of the protein molecules was responsible for facilitated diffusion, the molecules functioning somewhat after the fashion of a bucket brigade. Calculations by Wyman (121) show that this could not be the case; the contribution by rotatory diffusion is orders of magnitude too small to account for the results.

Translational diffusion of oxymyoglobin molecules is sufficient to explain facilitated diffusion. Proof comes from the experiment of Riveros-Moreno and Wittenberg (77), who determined the translational self-diffusion coefficients of hemoglobin and myoglobin to the highest achievable protein concentrations. The facilitated component of the oxygen flux through a hemoglobin or myoglobin solution was found to be proportional to the self-diffusion coefficient of the protein at all concentrations (Fig. 3).

It is convenient to regard the total flux of

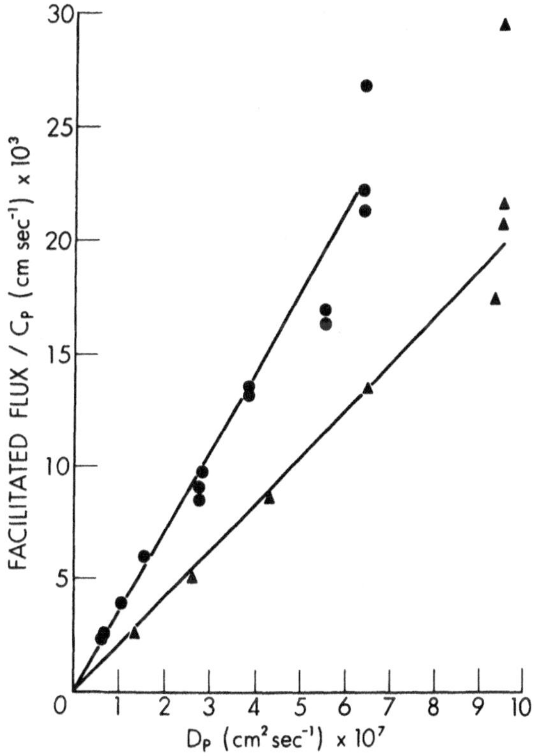

Figure 3. Facilitated flux per unit protein concentration as a function of self-diffusion coefficient of protein. *Solid circles*, hemoglobin; *solid triangles*, myoglobin. From Wittenberg (111).

oxygen through a myoglobin solution from a region of high to a region of very low oxygen pressure as the sum of two separate fluxes which are additive. These are a simple diffusive flux of dissolved oxygen (the light arrow of Fig. 2), and a myoglobin-facilitated oxygen flux (the heavy arrow of Fig. 2). The myoglobin-facilitated flux is actually a flux of oxymyoglobin molecules undergoing translational diffusion in a gradient of myoglobin oxygenation. This is expressed in the integrated form of Wyman's equation, where the total oxygen flux, $-F$, is equal to the sum of two terms:

$$-F = mD_pC_p\frac{dY}{dx} + D_c\frac{dC}{dx} \qquad (1)$$

Here D_p and D_c are the diffusion coefficients of protein and oxygen respectively; mC_p is the product of the protein concentration C_p and the number of binding sites, m, per molecule; x is the distance through the membrane and $(dY)/(dx)$ and $(dC)/(dx)$ are the gradients, respectively, of myoglobin oxygenation and oxygen concentration. In the case of hemoglobin, the assumption that the protein is essentially saturated with oxygen at the high pressure face of the slab and essentially desaturated at the low pressure face ($\Delta Y \cong 1$) suffices; theory and experiment agree within 15%. This is not true for myoglobin; under the conditions of Wittenberg's experiment ΔY was much less than unity. To consider this case and to learn more about the behavior of the system it is necessary to solve the equations in detail.

The governing equations were solved by Murray (70) using the method of singular perturbations. One condition, which causes the phenomenon to be singular in the mathematical sense and thereby solvable, is that the relaxation times of the myoglobin-oxygen reaction are small compared with the diffusion time of the myoglobin–oxygen complex (71). Two important aspects of facilitation at once become apparent. First, the myoglobin or hemoglobin is everywhere nearly at equilibrium with the oxygen in solution; it is often not necessary to examine the individual rate constants for oxygen combination and dissociation to predict facilitation [for a contrary point of view see Kreuzer and Hoofd (57)]. Second, the principle emerges that hemoglobin, myoglobin, or any macromolecule can function as a carrier only if its saturation with its ligand is incomplete somewhere in the system (18,71,72). We have already noted that myoglobin and leghemoglobin function in states of partial oxygenation in the tissue.

It might at first be supposed that the flux of myoglobin-bound oxygen would be trifling compared to the flux of free oxygen. This turns out not to be so. The explanation is found in the great insolubility of oxygen in water and in the abundance of myoglobin. The concentration of oxygen in air-equilibrated water at 20°C is 270 μM; in muscle tissue water at an assumed venous P_{O_2} of 20 torr and at 37°C, it is 35 μM. The concentration of myoglobin in many muscles, 100–500 μmoles per kg wet weight (81), is of the order of 10–15-fold greater than the oxygen concentration. On the other hand, the diffusivity of myoglobin is only about 20-fold less than the diffusivity of free oxygen. The sheer abundance of myoglobin in muscle, relative to dissolved oxygen, thus compensates for its lesser diffusion coefficient, and it is apparent from Eq. (1) that the fluxes into muscle of free oxygen and of myoglobin-bound oxygen will be of the same order. In the extreme instance of leghemoglobin in the functional root nodule, diffusion of free dissolved oxygen may be neglected, the second term of Eq. (1) vanishes, and essentially the entire flux is leghemoglobin facilitated.

The profiles of oxygen concentration and of myoglobin saturation through the thickness of the slab of hemoglobin or myoglobin solution can be specified using Murray's solution of the governing equations. This is done in Fig. 4A using the conditions of Wittenberg's experiments. The profiles are nonlinear and the fractional saturation, Y, can be far from zero or unity as the boundaries are approached. When, as in Fig. 4, the rate of oxygen removal at the low pressure boundary is large, the fraction of the total flux carried

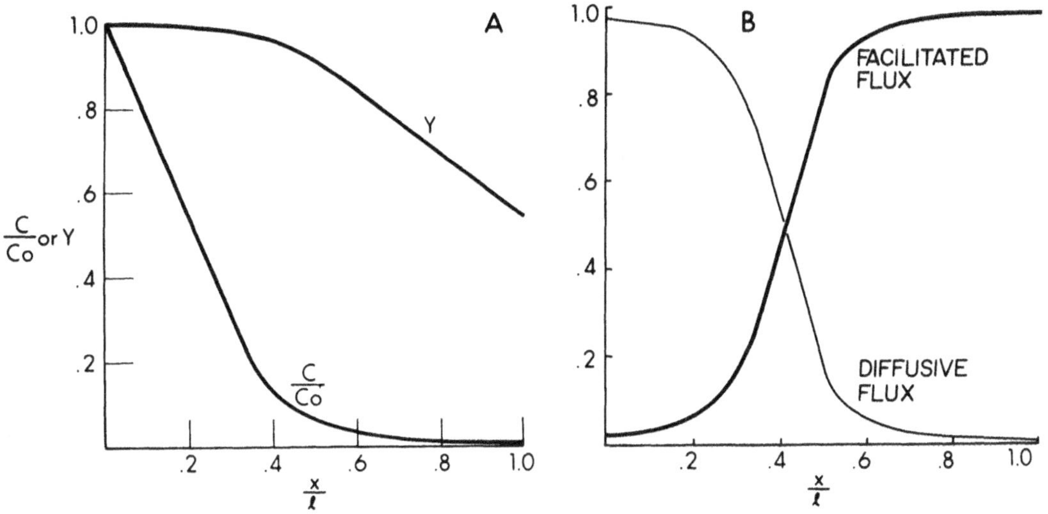

Figure 4. ***A.*** Dimensionless profiles of oxygen concentration (C/C_0) and myoglobin saturation (Y) as functions of the distance (x/l) through slabs of hemoglobin or myoglobin solution. From Murray (70). ***B.*** Dimensionless profiles of the fraction of total steady-state flux through a layer of myoglobin solution mediated by myoglobin-bound oxygen (*heavy line*) and by dissolved oxygen (*light line*). From Wittenberg (111).

by protein-bound oxygen near that boundary becomes very large, Fig. 4B.

This calculation extended to respiring muscle (71) predicts the oxygen supply within a typical muscle fiber. As a first approximation it is assumed that muscle respiration is independent of mitochondrial oxygen pressure. On this basis, the center of the fiber would be anaerobic were it not for the presence of myoglobin. In the presence of myoglobin, oxygen is available throughout the fiber. A useful way of looking at the situation is to estimate the maximum metabolism which a muscle fiber can sustain before an anaerobic center just appears. First approximated by Wyman (121), Murray's calculation shows that facilitated diffusion will contribute about half of the total oxygen influx. The calculation is in remarkable agreement with the experiments of Wittenberg et al. (106) using pigeon breast muscle. Assuming Michaelis–Menten kinetics for mitochondrial oxygen consumption, the central anaerobic region becomes smaller by a factor of about 2 (95). Profiles of oxygen concentration and myoglobin saturation for these two cases are presented in Fig. 5. At the rate of oxygen consumption characteristic of red muscle performing sustained work, the concentrations of free and myoglobin-bound oxygen remain significantly above zero all the way to the center of the muscle fiber.

Carbon monoxide, present even at relatively minute concentration compared to oxygen, by binding preferentially to myoglobin, will seriously impair facilitated diffusion (18,58,72,92). This suggests the possibility that traces of carbon monoxide from cigarette smoking may prove literally the last straw to persons with impaired cardiac function. The wisdom of smoking at low oxygen pressure, for instance by airplane pilots, is controversial.

Many other authors have presented solutions of the equations of diffusion accompanied by chemical reaction. Among these those of Kreuzer and Hoofd (43,56–58) are particularly addressed to the question of muscle respiration. The status of these efforts, to 1974, has been reviewed (82). The problem of the boundary conditions has been difficult, and fundamentally different conditions have been chosen by different authors. Numerical solutions which bystep the problem have also been developed. Murray and associates [see particularly Mitchell and

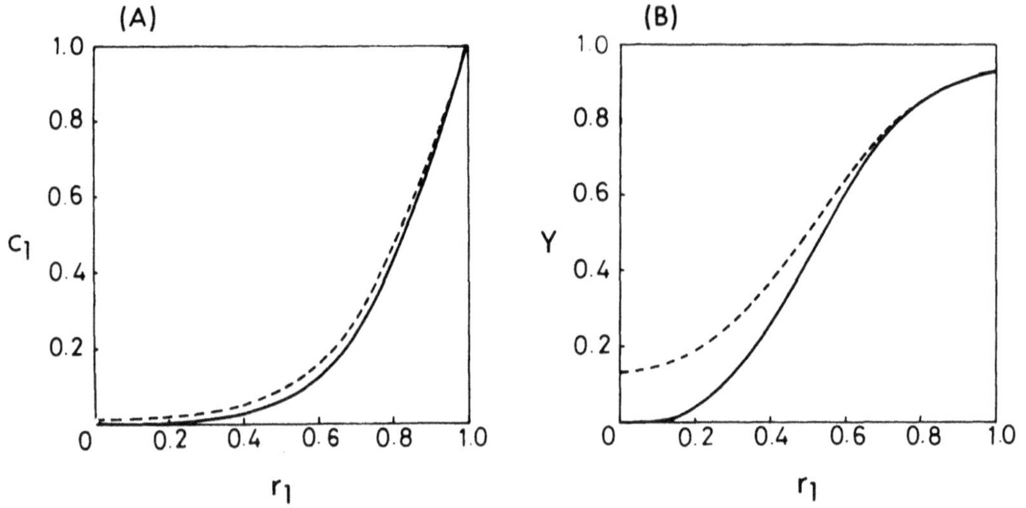

Figure 5. ***A.*** Dimensionless profiles of oxygen concentration (C_1) and ***B.*** dimensionless profiles of myoglobin saturation (Y) for a cylindrical cell respiring at a rate expected for muscle working at a steady state. *Solid line*, oxygen consumption independent of oxygen pressure. *Broken line*, oxygen consumption following a Michaelis-Menten type dependence on oxygen pressure. From Taylor and Murray (95).

Murray (69) and Taylor and Murray (95)] writing from a biologist's point of view, meticulously compare their solutions with those of other authors. The upshot is that the different boundary conditions are equivalent for practically all situations of biological interest, and the simple algebraic solution gives results in good accord with numerical computation of the full problem. Subsequent to the review of Schultz et al. (82), several mathematical treatments of facilitated diffusion have appeared (79,86,99). Some extension of this theory has been made for whole tissue as well (90,91). A role in whole body respiration has been calculated (37).

Accurate values of some of the constants of Eq. 1 have recently become available. These included the diffusion coefficients of oxygen (48,66), ions (20,21,60), water (22), inert gases (23,47), and carbon dioxide (48) in muscle. The volume of water available for diffusion is estimated at 80% of the muscle water (22). Krogh's diffusion coefficient for gases in muscle is about half that in water (47,48,66). The diminished diffusion coefficients compared to those in pure water are accounted for by the tortuosity of the diffusion path as small molecules wend their way around bulky protein molecules (101). The diffusion of carbon dioxide in muscle is facilitated (49). The solubility of gases in muscle is complicated by partition between aqueous and lipid phases of the tissue (23). The diffusivity of proteins in cytoplasm remains unknown. The self-diffusion coefficients of proteins in concentrated solution, however, have been determined (50,77,100).

Facilitated diffusion requires mobility of the oxygen-binding macromolecule within the cytoplasm. This is not proved. Accordingly, an experimental indication of facilitated diffusion in a muscle tissue is welcome. Kreuzer and colleagues (55) have attempted to demonstrate myoglobin-mediated oxygen transport across slabs of respiring chicken gizzard interposed between two chambers of differing oxygen pressure. Chicken gizzard is a smooth muscle whose myoglobin content, 500–1000 μmoles per kg wet weight, exceeds severalfold that of most heart or skeletal muscle (55,81). They find a small, carbon monoxide-inhibitable component of the oxygen flux, which is independent of the oxygen pressure on the upstream side of the slab. This, they calculate, is of the order expected for facilitated diffusion.

Profiles of myoglobin saturation through slabs of respiring chicken gizzard have been constructed for slabs frozen rapidly while exposed to known oxygen pressure on their opposing faces (83). In light of the probably very low respiratory rate (55), the long time course of the experiment, and the specialized structure of gizzard smooth muscle, these findings may have only limited relevance to facilitated diffusion within skeletal muscle cells.

Claims that cytochrome *P*-450 accelerates the flow of oxygen through tissue (19,38, 63,64) fall outside the scope of our discussion.

V. Oxygen Delivery in Saline-Perfused Muscle

Blood hemoglobin interferes in spectrophotometric or fluorometric measurements and, for the study of whole organs, perfusion with balanced saline solutions offers an obvious advantage.

Tamura et al. (94) present the outstanding study addressed to oxygen gradients in the saline-perfused rat heart. In contradiction of the earlier findings of Fabel and Lübbers (35), they find that the degree of myoglobin oxygenation changes with each beat. The time-averaged degree of deoxygenation of myoglobin and reduction of cytochrome oxidase go hand in hand, with direct proportionality between them. This is a surprising result because the oxygen pressures for half-reaction of the purified components are far apart, and in artificial mixtures they react independently (94). The result is confirmed in experiments from Lübbers' laboratory using the saline-perfused guinea pig heart (36). The explanation probably lies in the heterogeneity of tissue oxygenation (76,84,88).

With mild reduction of arterial oxygen content, the myocardium of the saline-perfused heart remains homogeneously fully aerobic and the coronary blood flow responds smoothly to decreased arterial oxygen content (88). With more profound hypoxia, the oxygen supply becomes heterogeneous with sharply bounded zones of fully aerobic and completely anaerobic tissue, each 200–500 μm wide, spanning tens of capillaries (26,87,88). Very little tissue is partially anaerobic. The anaerobic zones apparently correspond to the capillary beds supplied by individual arterioles (88). Optical absorbance changes in perfused myocardium (76,88) or liver (84) therefore represent the sum of fully aerobic and fully anaerobic zones. They cannot be interpreted in terms of intracellular gradients.

Another important attempt to study myoglobin function is that of Cole et al. (30) using the fluorocarbon-perfused dog heart. The oxygen-binding function of myoglobin, in this study, was blocked by nitrite introduced into the perfusion fluid. This was without effect either on the steady-state oxygen consumption or the mechanical performance of the isovolumically contracting ventricle, even under conditions of limiting oxygen supply. In addition to the problem of patchy hypoxia, another equally severe difficulty complicated this experiment. The heart became edematous, and although the degree of edema was slight, electron microscopy showed that the edema was mainly localized around the capillaries (Cole, private communication). This extracellular barrier to oxygen diffusion might easily dominate the total oxygen flow and override intracellular mechanisms.

VI. Oxygen Gradients in Isolated Cardiac Cells

Cells isolated from the tissues of adult animals are particularly suited to the study of intracellular oxygen gradients. The extracellular oxygen pressure can be controlled without the ambiguity which plagues experiments using perfused tissues. The diffusion path for oxygen is homogeneous and accurately defined. Fortunately optical spectra of cell suspensions are crisply resolved and free from some of the artifacts which distort spectra of slabs of tissue. Adult heart cells are best because intracellular myoglobin content increases with age. Freshly isolated cells are preferred to cultured cells because the metabolic requirements of heart cells are known to change in culture.

The diffusion path for oxygen in isolated cells will be different from that in tissue. Quantitative electron microscopy, morphometry, has led to a restatement of the diffusion path for oxygen in muscle. On the basis of limited data, Weibel (102) suggests tentatively that the surface area of the capillaries supplying a volume of red muscle may be proportional to the density of mitochondria. In diverse muscles, rat heart, diaphragm, and soleus, he finds that 1 μm^2 of capillary surface supplies in each case 5 μm^3 of mitochondria. The capillary wall offers by far the smallest area in the cellular diffusion path. (Table 1); the oxygen flux per unit area (Table 2) is greatest at this point. The proper model for diffusion into a red muscle cell therefore is a sector of a cylinder, truncated at the apex where the capillary lies, with the oxygen flux per unit area, even without correction for oxygen consumed, decreasing as the square of the radial distance from the capillary. This is essentially the classic Krogh cylinder. There is nothing, however, in the structure of muscle or in the known physiology of red muscle (111) to suggest that the central mitochondria of the myocyte are starved for oxygen, and we must suppose that the oxygen flux per unit area mitochondria is everywhere the same.

In contrast the entire surface area of the isolated myocyte is available for oxygen entry. This is five times the area of capillary surface available to each myocyte in tissue, Table 1. In order not to flood the cells with oxygen, the experimenter handling isolated myocytes must work in the technically awkward range of very low oxygen pressure in the suspending medium.

In order to explore the role of myoglobin in intracellular oxygen transport, B. A. Wittenberg (105,107) has developed a preparation of isolated adult ventricular rat heart cells in which a considerable fraction of the cells meets rigid criteria of cellular integrity. Normal morphology is retained; individual cells exhibit clear cross striations. Scanning electron microscopy (Nagel and Wittenberg, unpublished observations) reveals individual cells whose surface membranes appear vel-

Table 1. *Areas of Surfaces through Which Oxygen Consumed by Cardiac or Hepatic Cells Must Diffuse*[a]

	Area (μm^2/cell)		
	Cardiac myocytes		
Membrane component	Endocardial	Epicardial	Hepatocytes
Capillary lumenal surface	1,540	2,440	700
Cell membrane or sarcolemma	7,540	10,700	2,430 (1,200)
Mitochondrial outer membrane	66,800	74,000	5,170
Endoplasmic reticulum			
Total			22,500
Rough	1,000	1,220	13,800
Smooth			9,580

[a]Calculations for left ventricular myocytes of the adult rat are based on the data of Anversa et al. (1). These data are expressed by the authors on the basis of a "typical mononuclear myocyte." Isolated myocytes are binuclear and twice as large as mononuclear myocytes (16). Accordingly we have doubled their values [see Stewart and Page (89)]. On this basis, typical cell volumes are 20,700 μm^3 (endocardial) and 25,200 μm^3 (epicardial) (65). The area of the sarcolemma does not include the intercalated disc. Data for hepatocytes are recalculated from Bolender et al. (17) assuming 2.07×10^8 hepatocytes per gram liver (Bolender, private communication). The area of the hepatocyte cell membrane includes that of the microvilli. The bracketed value is an approximation considering the hepatocyte as a sphere without microvilli. The area of the liver capillary lumenal surface is estimated from data given by Weibel (102) taking the hepatocyte volume as 3750 μm^3 (Bolender, private communication).

Table 2. Fluxes of Consumed Oxygen per Unit Area Membrane in Isolated Cardiac or Hepatic Cells[a]

Membrane component	Oxygen flux (nmol cm^{-2} hr^{-1})			
	Cardiac myocytes		Hepatocytes	
	Coupled	Uncoupled	Cytochrome *c* oxidase	Cytochrome *P*-450
Capillary lumenal surface	60	448	129	19
Cell membrane or sarcolemma	13	98	75	11
Mitochondrial outer membrane	2	13	17	
Endoplasmic reticulum				
Total				0.6
Smooth				1.4

[a]Oxygen consumption of isolated cardiac myocytes (1200 nmol/10^6 cells/hr, coupled, and 8,900 nmol/10^6 cells/hr, uncoupled) from Wittenberg (105). Reaction rates in isolated hepatocytes (900 nmol/10^6 cell/hr, mitochondrial, and 134 nmol/10^6 cell/hr, microsomal) from Jones and Mason (45); endoplasmic reticulum oxygen uptake is equivalent to oxidation of 0.5 m*M* hexobarbital, noninduced cells. Membrane areas from Table 1: in order not to consider the large area of microvilli, the hepatic cell is treated as a sphere of 3750 μm^3 volume. Fluxes are given for total, rough plus smooth, endoplasmic reticulum, and for the least favorable case in which cytochrome *P*-450 is assumed to be restricted to the smooth endoplasmic reticulum.

vety and smooth, as do those of intact muscle, suggesting that the cell coat is intact. Transmission electron microscopy shows intact sarcolemma and surface coat (105A). The sarcolemma of individual cells is functionally intact; normal excitation–contraction coupling can be demonstrated. The sarcolemma retains its normal impermeability to ions, as proved by the tolerance of the cells to extracellular calcium ion. Intracellular mitochondria retain respiratory control and coupling, as evidenced by a fivefold increase in oxygen consumption following treatment with the uncoupling agent, CCCP. Interestingly, undamaged cells are quiescent; they do not beat. These cells can contract repeatedly, more than 100 times, with uniform, synchronous sarcomere shortening in response to electrical stimulation (59). Myoglobin is retained in these isolated cells and dominates their optical spectra (105). Oxygen saturation of cell myoglobin can be monitored. The oxygen affinity of myoglobin in situ in the cytoplasm of isolated cells is identical to that of myoglobin free in solution (unpublished observations).

The steady-state oxygen uptake of our preparation of isolated cardiac cells is oxygen limited only at very low oxygen pressure, with half maximal rate near 0.1 torr. Oxygen uptake of isolated mitochondria, measured in the same way, is half maximal at a very much lower oxygen pressure still. This unexpected finding implies that the flow of oxygen to the mitochondria fails only when the oxygen pressure outside of the cell is too small to saturate a significant fraction of myoglobin near the periphery of the cell.

VII. Oxygen Gradients in Isolated Hepatic Cells

The oxygen supply to hepatic cells may be less changed by isolation from the tissue than was the case with cardiac myocytes. A substantial fraction, 37% (103), of the hepatic cell surface faces the sinusoid, and the capillary surface area per unit cytoplasmic volume is perhaps twice that in muscle. Accordingly a homogeneous medium of low oxygen pressure may offer a reasonable model for oxygen supply to a liver cell.

In contrast to muscle, in which nearly all the oxygen taken up is consumed by mitochondrial cytochrome oxidase, the liver consumes oxygen in several subcellular organelles each having its own complement of terminal oxidase. Jones and Mason (45) have addressed the questions: What is the oxygen gradient to the mitochondria? And separately, what is the gradient to the endoplasmic reticulum of fully functional isolated hepatocytes? Their experiments were done in steady states of extracellular oxygen pressure.

The oxygen concentration required for half-maximal cellular oxygen uptake was 1.90 μM at 37°C, which is an order of magnitude greater than that required for half-maximal rate of oxygen uptake by isolated mitochondria. This is evidence for a gradient of oxygen concentration from outside the cell to the mitochondrion.

Several functions were used as indicators of the steady state level of oxidation–reduction of mitochondrial cytochrome oxidase and other components of the electron transport chain. The extracellular oxygen concentration required for half-maximal change of each of these functions in isolated hepatocytes was about an order of magnitude greater than that required for the corresponding function in isolated mitochondria. Jones and Mason conclude that a substantial intracellular oxygen gradient exists between the outer cellular membrane and the mitochondrial inner membrane at oxygen pressure below the critical oxygen pressure (that is, the oxygen pressure where oxygen uptake becomes oxygen-dependent). In sharp contrast, the oxygen partial pressures required for half-maximal rate of metabolism of two substrates of cytochrome *P*-450 (a microsomal terminal oxidase) by intact hepatocytes were found very similar to the corresponding values for isolated microsomes. Therefore, no detectable gradient of oxygen concentration exists between extracellular space and endoplasmic reticulum of hepatocytes at the relatively low extracellular oxygen pressure studied.

The experiment is open to the objection that microsomal functions were measured as rates of reaction, while mitochondrial functions were measured as steady state levels of oxidation–reduction and closely dependent indicators of these levels. Sensu strictu these two sets of measurements are not comparable. Calculation (104) shows that, in the instance of mitochondrial oxygen uptake by hepatocyes, the gradient of oxygen concentration may extend outside of the cell boundary into an unstirred layer surrounding each cell.

Mitochondria and endoplasmic reticulum are evenly distributed within the hepatocyte volume. A gradient of oxygen pressure from the cell periphery to a central point will encounter and pass through many mitochondria and many sheets of endoplasmic reticulum. On this model, both organelles experience the same gradient.

The oxygen flux across the outer surface of the respiring mitochondrion is of the same order in myocytes and hepatocytes and is 30-fold greater than that required to support cytochrome *P*-450-mediated oxidations in the endoplasmic reticulum (Table 2). This suggests that the apparent oxygen limitation to oxygen uptake by hepatocyte mitochondria may reflect a very local steep oxygen gradient, immediately adjacent to the mitochondrial surface. Another possibility, equally deserving consideration, is that the steepest part of the apparent gradient lies entirely within the mitochondrion where the great concentration of cytochrome oxidase, arrayed on the cristae, exhausts the oxygen locally to a very low pressure. The oxygen-dependent enzyme monoamine oxidase (96) which is localized in the mitochondrial outer membrane might provide a useful probe of the oxygen pressure at this point in the diffusion path and help to distinguish these two possibilities.

Two other enzymes have been used as probes of intracellular oxygen pressure gradients. Peroxisomal uric acid oxidase has a very low apparent affinity for oxygen (half-maximal rate near the partial pressure of oxygen in air), which depends on local substrate availability. Urate oxidation by isolated hepatocytes is oxygen limited (84). Studies on glycolate oxidase activity in perfused liver showed that under some conditions, no significant oxygen concentration gradient exists between the perfusate and the peroxisomes (73,74).

VIII. Oxygen Transfer to Mitochondria and Bacteroids

It is commonly assumed that cytochrome oxidase reacts with oxygen arriving at the mitochondrial surface in free solution in the cytoplasm. This is reasonable for liver cells, but may be open to question in cardiac and red skeletal muscle where we are seeking to find functions for myoglobin. Cole et al. (31) have made a careful study of the simultaneous rates of oxygen uptake and ATP production by isolated mitochondria in steady states of oxygen pressure from 150 μM oxygen to levels where ATP production was oxygen dependent. ATP production at any oxygen pressure (in the presence of abundant ADP) was not changed by the presence of myoglobin.

A very different situation obtains with bacteroids isolated from nitrogen-fixing nodules of soybean roots (7,115). Here the rate of ATP generation is strongly enhanced by oxygen delivered in combination with oxygen binding proteins (118). Oxygen arriving in free solution is consumed but supports ATP production only weakly. We have already noted that bacteroids have two different terminal oxidase systems serving these two functions: an *effective* oxidase system accepting leghemoglobin-bound oxygen and phosphorylating ADP to ATP, and an *ineffective* oxidase accepting free dissolved oxygen. Certainly leghemoglobin by facilitating the diffusion of oxygen to the very surface of the bacteroid (see Fig. 4B) insures that the oxygen concentration delivered to the bacteroids is optimal for the operation of the effective terminal oxidase (13,14,115). In the absence of leghemoglobin the oxygen pressure at the bacteroid surface could become very small.

IX. Oxymyoglobin as an Electron Acceptor

Although facilitated diffusion offers a sufficient explanation of the mechanism by which leghemoglobin supports bacteroid ATP production, a body of suggestive evidence forces us to question whether the explanation is complete (115). We ask whether chemical events, other than reversible dissociation and combination with oxygen, may not participate in oxygen transfer to intracellular organelles. In particular we note that among the several proteins which support the action of the effective terminal oxidase, two have oxygen dissociation rates which appear inappropriately small for an oxygen transport function (118). We note also that in both kinetic and steady-state experiments ATP production by isolated bacteroids is maximal when the oxygen-carrying protein is partially saturated, whether the protein is leghemoglobin (half saturated at 70 nM oxygen) or myoglobin (half saturated at 1000 nM oxygen) (11,13, 14,61,115). Perhaps the deoxy protein is a required intermediate.

It may be relevant that myoglobin in intact surviving molluscan and annelid nerves and muscle can undergo a chemical reaction other than oxygenation–deoxygenation (108,119). Under some conditions, when the tissue is denied oxygen and the oxygen present is removed by metabolic consumption, all, or nearly all, of the tissue myoglobin is converted to a spectral species identified with reasonable certainty as the higher oxidation state of myoglobin. The reaction is reversed when air is readmitted to the tissue. The well-known higher oxidation state, myoglobin(IV), is one oxidizing equivalent more oxidized than ferric myoglobin; two oxidizing equivalents less oxidized than oxymyoglobin. The corresponding form of leghemoglobin, leghemoglobin(IV), is particularly well characterized thanks to its notable stability (10, 114).

Oxyhemeproteins, for instance, the prosthetic groups of oxygenases or mixed function oxidases, may serve as electron acceptors. Oxyleghemoglobin or oxymyoglobin also will serve this function and will oxidize the arbitrarily chosen reductant, 2-hydroxyethylhydrazine, undergoing several hundred cycles of reduction and reoxygenation before the dissolved oxygen is exhausted and the reaction terminates with formation of the ferric protein (Wittenberg and Wittenberg, unpublished results). Horseradish peroxidase undergoes a similar cycle of reactions (51,52, 93), and, although the reaction sequence in this case is also unknown, one can show that oxygenated peroxidase accepts electrons from other hemeproteins (120). The possibility that oxymyoglobin acts in vivo as an electron acceptor must be included in our speculations.

X. Myoglobin-Associated Iron Protein

Myoglobin in no known instance functions alone. In all myoglobin-rich tissues examined, including muscles and nerves from six phyla of invertebrate animals (112), vertebrate red muscle (Wittenberg, unpublished observations), and soybean root nodules (115), myoglobin is accompanied by additional iron in the ratio: one gram atom additional iron per mole intracellular hemoglobin. This iron-containing substance is isolated and purified most readily from soybean root nodules. In this tissue leghemoglobin occurs at an effective concentration of 1.0–1.5 m*M* and constitutes about 40% of the soluble protein of the plant cytoplasm. The iron-containing substance is equally abundant. It is not an iron–sulfur protein. A red chromophore containing the leghemoglobin-associated iron may be brought into organic solvents. Optical spectra suggest that it may be an iron tetrapyrrole other than iron–protoporphyrin IX. A red, partially purified iron protein of molecular weight 31,000 has been isolated; it accounts for nearly all of the leghemoglobin-associated iron in the nodules from which it came. No function is known for the leghemoglobin (myoglobin) associated chromophore.

XI. The Myoglobin Molecule

In closing, we look at the structure of the myoglobin molecule to see what can be inferred about its function. The molecule is small, 153 amino acid residues in a compactly folded polypeptide chain, and slippery in the sense that myoglobin molecules in concentrated solution slide past each other with minimal frictional interaction (77,100). This no doubt is a consequence of the fact that the outside of the molecule is composed mainly of hydrophilic residues (75). The molecular size and the placement of the polypeptide chain in space are constant over (probably) all animal phyla and are very nearly the same even in the plant protein, leghemoglobin (98). This remarkable constancy suggests that the myoglobin molecule may represent a minimal structure required to provide an environment suited to reversible oxygen binding by the heme. The requirements for reversible oxygen binding can now be specified with confidence (46). It is a truism that myoglobin is intermediate in its affinity for oxygen between blood hemoglobin and tissue cytochrome oxidase. Allosteric interactions lowering the oxygen affinity of the tetrameric hemoglobin molecule help bring this about. Leghemoglobin offers the uniquely convincing instance of a monomeric cytoplasmic hemoglobin whose oxygen affinity can be shown to be adapted to its function, in this case oxygen transport in an environment of vanishingly low oxygen pressure. The extraordinary oxygen affinity of leghemoglobin, as we have noted, is a consequence of rapid, almost diffusion limited combination with oxygen. We speculate that the large, easily accessible heme pocket of leghemoglobin (reviewed in 6,116) and the flexibility which permits the distal histidine to move close to or away from the heme are molecular adaptations to allow oxygen unhindered access to its binding site.

Romero-Herrera et al. (78) have considered the amino acid sequences of myoglobins of the 29 mammals and two birds so far investigated. Throughout some 300 million years of evolution (since the ancestors of the mammals and birds diverged) two regions of the myoglobin molecule have remained remarkably conservative. One, of particular interest here, is a flat plane formed by the C helix and the CD interhelical bend in the vicinity of the heme group (Fig. 6). This segment of the molecule contains, on the surface, four residues which are invariant and two which are nearly so, with only very conservative substitutions. Five of the six nearly invariant residues in this plane are ionized. Romero-Herrera et al. (78) suggest: "It seems possible that the invariant charged residues on this flat plane are involved in an interaction between myoglobin and another molecule, functioning as a 'docking site' facilitating the release of oxygen in the vicinity of the mitochondria. Another possibility is that this 'docking site,' being close to the iron atom, facilitates the interaction with metmyoglobin reductase."

We suggest the additional possibility that the "docking site" may be involved in interaction with the iron chromophoric protein which we suggest accompanies myoglobin in one-to-one molecular proportion throughout the animal and plant kingdoms.

In conclusion the general question whether oxygen gradients exist within cells is resolved into particular questions about the geometry of very local oxygen pressure gradients and into questions about the molecular mech-

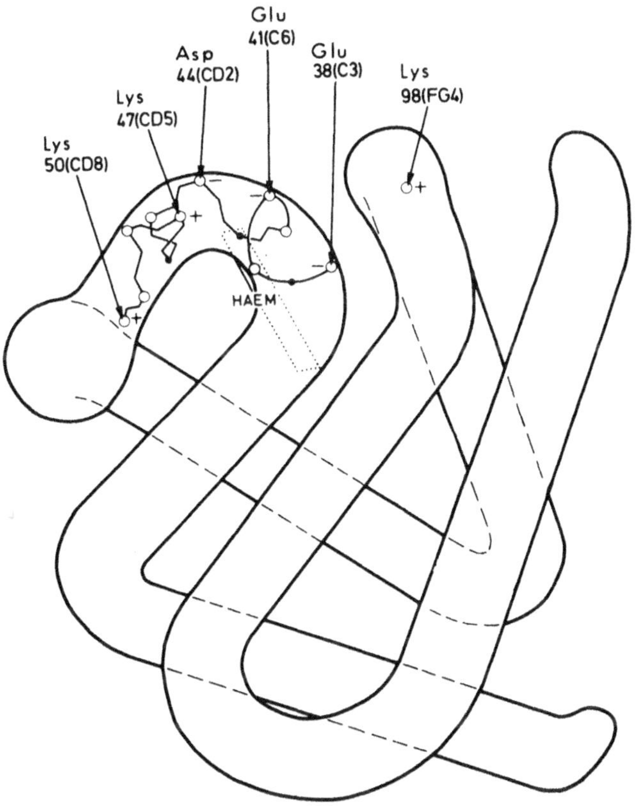

Figure 6. An outline drawing of the sperm whale myoglobin molecule showing the location of a conservative region, the "docking site," discussed in the text. After Romero-Herrera et al. (78).

anisms of oxygen transport and oxygen reduction.

References

1. Anversa, P., Loud, A. V., Giacomelli, F., and Wiener, J. (1978). Absolute morphometric study of myocardial hypertrophy in experimental hypertension. II. Ultrastructure of myocytes and interstitium. Lab. Invest. 38:597–609.
2. Appleby, C. A. (1969). Electron transport systems of *Rhizobium japonicum*. I. Hemeprotein P-450, other CO-reactive pigments, cytochromes and oxidases in bacteroids N_2-fixing root nodules. Biochim. Biophys. Acta 172:71–87.
3. Appleby, C. A. (1969). Electron transport systems of *Rhizobium japonicum*. II. Rhizobium hemoglobin, cytochromes and oxidases in free-living (cultured) cells. Biochim. Biophys. Acta 172:88–105.
4. Appleby, C. A. (1969). Properties of leghemoglobin *in vivo* and its isolation as ferrous oxyhemoglobin. Biochim. Biophys. Acta 188:222–229.
5. Appleby, C. A. (1977). Function of P-450 and other cytochromes in *Rhizobium* respiration. pp. 11–20. In: Functions of Alternative Terminal Oxidases. H. Degn, D. Lloyd, G. C. Hill (editors), vol. 49, Colloquium B 6, Federation of European Biochemical Societies, 11th Meeting. New York: Pergamon Press.
6. Appleby, C. A. (1979). The structure and reactivity of leghemoglobin, a monomeric hemoglobin. Proceedings of the 11th IUPAC International Symposium on the Chemistry of Natural Products. R. Vlahov (editor). In press.
7. Appleby, C. A., Bergersen, F. J., Macnicol, P. K., Turner, G. L., Wittenberg, B. A., and Wittenberg, J. B. (1976). Role of leghemoglobin in symbiotic nitrogen fixation. In: Proceedings of the First International Symposium on Nitrogen Fixation, W. E. Newton and C. J. Nyman (editors). Pullman, Washington: Washington State University Press, pp 274–292.
8. Appleby, C. A., Turner, G. L., and Macnicol, P. K. (1975). Involvement of oxyleghemoglobin and cytochrome P-450 in an efficient oxidative phosphorylation pathway which supports nitrogen fixation in *Rhizobium*. Biochim. Biophys. Acta 387: 461–474.
9. Åstrand, I., Åstrand, P., Christensen, E. H., and Hedman, R. (1960). Myohemoglobin as an oxygen store in man. Acta Physiol. Scand. 48:454–460.
10. Aviram, I., Wittenberg, B. A., and Wittenberg, J. B. (1978). The reaction of ferrous leghemoglobin with hydrogen peroxide to form leghemoglobin(IV). J. Biol. Chem. 253:5685–5689.
11. Bergersen, F. J. (1978). Leghemoglobin, oxygen supply and nitrogen fixation: Studies with soybean nodules. In: J. Döbereiner, R. H. Burris, A. Hollaender (editors), Limitations and Potentials for Biological Nitrogen Fixation in the Tropics. New York: Plenum Press, pp 247–261.
12. Bergersen, F. J., and Goodchild, D. J. (1973). Aeration pathways in soybean root nodules. Aust. J. Biol. Sci. 26:229–240.
13. Bergersen, F. J., and Turner, G. L. (1975). Leghemoglobin and the supply of O_2 to nitrogen-fixing root nodule bacteroids: Studies of an experimental system with no gas phase. J. Gen. Microbiol. 89:31–47.
14. Bergersen, F. J., and Turner, G. L. (1975). Leghemoglobin and the supply of oxygen to nitrogen-fixing root nodule bacteroids: Presence of two oxidase systems and ATP production at low free oxygen concentration. J. Gen. Microbiol. 91:345–354.
15. Bergersen, F. J., Turner, G. L., and Appleby, C. A. (1973). Studies on the physiological role of leghemoglobin in soybean root nodules. Biochim. Biophys. Acta 292:271–282.
16. Bishop, S. P., and Drummond, J. L. (1979). Surface morphology and cell size measurement of isolated rat cardiac myocytes. J. Mol. Cell. Cardiol. 11:423–433.
17. Bolender, R. P., Paumgartner, D., Losa, G., Muellener, D., and Weibel, E. R. (1978). Integrated stereological and biochemical studies on hepatocyte membranes. I. Membrane recoveries in subcellular fractions. J. Cell Biol. 77:565–583.
18. Britton, N. F., and Murray, J. D. (1977). The effect of carbon monoxide on haem-facilitated oxygen diffusion. Biophys. Chem. 7:159–167.
19. Buerk, D. G., and Longmuir, I. S. (1977). Evidence for nonclassical respiratory activ-

ity from oxygen gradient measurements in tissue slices. Microvasc. Res. 13:345–353.
20. Caillé, J. P., and Hinke, J. A. M. (1972). Evidence for Na sequestration in muscle from Na diffusion measurements. Can. J. Physiol. Pharmacol. 50:228–237.
21. Caillé, J. P., and Hinke, J. A. M. (1973). Evidence for K and Cl binding inside muscle from diffusion studies. Can. J. Physiol. Pharmacol. 51:390–400.
22. Caillé, J. P., and Hinke, J. A. M. (1974). The volume available to diffusion in the muscle fiber. Can. J. Physiol. Pharmacol. 52:814–828.
23. Carles, A. C., Kawashiro, T., and Piiper, J. (1975). Solubility of various inert gases in rat skeletal muscle. Pfluegers Arch. 359: 209–218.
24. Chalazonitis, N. (1968). Intracellular pO_2 control on excitability and synaptic activability, in *Aplysia* and *Helix* identifiable giant neurons. Ann. N.Y. Acad. Sci. 147: 419–459.
25. Chalazonitis, N., and Arvanitaki, A. (1970). Neuromembrane electrogenesis during changes in pO_2, pCO_2, and pH. Adv. Biochem. Psychopharmacol. 2:245–284.
26. Chance, B., Barlow, C., Nakase, Y., Takeda, H., Mayevsky, A., Fischetti, R., Graham, N., and Sorge, J. (1978). Heterogeneity of oxygen delivery in normoxic and hypoxic states: A fluorometer study. Am. J. Physiol. 235:H809–H820.
27. Coburn, R. F., and Mayers, L. B. (1971). Myoglobin oxygen tension determined from measurements of carboxymyoglobin in skeletal muscle. Am. J. Physiol. 220:66–74.
28. Coburn, R. F., and Pendleton, M. (1979). Effects of norepinephrine on oxygenation of resting skeletal muscle. Am. J. Physiol. 236: H307–H313.
29. Coburn, R. F., Ploegmakers, F., Gondrie, P., and Abboud, R. (1973). Myocardial myoglobin oxygen tension. Am. J. Physiol. 224:870–876.
30. Cole, R. P., Wittenberg, B. A., and Caldwell, P. R. B. (1978). Myoglobin function in the isolated fluorocarbon-perfused dog heart. Am. J. Physiol. 234:H567–H572.
31. Cole, R. P., Wittenberg, J. B., and Wittenberg, B. A. (1979). Mitochondrial function in the presence of myoglobin. Physiologist 22:21.
32. Colton, C. K., Stroeve, P., and Zahka, J. G. (1973). Mechanism of oxygen transport augmentation by hemoglobin. J. Appl. Physiol. 35:307–309.
33. Criswell, J. G., Havelka, U. D., Quebedeaux, B., and Hardy, R. W. F. (1976). Adaptation of nitrogen fixation by intact soybean nodules to altered rhizosphere pO_2. Plant Physiol. 58:622–625.
34. Douglas, C. G., and Haldane, J. S. (1922). The regulation of the general circulation rate in man. J. Physiol. (London) 56:69–100.
35. Fabel, H., and Lübbers, D. W. (1965). Measurements of reflection spectra of the beating rabbit heart *in situ*. Biochem. Z. 341:351–356.
36. Figulla, H. R., Wodick, R., Hoffman, J., and Lübbers, D. W. (1979). The oxygen saturation of myoglobin and cytochrome a a_3 during high flow hypoxia and low flow hypoxia in the beating, hemoglobin-free perfused Langendorf guinea pig heart. Pfluegers Arch. 379:R3.
37. Garby, L., and Meldon, J. (1977). The Respiratory Functions of the Blood. New York: Plenum Press, pp. 179–183.
38. Gurtner, G. H., Burns, B., Peary, H. H, Mendoza, C. J., Traystman, R. J., Summer, W., and Sciuto, A. M. (1978). Specific mechanisms for oxygen and carbon monoxide transport in the lung and placenta. In: Regulation of Ventilation and Gas Exchange. D. G. Davies and C. D. Barnes (editors) New York: Academic Press.
39. Haldane, J. S., and Priestley, J. G. (1935). Respiration. New Haven: Yale University Press.
40. Hemmingsen, E. A. (1965). Accelerated transfer of oxygen through solutions of heme pigments. Acta Physiol. Scand. 64, Suppl. 246:1–53.
41. Hill, R. (1936). Oxygen dissociation curves of muscle hemoglobin. Proc. Roy. Soc. London, Ser. B 120:472–483.
42. Holloszy, J. O. (1975). Adaptation of skeletal muscle to endurance exercise. Med. Sci. Sports 7:155–164.
43. Hoofd, L., and Kreuzer, F. (1978). Calculation of the facilitation of oxygen or carbon monoxide by hemoglobin or myoglobin by means of a new method for solving the carrier diffusion problem. Adv. Exp. Med. Biol. 94:163–168.
44. Imamura, T., Riggs, A., and Gibson, A. H. (1972). Equilibria and kinetics of ligand binding by leghemoglobin from soybean root nodules. J. Biol. Chem. 247:521–526.

45. Jones, D. P., and Mason, H. S. (1978). Gradients of oxygen concentration in hepatocytes. J. Biol. Chem. 253:4874–4880.
46. Jones, R. D., Summerville, D. A., and Basolo, F. (1979). Synthetic oxygen carriers related to biological systems. Chem. Rev. 79:139–179.
47. Kawashiro, T., Carles, A. C., Perry, S. F., and Piiper, J. (1975). Diffusivity of various inert gases in rat skeletal muscle. Pfluegers Arch. 359:219–230.
48. Kawashiro, T., Nüsse, W., and Scheid, P. (1975). Determination of diffusivity of oxygen and carbon dioxide in respiring tissue. Results in rat skeletal muscle. Pfluegers Arch. 359:231–251.
49. Kawashiro, T., and Scheid, P. (1976). Measurement of Krogh's diffusion constant of CO_2 in respiring muscle at various CO_2 levels: Evidence for facilitated diffusion. Pfluegers Arch. 362:127–133.
50. Keller, K. H., Canales, E. R., and Yum, S. I. (1971). Tracer and mutual diffusion coefficients of proteins. J. Phys. Chem. 75: 379–387.
51. Klapper, M. H., and Hackett, D. P. (1963). The oxidative activity of horseradish peroxidase. I. Oxidation of hydro- and naphthohydroquinones. J. Biol. Chem. 238: 3736–3742.
52. Klapper, M. H., and Hackett, D. P. (1963). The oxidative activity of horseradish peroxidase. II. Participation of ferroperoxidase. J. Biol. Chem. 238:3743–3749.
53. Klippenstein, G. L., Van Riper, D. A., and Oosterom, E. A. (1972). A comparative study of the oxygen transport proteins of *Dendrostomum pyroides*. Isolation and characterization of hemerythrins from muscle, the vascular system and the coelom. J. Biol. Chem. 247:5959–5963.
54. Kreuzer, F. (1970). Facilitated diffusion of oxygen and its possible significance, a review. Respir. Physiol. 9:1–30.
55. Kreuzer, F., de Koning, J., van Haren, R., and Hoofd, L. J. C. (1977). Oxygen diffusion facilitated by myoglobin in the chicken gizzard smooth muscle. 9th Eur. Conf. Microcirc., Antwerp. Bibl. Anat. 15:380–385. Karger, Basel.
56. Kreuzer, F., and Hoofd, L. J. C. (1970). Facilitated diffusion of oxygen in the presence of hemoglobin. Respir. Physiol. 8: 280–302.
57. Kreuzer, F., and Hoofd, L. J. C. (1972). Factors influencing facilitated diffusion of oxygen in the presence of hemoglobin and myoglobin. Respir. Physiol. 15:104–124.
58. Kreuzer, F., and Hoofd, L. J. C. (1976). Facilitated diffusion of carbon monoxide and oxygen in the presence of hemoglobin or myoglobin. Adv. Exp. Med. Biol. 75: 207–215.
59. Krueger, J. W., Forletti, D., and Wittenberg, B. A. (1980). Uniform sarcomere shortening in isolated cardiac muscle cells. J. Gen. Physiol. 76:587–607.
60. Kushmerick, M J., and Podolsky, R. J. (1969). Ionic mobility in muscle cells. Science 166:1297–1298.
61. Laane, C., Haaker, H., and Veeger, C. (1978). Involvement of the cytoplasmic membrane in nitrogen fixation by *Rhizobium leguminosarum* bacteroids. Eur. J. Biochem. 87:147–153.
62. Lawrie, R. A. (1953). The activity of the cytochrome system in muscle and its relation to myoglobin. Biochem. J. 55:298–305.
63. Longmuir, I. S. (1976). The measurement of the fraction of oxygen carried by facilitated diffusion. Adv. Exp. Med. Biol. 75: 217–223.
64. Longmuir, I. S., Martin, D. C., Gold, H. J., and Sun, S. (1971). Nonclassical respiratory activity of tissue slices. Microvasc. Res. 3:125–141.
65. Loud, A. V., Anversa, P., Giacomelli, F., and Wiener, J. (1978). Absolute morphometric study of myocardial hypertrophy in experimental hypertension. I. Determination of myocyte size. Lab. Invest. 38:586–596.
66. Mahler, M. (1978). Diffusion and consumption of oxygen in the resting frog sartorius muscle. J. Gen. Physiol. 71:533–557.
67. Millikan, G. A. (1937). Experiments on muscle haemoglobin *in vivo*; the instantaneous measurement of muscle metabolism. Proc. Roy. Soc. London, Ser. B 123: 218–241.
68. Millikan, G. A. (1939). Muscle hemoglobin. Physiol. Rev. 19:503–523.
69. Mitchell, P. J., and Murray, J. D. (1973). Facilitated diffusion: The problem of boundary conditions. Biophysik 9:177–190.
70. Murray, J. D. (1971). On the molecular mechanism of facilitated oxygen diffusion by haemoglobin and myoglobin. Proc. Roy. Soc. London, Ser. B 178:95–110.

71. Murray, J. D. (1974). On the role of myoglobin in muscle respiration. J. Theor. Biol. 47:115–126.
72. Murray, J. D., and Wyman, J. (1971). Facilitated diffusion, the case of carbon monoxide. J. Biol. Chem. 246:5903–5906.
73. Oshino, N., Jamieson, D., and Chance, B. (1975). The properties of hydrogen peroxide production under hyperoxic and hypoxic conditions of perfused rat liver. Biochem. J. 146:53–65.
74. Oshino, N., Jamieson, D., Sugano, T., and Chance, B. (1975). Optical measurements of the catalase-hydrogen peroxide intermediate (Compound I) in the liver of anaesthetized rats and its implication to hydrogen peroxide production in situ. Biochem. J. 146:67–77.
75. Perutz, M. F., Kendrew, J. C., and Watson, H. C. (1965). Structure and function of hemoglobin II. Some relations between polypeptide chain configuration and amino acid sequence. J. Mol. Biol. 13:669–678.
76. Rich, T. L., and Williamson, J. R. (1978). Optical evidence for steep oxygen gradients and sharp border zones in hypoxic cardiac tissue. Fed. Proc. 37:780.
77. Riveros-Moreno, V., and Wittenberg, J. B. (1972). The self-diffusion coefficients of myoglobin and hemoglobin in concentrated solutions. J. Biol. Chem. 247:895–901.
78. Romero-Herrera, A. E., Lehmann, H., Joysey, K. A., and Friday, A. E. (1978). On the evolution of myoglobin. Phil. Trans. Roy. Soc. London, Ser. B 283:61–163.
79. Rubinow, S. I., and Dembo, M. (1977). The facilitated diffusion of oxygen by hemoglobin and myoglobin. Biophys. J. 18:29–42.
80. Scholander, P. F. (1960). Oxygen transport through hemoglobin solutions. Science 131: 585–590.
81. Schuder, S., Wittenberg, J. B., Haseltine, B., and Wittenberg, B. A. (1979). Spectrophotometric determination of myoglobin in cardiac and skeletal muscle: Separation from hemoglobin by subunit exchange chromatography. Anal. Biochem. 92:473–481.
82. Schultz, J. S., Goddard, J. D., and Suchdeo, S. R. (1974). Facilitated transport via carrier-mediated diffusion in membranes. Part I. Mechanistic aspects, experimental systems and characteristic regimes. AIChE J. 20:417–445.
83. Schwarzmann, V., and Grunewald, W. A. (1978). Myoglobin oxygen saturation profiles in muscle sections of chicken gizzard and the facilitated O_2-transport by myoglobin. Adv. Exp. Med. Biol. 94:301–310.
84. Sies, H. (1977). Oxygen gradients during hypoxic steady states in liver. Urate oxidase and cytochrome oxidase as intracellular oxygen indicators. Hoppe-Seyler's Z. Physiol. Chem. 358:1021–1032.
85. Smith, J. D. (1949). Haemoglobin and the oxygen uptake of leguminous root nodules. Biochem. J. 44:591–598.
86. Smith, K. A., Meldon, J. H., and Colton, C. K. (1973). An analysis of carrier mediated transport. AIChE J. 19:102–111.
87. Steenbergen, C., Deleeuw, G., Barlow, C., Chance, B., and Williamson, J. R. (1977). Heterogeneity of the hypoxic state in perfused rat heart. Circ. Res. 41:606–615.
88. Steenbergen, C., Williamson, J. R., Deleeuw, G. J. (1978). Nature of flow and oxygen border zones in hypoxic and ischemic myocardium. pp. 1542–1550 in: Frontiers of Biological Energetics, Vol. 2, Electrons to tissues. P. L. Dutton, J. S. Leigh and A. Scarpa (editors). New York: Academic Press.
89. Stewart, J. M., and Page, E. (1978). Improved stereological techniques for studying myocardial cell growth: Application to external sarcolemma, T-system, and intercalated discs of rabbit and rat hearts. J. Ultrastruct. Res. 65:119–134.
90. Stroeve, P. (1977). The facilitated transport of oxygen into muscle tissue. In: 9th Eur. Conf. Microcirc., Antwerp 1976. Bibl. Anat. 15:429–432.
91. Stroeve, P., and Eagle, K. (1979). An analysis of diffusion in a medium containing dispersed reactive cylinders. Chem. Eng. Commun. 3:189–198.
92. Suchdeo, S. R., Goddard, J. D., and Schultz, J. S. (1973). An analysis of the competitive diffusion of O_2 and CO through hemoglobin solutions. Adv. Exp. Med. Biol. 37B:951–961.
93. Swedin, B., and Theorell, H. (1940). Dioxoimaleic acid oxidase action of peroxidases. Nature 145:71–72.
94. Tamura, M., Oshino, N., Chance, B., and Silver, I. A. (1978). Optical measurements of intracellular oxygen concentration of rat heart *in vitro*. Arch. Biochem. Biophys. 191: 8–22.

95. Taylor, B. A., and Murray, J. D. (1977). Effect of the rate of oxygen consumption on muscle respiration. J. Math. Biol. 4:1–20.
96. Tipton, K. F. (1972). Some properties of monoamine oxidase. Adv. Biochem. Psychopharmacol. 5:11–24.
97. Tjepkema, J. D., and Yocum, C. S. (1973). Respiration and oxygen transport in soybean nodules. Planta 115:59–72.
98. Vainshtein, B. K., Arutyunian, E. G., Kuranova, I. P., Borisov, V. V., Sosfenov, N. I., Pavlovskii, A. G., Grebenko, A. I., Konareva, N. V., and Nekrasov, Y. V. (1977). Three-dimensional structure of lupine leghemoglobin at 2.8 Angstrom resolution. Dokl. Akad. Nauk SSSR 233:238–241.
99. Van Ouwerkerk, H. J. (1977). Facilitated diffusion in a tissue cylinder with an anoxic region. Pfluegers Arch. 372:221–230.
100. Veldkamp, W. B., and Votano, J. R. (1976). Effects of intermolecular interaction on protein diffusion in solution. J. Phys. Chem. 80:2794–2801.
101. Wang, J. H. (1954). Theory of self-diffusion of water in protein solutions. A new method for studying the hydration and shape of protein molecules. J. Am. Chem. Soc. 76:4755–4763.
102. Weibel, E. R. (1979). Oxygen demand and the size of respiratory structures in mammals. pp. 289–346. In: Evolution of Respiratory Processes. S. C. Wood and C. Lenfant (editors). Vol. 13. New York: M. Dekker.
103. Weibel, E. R., Stäubli, W., Gnägi, H. R., and Hess, F. A. (1969). Correlated morphometric and biochemical studies on the liver cell. I. Morphometric model, stereological methods and normal morphometric data for rat liver. J. Cell Biol. 42:68–91.
104. Weisz, P. B. (1973). Diffusion and chemical transformation. An interdisciplinary transformation. Science 179:433–440.
105. Wittenberg, B. A. (1979). Myoglobin in isolated adult heart cells. pp. 35–51. In: Oxygen: Biochemical and Clinical Aspects. W. S. Caughey (editor). Academic Press, New York.
105A. Wittenberg, B. A., and Robinson, T. F. (1981). Oxygen requirements, morphology, cell coat and membrane permeability of calcium-tolerant myocytes from hearts of adult rats. Cell and Tissue Res. In press.
106. Wittenberg, B. A., Wittenberg, J. B., and Caldwell, P. R. B. (1975). Role of myoglobin in the oxygen supply to red skeletal muscle. J. Biol. Chem. 250:9038–9043.
107. Wittenberg, B. A., Wittenberg, J. B., and Krueger, J. W. (1978). Myoglobin in oxygen economy of isolated cardiac cells. Fed. Proc. 37:1608.
108. Wittenberg, B. A., Wittenberg, J. B., Stolzberg, S., and Valenstein, E. (1965). A novel reaction of hemoglobin in invertebrate tissues. II. Observations on molluscan muscle. Biochim. Biophys. Acta 109:530–535.
109. Wittenberg, J. B. (1959). Oxygen transport: A new function proposed for myoglobin. Biol. Bull. 117:402.
110. Wittenberg, J. B. (1966). The molecular mechanism of hemoglobin-facilitated oxygen diffusion. J. Biol. Chem. 241:104–114.
111. Wittenberg, J. B. (1970). Myoglobin facilitated oxygen diffusion: Role of myoglobin in oxygen entry into muscle. Physiol. Rev. 50:559–636.
112. Wittenberg, J. B. (1972). An ubiquitous association between myoglobin and equimolar quantities of iron protein(s). Fed. Proc. 31:923.
113. Wittenberg, J. B. (1976). Facilitation of oxygen diffusion by intracellular leghemoglobin and myoglobin. pp. 228–246. In: Oxygen and Physiological Function. F. F. Jöbsis (editor). Professional Information Library, Dallas, Texas.
114. Wittenberg, J. B. (1978). Leghemoglobin. Low temperature optical spectra of acid and alkaline forms of leghemoglobin(IV). Configuration of the heme. J. Biol. Chem. 253: 5690–5693.
115. Wittenberg, J. B. (1980). Utilization of leghemoglobin-bound oxygen by Rhizobium bacteroids. pp. 53–67. In: Nitrogen Fixation, Vol. II, W. H. Orme-Johnson and W. E. Newton (editors). Baltimore: University Park Press.
116. Wittenberg, J. B. (1979). Reactivity and function of leghemoglobin. pp. 53–68. In: Oxygen: Biochemical and Clinical Aspects. W.S. Caughey (editor). New York: Academic Press.
117. Wittenberg, J. B., Appleby, C. A., and Wittenberg, B. A. (1972). The kinetics of the reactions of leghemoglobin with oxygen and carbon monoxide. J. Biol. Chem. 247:527–531.
118. Wittenberg, J. B., Bergersen, F. J., Appleby, C. A., and Turner, G. L. (1974). Facilitated oxygen diffusion. The role of leghemoglobin

in nitrogen fixation by bacteroids isolated from soybean root nodules. J. Biol. Chem. 249:4057–4066.

119. Wittenberg, J. B., Brown, P. K., and Wittenberg, B. A. (1965). A novel reaction of hemoglobin in invertebrate nerves. I. Observations on annelid and molluscan nerves. Biochim. Biophys. Acta 109:518–529.

120. Wittenberg, J. B., Noble, R. W., Wittenberg, B. A., Antonini, E., Brunori, M., and Wyman, J. (1967). Studies on the equilibria and kinetics of the reactions of peroxidase with ligands. II. The reaction of ferroperoxidase with oxygen. J. Biol. Chem. 242: 626–634.

121. Wyman, J. (1966). Facilitated diffusion and the possible role of myoglobin as a transport mechanism. J. Biol. Chem. 241:115–121.

10

The Reaction of Oxygen with Cytochrome Oxidase: The Role of Sequestered Intermediates

BRITTON CHANCE

I. Introduction

Cytochrome oxidase is specifically served by oxygen delivery from the blood, and affords the main energy conservation pathway of the eukaryotic cell. The mechanism by which cytochrome oxidase reduces to water the total oxygen delivery and affords efficient production of ATP without production of toxic intermediates of oxygen reduction is one of the principal problems of modern biochemistry, on which significant advances have been made recently. The quantitation of structural and functional features of cytochrome oxidase has been obtained by low dose electron microscopy, low angle diffraction from oriented bilayers, and most recently anomalous scattering. These approaches are complemented by low temperature kinetics and sensitive spectrophotometry to afford a "first step" toward the correlation of structural and functional aspects of cytochrome oxidase and the cytochrome oxidase–oxygen reaction mechanism. Salient among the findings is the identification of several types of oxygen compounds formed at very low temperatures by a flash photolysis procedure. The optical and magnetic properties of these compounds have been studied in detail and lead to a new insight on how the four electrons of cytochrome oxidase can be transferred to oxygen at a sequestered active site from which the end product is water, the oxidation of metalloenzyme components, and the formation of an asymmetric hydrogen ion gradient. In addition to oxygen reduction to the peroxide level by the iron and copper binuclear complex of cytochrome a_3–Cu_{a_3}, oxygen is further reduced to water by electron transfer from an electron reservoir—cytochrome a—Cu_a— and from cytochrome c. The pathways of electron transfer operate by nuclear assisted electron transfer among widely spaced porphyrin centers, in distinct contrast to the simplistic notion of a vectorial transmembrane process of the chemiosmotic theory. The following describes the current state of our research on the mechanism of oxygen reduction by cytochrome oxidase.

II. A Possible Mechanism for Oxygen Reduction

The chemistry of oxygen reduction by cytochrome oxidase is summarized in Fig. 1, which represents a sequence of reactions

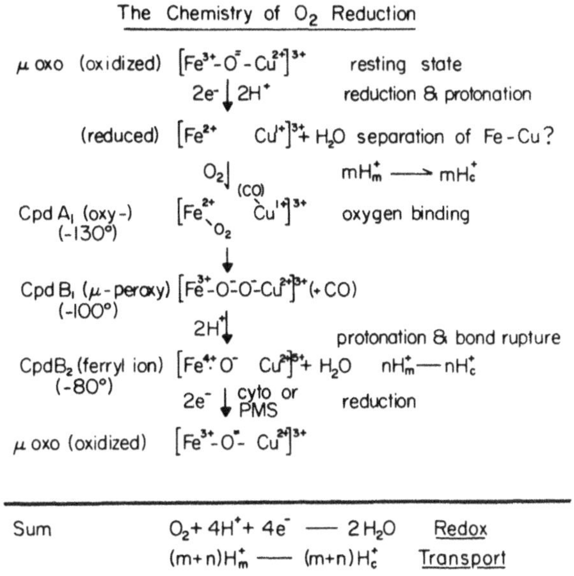

Figure 1. Chemistry of oxygen reduction by cytochrome oxidase.

based on our research (1–9), starting from the resting μoxo form of the enzyme, and cycling through two oxygen intermediates and protonation steps back to the μoxo form. This is a characteristic cycle of enzyme activity when viewed from the oxygen reaction only: details of electron donation are included in Fig. 2. The proposal for μoxo configuration for the resting enzyme is based on the studies of iron porphyrin models in which the oxygen atom of the water molecule is the bridging ligand between the two porphyrins. In this case, the μoxo bridge is between the iron and copper atoms of the binuclear complex, heme a_3–Cu_{a_3} of cytochrome oxidase. In order to create the μoxo form, the Fe–Cu configuration must be appropriate to the μoxo bond length of 1.7 Å on each side of the cytochrome oxidase atom (total Fe–Cu distance ~3.5 Å).

Reaction with oxygen requires reduction of iron and copper by a pair of electrons, and dissociation of the oxygen atom from the μoxo structure, illustrated in step 2. In steps 2 or 3, a separation of the iron and copper to ~3.5 Å may occur in order better to accommodate dioxygen as a ligand of the iron, as suggested by EXAFS studies of the binding of oxygen to hemocyanin (10). Furthermore, since infrared spectral studies suggest that CO may be bound to the copper or to the iron (11), the increased spacing of the Fe–Cu couple is required. This configurational change and the associated reduction of the Fe–Cu couple are considered to be energetic, i.e., one of the energy-yielding steps of an oxidation–reduction cycle (12–14) and therefore, coupling of the structural change due to the iron copper separation and oxidoreduction to the p*K*'s of the pH-linked groups (carboxyl, amino, etc.) located within the channel system described in Fig. 2 is proposed at this point. This causes a H^+ change in the stoichiometric ratio of *m* per electron, as indicated in the side of the diagram of Fig. 1. This energy coupling, could, however, occur at other energy-yielding points in the redox cycle.

The third step is indicated to be bonding of the ferrous iron to oxygen to form oxycytochrome oxidase. The reaction can best be demonstrated by low temperature photolysis (−130°C) of the CO compound in the presence of oxygen previously added at higher temperature (usually −20°C). Electron delocalization and transfer to the oxygen does not occur to a great extent in cytochrome oxidase as judged by the spectrum of the

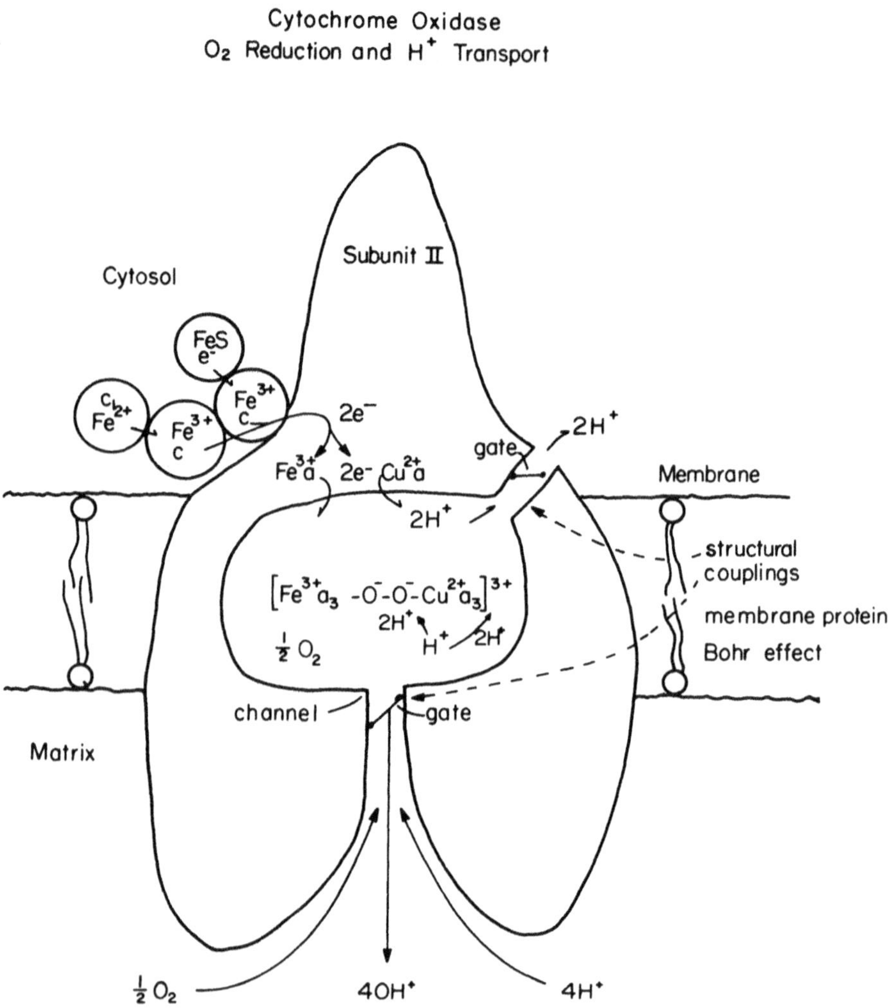

Figure 2. Structure of the chemistry of oxygen reduction by cytochrome oxidase, indicating association of chemical intermediates, transport pathways, and structural couplings within the cytochrome oxidase structure.

oxygen intermediate, and from attempts to identify an iron-bound superoxide anion by ESR techniques, but the EPR sensitivity is low due to the magnetic effect of the iron atom to which oxygen is bonded. Alben's recent data suggest that the photolyzed CO molecule may migrate to the adjacent copper (11). This appears not to affect the binding of the iron, as judged from the similarity of rate constants for the oxygen reaction determined in the presence and absence of CO. However, an appropriate spacing of the Fe and Cu atoms is required to accommodate both ligands.

The concentration of the oxygen compound can be maximized, depending upon the oxygen concentration at temperatures between $-130°$ and $-100°C$. Above these temperatures, electron transfer processes compete with the oxygen reaction and diminish the steady-state concentration of the oxygen compound to the relatively low values due to limitation on the maximal concentration of oxygen in water (~ 1 m*M*). Thus, oxycytochrome appears only as a transient species at temperatures above $-90°C$, and can be observed only with rapid kinetic methods at $-30°C$. At temperatures above these, the oxygen compound has not been identified spectroscopically, but can be iden-

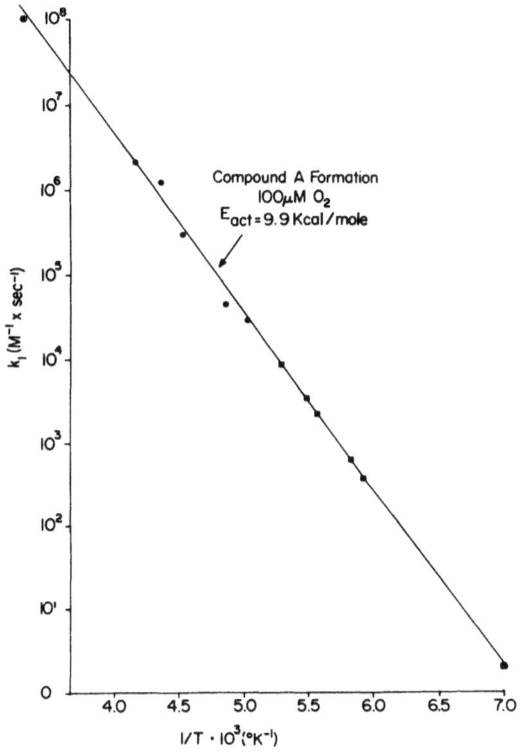

Figure 3. Determination of the energy of activation for the reaction of cytochrome oxidase over the temperature range of roughly −43° to −93°, and including the range for a transition in the binding of CO to O_2 (14CO to copper) (27).

tified kinetically from the continuity of the Arrhenius profile for the oxygen reaction, which appears as an unbroken curve (Fig. 3) from −130°C to room temperature with a constant Arrhenius activation energy of 10 kJ. This suggests further that the oxygen reaction is rate limiting throughout the higher range of temperatures (>−50°C). However, under the conditions where Gibson and I labored (25°), electron donation was faster than the oxygen reaction, and no significant amounts of the functional oxygen intermediates were detectable (15). At the lower temperatures (<−50°C), intermediates can be identified; the oxygen reaction is more rapid than the electron donation, and significant amounts of the oxygen intermediate are detectable.

Above −100°C, electron donation to the oxygen compound from both Fe and Cu (heme a_3 and Cu_{a_3}) is most probable from the thermodynamic standpoint, and is supported by observations of the disappearance of α, β, γ bands of reduced heme and the appearance of absorption bands in the infrared region attributable to oxidized copper. This leads us to formulate a μperoxo intermediate termed compound B, readily detectable at −100°C and stable to ~−50°C. To accommodate the μperoxo bridge, ~3.5 Å between the two metal atoms centers is needed, possibly requiring some increase of the distance between the metal atoms that was appropriate to the μoxo state, as mentioned above. At this time, any CO bound to the copper would be ejected. Identification of the iron of heme a_3 involvement in this complex seems relatively secure, based not only on the photochemical action spectrum of Warburg, but also as obtained more precisely by Castor and Chance, since this reaction can be observed in forms of cytochrome oxidase found in *Candida utilis*, in which separation of the α bands of the cytochromes *a* and a_3 is sufficient so that the absorption attributable to the reduced cytochrome *a* can be observed to remain at fixed level, while that attributable to the reduced form of cytochrome a_3 is observed to disappear in an independent oxidation reaction, presumably the formation of the μperoxo intermediate (Fig. 4). Resolution of the optical contributions of *a* and a_3 is more difficult in the case of copper, where the majority of the absorption in the 830 nm region is attributable to copper associated with heme a_3 (Cu_{a_3}). However, the band position for $Cu_{a_3}^{2+}$ (790 nm) is different from that for the sum of $Cu_{a_3+}^{2+}$ and Cu_a^{2+} (830 nm), and the binuclear nature of the iron–copper complex suggests a function of the copper in electron donation to oxygen. Thus, Cu_{a_3} is identified as the copper that reduces oxygen. A model that supports the interpretation of a μperoxo intermediate is given by the studies of Brown et al. on hemocyanin, where a cuprous–cuprous binuclear complex with a spacing of ~3.5 Å reacts with oxygen and transfers two electrons to oxygen to form a μperoxo intermediate. In this case, the reaction is a reversible equilibrium and the re-

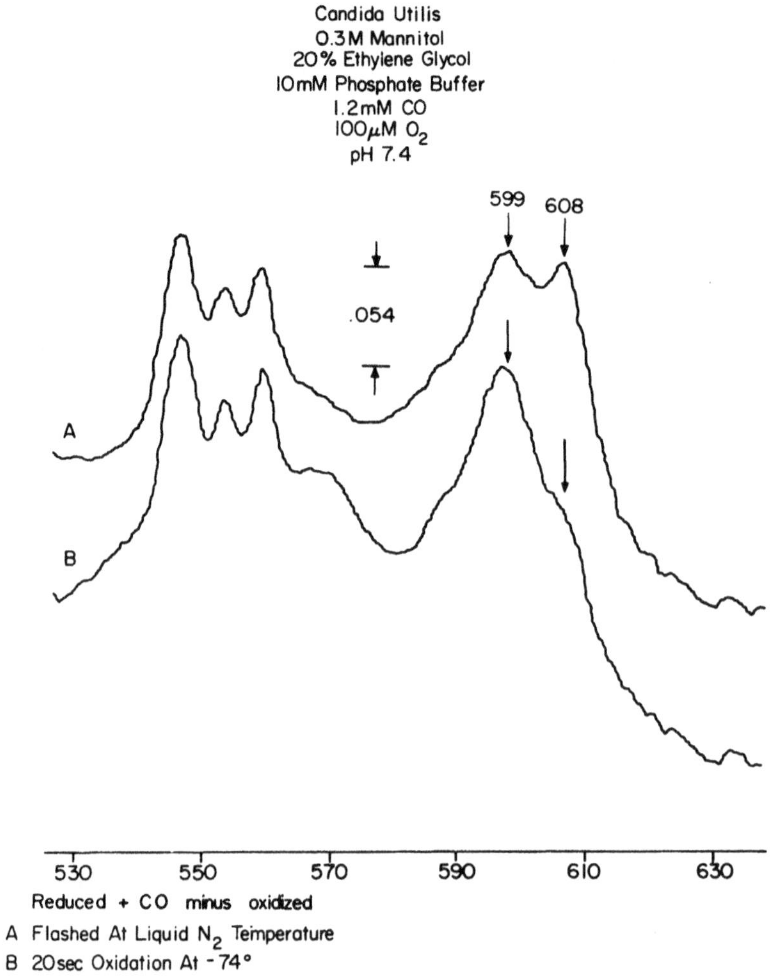

Figure 4. Formation of the μperoxo intermediate in Candida utilis cytochrome oxidase causes absorption attributable to the reduced form of cytochrome a_3 to disappear. Separation of the α bands of cytochromes a and a_3 is such that absorption attributable to reduced cytochrome a remains at a fixed level.

duced oxy form is readily regenerated in contradistinction to cytochrome oxidase. Presumably, the spacing and reactivity of the metal atoms of the iron–copper complex is similar to that of the two Cu atoms of hemocyanin.

Protonation of the μperoxo intermediate and electron transfer may give the ferryl ion familiar in studies of peroxidases and myoglobins (16). Such a transition is possible in cytochrome oxidase, although it has not been spectroscopically identified. Nevertheless, the possibility suggests that cytochrome oxidase can act in a peroxidatic way toward its electron donor substrates, just as the well-known peroxidases do, particularly cytochrome c peroxidase of yeast cells (17).

Reduction of compounds B_1 or B_2 can occur by electron tunneling reactions from cytochrome c or PMS. Such tunneling reactions are likely to occur at both higher and lower temperatures, but are usually easier to identify at the lower temperatures. Presumably, a variety of electron donors are appropriate to peroxidase-like intermediates. At higher temperatures, electron donation from cytochrome a becomes the most rapid reaction, and the sequential function of the respiratory chain can be observed to occur as the temperatures are raised toward physiological level, as has previously been found in kinetic and equilibrium studies (18).

The reaction product of two electron reduction of compounds B_1 or B_2 is the μoxo form itself, in which the proximation of the iron and copper, in order to accommodate the oxygen bridge optimally, is proposed. The sum of redox reactions is the usual stoichiometry for oxygen reduction with $4H^+$ and $4e^-$. The sum of the transport reactions is $m + n$ per electron from the matrix phase to the cytosolic phase.

III. Large-Scale Structural Aspects

The structural import of the foregoing reaction steps is indicated in Fig. 2, where the association of the chemical intermediates, the transport pathways, and the structural couplings are indicated within the framework of the cytochrome oxidase structure, as identified by Capaldi, Frey, Blasie, and others, who have used low dose optical processing of electron microscope and low angle X-ray diffraction data (19–21). The identification of the metal atoms with a particular one of the seven subunits is difficult at the present time, although the existence of the amino acid sequence appropriate to one of the two coppers of cytochrome oxidase—presumably copper a_3, the more ionic type (8)—suggests the location of both elements on the binuclear complex heme a_3 and Cu_{a_3} in subunit II. Electron transfer reactions that can occur by tunneling seem possible over distances of 20–25 Å, and therefore permit the location of heme a and Cu relatively distant from heme a_3 and Cu_{a_3}, as for example, cytochrome c is distant from its electron acceptor heme a and its associate Cu_1,Cu_a. Cytochrome c itself with its appropriate electron donors, the Rieske iron–sulfur protein, and cytochrome c_1 are also indicated, although their association with a neighboring macromolecule, the b–c_1 complex, is well recognized (22).

The approximate location of the components with respect to the plane of the membrane is known from the broading of their epr signals caused by paramagnetic ions such as dysprosium. More accurate effects are being studied by anomalous X-ray scattering using Synchrotron radiation (23). Thus, compound B_1 is located within tunneling distance of the remainder of the electron donors, which places it on the cytosolic side of the membrane.

The reactive intermediate included in the diagram is the μperoxo intermediate (B_1), and it can, according to the previous diagram, accept two protons and two electrons to make the μoxo form. The entry of the protons and indeed, the oxygen from which the μoxo is formed, is indicated through a gated channel, because of the observation of a rate limitation in the oxygen reaction, readily demonstrable at low temperatures by a break in the pseudo-first-order rate oxygen concentration relationship. The idea that protons may enter the reactive space through the same channel as oxygen is mainly one of simplification: the proton channel could be located in a different subunit from that of heme a_3 (Cu_{a_3}). However, Ludwig and Sone both demonstrate proton translocation in the one and two subunit oxidases (24,25). Thus, the proton channel may be in the same subunit as contains the hemes (however, the latter oxidase appears to lack a functional heme a_3 (Cu_{a_3}), and instead, relies upon cytochrome o, a protoporphyrin IX oxidase. In any case, the oxygen reduction protons are most likely to enter through the oxygen channel, and the diagram itself is simplified by indicating that the proton channel-translocated protons behave similarly. The exit of the protons to the cytosolic side is also indicated to have a channel and an appropriate gate. While two gates are not essential, structural couplings may occur at various places within the subunit bearing heme a_3 and Cu_{a_3}. These couplings may arise from forces and displacements generated in the redox reactions, particularly those involving alterations of the spacings of the iron–copper binuclear complex. These changes are coupled to the structural factors controlling the pK's of carboxyl and amino acid groups within the channel, and controlling as well the function of the gates in allowing protons to pass when their concentration had reached a maximum, and preventing backflow of protons at other times during the reaction cycle. The hydroxyl ions are of course ejected back into the matrix side through the same or similar channels as that through which the hydrogen ions entered as water molecules.

The mechanism that affords structural coupling between redox changes and protein structures is the same as originally postulated by Wyman in 1949 as "indirectly-linked functions" that cause the release of four protons when four oxygen molecules react with hemoglobin. This is referred to in recent "membrane Bohr effect" papers (26); to identify the current ideas with the older ones, we use this term for the structurally reactive link between proton movement and the membrane protein and presume that the basic mechanism of coupling of redox energy changes to structural changes is similar.

IV. Short-Range Structural Features

Particular interest is focused upon the nature of the iron and copper atoms that cooperate in combining with and reducing oxygen (Figs. 1 and 2). This must be a special configuration of iron and copper, different from that in hemoglobin and myoglobin, and indeed, from hemocyanin. Iron, alone, as in hemoglobin and myoglobin, shows a slight tendency to reduce oxygen to the superoxide anion stage, a readily reversible reaction when the need for dissociation of oxygen from hemoglobin or myoglobin occurs. The reaction proceeds farther with the binuclear copper–copper of hemocyanin; there oxygen is reduced rapidly to the peroxide stage but no further electron donation is possible (the "other half" of the oxidase, such as cytochrome oxidase, where four electron donors are present, is absent in hemocyanin) and oxygen can be reversibly dissociated. In the case of the iron–copper binuclear complex of cytochrome oxidase, the oxygen binding properties resemble oxyhemoglobin and myoglobin when the temperature is low enough to slow electron donation from the copper and iron atoms. As the temperature is raised, electrons are donated from iron and copper to make a peroxo form, as in the case of hemocyanin. The difference arises on raising the temperature farther to the point where electrons can be donated by the second part of cytochrome oxidase, the additional iron and copper components, and from cytochrome c, c_1, and Fe–S protein. Thus, it is of particular interest to learn the nature of the iron atom, expected to be similar to that of hemocyanin, the nature of the copper atom, expected to be similar to that of hemocyanin, and the distance between the two in order to ascertain whether stretching of peroxide between iron and copper is feasible.

Most optical and magnetic methods for exploring the copper atom combined with iron in a binuclear complex of cytochrome oxidase fail since the spin pairing or antiferromagnetic coupling existing between the two ablates the magnetic signals and alters the optical signals so they cannot be distinguished readily from those of the second copper atom. One method which does give unaltered signals from the spin-paired copper is X-ray synchrotron K edge absorption, made possible by the high intensity tunable X-ray beams obtained from synchrotron light sources. Such beams, with intensity of 10^{11} photons/sec, can stimulate fluorescence emission from the copper atom at up to 50,000 ct/sec, a rate which makes feasible a recording of the edge absorption spectrum and the extended edge absorption fine structure (EXAFS) in several minutes for exploratory studies and in 8–10 hr for highly accurate studies. Such work, at the Stanford Synchrotron Radiation Laboratory (SSRL), has led to the discovery of significant properties of the copper atom itself and to preliminary studies of the distance between the iron and copper atoms of the oxygen reaction portion of cytochrome oxidase.

Considering first the nature of the copper

atom, its edge absorption is ahown in Fig. 5, which illustrates the characteristic 1s–3d, 1s–4p, and 1s–4s transitions at appropriately spaced energies. Since the X-ray synchrotron treats both coppers equally, differences in the two would be expected to occur upon reduction if the two copper environments are different. Reduction indeed shows that they are different: the high energy portion shifts more than the low energy portion. Based on appropriate model studies, this immediately suggests that of the two coppers, one is more ionic and shows a larger charge density shift on reduction, and the other is more covalent and shows a correspondingly smaller shift. Furthermore, the more highly charged environment would give edge absorption at higher energy and the less highly charged environment at low energy. Thus, the attribution of the large shift and the higher energy to the complex copper and the low energy, smaller shifted portion to the uncomplex, second copper atom seems appropriate and has been ingenuously verified by comparison of the two portions of the spectrum to appropriate model compounds, the higher energy portion to the model copper proteins, stellacyanin–azurin, and plastocyanin, and the lower one to a host of covalently bound copper compounds, of which copper oxalate bisimidazole is an example. The resulting deconvoluted spectra can be compared to portions of the oxidase in the half-oxidized–half-reduced form; thereby, the observations are adequately controlled (8).

The matching of the copper spectrum to that of a Type I blue copper, such as stellacyanin, azurin, or plastocyanin, affords the first evidence of the kind of environment appropriate for a function as an iron–copper binuclear complex in oxygen and peroxide binding.

The extension of the X-ray spectrum to higher energies gives characteristic oscillations related to interference patterns generated by the emission of a photon from the copper/iron atom and its reflection from neighboring groups, which in the case of iron would be the nitrogen of the pyrrole rings, and the copper atom itself if it is near enough. Current cytochrome oxidase studies suggest that this is the case: the copper atoms can be shown by a type of quantitative structural study to be as close to the iron atom as conjectures based upon antiferromagnetic coupling would have it be.

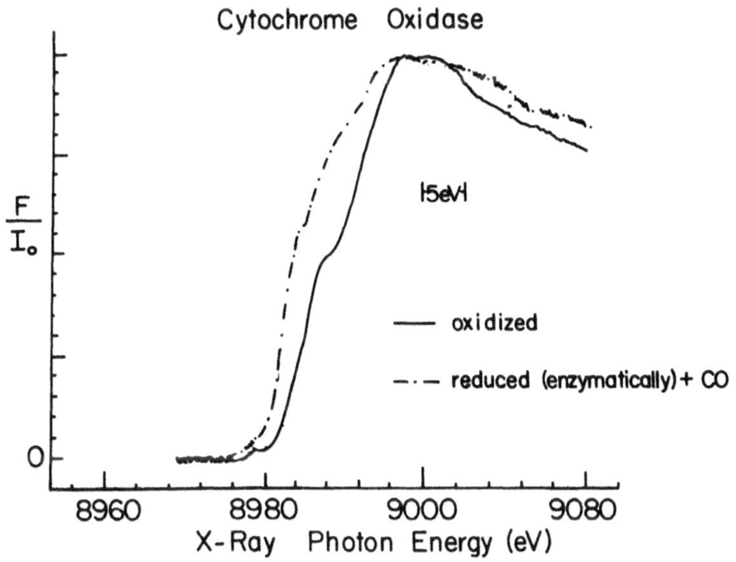

Figure 5. Experimental data obtained for the edge structure of oxidized, reduced, and CO-mixed valence states of cytochrome oxidase.

V. Summary

It has been our purpose to present a view of how cytochrome oxidase, the principal enzyme metabolism, deals with the transfer of electrons to oxygen in the energy coupled oxidoreduction reaction. This represents the most highly developed and specialized system for the nontoxic conservation of energy from intracellular oxidations. The matching of the electron donating capacity of cytochrome oxidase to its four electron accepting property suggests a specificity of design of electron donor and acceptor. However, our experimental results show that the reaction can be trapped in three distinct states (an oxy, a peroxy, and a fully oxidized state); a concerted process is not possible. Instead, a sequential process apparently follows a two electron path rather than Michaelis's idealized one electron path. Many other oxidases are less ideal and may be deficient in the number of electrons readily available for oxygen reduction in the tight binding of reactive intermediates characteristic of cytochrome oxidase. Among those are an interesting and recently discovered hybrid of cytochrome oxidase in which the Cu_a part is intact, while the a_3 part is replaced by the protoporphyrin IX oxidase known as cytochrome *o*. The way in which this oxidase serves its reactive intermediates will be of great interest to determine. Furthermore, cytochrome *o* exists as an oxidase only, without the benefit of heme *a* and Cu_a as ready electron donors for oxygen reduction. Thus, we visualize that nature has proceeded through a series of oxidases from a single iron porphyrin modeled upon myoglobin through various intermediate stages to the final four electron iron copper hybrid oxidase, cytochrome oxidase.

Acknowledgment

Support was generously provided by USNIH GM 12202, GM 27308, and HL 18708.

References

1. Chance, B., Waring, A., and Saronio, C. (1978). Low temperature transport in cytochrome *c* in the cytochrome *c*-cytochrome oxidase reaction: Evidence for electron tunneling. 29 Colloquium der Gesellschaft für Biologische Chemie 6–8 April 1978, Mosbach, Baden. In: Schafer, G., and Klingenberg, M. (Eds.). Energy Conservation in Biological Membranes. Berlin: Springer-Verlag, pp. 56–73.
2. Chance, B., Waring, A., and Yang, E. (1979). Interaction of electron doners and soluble cytochrome oxidase at intermediate-low temperatures. Biophys. J. 25:44a.
3. Chance, B. Waring, A., and Powers, L. (1979). The role of peroxidase-like intermediates in the enzymatic function of cytochrome oxidase over a variety of temperatures. Proceedings of the Japanese-American Seminar on Cytochrome Oxidase, Kobe, Japan, 1978. New York: Elsevier/North Holland, pp. 353–360.
4. Powers, L. Blumberg, W., Ching, Y., Eisenberger, P., Chance, B., Barlow, C., Leigh, J. S., Jr., Smith, J., Yonetani, T., Peisach, J., Vik, S., Hastings, B., and Perlman, M. (1979). X-ray absorption edge and extended fine structure (EXAFS) studies on cytochrome *c* oxidase. Biophys. J. 25(2):43a.
5. Hu, V. W., Chan, S. I., and Brown, G. (1977). X-ray absorption edge studies on cyanide-bound cytochrome *c* oxidase. FEBS Lett. 84:287–290.
6. Chance, B., and Leigh, J. S., Jr. (1977). Oxygen intermediates and mixed valence states of cytochrome oxidase: Infrared absorption difference spectra of compounds A, B, and C of cytochrome oxidase and oxygen. Proc. Natl. Acad. Sci. U.S.A. 74: 4777–4780.
7. Chance, B., Saronio, C., and Leigh, J. S., Jr. (1979). Compound C_2—a product of the reaction of oxygen and the mixed valence state of cytochrome oxidase: Optical evidence for a type I copper. Biochem. J. 177: 931–941.
8. Powers, L., Blumberg, W., Chance, B., Barlow, C., Leigh, J. S., Jr., Smith, J., Yonetani, T., Vik, S., and Peisach, J. (1979). The nature of copper atoms of cytochrome *c* oxidase as studied by optical and x-ray

absorption edge spectroscopy. Biochem. Biophys. Acta. 546:520–538.
9. Brown, J. M. (1978). Dissertation, Department of Chemistry. Princeton University.
10. Brown, J. M., Powers, L., Kincaid, B., Larrabee, J. A., and Spiro, T. (1980). Structural studies of the hemocyanin active site. I. EXAFS (extended x-ray absorption fine structure) analysis. J. Am. Chem. Soc. 102(2):4210–4216.
11. Alben, J. O., Altschuld, F., and Moh, P. (1981). Structure of the cytochrome oxidase (a_3) heme pocket. Low temperature FTIR spectroscopy of the photolyzed CO complex. In: Ho, Chien (Ed.). Interaction between Iron and Proteins in Oxygen and Electron Transport. New York: Elsevier/North Holland. In press.
12. Chance, B., Angiolillo, P., Yang, E., and Powers, L. (1980). Identification and assay of synchrotron radiation-induced alterations on metalloenzymes and proteins. FEBS Lett. 112(2):178–182.
13. DeVault, D., and Chance, B. (1966). Studies of photosynthesis using a pulsed laser. Biophys. J. 6(6):825–847.
14. Wikstrom, M. (1980). Mechanism of proton translocation by mammalian cytochrome oxidase. First European Bionergetics Conference, Urbino, Italy.
15. Chance, B., Gibson, Q., Eisenhardt, R., and Lonberg-Holn, K. (1964). Rapid mixing and sampling techniques as applied to biochemical kinetics. Science 146:3652.
16. Chance, B. Possible structures of cytochrome oxidase-oxygen intermediates and their reactivity toward cytochrome *c*. In: Lee, C. P. Schatz, G., Ernster, L. (Eds.). Membrane Bioenergetics. Reading, Massachusetts: Addison Wesley, pp. 1–12.
17. Chance, B. (1943). The kinetics of the enzymatic substrate compound of peroxidase. J. Biol. Chem. 151:553–557.
18. Chance, B., Azzi, A., Lee, I., Lee, C. P., and Mela, L. (1968). In: Ernster, L., and Drahota, Z. (Eds.). Mitochondria: Structure and Function. London: Academic Press, FEBS Symp. 17:233.
19. Capaldi, R. (1981). Structure and function of cytochrome oxidase. In: Ho, Chien (Ed.). Interaction between Iron and Proteins in Oxygen and Electron Transport. New York: Elsevier/North Holland. In press.
20. Frey, T. (1978). Structure and orientation of cytochrome *c* oxidase in crystalline membranes. J. Biol. Chem. 253:4389–4395.
21. Henderson, R., Capaldi, R., and Leigh, J. S., Jr. (1977): Arrangement of cytochrome oxidase molecules in two dimensional vesicle crystals. J. Mol. Biol. 112:631–648.
22. Weiss, H., Wingfield, P., Leonard, K., Winkler, F., Perkins, S., and Miller, A. (1980). Structure of ubiquinone: Cytochrome *c* reductase from neurospora mitochondria. First European Biogenetics Conference, Urbino, Italy.
23. Blasie, J. K., and Stamatoff, J. (1981). The use of resonance x-ray scattering for determining spatial relationships among metal atoms within macromolecules in noncrystalline state. Ann. Rev. Biophys. Bioeng. In press.
24. Ludwig, B., and Schatz, G. (1980). Two subunit cytochrome *c* oxidase (cytochrome aa_3) from Paracoccus denitrificans. Proc. Natl. Acad. Sci. U.S.A. 77(1):196–200.
25. Sone, N., Ohyama, T., and Kagaway, Y. (1979) Thermostable single-band cytochrome oxidase. FEBS Lett. 106(1):39–42.
26. Chance, B., Crofts, A., Nishimura, N., and Price, B. (1970), Fast membrane H+ binding in the light-activated state of Chromatium chromatophores. Eur. J. Biochem. 13: 364–374.
27. Chance, B. (1981). Oxygen transport and oxygen reduction. In: Ho, Chien (Ed.). Interaction between Iron and Proteins and Electron Transport. New York: Elsevier/North Holland. In press.

11

Biochemical Aspects of Oxygen Toxicity in the Metazoa

GYULA B. KOVACHICH AND NIELS HAUGAARD

The atmosphere of our planet contains about 20% oxygen, an amount of this vital substance that has made life possible in all its manifestations. To what extent life as we know it on earth has developed to accommodate to this particular oxygen tension, and to what extent life would have developed under conditions of lower or higher oxygen tensions, are matters for speculation. Some organisms, such as anaerobic bacteria, cannot tolerate the oxygen present in the air and thrive only when oxygen is sparse or absent. However, the great majority of animals and plants have adapted to live under existing atmospheric conditions. They have taken advantage of oxidative processes that produce much greater amounts of energy per mole of metabolite utilized than glycolysis or other half-reactions used by many lower organisms or by higher organisms during stress. Nonetheless, oxidative reactions are associated with potentially harmful effects on numerous reactions and enzymes in the cell. It is clear that during evolution animal and plant cells developed special mechanisms to control the oxidizing power of molecular oxygen. In a sense the answer to the question of the mechanism of oxygen toxicity can be found in the many ways in which cells have developed defenses against the toxic side effects of oxygen on cell function and metabolism.

When organisms are exposed to an atmosphere in which the concentration of oxygen is higher than that of air at sea level, alterations occur in cell metabolism and function. Such changes may be concerned primarily with adaptations to increased oxygen pressure and may not lead to serious damage or symptoms, except when the exposure to increased oxygen becomes extreme, either with respect to duration of exposure or concentration of molecular oxygen. An obvious example of an effect of increased molecular oxygen in the mammalian organism is the loss of the dual function of hemoglobin. This effect becomes progressively larger as the concentration of oxygen in inspired gas is increased. In venous blood the fraction of hemoglobin in reduced form decreases steadily until in man at about 2.5 atm oxygen there is virtually no reduction of hemoglobin, since the body requirements for oxygen are fully satisfied by dissolved oxygen. Since oxyhemoglobin is a stronger acid than reduced hemoglobin, this leads to a loss of buffering capacity of the blood (dual function of hemoglobin), to a decrease of pH, and to an increase in P_{CO_2} of the venous blood. This phenomenon was at one time considered an important factor in the development of oxygen toxicity (Gesell, 1925), but has been shown since to play only a minor role (Behnke et al., 1934; Stadie et al., 1944). The concept, however, remains that biochemical changes can occur in the body that are directly related to an increase in oxygen tension of the ambient atmosphere. Other examples, such as changes in ratios of oxidized and reduced forms of enzymes or coenzymes, may be found in the future.

One of the major manifestations of oxygen toxicity in mammals concerns the lung and can be seen during prolonged breathing of oxygen-enriched gas mixtures at sea level (Clark and Lambertsen, 1971). With hyperbaric oxygenation, effects on the central nervous system predominate, giving rise to the familiar oxygen convulsions and death. Changes in retinal function have also been observed during hyperoxygenation (Nichols and Lambertsen, 1969).

There are numerous review articles concerning the general aspects of oxygen poisoning (Bean, 1945; Haugaard, 1968, 1974; Clark and Lambertsen, 1971) and other chapters in this book deal with important recent developments. This chapter will emphasize biochemical changes in hyperoxygenated brain tissue and will also discuss certain factors that increase or decrease the susceptibility of an organism to the toxic effect of oxygen.

I. Molecular Mechanisms of Oxygen Toxicity

A. The Role of Free Radicals

A general theory of oxygen toxicity was formulated by Gerschman (1959), who proposed that oxygen-induced damage is caused by free radical intermediates generated in excessive amounts during exposure of animals to increased pressures of oxygen. The theory was based on the similarities between cellular alterations produced by oxygen and radiation, and on the protective role of sulfhydryl compounds and other agents against both types of cellular damage. Strong support for this theory came from findings that synergistic effects between hyperbaric oxygen and radiation have been observed with many biological systems.

The discovery of the enzyme superoxide dismutase (McCord and Fridovich, 1969) gave further strength to Gerschman's theory and added an exciting new dimension to it. Superoxide dismutase is an enzymatic "scavenger" of the highly reactive superoxide free radical that is produced during reduction of molecular oxygen by cellular oxidative reactions (Fridovich, 1974, 1975). Since superoxide dismutase has been found in all aerobic organisms and could not be detected in anaerobes (McCord et al., 1971), it appears that this enzyme is necessary for life even at the oxygen pressure found in ambient air. Thus oxygen toxicity, which once seemed only an interesting but somewhat esoteric problem, is now emerging as one of the most fundamental phenomena in biological sciences.

Superoxide free radicals, however, also serve a useful purpose in the life of higher organisms. The bacteriocidal action of the granulocyte commences with a sudden increase in respiration (Sbarra and Karnovsky, 1959; Baehner et al., 1970), followed by release of superoxide and hydrogen peroxide (Iyer et al., 1961; Babior et al., 1973; Curnutte and Babior, 1974; Homan-Muller et al., 1975). Superoxide, formed by superoxide synthase (McCord and Salin, 1975; Babior et al., 1976; Babior and Kipnes, 1977), is the precursor of the actual bacteriocidal agent, the hydroxyl radical (Johnston et al., 1975).

Biological systems that utilize oxygen are endowed with antioxidants, which protect against the deleterious effects of superoxide and superoxide-generated free radical intermediates. When cellular protective mechanisms are overwhelmed, free radicals initiate chain reactions leading to either slowly reversible or irreversible changes in cell structure. Among the cellular effects that may be initiated by an increase in the concentration of free radicals are oxidation of sulfhydryl groups and peroxidation of lipids.

B. Oxidation of Sulfhydryl Groups

Barron and Singer (1943) proposed that enzymes which require reduced —SH groups for their catalytic action should be called sulfhydryl enzymes. The enzyme inhibitory theory of oxygen toxicity advanced by Stadie et al. (1944) and by Dickens (1946b) involved this group of enzymes. The theory is based on the susceptibility of these enzymes to inactivation by oxidation of their essential sulfhydryl groups. Tjioe and Haugaard (1972) demonstrated this action of oxygen experimentally. They correlated inactivation of crystalline glyceraldehyde phosphate dehydrogenase by oxygen at 5 atm with disappearance of enzyme sulfhydryl groups, in the presence of trace amounts of heavy metals. It is important to point out, however, that some biologically active substances, such as insulin, require —S—S—bridges for activity, and are inactivated by reduction of the disulfide links (Tomizawa and Halsey, 1959). There is ample evidence that high-pressure oxygen inactivates sulfhydryl enzymes in various in vitro preparations (Stadie et al., 1945a,b,c; Stadie and Haugaard, 1945; Dickens, 1946b; Haugaard, 1946), but attempts to demonstrate oxidation of sulfhydryl active sites in vivo have met with difficulty (Teng and Harris, 1970; Begin-Heick et al., 1969). A significant reduction in the —SH/S—S ratio and inhibition of succinic dehydrogenase, however, has been demonstrated in rat lung tissue after the experimental animals had been exposed to 5 atm oxygen for 45 min (Jamieson et al., 1963).

Barron (1955) proposed that reduced glutathione plays an important role in maintaining essential —SH groups in the reduced state. Indeed, an increase in gluthathione turnover has been shown in frog bladder preparation (Allen et al., 1973) and in the isolated lung preparation (Nishiki et al., 1976) immediately upon raising oxygen pressure. Negative findings in measurements of tissue content of —SH groups, therefore, do not allow one to conclude with certainty that sulfhydryl oxidation is not a significant effect of hyperoxygenation in vivo. Regeneration of —SH groups may occur and significant decreases in cellular sulfhydryl content could be a late event in oxygen toxicity. Nishiki et al. (1976) postulated that the main defense mechanism against oxygen toxicity is an early increase in the production of reduced glutathione. When production of reducing equivalents cannot meet the demand, first membrane damage occurs by lipid peroxidation, followed by oxidation of sulfhydryl enzymes.

A comparatively little known area of sulfhydryl chemistry deals with the role of —SH groups in membrane function. There is evidence that sulfhydryl agents in low concentrations affect the properties of mitochondrial (Haugaard et al., 1969), microsomal (Robinson, 1965a), and nerve plasma membranes (Smith, 1958; Spyropoulos et al., 1960; Huneeus-Cox et al., 1966). For example, diamide, an —SH oxidizing agent, highly specific for glutathione (Kosower and Kosower, 1969; Kosower et al., 1972), has been shown to decrease reduced glutathione levels and inhibit cation transport of isolated rat lens (Epstein and Kinoshita, 1970). Since oxygen at increased pressure interferes with normal membrane excitability (Rudin and Eisman, 1952; Perot and Stein, 1959; Cymerman and Gottlieb, 1970; Rossier, 1970; Ochs and Hackaman, 1972), it is likely that oxidation of membrane sulfhydryl groups plays a part in these effects.

C. Lipid Peroxidation

Lipid peroxidation is a form of oxidative degeneration of polyunsaturated lipids (Tappel, 1973). Unsaturated fatty acids are liable to peroxidation because the presence of carbon–carbon double bonds which weaken the carbon–hydrogen bond on the adjacent carbon atom. The interaction of oxygen and such lipids generates free radical intermediates, such as the highly reactive ROO· form, which in turn initiates chain reactions (Peters and Foote, 1976; McCay and Poyer, 1976). There are indications that lipid peroxidative intermediates are also powerful inactivators of sulfhydryl enzymes (Chio and Tappel, 1969).

Evidence is accumulating for the importance of lipid peroxidation as a pathogenic process in oxygen toxicity, as shown by measurements involving brain tissue (Becker and Galvin, 1962; Wollman, 1962; Kahn et al., 1964; Jerrett et al., 1973), red blood cells (Tsen and Collier, 1960; Zalkin and Tappel, 1960; Hochstein, 1966; Mengel and Kann, 1966), the isolated rat lung (Nishiki et al., 1976), and the frog bladder preparation (Allen et al., 1973). Among cellular structures known to be affected are mitochondria (Hunter et al., 1964; McKnight et al., 1965) and microsomes (Robinson, 1965b).

An oxygen-dependent inactivation of the sodium transport mechanism has been shown in the frog bladder preparation by Allen et al. (1973), and they concluded that lipid peroxidation was the most likely molecular mechanism to cause this. There is experimental evidence that brain NaK-ATPase may be inactivated by lipid peroxidation. An ascorbic acid-dependent, metal ion-catalyzed inactivation of microsomal NaK-ATPase activity was found to be due to lipid peroxidation (Schaeffer et al., 1975), and it was shown that synaptosomal NaK-ATPase is inactivated by lipoxygenase–H_2O_2 treatment (Sun, 1972).

II. Oxygen Toxicity in Brain Tissue

A. *The Effect of Metabolic Rate*

The effect of high-pressure oxygen on respiration in rat brain slices has been investigated extensively by Dickens (1946a) and by Stadie and colleagues (1945a). They found that tissue respiration is depressed by elevated oxygen pressures and that the extent of the depression is a function of oxygen pressure and the duration of hyperoxygenation. The relationship was quantitated by Dickens (1962) in an empirical equation expressing the duration of hyperoxygenation (T) necessary for 50% depression of oxygen consumption at a given atmospheric pressure of oxygen (P). For brain slices incubated in normal Krebs–Ringer phosphate medium the following equation was proposed:

$$\log T = 2.56 - 0.77 \log P$$

Another expression (Dickens, 1962), derived from findings of several independent laboratories (Prekladovizkii, 1936; Jenkinson, 1940; Haldane, 1941), predicts the probability of the onset of in vivo neurological symptoms of oxygen toxicity as a function of the same parameters T and P:

$$\log T = 3.0 - 2.6 \log P$$

These equations are the linear forms of a rectangular hyperbola (Fig. 1). It is evident that the progression of oxygen toxicity with time in vivo and in brain tissue in vitro can be expressed by the same mathematical function but that the slope of this function is different under the two conditions. Another difference between in vivo and in vitro effects of high-pressure oxygen is that brain slices under standard incubation conditions are much more resistant to the toxic effects of oxygen than brain tissue in situ.

The problems involved in the study of oxygen toxicity with brain slices were reexamined in this laboratory, based on the following considerations. In normal Krebs–Ringer or similar media, the brain tissue is in a resting condition with the plasma membrane polarized and the oxygen consumption about 50% below that found in the intact brain (Geiger and Magnes, 1947; McIlwain, 1953; Kety, 1957; Hertz and Schousboe, 1975). It has been suggested (Quastel and Quastel, 1961; McIlwain, 1966) that the in vitro conditions are a better simulation of the metabolic state in situ when oxygen consumption of brain slices is increased by

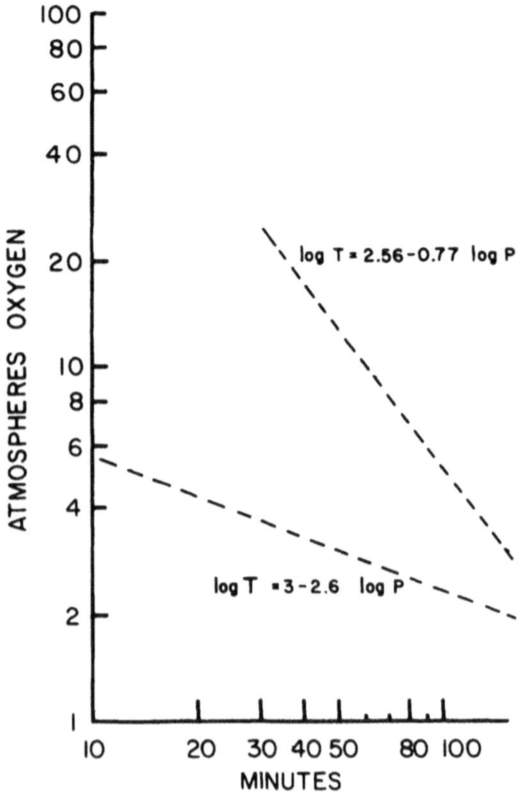

Figure 1. Comparison of the empirical equations (Dickens, 1962) expressing the onset of neurological symptoms (log $T = 3–2.6 \log P$) and ED_{50} for depression of oxygen consumption in brain slices (log $T = 2.56–0.77 \log P$). These two parameters of toxicity are clearly two different functions of oxygen pressure. However, when the metabolic rate of brain slices is increased (Figs. 2 and 3), the slope of the pressure function for 25% depression of $^{14}CO_2$ production approaches the slope of the equation for in vivo oxygen toxicity (log $T = 3–2.6 \log P$).

potassium depolarization (Ashford and Dixon, 1935; Dickens and Greville, 1935) or by electrical stimulation (McIlwain, 1953). We investigated the effects of oxygen at elevated pressures on the glucose metabolism of brain slices under conditions at which the rate of respiration was altered by the presence of high concentrations of potassium ions or by changing calcium concentration in the incubation medium or by addition of drugs that affect CO_2 production in a variety of ways (Kovachich et al., 1977; Kovachich, 1978a, 1979a, 1980). These experiments showed that when the metabolic rate of brain slices is increased the toxic effects of oxygen become more pronounced.

The experiments presented in Fig. 2 show the effects of varying the pressure of oxygen from 1 to 10 atm on the formation of $^{14}CO_2$ from uniformly labeled [^{14}C]glucose by brain slices at increasing concentrations of KCl in the incubation medium. At 1 atm of oxygen it is clear that glucose oxidation is markedly increased as the potassium concentration in the medium is raised. It is also apparent that as the rate of glucose metabolism is increased, the tissue becomes more sensitive to the toxic effects of oxygen. Compare, for example, the steep depression of CO_2 production from glucose at 100 m*M* KCl at 60-min incubation with the much more modest decrease seen in normal Krebs–Ringer phosphate solution for the same incubation time.

Similar results were obtained when respiration and glucose oxidation were increased by the addition of the depolarizing agent veratridine to the incubation medium (Fig. 3). Veratridine, like excess potassium, increased glucose oxidation as measured by the production of CO_2. This drug also made the brain much more sensitive to the toxic effects of hyperbaric oxygen.

Figure 4 shows the combined results of a number of experiments in which the effects of elevated oxygen pressures on brain glucose oxidation have been tested both under conditions where the metabolic rate has been increased and where it was lower than that of brain slices incubated in Krebs–Ringer phosphate at 37°C. Veratridine at 15 μM greatly stimuated CO_2 production from glucose and under these conditions the brain slices were very sensitive to the toxic effects of oxygen as indicated by the steep slope of the curve representing CO_2 production from glucose per hour as a function of the pressure of oxygen. This was also true, although to a lesser extent, when glucose oxidation was increased by raising the KCl concentration to 75 m*M*, adding dinitrophenol (DNP, 5 μM) to the medium, or omitting calcium ions from the solution. The effect of oxygen on glucose oxidation in the usual Krebs–Ringer phos-

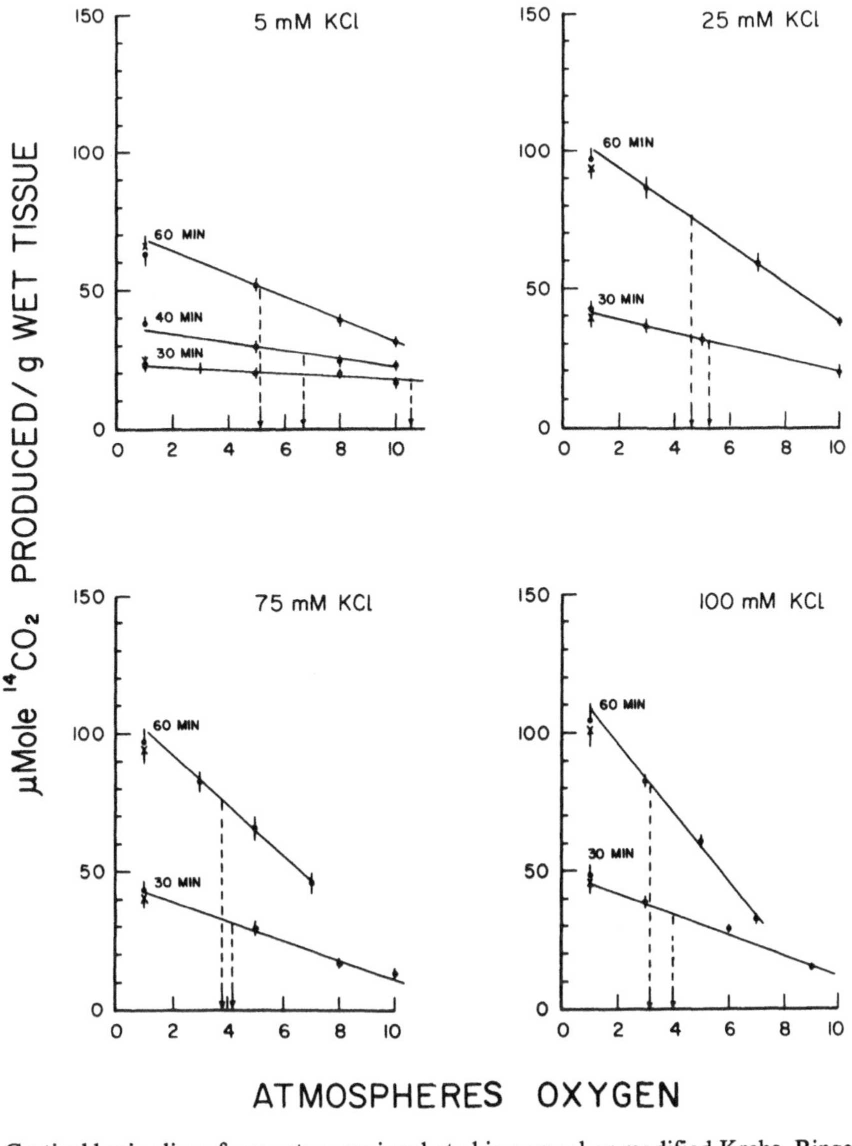

Figure 2. Cortical brain slices from rats were incubated in normal or modified Krebs–Ringer phosphate medium: 5 m*M* KCl, 120 m*M* NaCl, 1.3 m*M* $CaCl_2$, 1.2 m*M* $MgCl_2$, sodium phosphate 16 m*M*, pH 7.4, with 10 m*M* [U-^{14}C]glucose. In media containing elevated levels of potassium, the sodium concentration was correspondingly reduced. The incubation vessels were Erlenmeyer flasks, sealed by a rubber stopper holding a 26g hypodermic needle to allow compression of the gas phase inside the flasks. The hyperbaric chamber was a Bethlehem Corporation Model G15-APSP, mounted on a shaker and fitted with a temperature control system. The reaction was terminated by tilting the chamber to transfer the trichloroacetic acid from the side arm of the reaction vessel into the main compartment. The evolved $^{14}CO_2$ was absorbed by paper wicks soaked in 100 μl phenylethylamine in the center well, prior to decompression. Each point on the plot represents the mean ± SEM of 4–6 experiments. The broken vertical lines indicate oxygen pressure necessary for 25% depression of $^{14}CO_2$ production. Additional information is available in Kovachich et al. (1977) and Kovachich (1980).

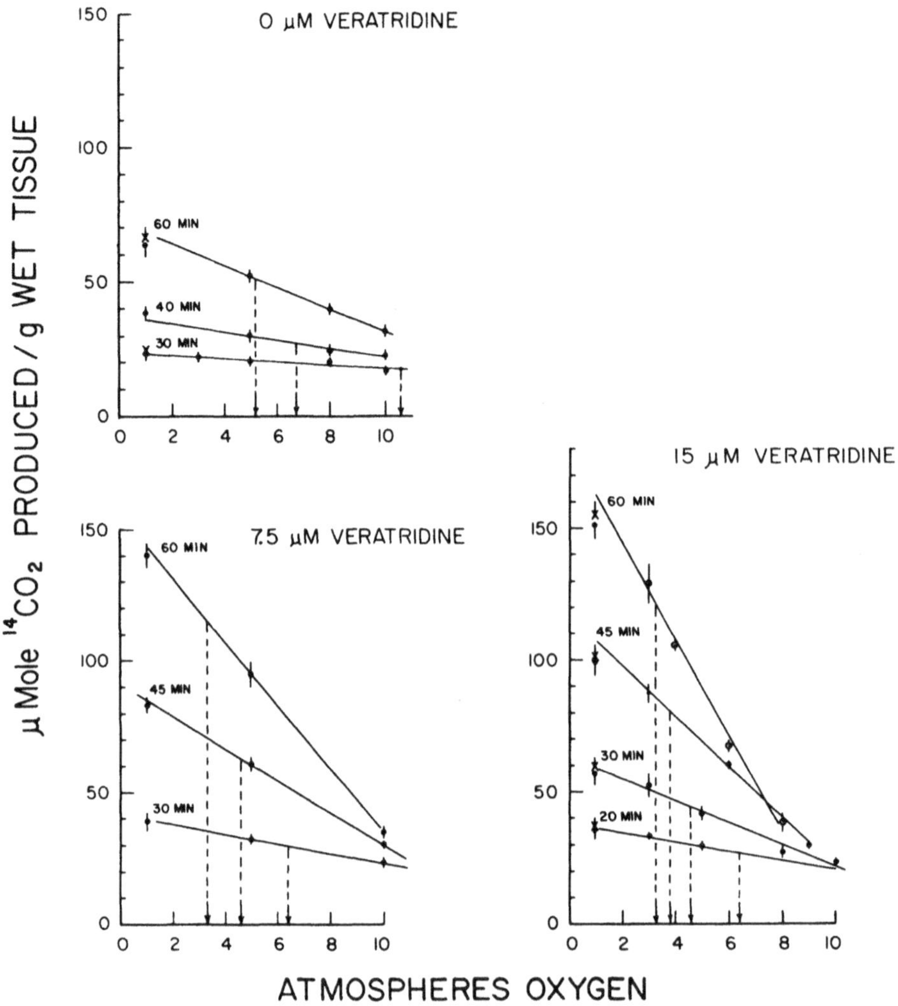

Figure 3. Experimental conditions are described in legend for Fig. 2. The medium was normal Krebs–Ringer phosphate with 7.5 or 15 μM veratridine.

phate medium is shown in the graph as the curve labeled 5 mM KCl-P_i buffer. Under these conditions the toxic action of oxygen is small although significant as shown by many earlier measurements. When TRIS buffer was substituted for phosphate or when a high concentration of calcium ions was present in the medium, CO_2 production from glucose was depressed at 1 atm O_2 and the inhibition of glucose oxidation by hyperbaric oxygen was very small. Finally, when brain slices were incubated at 32°C in Krebs–Ringer phosphate medium the rate of glucose oxidation was low at 1 atm; under these conditions of depressed metabolism, oxygen at pressures up to 10 atm had no significant effect.

It is not known why brain tissue is more sensitive to the toxic effects of oxygen when the rate of respiration is increased. There are two possibilities that should be considered. First, when electron transport is increased additional free radicals may be formed, and these, in the face of the action of hyperbaric oxygen, could lead to an excessive rise in highly reactive molecular species. Second, during heightened metabolic activity, the cellular membranes that are targets for the harmful effects of oxygen may shift into configurations that are more vulnerable to the oxidizing effect of free radicals. The latter possibility, although highly speculative, is not entirely unreasonable, since it has been

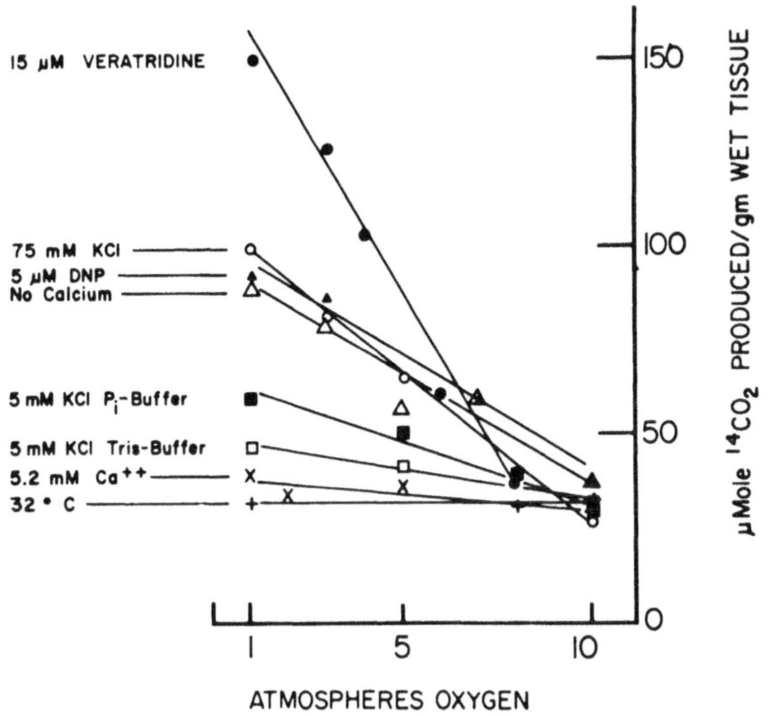

Figure 4. Experimental conditions were the same as described in legend for Fig. 2. Metabolic rate was altered by the following changes in the normal Krebs–Ringer phosphate medium (5 m*M* KCl): addition of 15 μ*M* veratridine, raising potassium concentration to 75 m*M*; addition of 5 μ*M* dinitrophenol, omission of calcium, replacement of phosphate buffer by TRIS buffer, raising Ca concentration to 5.2 m*M*. In one experiment the temperature was lowered from 37° to 32°C without changing the composition of the medium. Duration of all experiments was 1 hr. The points represent the mean of 4–6 experiments.

shown, for example, that mitochondrial membranes undergo extensive alterations in structure when the respiration changes from the resting state-4 to the active state-3 (Hackenbrock et al., 1971). There is considerable evidence that free radicals may be formed during cell respiration (Dionisi et al., 1975; Loschen and Azzi, 1976) and that when the rate of electron transport is increased there is an increase in the formation of these molecules. These free radicals may directly react with vital cell constituents, or may be transformed into more reactive species.

B. *Changes in Pathways of Glucose Oxidation at Elevated Oxygen Pressure*

In brain slices glucose oxidation is altered to a considerable extent in the presence of elevated oxygen pressures (Kovachich, 1978b) as shown in Fig. 5.

In experiments in which C-1- and C-6-labeled glucose was used, in the presence of Krebs–Ringer solutions containing 5 m*M* KCl there was a significant although relatively small decrease in CO_2 formation from the 6-position of the glucose molecule as oxygen pressure was raised. In contrast, CO_2 formation from the glucose 1-position increased with the pressure of oxygen, indicating that a rise in pentose-shunt activity occurred. In a medium containing 75 m*M* KCl, which caused depolarization of the cell membrane, glucose metabolism was much greater than in normal Krebs–Ringer phosphate. Oxidation of glucose, as measured by CO_2 formation from the 6-position of glucose, was extremely sensitive to inhibition by oxygen under these conditions. Formation of CO_2 from the 1-position of glucose was unaffected by oxygen until a pressure of 4 atm had been reached, indicating a relative increase in the contribution of the pentose-shunt pathway to the total oxidation of

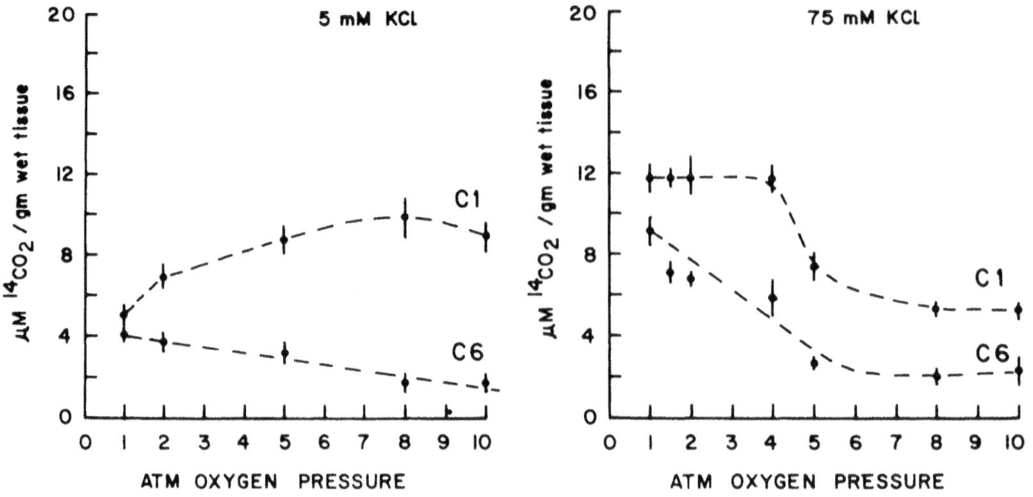

Figure 5. Incubation conditions were the same as described in Fig. 2. The substrate was 10 m*M* [1−^{14}C]glucose or [6−^{14}C]glucose; incubations were 60 min.

glucose. As the oxygen pressure was increased above 4 atm, CO_2 formation from the 1-position became markedly depressed.

While the measurements of CO_2 from C-1- and C-6-labeled glucose do not give strictly quantitative estimates of the two pathways of glucose oxidation, the differences in the effects of elevated oxygen pressures are so large as to lead to the conclusion that oxygen toxicity occurs predominantly in the main glucose oxidation pathway and that the pentose shunt activity is little affected or increased as oxygen pressure is raised. The same conclusion was reached by Brue et al. (1978) who measured $^{14}CO_2$ formation from [1−^{14}C]- and [6−^{14}C)glucose in the isolated retina under high-pressure oxygen.

The above experiments indicate that the entrance of pyruvate into the tricarboxylic acid cycle is inhibited by high-pressure oxygen. This conclusion is supported by further results. Pyruvate metabolism is rate limiting in glucose oxidation in potassium depolarized brain slices (Kini and Quastel, 1959; Bilodeau and Elliott, 1963). This is apparently caused by the relatively low activity of pyruvate dehydrogenase itself (Kovachich and Haugaard, 1977). The activity of this enzyme is reduced in homogenates of brain slices exposed to high oxygen pressures (Kovachich et al., 1977). Furthermore, in brain slices incubated with glucose as a substrate there is an accumulation of alanine under hyperbaric oxygenation (Kovachich, 1979b). Since alanine formation is an alternative pathway of pyruvate metabolism, such an increase would be expected if pyruvate entrance into the tricarboxylic acid cycle became limited. These observations indicate an alteration in glucose metabolism under high-pressure oxygen, where complete oxidation of glucose is hindered. The following section will show that these conclusions are compatible with the results of extensive earlier experiments on the effects of high-pressure oxygen on cerebral GABA metabolism.

C. *The Role of γ-Aminobutyric Acid (GABA) and Other Metabolites in CNS Toxicity*

Wood and colleagues demonstrated a striking correlation between depression of cerebral GABA levels and the onset of oxygen-induced seizures (Wood and Watson, 1962; Wood et al., 1963, 1965, 1967, 1969; Wood, 1970). The decrease in GABA was determined during the first 20 min of hyperoxygenation and compared to the exposure time needed to produce convulsions in 50% of the animals (CT_{50}) at the same pressure. The results showed that the greater the decrease in brain GABA concentration, the shorter the preconvulsion time. Regardless of the magnitude of CT_{50} this relationship was manifested in mice, hamsters, rats, guinea pigs, and rabbits at 6.1 atm; in chicks of 1 day, 1 week, and 2 weeks of age at 5.1 atm; and in mice in the 3.7 to 6.4 atm range with 100% oxygen and with gas mixtures containing oxygen and 0.5%, 1%, or 5% CO_2. Lowering the cerebral GABA level was also directly related to oxygen pressure in mice treated with oxygen in the 3.7 to 6.1 atm range (Wood et al., 1969). When the decrease in GABA, as obtained in these experiments, is represented on the y-axis and oxygen pressures above 1 atm are shown on the x-axis, then on extrapolation, the decrease in GABA would be minimal at about 3 atm oxygen. This evidence further supports the importance of the GABA content of brain in the development of oxygen-induced seizures, since it is well known that seizures rarely develop below 3 atm oxygen, even during prolonged exposures to excess oxygen (Shilling and Adams, 1933; Lambertsen, 1955; Foster, 1965).

There is no doubt that the reduction of GABA levels is a manifestation of neurochemical changes associated with oxygen-induced seizures, but it is not yet clear how these two phenomena are connected. GABA is formed by the GABA shunt, a bypass in the citric acid cycle. It is not only a major metabolite of intermediary metabolism in the brain, but it also acts as an inhibitory transmitter in certain CNS pathways (Curtis and Watkins, 1960, 1965). It is possible that a cause and effect relationship could exist between lowered levels of GABA and the onset of seizures, in view of the transmitter function of this substance. As an inhibitory transmitter, GABA is necessary for maintaining a proper balance of excitatory and inhibitory influences in certain areas of the brain. It is reasonable to assume, therefore, that hyperexcitability will develop when the concentration of this substance is lowered by oxygen below a certain point. This view, however, would have to be reconciled with some recent findings. For example, disulfiram prevents development of seizures without preventing a reduction in GABA content (Alderman et al., 1974; Faiman et al., 1977). Furthermore, oxygen-induced convulsions may ensue while GABA levels are higher than normal. GABA levels may be elevated by hydrazinopropionic acid (HPA) (Van Gelder, 1968, 1969) or by amino-oxyacetic acid (AOAA) (Baxter and Roberts, 1961; Roberts and Simonsen, 1963; Haber, 1965). These agents interfere with the action of the GABA-metabolizing enzyme, GABA-transaminase (GABA-T). However, treatment of rats with HPA or AOAA has been reported to actually shorten preconvulsion time and increase intensity and frequency of convulsions under high-pressure oxygen, while cerebral GABA levels are increased by 10–200% (Alderman et al., 1974).

Depression of brain GABA concentrations by high-pressure oxygen, therefore, may be a manifestation of an overall derangement in brain metabolism and the seizures are generated by a so far unknown mechanism. This contention is in line with the interpretation of several experiments involving the GABA system. GABA is formed from glutamic acid by the action of glutamic acid decarboxylase (GAD) (Roberts and Frankel, 1950, 1951a,b) and is converted to succinic semialdehyde by the enzyme GABA-T (Roberts et al., 1958; Lajtha et al., 1959). GABA-T is quite stable under increased oxygen pressure (Wood et al., 1971), but there is evidence that GAD, the GABA-forming enzyme, is sensitive to inhibition by oxygen both in vitro (Roberts and Simonsen, 1963; Wood and Watson, 1964) and in vivo (Scherbakova, 1962; Wood et al., 1966). However, sensitivity of GAD to inhibition by oxygen is independent of the age of the animals, while depression of GABA by oxygen becomes more marked during postnatal development (Wood et al., 1971). Furthermore, no evidence is available that inhibition of GAD is reversible, whereas GABA depression is reversible (Wood and Watson, 1963). In view of these findings, Radomski and Watson

(1973) speculated that changes in GABA metabolism in hyperoxygenated brain tissue "may in fact be secondary to oxygen-induced changes in ionic gradients."

Let us therefore consider the effect of oxygen on brain content of glutamic acid, the immediate precursor of GABA. This substance is found in the CNS in abundance, and is not only a metabolite of intermediary metabolism, but is also considered an excitatory neurotransmitter (Krnjevic and Phillis, 1963; Johnson, 1972; Crawford and Connor, 1973). There are contradictory reports about the in vivo effect of high-pressure oxygen on glutamic acid in whole brain; according to some reports there is no change (Wood and Watson, 1963; Faiman et al., 1977) while according to others there is a decrease (Banister et al., 1976; Schafer, 1978) or increase (Gershenovich and Krichevskaya, 1956) in cerebral glutamate levels. Some of these contradictions may be explained by differences in experimental conditions, such as the application of different oxygen pressures for different amounts of time causing various degrees of CNS damage prior to analysis. Even so, these measurements only reflect steady-state levels without regard to turnover rates; thus, even negative results do not necessarily indicate that there are no alterations in glutamic acid metabolism. Furthermore, these results do not reflect regional variations. Since brain tissue is heterogeneous in terms of tissue types (i.e., neuronal and glial elements), in functional and biochemical organization, and most likely in susceptibility to oxygen as well, whole brain determinations are of limited value.

Experiments with cortical brain slices produced results that are easier to interpret. Kaplan and Stein (1957) found that incubation for 90 min at 6 atm oxygen impaired the ability of guinea pig brain cortical slices to accumulate glutamate from the medium and to retain tissue potassium. The latter observation was confirmed by Joanny et al. (1970). It is well documented that glutamate uptake is dependent on a transport mechanism (Tsukada et al., 1963; Margolis and Lajtha, 1968; Logan and Snyder, 1972; Bennett et al., 1973; Takagaki, 1976a), which is coupled to transport of potassium (Stern et al., 1949; Terner et al., 1950; Pappius and Elliott, 1956; Takagaki et al., 1959; Tsukada et al., 1963), and that both processes are inhibited by ouabain (Gonda and Quastel, 1962; Nakazawa and Quastel, 1968; Takagaki, 1976b). These results therefore suggest that high-pressure oxygen impairs membrane-bound transport systems, or otherwise interferes with glutamate and electrolyte homeostasis. Such effects, if occurring in vivo, would have far-reaching consequences in brain function. For example, neuronal excitability could be increased if the extracellular space accumulated glutamate, an excitatory transmitter, and potassium, a depolarizing agent. Furthermore, it is now generally accepted that glutamate is distributed in at least two metabolic compartments, probably corresponding to neuronal and glial tissues and that cellular intermediary metabolism requires transport of glutamate between these compartments (Van den Berg et al., 1969; Berl and Clarke, 1969; Benjamin and Quastel, 1972; Balazs and Cremer, 1973).

Experiments in this laboratory have shown that glutamate levels in cortical brain slices are lowered by 10 atm oxygen within 30 min (Kovachich, 1979b). This may be due to increased glutamate release or stimulation of glutamate catabolism. The latter possibility is likely, since aspartate is accumulated in these tissue samples, suggesting transamination of glutamate with oxaloacetate. It has been shown that glutamate transamination will occur when entrance of pyruvate into the Krebs cycle is depressed (Balazs, 1964), a condition which seems to prevail in brain slices under high-pressure oxygen (Kovachich et al., 1977; Kovachich, 1978b, 1979b).

In summary, experiments with brain slices indicate changes in cortical glutamate metabolism under high-pressure oxygen. This phenomenon may be important for a better understanding of the mechanism by which oxygen lowers GABA levels in brain. The changes in glutamate and GABA levels and the ability of succinate to delay the onset of CNS oxygen toxicity (Currie et al., 1971;

Mayevsky et al., 1974) may be indicative of impaired GABA shunt activity. Although these implications are only speculative, it is well to keep in mind that reduction in cortical glutamate levels has been associated with several types of convulsions, such as human epilepsy (Van Gelder et al., 1972), hypoglycemic seizures (Cravioto et al., 1951; Dawson, 1953; Tews et al., 1965; Lewis et al., 1974), and in seizures induced by cobalt (Koyama, 1972; Van Gelder, 1972), mescaline (Mison-Crighel et al., 1964), methionine sulphoximine (Peters and Tower, 1959), local freezing (Berl et al., 1959) and pentelene tetrazol (Tews et al., 1963; Whisler et al., 1968).

III. Factors Influencing Susceptibility to Oxygen Poisoning in Vivo

Experiments carried out over many years have led to the realization that resistance to the toxic effects of oxygen may change during the life of a multicellular organism, as a result of hormonal or environmental influences but also as a function of age, with wide differences in resistance to hyperbaric oxygen among species.

A. Metabolic and Hormonal Conditions and Oxygen Toxicity

When Gerschman et al. (1954) proposed a general theory of oxygen toxicity based on the action of free radicals they pointed out that increased metabolism may enhance the formation of these toxic agents. At this time Stadie et al. (1944) and Bean (1945) had already elaborated on experimental evidence linking metabolic rate to the rate of development of oxygen toxicity.

It has been known for a long time that man is more resistant to the toxic effects of oxygen than, for example, small laboratory animals with considerably higher basal metabolism rates, and that exercise reduces man's relative tolerance to increased oxygen pressure (Bornstein and Stroink, 1912; Donald, 1947). Certain natural changes in the metabolic state of organisms during their life cycle will effect susceptibility to oxygen. Continuous exposure of insects (Lipidopterans) to 100% oxygen during morphogenesis is detrimental to their development, but they are most susceptible to oxygen poisoning during early larval development and during pupation (Brown and Hines, 1976). These stages are characterized by the most intense biosynthetic and metabolic activity. Ground squirrels have increased resistance to the effects of high-pressure oxygen during hibernation, when the metabolic rate is markedly reduced (Popovic et al., 1964). Cold-blooded animals are less susceptible to oxygen poisoning than are mammals, but it was shown that when turtles are exposed to high-pressure oxygen at 37°C, symptoms of oxygen poisoning appear as rapidly as in mammals (Faulkner and Binger, 1927).

Hormonal changes may also significantly affect the oxygen tolerance of the organism. The role of the thyroid hormone has been repeatedly examined and it is generally accepted that the hyperthyroid state reduces oxygen tolerance (Campbell, 1937; Grossman and Penrod, 1949b; Bean and Bauer, 1952; Smith et al., 1960; Szilagyi et al., 1969). Grossman and Penrod (1949b) demonstrated that rats fed a diet supplemented with thyroid extract had increased oxygen consumption, and those given propylthiouracil to produce hypothyroidism had decreased oygen consumption compared to controls. When these rats were exposed to 5.5 atm oxygen for 1 hr, 59% of the control group survived, while 29% of the hyperthyroid and 80% of the hypothyroid animals survived.

The adrenergic system appears to play a role in the development of oxygen-related tissue damage. The increased adrenergic tone during hyperoxygenation was found to be detrimental to the animals' defenses against a hyperoxic challenge (Bean and Johnson, 1955), and it has been suggested that CNS catecholamine levels play a part in determining the threshold for seizure development (Hof et al., 1972; Ngai et al., 1977).

The importance of metabolic rate in determining sensitivity to the toxic effects of oxygen is also manifested in vitro. In a comparative study using retina, brain, and liver of teleost, amphibian, and mammal, Baeyens et al. (1973) concluded that the higher the metabolic rate, the more rapidly oxygen toxicity develops. This concept is further strengthened by the results of experiments showing that lowered temperature has a protective effect against the toxic action of oxygen in whole animals (Grossman and Penrod, 1949a; Giretti et al., 1969), cortical brain slices (Kovachich, 1979a, 1980), and liver homogenates (Norman et al., 1966).

B. *Effect of Age*

It is generally accepted that newborns are more resistant to oxygen-induced lung damage and CNS toxicity than their adult counterparts (Faulkner and Binger, 1927; Smith et al., 1932a, 1932b; Williams and Beacher, 1944; Gerschman et al., 1955; Kyle, 1965; Polgar et al., 1966; Craig et al., 1968; Shanklin, 1969; Sperling, 1970; Wood, 1970; Smith and Usubiaga, 1971; Balentine, 1973; Chen and Fang, 1975; Bonikos et al., 1976; Berry et al., 1977). In experiments with mice and rats a large number of newborns survived for 6 weeks in 100% O_2 at 1 atm (Bonikos et al., 1976) and typical seizures rarely developed even at 4.4 or 5 atm absolute oxygen before the second postnatal week (Sperling, 1970; Balentine, 1973). Although neonates have a far better chance of surviving otherwise lethal doses of oxygen and the usual pattern of symptoms is delayed or absent, other manifestations of oxygen toxicity have been noticed. Raising rats at 70–80% oxygen at atmospheric pressure for the first 9 days of life significantly inhibited the growth of the brain and reduced accumulation of DNA, RNA, total protein, and lipoproteins (Grave et al., 1972); and at 1 atm, 100% oxygen inhibited DNA synthesis in the lungs of newborn mice (Northway et al., 1973). In most of these experiments special arrangements were made to provide maternal care for the newborns, since the adult species would not survive the effects of continuous hyperoxygenation. Dams were periodically shifted between control and treated groups for this purpose and also to minimize the influence of natural mother and natural litter interaction in each group.

Although development of newborns is adversely affected by increased oxygen tension, it is nevertheless clear that during the first neonatal weeks most species have a greater flexibility in adapting to increased environmental oxygen pressure. How is this adaptability lost during development? It is possible that in the young, tissue production of oxygen-generated toxic intermediates proceeds at a slower rate, or that the natural defenses of cells are better able to handle the oxidizing challenge. If the latter alternative is correct, it could mean that cellular structures or enzymes are less susceptible to oxidative damage because enzymes such as superoxide dismutase or glutathione reductase are particularly active in the young animal.

The decreased resistance to pulmonary oxygen toxicity observed with age may have an enzymatic basis. It has been demonstrated, for example, that exposure to a raised O_2 tension increases superoxide dismutase activity in the lung in certain species (Frank et al., 1978). Species which do not manifest this adaptive response to increased oxygen tension in the neonatal stage succumb to oxygen more rapidly, as do adults. There is no evidence that the fully developed lung tissue increases superoxide dismutase activity after exposure to high-pressure oxygen.

The explanation for the greater resistance of the immature CNS is speculative. Wood (1970) observed gradually increasing susceptibility to oxygen-induced seizures in chicks during the first weeks following hatching and found a high correlation between the rate at which cerebral GABA levels decreased and the time of onset of seizures. The activities of the two enzymes in GABA metabolism, glutamate decarboxylase and GABA-α-ketoglutarate transaminase, were found to increase with age; however, no age-dependent oxygen effect could be demonstrated on either of these two enzymes in vitro, or on GABA transport in brain slices (Wood et al., 1971). The age-related increase in susceptibility, therefore, could not be explained in terms of direct effects of oxygen on the enzymatic steps in GABA metabolism.

Any proposed mechanism for the development of seizures should account for the fact that normal cell excitability depends on the proper distribution of electrolytes across the cell membrane. A primary role for maintaining electrolyte balance is assigned to the NaK-ATPase or sodium pump (Skou, 1965; Harvey and McIlwain, 1968), and in the adult brain the bulk of cellular energy production is geared toward the operation of this system (Whittam, 1964). In newborn rats brain-tissue NaK-ATPase activity is low, with sharp increases during the second and third postnatal weeks (Abdel-Latif et al., 1967; Samson and Quinn, 1967; Medzihradsky et al., 1972; Nagata et al., 1974). By the third week the adult level of NaK-ATPase activity is reached, and oxygen consumption in the brain approaches the adult level from the slow metabolic rate characteristic of newborns (Himwich, 1951). During this time, simultaneously with the development of the sodium pump in vivo, isolated brain tissue from these animals shows an increased response to stimulation of respiration by potassium (Himwich, 1951). The increase in respiration produced by high extracellular potassium is believed to be mediated through the activation of the sodium pump (Ruscak and Whittam, 1967). The potassium-activated brain slice has a markedly increased sensitivity to oxygen toxicity (Kovachich et al., 1977), which appears to be related to the increase in metabolic rate under this condition (Kovachich, 1979a, 1980). It is tempting to speculate, therefore, that the appearance of seizures during neonatal development and the increased susceptibility of adult brain to toxic effects of oxygen is related to the full development of the excitable membrane, and the appearance of the sodium pump and simultaneous increase in respiration.

C. *Cyclic Variations in Sensitivity to Oxygen Toxicity*

Rodents such as rats and mice are nocturnal animals and show increased spontaneous activity during the night. Saelens et al. (1968) quantitated the spontaneous locomotor activity of mice in circular 6-beam photocell activity cages and found a daily cycle of high and low activity periods with peak activity between 8 and 12 P.M., in agreement with the findings of Mount and Willmott (1967). A single dose of 200 mg/kg pargyline, a monoamine oxidase inhibitor, markedly potentiated, and a single dose of 400 mg/kg α-methylmetatyrosine, a depletor of norepinephrine, abolished the night-time increment in locomotor activity. Although both drugs were injected at 10 A.M., the treatments had no effect on the rest of the daily cycle, indicating that the adrenergic system plays an especially important role in the nocturnal high-activity period. Hof et al. (1972) demonstrated the existence of a circadian rhythm in susceptibility to oxygen-induced seizures in phase with the above cycles: rats manifest the longest latency period before convulsion between 7 A.M. and 11 A.M., coinciding with their least active period. When young rats were raised in a reverse dark–light schedule (dark from 7 A.M. until 7 P.M.), their sleep and wake periods as well as their oxygen sensitivity patterns were reversed (Dexter et al., 1972).

Seizure time is affected by the extent of environmental illumination even in the same stage of circadian rhythm. Ngai et al. (1977) reported that when mice were kept in a normal light–dark schedule prior to experimentation and exposed to 4 atm oxygen, CT_{50} was 16.6 min for mice exposed in dark (4 ft candles) and 24.4 min for light-exposed mice (130 ft candles). The protection by light was not manifested in blinded mice; treatment with norepinephrine depletors or with propranalol offered protection to the dark-exposed mice. Since the pineal gland is believed to be the generator of light–dark as well as diurnal cycles in brain metabolism (Axelrod, 1974), the authors speculated that the relationship between illumination and seizure time originates in variations in the activity of the eye–pineal gland axis. Signals from the eye are transmitted to the pineal gland through a sympathetic pathway with norepinephrine as a final transmitter acting on beta receptors. In the dark, for example, the pineal norepinephrine turnover rate is nearly tripled. The abolishment of night-time peak activity in mice by α-methylmetatyrosine agrees with this proposed mechanism (Saelens et al., 1968). Ngai et al. (1977) concluded that increased susceptibility of mice to seizures in a dark environment is related to increased pineal norepinephrine tone and subsequent neuroendocrine changes which may be manifested by increased spontaneous activity.

D. The Role of CO_2 and THAM

There has been considerable discussion concerning the role of CO_2 in the etiology of oxygen toxicity, since increased CO_2 accelerates the development of lung and CNS damage at increased oxygen pressure (Behnke et al., 1934; Marshall and Lambertsen, 1961; Bean, 1961; Van den Brenk and Jamieson, 1964; Mayevsky et al., 1974). It is still uncertain, however, how CO_2 exerts its effect. It has been proposed, for example, that the action of CO_2 is a result of vasodilation facilitating delivery of oxygen to tissues, or that it is related to the lowered pH of body fluids and cells (Ligou and Nahas, 1960; Bean, 1961; Sanger et al., 1961; Gottlieb and Jagodzinski, 1963; Van den Brenk and Jamieson, 1964; Nahas, 1965; McSherry and Veith, 1968). Experimentation in response to these questions led to the use of the buffer tris(hydroxymethyl)aminomethane (THAM), and the results showed significant protection of this agent against oxygen toxicity. In view of its dissociation properties, THAM is an ideal physiological buffer. At the normal pH of body fluids, 30% of this substance is in the un-ionized form; it is able to penetrate into the cytosol to a considerable extent (Ligou and Nahas, 1960; Brown and Goot, 1963; Lambotte et al., 1971), where it will act to increase pH. It is likely that the protective effect of THAM against O_2 toxicity is related to its buffering capacity.

Additional mechanisms should also be considered, however, since THAM exerts a variety of other actions even when it is used as a buffering agent. For example, it interferes with some enzyme reactions requiring monovalent cations (Betts and Evans, 1968), with the enzymes acting on carbonyl-containing substrates such as pyruvate (Mahler, 1961). THAM also influences electrolyte homeostasis in dogs (Samiy et al., 1961) and in unicellular organisms (Packer et al., 1961) and produces negative inotropic effects on the isolated rabbit heart (Gillespie and McKnight, 1976). These are only a few of the reported effects. For a review of the pharmacology of THAM see Nahas (1962).

Effects of THAM on the adrenergic system and on mitochondrial respiration may be relevant to its protective action against oxygen toxicity. There is considerable evidence that antiadrenergic substances increase resistance to O_2 toxicity, and that certain cellular reactions dependent on the adrenergic system are inhibited in THAM-buffered media. Gonzalez et al. (1979) incubated samples of rabbit vas deferens, rabbit superior cervical ganglion, and rat striatum in Tyrode's solution with tritiated tyrosine as the catecholamine precursor. When THAM was also present in the medium, tritiated norepinephrine and total catecholamine levels were reduced to less than 50% of control values within 2 hr. Since they found no evidence of increased release, the authors speculated that THAM reduced tissue catecholamine content by facilitating intracellular oxidation, secondary to alkalinization of intragranular space, although it also seems possible that the rate of synthesis was affected by THAM.

The findings of Gillespie and McKnight (1976) are related to these observations. They found that 10 mM THAM depressed motor responses to adrenergic stimulation in several in vitro preparations and suggested that the effects are mainly presynaptic and result from the metabolic action of THAM.

Stinson and Spencer (1968) compared the effect of various buffering systems on the respiratory parameters of isolated mitochondrial preparations. The rate of mitochondrial respiration was markedly lowered in THAM buffered media, compared to respiration in media buffered with HEPES, inorganic phosphate, *N*-tris(hydroxymethyl)methylglycine or *N*-tris(hydroxymethyl)-methyl-2-aminoethanesulfonic acid. Reduction of metabolic rate appears to increase resistance to toxic effects of oxygen, as discussed earlier in this chapter. Since THAM is able to penetrate into cells, reduction of mitochondrial respiration may be one component of its protective action against oxygen toxicity in intact tissues and in whole animals as well.

Median convulsion time at 3 atm oxygen is doubled in mice injected with 1 ml of 0.3 M THAM per 30 g body weight (Nahas, 1965). These animals had 25% lower oxygen consumption during hyperoxygenation, compared to saline-injected controls. A reduction in metabolic rate in THAM-treated animals was also noted by Sanger et al. (1961) and by Gottlieb and Jagodzinski (1963). In THAM-buffered Krebs–Ringer medium the metabolic rate of brain slices is lower than in Krebs–Ringer phosphate medium, and the inhibition of $^{14}CO_2$ production from [U-^{14}C] glucose by high-pressure oxygen develops at a slower rate (Kovachich, 1979a, 1980). It remains to be seen, however, whether there is a cause-and-effect relationship between these two phenomena brought about by THAM.

IV. Concluding Remarks

It is clear that the mechanisms by which molecular oxygen in excess exerts its many toxic effects on cells are exceedingly complex and are only beginning to be understood.

More and more evidence has accumulated in support of the concept first stated by Gerschman that free radicals play an essential part in the production of cellular damage by hyperbaric oxygen. The discovery by McCord and Fridovich (1969) of superoxide dismutase led to the realization that this enzyme may act as the most important defense developed by cells to protect against the ever present oxidizing potential of molecular oxygen. Glutathione and glutathione reductase as well as catalase probably also play important parts in protecting cells against the toxic effects of oxygen. Haugaard (1968, 1974) has pointed out that a chelating agent such as EDTA, that combines strongly with trace amounts of heavy metals (e.g., Cu^{2+}, Fe^{2+}), has a remarkable ability to diminish or prevent oxygen toxicity in tissue preparations in vitro. It is possible that in the cell in situ ATP and other cell constituents chelate trace metals. This can perhaps occur to such an extent that the concentrations of free metal ions under normal circumstances are too low to catalyze formation of free radicals to a significant extent. The binding of trace heavy metals to tissue constituents can also be considered a cellular defense against oxygen toxicity.

In multicellular organisms a great many factors influence the action of excess oxygen on particular cells or organs. It is obvious that when an animal is exposed to increased pressures of oxygen, some cells in close contact with the gaseous environment or arterial blood such as those in the lung are exposed to a greater extent to an increased pressure of oxygen than other cells in the body. Changes in the rate of blood flow to an organ, caused directly or indirectly by an increase in the pressure of oxygen in the ambient atmosphere, can obviously influence profoundly the increase in intracellular oxygen tension occurring in an individual cell and, therefore, the toxic effects produced.

Finally, we would like to express our conviction that there is no simple answer to the problem to oxygen toxicity. Numerous reactions in the cell are affected during hyperbaric oxygenation, and harmful effects of excess oxygen on cell function and metabolism may vary widely in different organs and cell types.

With the increasing interest by scientists of different backgrounds in oxygen toxicity, we can expect considerable progress in this important field in the next decade.

Acknowledgment

Preparation of this work was supported by grants from the National Institutes of Health (HL-01813, HL-08899-15) and from the Office of Naval Research (contract N00014-76-C-0248).

References

Abdel-Latif, A. A., Brody, J., and Ramahi, H. (1967). Studies on sodium-potassium adenosine triphosphatase of the nerve endings and appearance of electrical activity in developing rat brain. J. Neurochem. 14:1133–1141.

Alderman, J. L., Culwer, B. W., and Shellenberger, M. K. (1974). An examination of the role of γ-aminobutyric acid (GABA) in hyperbaric oxygen induced convulsions in the rat. I. Effects of increased gamma-aminobutyric acid and protective agents. J. Pharmacol. Exp. Ther. 190:334–340.

Allen, J. E., Goodman, D. E. P., Besarab, A., and Rasmussen, H. (1973). Studies on biochemical basis of oxygen toxicity. Biochim. Biophys. Acta 320:708–728.

Ashford, C. A., and Dixon, K. C. (1935). The effect of potassium on glycolysis of brain tissue with reference to the Pasteur effect. Biochem. J. 29:157–168.

Awapara, J., Landua, A. J., Fuerst, R., and Seale, B. (1950). Free γ-aminobutyric acid in brain. J. Biol. Chem. 187:35–39.

Axelrod, J. (1974). Pineal-gland-neurochemical transducer. Science 184:1341.

Babior, B. M., and Kipnes, R. S. (1977). Superoxide-forming enzyme from human neutrophils; evidence for flavin requirement. Blood 50: 517–524.

Babior, B. M., Curnutte, J. T., and McMurrich, B. J. (1976). The particulate superoxide-forming system from human neutrophils. Properties of the system, and further evidence supporting its participation in the respiratory burst. J. Clin. Invest. 58:989–996.

Babior, B. M., Kipnes, R. S., and Curnutte, J. T. (1973). Biological defense mechanisms: The production by leukocytes of superoxide, a potential bactericidal agent. J. Clin. Invest. 52: 741–744.

Baehner, R. L., Gilman, N., and Karnovsky, M. L. (1970). Respiration and glucose oxidation in human and guinea pig leukocytes: Comparative studies. J. Clin. Invest. 49: 692–700.

Baeyens, D. A., Hoffert, J. R., and Fromm, P. O. (1973). Comparative study of oxygen toxicity in retina, brain and liver of teleost, amphibian and mammal. Comp. Biochem. Physiol. 45A: 925–932.

Balazs, R. (1964). Control of glutamate metabolism. The effect of pyruvate. J. Neurochem. 12:63–76.

Balazs, R., and Cremer, J. E., eds. (1973). Metabolic Compartmentation in the Brain. London: Macmillan.

Balentine, J. D. (1973). Comparative responses of newborn and adult rats to hyperbaric oxygen exposure. Am. J. Pathol. 70:A21.

Banister, E. W., Bhakthan, N. M. G., and Singh, A. K. (1976). Lithium protection against oxygen toxicity in rats: Ammonia and amino acid metabolism. J. Physiol. (London) 260:587–596.

Barron, E. S. G. (1955). Oxidation of some oxidation-reduction systems by oxygen at high pressure. Arch. Biochem. Biophys. 59: 502–510.

Barron, E. S. G., and Singer, T. P. (1943). Enzyme systems containing active sulfhydryl groups. The role of glutathione. Science 97: 356.

Baxter, C. F., and Roberts, E. (1961). Elevation of γ-aminobutyric acid in brain: Selective inhibition of γ-aminobutyric-α-ketoglutaric acid transaminase. J. Biol. Chem. 236:3287–3294.

Bean, I. W., and Bauer, R. (1952). Thyroid in pulmonary injury by O_2 in high concentration at atmospheric pressure. Proc. Soc. Exp. Biol. Med. 81:693–694.

Bean, I. W., and Johnson, P. C. (1955). Epinephrine and neurogenic factors in pulmonary edema and CNS reaction induced by O_2 at high pressure. Am. J. Physiol. 160:438–444.

Bean, J. (1945). Effects of oxygen at increased pressure. Physiol. Rev. 25:1–147.

Bean, J. (1961). Tris buffer, CO_2, and sympathoadrenal system in reactions to O_2 at high pressure. Am. J. Physiol. 201:737–739.

Becker, N. H., and Galvin, J. F. (1962). Effect of oxygen-rich atmospheres on cerebral lipid peroxides. Aerosp. Med. 33:985.

Begin-Heick, N., Hochstein, P., and Hill, G. B. (1969). Investigations on the effects of hyperbaric oxygen on enxyme activity. Can. J. Physiol. Pharmacol. 47:400.

Behnke, A. R., Shaw, L. A., Shilling, C. W. Thomson, R. M., and Messer, A. C. (1934). Studies on the effects of high oxygen pressure. I. Effect of high oxygen pressure upon the carbon-dioxide and oxygen content, the acidity, and the carbon-dioxide combining power of the blood. Am. J. Physiol. 107:13–28.

Benjamin, A. M., and Quastel, J. H. (1972). Locations of amino acids in brain slices from the rat. Tetrodotoxin-sensitive release of amino acids. Biochem. J. 128:631–646.

Bennett, J. P., Jr., Logan, W. J., and Snyder, S. H. (1973). Amino-acids as central nervous transmitters—influence of ions, amino-acid analogs and ontogeny on transport systems for L-glutamic and L-aspartic-acids and glycine into central nervous synaptosomes of rat. J. Neurochem. 21:1533–1550.

Berl, S., and Clarke, D. D. (1969). Compartmentation of amino acid metabolism. In: Handbook of Neurochemistry, ed. A. Lajtha, vol. 2, pp. 447–472. New York: Plenum Press.

Berl, S., Purpura, D. P., Girado, M., and Waelsch, H. (1959). Amino acid metabolism in epileptogenic and non-epileptogenic lesions of the neocortex (cat). J. Neurochem. 4:311–317.

Berry, S., Fitch, J. W., and Schatte, C. L. (1977). Influence of sex and age on the susceptibility of mice to oxygen poisoning. Aviat. Space Environ. Med. 48:37–39.

Betts, G. F., and Evans, H. J. (1968). The inhibition of univalent cation activated enzymes by tris(hydroxymethyl)aminomethane. Biochim. Biophys. Acta 167:193–196.

Bilodeau, F., and Elliott, K. A. C. (1963). The influence of drugs and potassium on respiration and potassium accumulation by brain tissue. Can. J. Biochem. Physiol. 41:779–792.

Bonikos, D. S., Bensch, K. G., and Northway, W. H. (1976). Oxygen toxicity in the newborn. Am. J. Pathol. 85:623–650.

Bornstein, A., and Stroink. (1912). Ueber Sauerstoffvergiftung. Deut. Med. Wochensch. 38: 1495–1497.

Brown, E. B., Jr., and Goot, B. (1963). Intracellular hydrogen changes and potassium movement. Am. J. Physiol. 204:765–770.

Brown, O. R., and Hines, M. B. (1976). Selective toxicity of 1 atmosphere of oxygen during morphogenesis of two Lepidopterans. Aviat. Space Environ. Med. 47:954–957.

Brue, F., Joanny, P., Morcellet, J. L., and Corriol, J. H. (1978). Metabolic defense mechanisms at the cellular level against oxygen toxicity. Bull. Europeen Physiopathol. Respiratoire 14:152 (abstract).

Campbell, J. A. (1937). Oxygen poisoning and the thyroid gland. J. Physiol. (London) 90: 91–92.

Chen, C. F., and Fang, H. S. (1975). High oxygen pressure induced convulsions in suckling mice. Chin. J. Physiol. 22:1–5.

Chio, K. S., and Tappel, A. L. (1969). Inactivation of ribonuclease and other enzymes by peroxidizing lipid and by malonaldehyde. Biochemistry 8:2827–2832.

Clark, J. M., and Lambertsen, C. J. (1971). Pulmonary oxygen toxicity: A review. Pharmacol. Rev. 23:37–133.

Craig, J., Rev-Kury, L. H., and Kury, G. (1968). Oxygen toxicity in newborn rats. Pediat. Res. 2:316.

Cravioto, R. O., Massieu, G., and Izquierdo, J. J. (1951). Free amino acids in rat brain during insulin shock. Proc. Soc. Exp. Biol. Med. 78:856–858.

Crawford, I. L., and Connor, J. D. (1973). Localization and release of glutamic-acid in relation to hippocampal mossy fiber pathway. Nature (London) 244:442–443.

Curnette, J. T., and Babior, B. M. (1974). Biological defense mechanisms: The effect of bacteria and serum on superoxide production by granulocytes. J. Clin. Invest. 53:1662–1672.

Currie, W. D., Kramer, R. C., and Sanders, A. P. (1971). Effects of hyperbaric oxygenation on metabolism. VII. Succinate protection against oxygen toxicity in large animals. Proc. Soc. Exp. Biol. Med. 136:630–631.

Curtis, D. R., and Watkins, J. C. (1960). In: Inhibition in the Nervous System and Gamma-aminobutyric Acid, eds. E. Roberts et al., pp. 424–444. Oxford: Pergamon.

Curtis, D. R., and Watkins, J. C. (1965). The pharmacology of amino acids related to gamma-aminobutyric acid. Pharmacol. Rev. 17:347–391.

Cymerman, A., and Gottlieb, S. F. (1970). Effects of increased oxygen tension on bioelectric properties of frog sciatic nerve. Aerosp. Med. 41:36.

Dawson, R. M. C. (1953). Cerebral amino acids in fluoroacetate-poisoned, anaesthetised and hypoglycaemic rats. Biochim. Biophys. Acta 11:548–552.

Dexter, J. D., Hof, D. G., and Mengel, C. E. (1972). The effect of sleep-wake reversal and sleep deprivation on circadian rhythm of oxygen toxicity seizure susceptibility. Aerosp. Med. 43:1075–1078.

Dickens, F., and Greville, G. D. (1935). The metabolism of normal and tumour tissue. XIII. Neutral salt effects. Biochem. J. 29: 1468–1483.

Dickens, F. (1946a). The toxic effects of oxygen on brain metabolism and on tissue enzymes. 1. Brain metabolism. Biochem. J. 40:145–171.

Dickens, F. (1946b). The toxic effects of oxygen on brain metabolism. 2. Tissue enzymes. Biochem. J. 40:171–186.

Dickens, F. (1962). The toxic effects of oxygen on nervous tissue. In: Neurochemistry, eds. K. A. C. Elliott, I. H. Page, and F. H. Quastel, pp. 851–869. Springfield, Illinois: C. C Thomas.

Dionisi, O, Galeotti, T., Terranova, T., and Azzi, A. (1975). Superoxide radicals and hydrogen peroxide formation in mitochondria from normal and neoplastic tissues. Biochim. Biophys. Acta 403:292–300.

Donald, K. W. (1947). Oxygen poisoning in man. Brit. Med. J. 17:667–676.

Epstein, D. L., and Kinoshita, J. H., (1970). The effect of diamide on lens glutathione and lens membrane function. Invest. Ophthalmol. 9: 629–638.

Faiman, M. D., Nolan, R. J., Baxter, C. F., and Dodd, D. E. (1977). Brain gamma-aminobutyric acid, glutamic acid decarboxylase, glutamate, and ammonia in mice during hyperbaric oxygenation. J. Neurochem. 28:861–865.

Faulkner, J. M., and Binger, C. A. (1927). Oxygen poisoning in cold blooded animals. J. Exp. Med. 45:865–871.

Foster, C. A. (1965). In: Hyperbaric Oxygenation, ed. I. M. Ledingham, p. 380. Edinburgh: Livingstone.

Frank, L., Bucher, J. R., and Roberts, R. J. (1978). Oxygen toxicity in neonatal and adult animals of various species. J. Appl. Physiol. 45: 699–704.

Fridovich, I. (1974). Superoxide dismutases. Adv. Enzymol. 41:35–97.

Fridovich, I. (1975). Superoxide dismutases. Annu. Rev. Biochem. 44:147–159.

Geiger, A., and Magnes, J. (1947). The isolation of the cerebral circulation and the perfusion of the brain in the living cat. Am. J. Physiol. 149:517–537.

Gerschman, R. (1959). Oxygen effects in biological systems. Proc. Int. Congr. Physiol. Sci. 21st, Buenos Aires, pp. 222–226.

Gerschman, R., Gilbert, D. L., and Nye, S. W. (1955). Survival time of newborn rats and mice in high oxygen pressure (HOP). U.S.A.F. School of Aviation Medicine, Project No. 21-1201-0013. *Studies on Oxygen Poisoning*. Report No. 10:35–36.

Gerschman, R., Gilbert, D. L., Nye, S. W., Dwyer, P., and Fenn, W. O. (1954). Oxygen poisoning and X-irradiation: A mechanism in common. Science 119:623–626.

Gershenovich, Z. S., and Kricheyskaya, A. A. (1956). Activity of glutamine synthetase of the brain and liver following exposure of animals to high oxygen pressure. Biokhimiya 21:715–722.

Gesell, R. (1925). The chemical regulation of respiration. Physiol. Rev. 5:551–595.

Gillespie, J. S., and McKnight, A. T. (1976). Adverse effects of tris hydrochloride, a commonly used buffer in physiological media. J. Physiol. (London) 259:561–573.

Giretti, M. L., Rucci, F. S., and LaRocca, M. (1969). Effects of lowered body temperature on hyperoxic seizures. Electroencephalogr. Clin. Neurophysiol. 27:581–586.

Gonda, O., and Quastel, J. H. (1962). Effects of ouabain on cerebral metabolism and transport mechanisms in-vitro. Biochem. J. 84:394–406.

Gonzalez, C., Obeso, A., and Fidone, S. (1979). Tris buffer: Effects on catecholamine synthesis. J. Neurochem. 32:1143–1145.

Gottlieb, S. F., and Jagodzinski, R. V. (1963). Role of THAM in protecting mice against convulsive episodes caused by exposure to oxygen under high pressure. Proc. Soc. Exp. Biol. Med. 112:427–430.

Grave, G. D., Kennedy, C., and Sokoloff, L. (1972). Impairment of growth and development of the rat brain by hyperoxia at atmospheric pressure. J. Neurochem. 19:187–194.

Grossman, M. S., and Penrod, K. E. (1949a). Relationship of hypothermia to high oxygen poisoning. Am. J. Physiol. 156:177–181.

Grossman, M. S., and Penrod, K. E. (1949b). The thyroid and high oxygen poisoning in rats. Am. J. Physiol. 156:182–184.

Haber, B. (1965). The effect of hydroxylamine and amino-oxyacetic acid on the cerebral in vitro utilization of glucose, fructose, glutamic acid, and γ-aminobutyric acid. Can. J. Biochem. 43:865–876.

Hackenbrock, C. R., Rehn, T. G., Weinbach,

E. C., and Lemasters, J. F. (1971). Oxidative phosphorylation and ultrastructural transformation in mitochondria in intact ascites tumor cell. J. Cell. Biol. 51:123–137.

Haldane, J. B. S. (1941). Human life and death at high pressures. Nature 148:458–460.

Harvey, J. A., and McIlwain, H. (1968). Excitatory acidic amino acids and cation content and sodium ion flux of isolated tissues from brain. Biochem J. 108:269–274.

Haugaard, N. (1946). Oxygen poisoning XI. The relation between inactivation of enzymes by oxygen and essential sulfhydryl groups. J. Biol. Chem. 164:265–270.

Haugaard, N. (1968). Cellular mechanisms of oxygen toxicity. Physiol. Rev. 48:311–373.

Haugaard, N. (1974). The effects of high and low oxygen tensions on metabolism. In: Hayaishi, O. (Ed.). Molecular Oxygen in Biology: Topics in Molecular Oxygen Research, pp. 163–182. New York: American Elsevier.

Haugaard, N., Lee, N. H., Kostrzewa, R., Horn, R. S., and Haugaard, E. S. (1969). The role of sulfhydryl groups in oxidative phosphorylation and ion transport by rat liver mitochondria. Biochim. Biophys. Acta 172:198–204.

Hertz, L., and Schousboe, A. (1975). Ion and energy metabolism of the brain at the cellular level. Int. Rev. Neurobiol. 18:141–178.

Himwich, H. W. (1951). Brain Metabolism and Cerebral Disorders. Baltimore: Williams & Wilkins.

Hochstein, P. (1966). Antioxidant mechanisms and hemolysis associated with lipid peroxidation. In: Brown, I. W., and Cox, B. G. (Eds.). Proceedings of the Third International Conference on Hyperbaric Medicine, pp. 61–64. Washington, D. C.: Natl. Acad. Sci., Natl. Res. Council.

Hof, D. G., Cline, W. H., Dexter, J. D., and Mengel, C. E. (1972). CNS epinphrine tone, a possible etiology for the threshold in susceptibility to oxygen toxicity seizures. Aerosp. Med. 43:1194–1199.

Homan–Muller, J. W. T., Weening, R. S., and Roos, D. (1975). Production of hydrogen peroxide by phagocytizing human granulocytes. J. Lab. Clin. Med. 85:198–207.

Huneeus-Cox, F., Fernandez, H. L., and Smith, B. H. (1966). Effects of redox and sulfhydryl reagents on the bioelectric properties of the giant axon of the squid. Biophys. J. 6:675.

Hunter, F. E., Scott, A., Weinstein, J. et al. (1964). Effect of phosphate, arsenate, and other substances on swelling and lipid peroxide formation when mitochondria are treated with oxidized and reduced glutathione. J. Biol. Chem. 239:622–630.

Iyer, G. Y. N., Islam, M. F., and Quastel, J. H. (1961). Biochemical aspects of phagocytosis. Nature 192:535–541.

Jamieson, D., Ladner, K., and Van Den Brenk, H. A. S. (1963). Pulmonary damage due to high pressure oxygen breathing in rats. IV. Quantitative analysis of sulfhydryl and disulphide groups in rat lungs. Aust. J. Exp. Biol. Med. Sci. 41:491–497.

Jenkinson, S. (1940). Submarine salvage, the air in a submerged submarine: Means of exit when submerged, and disabilities of the survivors. Br. J. Surg. 27:767–780.

Jerrett, S. A., Jeferson, D., and Mengel, C. E. (1973). Seizures, H_2O_2 formation and lipid peroxides in brain during exposures to oxygen under high pressure. Aerosp. Med. 44: 40–44.

Joanny, P., Corriol, J., and Brue, F. (1970). Hyperbaric oxygen—effects of metabolism and ionic movement in cerebral cortex slices. Science 167:1505–1510.

Johnson, J. L. (1972). Glutamic acid as synaptic transmitter in nervous system. Brain Res. 37: 1–19.

Johnston, R. B., Jr., Keele, B. B., Jr., Misra, H. P., Webb, L. S., Lehmeyer, J. E., and Rajagopalan, K. V. (1975). Superoxide anion generation and phagocytic bactericidal activity. In: The Phagocytic Cell in Host Resistance, eds. J. A. Bellanti and D. H. Dayton, p. 51. New York: Raven Press.

Kahn, H. E., Jr., Mengel, C. E., Smith, W., and Horton, B. (1964). Oxygen toxicity and vitamin E. Aerosp. Med. 35:840.

Kaplan, S. A., and Stein, S. N. (1957). Effects of oxygen at high pressure on the transport of potassium, sodium and glutamate in guinea pig brain cortex. Am. J. Physiol. 190:157–162.

Kety, S. S. (1957). The general metabolism of the brain in vivo. In: Metabolism of the Nervous System, ed. D. Richter, pp. 221–237. Oxford: Pergamon.

Kini, N. N., and Quastel, J. H. (1959). Carbohydrate–amino acid interrelations in brain cortex in vitro. Nature 184:252–255.

Kosower, E. M., and Kosower, N. S. (1969). Lest I forget thee, glutathione. Nature 224:117–120.

Kosower, E. M., Correa, W., Kinon, B. J., and Kosower, N. S. (1972). Glutathione. VII. Differentiation among substrates by the thiol oxidizing agent, diamide. Biochim. Biophys. Acta 264:39–44.

Kovachich, G. B. (1978a). Depression of glucose

oxidation in rat brain slices under high oxygen as a function of membrane depolarization. Fed. Prod. 37:278.

Kovachich, G. B. (1978b). Increased pentose shunt activity in rat brain cortical slices at elevated oxygen pressures. Pharmacologist 20: 274.

Kovachich, G. B. (1979a). Depression of $^{14}CO_2$ formation in rat brain slices by high-pressure oxygen. Relationship between metabolic rate and tissue sensitivity. Pharmacologist 21:242.

Kovachich, G. B. (1979b) Alterations in levels of amino acid in cortical brain slices by high pressure oxygen. Physiologist. 22(4):72.

Kovachich, G. B. (1980). Depression of $^{14}CO_2$ production from [U$-^{14}$C] glucose in brain slices under high-pressure oxygen: Relationship between metabolic rate and tissue sensitivity to oxygen. J. Neurochem. 34:459–462.

Kovachich, G. B., and Haugaard, N. (1977). Pyruvate dehydrogenase activation in rat brain cortical slices by elevated concentrations of external potassium ions. J. Neurochem. 28: 923–927.

Kovachich, G. B., Schina, M. J., and Haugaard, N. (1977). Inhibitory effect of high oxygen pressure on potassium-induced activation of pyruvate dehydrogenase and glucose metabolism in rat brain slices. Biochim. Biophys. Acta 462:493–500.

Koyama, I. (1972). Amino acids in the cobalt induced epileptogenic and nonepileptogenic cat's cortex. Can. J. Physiol. Pharmacol. 50: 740–752.

Krnjevic, K., and Phillis, J. W. (1963). Actions of certain amines on cerebral cortical neurons. J. Physiol. (London) 165:274–304.

Kyle, J. D. (1965). The effects of 100% oxygen inhalation on adult and newborn rat lungs. South. Med. J. 58:1592.

Lajtha, A., Berl, S., and Waelsh, H. (1959). Amino acid and protein metabolism of the brain. IV. The metabolism of glutamic acid. J. Neurochem. 3:322–332.

Lambertsen, C. J. (1955). Respiratory and circulatory actions of high oxygen pressure. Proceedings of the Underwater Physiology Symposium, Publ. 377, p. 25. Washington, D.C.: Natl. Acad. Sci., Natl. Res. Council.

Lambotte, L., Kestens, P. J., and Haxhe, J. J. (1971). The effect of tris(hydroxymethyl) aminomethane on the potassium content and the membrane potential of liver cells. J. Pharmacol. Exp. Ther. 176:434–440.

Lewis, L. D., Ljunggren, B., Norberg, K., and Siesjo, B. K. (1974). Changes in carbohydrate substrates, amino acids, and ammonia in the brain during insulin-induced hypoglycemia. J. Neurochem. 23:659–671.

Ligou, J. C., and Nahas, G. G. (1960). Comparative effects of acidosis induced by acid infusion and CO_2 accumulation. Am. J. Physiol. 198: 1201–1206.

Logan, W. J., and Snyder, S. H. (1972). High affinity uptake systems for glycine, glutamate and aspartic acids in synaptosomes of rat central nervous tissue. Brain Res. 42:413–431.

Loschen, G., and Azzi, A. (1976). On the formation of hydrogen peroxide and oxygen radicals in heart mitochondria. Rec. Adv. Stud. Card. Struct. Metab. 7:3–12.

Mahler, H. R. (1961). The use of amine buffers in studies with enzymes. Ann. N.Y. Acad. Sci. 92: 426–440.

Margolis, R. K., and Lajtha, A. (1968). Ion dependence of amino acid uptake in brain slices. Biochim. Biophys. Acta 163:374–385.

Marshall, J. R., and Lambertsen, C. J. (1961). Interactions of increased pO_2 and pCO_2 effects in producing convulsions and death in mice. J. Appl. Physiol. 16:1–7.

Mayevsky, A., Jamieson, D., and Chance, B. (1974). Oxygen poisoning in the unanesthetized brain: Correlation of the oxidation-reduction state of pyridine nucleotide with electrical activity. Brain Res. 76:481–491.

McCay, P. B., and Poyer, J. L. (1976). Enzyme generated free radicals as initiators of lipid peroxidation in biological membranes. In: Enzymes of Biological Membranes, vol. 4, ed. A. Martonosi, pp. 250–275. New York: Plenum Press.

McCord, J. M., and Fridovich, I. (1969). Superoxide dismutase. An enzymic function for erythrocuprein (hemocuprein). J. Biol. Chem. 244:6049–6055.

McCord, J. M., and Salin, M. L. (1975). Free radicals and inflammation: Studies on superoxide mediated NBT reduction by leukocytes. In: Erythrocyte Structure and Function, ed. G. J. Brewer, p. 731. New York: A. R. Liss.

McCord, J. M., Keele, B. B., Jr., and Fridovich, I. (1971). An enzyme based theory of obligate anaerobiosis: The physiological function of superoxide dismutase. Proc. Natl. Acad. Sci. (U.S.A.) 68:1024–1027.

McIlwain, H. (1953). Glucose level, metabolism, and response to electrical impulses in cerebral tissues from man and laboratory animals. Biochem. J. 55:618–624.

McIlwain, H. (1966). Biochemistry and the Central Nervous System, pp. 49–77. Boston: Little Brown and Co.

McKnight, R. C., Hunter, F. E., Jr., and Oehlert, W. H. (1965). Mitochondrial membrane ghosts produced by lipid peroxidation induced by ferrous ion. J. Biol. Chem. 240:3439.

McSherry, K. C., and Veith, F. J. (1968). The relationship between the central nervous system and pulmonary forms of oxygen toxicity. Effect of THAM administration. Surg. Forum 19:33.

Medzihradsky, F., Sellinger, O. Z., Nandhasri, P. S., and Santiago, J. C. (1972). ATPase activity in glial cells in neuronal perikarya of rat cerebral-cortex during early postnatal-development. J. Neurochem. 19:543–545.

Mengel, C. E., and Kann, H. E., Jr. (1966). Effects of hyperoxia on erythrocytes: III. In vivo peroxidation of erythrocyte lipids. J. Clin. Invest. 45:1150–1158.

Mison-Crighel, N., Luca, N., and Crighel, E. (1964). The effect of an epileptogenic focus, induced by topical administration of mescaline, on glutamic acid, glutamine and GABA in the neocortex of the cat. J. Neurochem. 11: 333–340.

Mount, L. E., and Willmott, J. V. (1967). The relation between spontaneous activity, metabolic rate and the 24 hour cycle in mice at different environmental temperatures. J. Physiol. (London) 190:371–380.

Nagata, Y., Mikoshiba, K., and Tsukada, Y. (1974). Neuronal cell body enriched and glial cell enriched fractions from young and adult rat brains: Preparation and morphological and biochemical properties. J. Neurochem. 22: 493–503.

Nahas, G. G. (1962). The pharmacology of Tris(hydroxymethyl)amino-methane (THAM). Pharmacol. Rev. 14:447–472.

Nahas, G. G. (1965). Control of acidosis in hyperbaric oxygenation. Ann. N.Y. Acad. Sci. 117:774–786.

Nakazawa, S., and Quastel, J. H. (1968). Effects of inorganic salts and of ouabain on some metabolic responses of rat cerebral cortex slices to cationic and electrical stimulations. Can. J. Biochem. 46:355–362.

Ngai, S. H., Levy, A., Finck, A. D., Yang, J. C., and Spector, S. (1977). Central nervous system toxicity of hyperbaric oxygen—effects of light, norepinephrine depletion and beta-adrenergic blockade. Neuropharmacology 16:675–679.

Nichols, C. W., and Lambertsen, C. J. (1969). Effects of high oxygen pressures on the eye. N. Eng. J. Med. 281:25–30.

Nishiki, K., Jamieson, D., Oshino, N., and Chance, B. (1976). Oxygen toxicity in the perfused rat liver and lung under hyperbaric conditions. Biochem. J. 160:343–355.

Norman, J. N., Smith, G., and Douglas, T. A. (1966). The effect of oxygen at elevated atmospheric pressure and hypothermia on tissue metabolism. Surg. Gynecol. Obstet. 122: 778–784.

Northway, W. H., Rezeau, L., Petriceks, R., Dorsett, R., and Roberts, T. (1973). Pulmonary oxygen toxicity: Adaptive response in the newborn mouse. Invest. Radiol. 8:281–282.

Ochs, S., and Hackaman, P. (1972). Block of nerve excitability and fast axoplasmic transport by increased O_2 and CO_2. Biophys. Soc. Abstr. 12:192a.

Packer, E. L., Hunter, S. H., Cox, D., Mendelow, M. A., Baker, H., Frank, O., and Amsterdam, D. (1961). Use of amine buffers in protozoan nutrition. Ann. N.Y. Acad. Sci. 92:486–491.

Pappius, H. M., and Elliott, K. A. C. (1956). Factors affecting the potassium content of incubated brain slices. Can. J. Biochem. Physiol. 34:1053–1064.

Perot, P. L., Jr., and Stein, S. N. (1959). Conduction block in mammalian nerve produced by oxygen at high pressure. Science 123: 802–803.

Peters, E. L., and Tower, D. B. (1959). Glutamic acid and glutamine metabolism in cerebral cortex after seizures induced by methionine sulphoximine. J. Neurochem. 5:80–90.

Peters, J. W., and Foote, C. S. (1976). Chemistry of superoxide ion. II. Reaction with hydroperoxides. J. Am. Chem. Soc. 98:873–875.

Polgar, G., Antagnoli, W., Ferrigan, L. W., Marten, E. A., and Gregg, W. P. (1966). The effect of chronic exposure to 100% oxygen in newborn mice. Am. J. Med. Sci. 252: 580–587.

Popovic, V., Gerschman, R., and Gilbert, D. L. (1964). Effects of high oxygen pressure on ground squirrels in hypothermia and hibernation. Am. J. Physiol. 206:49–50.

Prekladovizkii, S. I. (1936). Toxic effect of high oxygen tension on the animal organism. Fiziol. Zh. (Moskva) 20:518–533.

Quastel, J. H., and Quastel, D. M. J. (1961). The Chemistry of Brain Metabolism in Health and Disease, p. 11. Springfield, Illinois: C. C Thomas.

Radomski, M. W., and Watson, W. J. (1973).

Effect of lithium on acute oxygen toxicity and associated changes in brain gamma-aminobutyric acid. Aerosp. Med. 44:387–392.

Roberts, E., and Frankel, S. (1950). γ-aminobutyric acid in brain: Its formation from glutamic acid. J. Biol. Chem. 187:55–63.

Roberts, E., and Frankel, S. (1951a). Glutamic acid decarboxylase in brain. J. Biol. Chem. 188: 789–795.

Roberts, E., and Frankel, S. (1951b). Further studies of glutamic acid decarboxylase in brain. J. Biol. Chem. 190:505–512.

Roberts, E., and Simonsen, D. G. (1963). Some properties of L-glutamic decarboxylase in mouse brain. Biochem. Pharmacol. 12:113–134.

Roberts, E., Rothstein, M., and Baxter, C. F. (1958). Some metabolic studies of γ-aminobutyric acid. Proc. Soc. Exp. Biol. 97:796–802.

Robinson, J. D. (1965a). Structural changes in microsomal suspensions. II. Studies with brain microsomes. Arch. Biochem. Biophys. 110: 475–484.

Robinson, J. D. (1965b). Structural changes in microsomal suspensions. III. Formation of lipid peroxides. Arch. Biochem. Biophys. 112: 170–179.

Rossier, B. (1970). Effets de l'oxygene hyperbare sur la conduction nerveuse et la transmission synaptique dans le ganglion sympathique isole du rat. Brain Res. 18:257–272.

Rudin, D. O., and Eiseman, G. (1952). Effects of oxygen at high pressure on central nervous system axons. Fed. Proc. 11:133.

Ruscak, M., and Whittam, R. (1967). Metabolic response of brain slices to agents affecting sodium pump. J. Physiol. (London) 190:595.

Saelens, J. K., Kovacsics, G. B., and Allen, M. P. (1968). The influence of the adrenergic system on the 24-hour locomotor activity pattern in mice. Arch. Int. Pharmacol. Pharmacodyn. 173:411–416.

Samiy, A. H., Oken, D. E., Rees, S. B., Robin, E. D., and Merrill, J. P. (1961). Effect of 2-amino-2-hydroxyl-1,3-propanediol on excretion. Ann. N.Y. Acad. Sci. 92:570–579.

Samson, F. E., and Quinn, D. J. (1967). Na^+-K^+-Activated ATPase in rat brain development. J. Neurochem. 14:421–427.

Sanger, C., Nahas, G. G., Goldberg, A. R., and D'Allesio, G. M. (1961). Effects of 2-amino-2-hydroxymethyl-1,3-propanediol on oxygen toxicity in mice. Ann. N.Y. Acad. Sci. 92: 710–723.

Sbarra, S. J., and Karnovsky, M. L. (1959). The biochemical basis of phagocytosis. I. Metabolic changes during ingestion of particles by polymorphonuclear leukocytes. J. Biol. Chem. 234: 1355–1362.

Schaeffer, A., Komlos, M., and Seregi, A. (1975). Lipid peroxidation as the cause of ascorbic acid induced decrease of adenosine triphosphate activities of rat brain microsomes and its inhibition by biogenic amines and psychotropic drugs. Biochem. Pharmacol. 24:1781–1786.

Schafer, G. (1978). Influence of hyperoxia (1 ATA) on mouse brain GABA, glutamate and glutamine. Aviat. Space Environ. Med. 49: 470–475.

Shanklin, D. R. (1969). On the pulmonary toxicity of oxygen. I. Relationship of oxygen content to the effect of oxygen on the lung. Lab. Invest. 21:439–448.

Shcherbakova, G. V. (1962). Activity of glutamic decarboxylase and the content of γ-aminobutyric acid in rat brain at various levels of functional state caused by increased pressure of oxygen. Dokl. Akad. Nauk SSSR 146:1213–1215.

Shilling, C. W., and Adams, B. H. (1933). A study of convulsive seizures caused by breathing oxygen at high pressures. U.S. Naval Med. Bull. 31:112–121.

Skou, J. C. (1965). Enzymatic basis for active transport of Na^+ and K^+ across cell membrane. Physiol. Rev. 45:596–617.

Smith, B. E., and Usibiaga, L. E. (1971). Oxygen toxicity: Increased resistance in newborn mice and progressive diminution with advancing age. Pharmacologist 13:240.

Smith, C. W., Bean, I. W., and Bauer, R. (1960). Thyroid influence in reaction to O_2 at atmospheric pressure. Am. J. Physiol. 199:883–888.

Smith, F. J. C., Heim, J. W., Thomson, R. M., and Drinker, C. K. (1932a). Bodily changes and development of pulmonary resistance in rats living under compressed air conditions. J. Exp. Med. 56:63–78.

Smith, F. J. C., Bennet, G. A., Heim, J. W., Thomson, R. M., and Drinker, C. K. (1932b). Morphological changes in the lungs of rats living under compressed air conditions. J. Exp. Med. 56:79–89.

Smith, H. M. (1958). Effects of sulfhydryl blockage on axonal function. J. Cell. Comp. Physiol. 51:161.

Sperling, D. R. (1970). Hyperbaric oxygen toxicity in newborn and developing mice. J. Appl. Physiol. 29:472.

Spyropoulos, C. S., Tasaki, I., and Brady, R. O. (1960). The effects of oxidizing agents upon the electrophysiological properties of the nerve fiber. Proc. 4th Int. Congr. Biochem, Vienna, Sept. 1958, Vol. XV, Index to Symp. and Colloquia and Abstr. of Sect. Papers. IUB Symp. Ser. 17:181.

Stadie, W. C., Riggs, B. C., and Haugaard, N. (1944). Oxygen poisoning. Am. J. Med. Sci. 207:84–114.

Stadie, W. C., and Haugaard, N. (1945). Oxygen poisoning. V. The effect of high oxygen pressure upon enzymes: Succinic dehydrogenase and cytochrome oxidase. J. Biol. Chem. 161: 175–180.

Stadie, W. C., Riggs, B. C., and Haugaard, N. (1945a). Oxygen poisoning. III. The effect of high oxygen pressures upon the metabolism of brain. J. Biol. Chem. 160:191–208.

Stadie, W. C., Riggs, B. C., and Haugaard, N. (1945b). Oxygen poisoning. IV. The effect of high oxygen pressures upon the metabolism of liver, kidney, lung and muscle tissue. J. Biol. Chem. 160:209–216.

Stadie, W. C., Riggs, B. C., and Haugaard, N. (1945c). Oxygen poisoning. VI. The effect of high oxygen pressure upon enzymes: pepsin, catalase, cholinesterase, and carbonic anhydrase J. Biol. Chem. 161:175–180.

Stern, J. R., Eggleston, L. V., Hems, R., and Krebs, H. A. (1949). Accumulation of glutamic acid in isolated brain tissue. Biochem. J. 44: 410–418.

Stinson, R. A., and Spencer, M. (1968). An evaluation of the effects of five buffers on respiratory parameters of isolated mitochondria. Can. J. Biochem. 46:43–50.

Sun, A. U. (1972). The effect of lipoxidation on synaptosomal ($NA^+ + K^+$)-ATPase isolated from the cerebral cortex of squirrel monkey. Biochim. Biophys. Acta 266:350–360.

Szilagyi, T., Toth, S., Miltenyi, L., and Jona, G. (1969). Oxygen poisoning and thyroid function. Acta Physiol. Acad. Sci. Hung. 35:59–61.

Takagaki, G. (1976a). Properties of the uptake and release of glutamic acid by synaptosomes from rat cerebral cortex. J. Neurochem. 27: 1417–1425.

Takagaki, G. (1976b). Characteristics of the uptake and release of glutamic acid in synaptosomes from rat cerebral cortex. Effects of ouabain. Adv. Exp. Med. Biol. 69:307–317.

Takagaki, G., Hirano, S., and Nagata, Y. (1959). Some observations on the effect of d-glutamate on the glutamate metabolism and the accumulation of potassium ions in brain cortex slices. J. Neurochem. 4:124–134.

Tappel, A. L. (1973). Lipid peroxidation damage to cell components. Fed. Proc. 32:1870.

Teng, S. S., and Harris, J. W. (1970). Effect of hyperbaric oxygen on cellular dehydrogenases and sulfhydryls. Exp. Cell. Res. 60:451.

Terner, C., Eggleston, L. V., and Krebs, H. A. (1950). The role of glutamic acid in transport of potassium in brain and retina. Biochem. J. 47:139–149.

Tews, J. K., Carter, S. H., Roa, P. D., and Stone, W. E. (1963). Free amino acids and related compounds in dog brain: Post-mortem and anoxic changes. Effects of ammonium chloride infusion, and levels during seizures induced by picrotoxin and by pentelenetetrazol. J. Neurochem. 10:641–653.

Tews, J. K., Carter, S. H., and Stone, W. E. (1965). Chemical changes in the brain during insulin hypoglycaemia and recovery. J. Neurochem. 12:679–693.

Tjioe, G., and Haugaard, N. (1972). Oxygen inhibition of crystalline glyceraldehyde phosphate dehydgrogenase and disappearance of enzyme sulfhydryl groups. Life Sci. 11:329–335.

Tomizawa, H. H., and Halsey, Y. D. (1959). Isolation of an insulin degrading enzyme from liver. J. Biol. Chem. 234:307–310.

Tsen, C. C., and Collier, H. B. (1960). The protective action of tocopherol against hemolysis of rat erythrocytes by dialuric acid. Can. J. Biochem. 38:957.

Tsukada, Y., Nagata, Y., Hirano, S., and Matsutani, T. (1963). Active transport of amino acid into cerebral cortex slices. J. Neurochem. 10:241–256.

Van den Berg, C. J., Krzalic, L., Mela, P., and Waelsch, H. (1969). Compartmentation of glutamate metabolism in brain. Evidence for the existence of two different tricarboxylic acid cycles in brain. Biochem. J. 113:281–290.

Van den Brenk, H. A. S., and Jamieson, D. (1964). Brain damage and paralysis in animals exposed to high pressure oxygen—pharmacological and biological observations. Biochem. Pharmacol. 13:165–182.

Van Gelder, N. M. (1968). Hydrazinopropionic acid. A new inhibitor of aminobutyrate transaminase and glutamate decarboxylase. J. Neurochem. 15:747–757.

Van Gelder, N. M. (1969). The action in vivo of a structural analogue of GABA: hydrazinopropionic acid. J. Neurochem. 16:1355–1360.

Van Gelder, N. M. (1972). Antagonism by taurine of cobalt induced epilepsy in cat and mouse. Brain Res. 47:157–165.

Van Gelder, N. M., Sherwin, A. L., and Rasmussen, T. (1972). Amino acid content of epileptogenic human brain: Focal versus surrounding regions. Brain Res. 40:385–393.

Whisler, K. E., Tews, J. K., and Stone, W. E. (1968). Cerebral amino acids and lipids in drug-induced status epilepticus. J. Neurochem. 15: 215–220.

Whittam, R. (1964). The interdependence of metabolism and active transport. In: Cellular Functions of Membrane Transport, ed. J. F. Hoffman, pp. 184–202. New York: Prentice-Hall, Inc.

Williams, C. M., and Beacher, H. K. (1944). Sensitivity of Drosophila to oxygen poisoning. Am. J. Physiol. 140:566–573.

Wollman, M. (1962). Study of the effects of hyperoxia on the central nervous system. In: The Selective Vulnerability of the Brain in Hypoxemia: A Symposium, eds. J. Schade and W. McMenemey, pp. 349–356. Oxford: Blackwell.

Wood, J. D. (1970). Seizures induced by hyperbaric oxygen and cerebral gamma-aminobutyric acid in chicks during development. J. Neurochem. 17:573–579.

Wood, J. D., and Watson, W. J. (1962). Protective action of γ-aminobutyric acid against oxygen toxicity. Nature 195:296.

Wood, J. D., and Watson, W. J. (1963). Gamma-aminobutyric acid levels in the brain of rats exposed to oxygen at high pressures. Can. J. Biochem. Physiol. 41:1907–1913.

Wood, J. D., and Watson, W. J. (1964). The effect of oxygen on glutamic acid decarboxylase and gamma-aminobutyric acid-alpha-ketoglutaric acid transaminase activities in rat brain homogenates. Can. J. Physiol. Pharmacol. 42: 277–279.

Wood, J. D., Watson, W. J., and Clydesdale, F. M. (1963). Gamma-aminobutyric acid and oxygen poisoning. J. Neurochem. 10:625–633.

Wood, J. D., Stacey, N. E., and Watson, W. J. (1965). Pulmonary and central nervous system damage in rats exposed to hyperbaric oxygen and protection therefrom by gamma-aminobutyric acid. Can. J. Physiol. Pharmacol. 43: 405–410.

Wood, J. D., Watson, W. J., and Ducker, A. J. (1967). Oxygen poisoning in various mammalian species and the possible role of gamma-aminobutyric acid metabolism. J. Neurochem. 14:1067–1074.

Wood, J. D., Watson, W. J., and Murray, G. W. (1969). Correlation between decrease in brain gamma-aminobutyric acid levels and susceptibility to convulsions induced by hyperbaric oxygen. J. Neurochem. 16:281–287.

Wood, J. D., Radomski, M. W., and Watson, W. J. (1971). A study of possible biochemical mechanisms involved in hyperbaric oxygen-induced changes in cerebral gamma-aminobutyric acid levels and accompanying seizures. Can. J. Biochem. 49:543–547.

Wood, J. D., Watson, W. J., and Stacey, N. E. (1966). A comparative study of hyperbaric oxygen-induced and drug-induced convulsions with particular reference to γ-aminobutyric acid metabolism. J. Neurochem. 13:361–370.

Zalkin, H., and Tappel, A. L. (1960). Studies of the mechanism of vitamin E action. IV. Lipid peroxidation in the vitamin E deficient rabbit. Arch. Biochem. Biophys. 88:113.

12

Antioxidant Defenses

HENRY J. FORMAN AND ARON B. FISHER

I. Introduction

Gerschman first proposed that the initial event of oxygen poisoning may be the generation of toxic free radicals (28). This has led to a recent focus on free radicals generated from oxygen itself, including superoxide anion and hydroxl radical and derivatives of these radicals including H_2O_2 and singlet oxygen. It is now known that these partially reduced species of O_2 are generated normally during cellular metabolism and, at least for some reactions, their rate of generation increases as a function of P_{O_2} (26,27,46). This supports the possibility that exposure to elevated oxygen tensions results in increased generation of toxic free radicals which may then be responsible for tissue damage.

This chapter concerns the mechanisms which have evolved to cope with the hazards of life in an oxidizing atmosphere. To start, we will briefly review the chemistry of the "activated" oxygen species that might produce cellular damage and then discuss the defense mechanisms available to the cell for its protection. Finally, we will review the experimental evidence for the role of these defense mechanisms. This chapter deals primarily with defense against O_2 and its metabolites, although defenses against other oxidants such as ozone and NO_2 probably utilize similar mechanisms.

A. Species of "Activated Oxygen"

"Electron activated oxygen." The evolution of enzymatically controlled reduction of oxygen to water in mitochondria allowed development of complex multicellular organisms. However, several enzymatic and nonenzymatic reactions of oxygen which occur in biological systems produce molecules which are intermediate in electrical charge between oxygen and water. Oxygen reacts more readily through one electron, or free radical processes, than through direct two electron transfer reactions because of the relatively long energy duration required to invert the spin of one electron. Sequential one electron reduction of oxygen produces superoxide anion, hydrogen peroxide, hydroxyl radical, and water as indicated in Fig. 1. The presence of these "electron activated" oxygen species because of their electrophilic nature presents a challenge to the integrity of cell constituents. The mechanisms for production of these molecules and the nature of

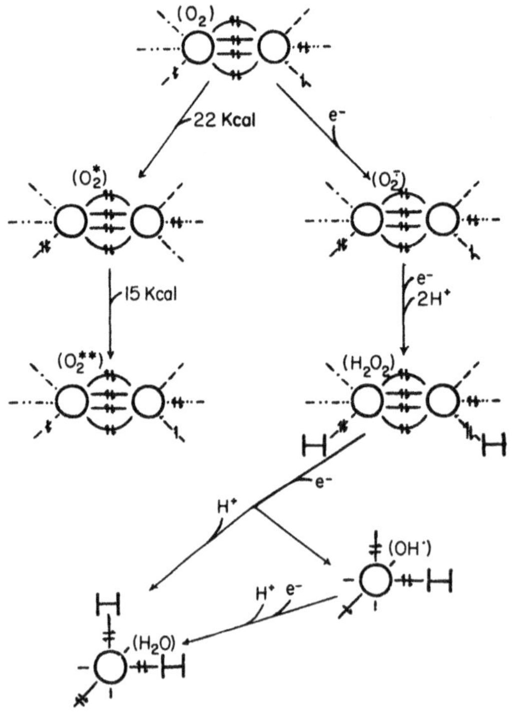

Figure 1. Schematic representation of oxygen and electron and energy activated forms. Indicated are the molecular orbital configuration of oxygen (O_2), its products by reduction, superoxide anion (O_2^-), hydrogen peroxide (H_2O_2), hydroxyl radical ($OH^{\cdot}$) and water (H_2O) and its higher energy singlet states (O_2^* and O_2^{**}). Small arrows indicate direction of spin of electrons, solid lines indicate bonding orbitals, and broken lines indicate antibonding orbitals.

their attack on biological components are presented elsewhere in this volume.

"Energy activated oxygen." Oxygen in its lowest energy state (ground state) can either accept electrons as described above or it can accept energy and rearrange its electronic configuration. Oxygen in biological systems is predominantly in the ground state but oxygen in higher energy states may possibly be present. Ground state oxygen is a triplet state with two unpaired electrons that may have the same spin and are necessarily in different orbitals. Singlet oxygen can exist in two forms characterized by different energy states. In one form (O_2^{**}) which is 37 kcal above ground state, the two unpaired electrons spin in the opposite direction but remain in separate orbitals. In the other form (O_2^*) which is 22 kcal above ground state, the outer shell electrons are in the same orbital and have opposite spins. Because reactions of organic molecules most easily proceed with transfer of pairs of electrons (which must have opposite spins), singlet oxygen is biologically more reactive than ground state oxygen (55,61). The increased reactivity of singlet oxygen presents challenges to the cell for defense against cellular damage.

B. General Mechanisms for Defense against Oxidative Damage

Protection against electron activated oxygen. The reactivity of ground state oxygen and water does not appear to present any danger to biological systems (61). A general mechanism for protection of biological systems against the potential hazards of superoxide, hydrogen peroxide, and hydroxyl radical is to convert them either to oxygen by oxidation or to water by reduction. In one series of reactions, the "electron activated" species react with themselves in dismutation reactions:

$$2O_2^- + 2H^+ \rightarrow H_2O_2 + O_2 \qquad \textbf{(1)}$$

$$2H_2O_2 \rightarrow 2H_2O + O_2 \qquad \textbf{(2)}$$

$$2OH^{\cdot} \rightarrow H_2O_2 \qquad \textbf{(3)}$$

The advantage of these reactions to the cell is that essential components are not sacrificed in removing the potentially dangerous species. In a second series of general reactions, a trapping agent (A or H_2A) is reduced or oxidized:

$$2O_2^- + 2H^+ + A \rightarrow 2O_2 + H_2A \qquad \textbf{(4)}$$

$$2O_2^- + 2H^+ + H_2A \rightarrow 2H_2O_2 + A \qquad \textbf{(5)}$$

$$H_2O_2 + H_2A \rightarrow 2H_2O + A \qquad \textbf{(6)}$$

$$2OH^{\cdot} + H_2A \rightarrow 2H_2O + A \qquad \textbf{(7)}$$

In a third group of reactions, the "electron activated" oxygen species forms peroxides, alcohols, ketones, sulfones, etc., by reacting with the trapping molecule itself.

Protection against "energy activated" oxygen. Two general defense mechanisms are possible for protection of cell constituents against oxidation by singlet oxygen (24). In the first general mechanism, singlet oxygen is converted into ground state oxygen through absorption of energy by a quencher (Q) which may then give up the energy in the form of heat or may undergo a transition to another compound (X).

$$O_2^* + Q \rightarrow O_2 + Q^* \quad (8)$$

$$Q^* \rightarrow Q + \text{heat} \quad (9)$$

$$Q^* \rightarrow X \quad (10)$$

The second general mechanism is the reaction of a quencher with singlet oxygen with incorporation of oxygen:

$$O_2^* + Q \rightarrow QO_2 \quad (11)$$

QO_2 may be a ketone, peroxide, etc. While the two forms of singlet oxygen (O_2^* and O_2^{**}) have different electronic configurations and energies which result in different reactions for each, the same general mechanisms for protection apply to both.

Expense of antioxidant defenses to biological systems. The reactions described above which use a trapping agent occur within biological systems both enzymatically and nonenzymatically. In some cases, the defense reactions are analogous to normal metabolism and the reaction products may be further metabolized or excreted. Other antioxidants exist in cells for the sole purpose of trapping activated oxygen species and are expendable. On the other hand, if the trapping agents are not expendable, the reaction can then be considered as damaging rather than protective. For example, a free amino acid that is "damaged" by oxidant interaction can be replaced more easily than a "damaged" protein containing that amino acid. The least potentially hazardous defense is elimination of the activated oxygen species through the dismutation reactions (Eq. 1–3) since no trapping is required. Enzymes have evolved which can catalyze the dismutation of $O_2^{\overline{\cdot}}$ and H_2O_2 with extreme rapidity thereby eliminating a requirement for reaction with nonexpendable tissue components. These enzymes form the first line of defense against oxidizing radicals by limiting their tissue concentrations. Other enzymes and a wide range of trapping agents (referred to here as quenchers or interceptors) form a second line of defense to limit the interactions of tissue oxidants with essential tissue components.

II. Mechanisms of Protection

A. Enzymes That Catalyze Reactions with Activated Species

Superoxide dismutases. Superoxide dismutases (a group of metalloproteins) are characterized by their catalysis of the reaction in Eq. 1. The importance of these enzymes in protection of biological molecules is great enough to merit consideration in another chapter. Superoxide dismutases catalyze a reaction which removes one potentially dangerous species ($O_2^{\overline{\cdot}}$) while producing another (H_2O_2). Other enzymes must then remove H_2O_2 from the cell.

Catalase. Catalase, a heme protein, catalyzes the dismutation reaction of H_2O_2 shown in Eq. 2. The rate enhancement provided by catalase is great ($>10^8$-fold) resulting in rapid H_2O_2 removal. Because the rate of decomposition of the enzyme substrate complex is so rapid, saturation can not occur in the range of biological concentrations.

Catalase can use alternate substrates in reactions which are peroxidatic in nature (58):

$$H_2O_2 + ROOH \rightarrow H_2O + ROH + O_2 \quad (12)$$

$$H_2O_2 + RCH_2OH \rightarrow 2H_2O + R\underset{\underset{O}{\|}}{C}H \quad (13)$$

$$H_2O_2 + H\underset{\underset{O}{\|}}{C}{-}OH \rightarrow 2H_2O + CO_2 \quad (14)$$

In these reactions, R is a short chain alkyl group (ethyl for example). These reactions are modifications or examples of reaction 6. Because of the low affinity of catalase for ROOH compared with H_2O_2, reaction 12 is unlikely to be important as a defense mechanism. However, the affinity of catalase for alcohols and formate is sufficiently high so that it could act as a peroxidatic catalyst (reactions 13 and 14). At the estimated steady-state in vivo concentration of H_2O_2 of 10^{-9} to 10^{-8} *M* (9), it is probable that catalase functions in both catalatic and peroxidatic modes.

Peroxidases. These enzymes remove H_2O_2 through peroxidatic mechanisms (reaction 6) where H_2A represents the hydrogen donor. The use of specific hydrogen donors differentiates the peroxidases from catalase. Peroxidases also differ from catalase in their use of long chain or branched chain alkyl hydroperoxides as substrates.

The peroxidase with the greatest potential protective action for the cell is glutathione peroxidase (14). This enzyme requires selenium for its activity (49) and catalyzes the reaction:

$$ROOH + 2GSH \rightarrow ROH + H_2O + GSSG \quad \textbf{(15)}$$

In this reaction, GSH is reduced glutathione, GSSG is oxidized glutathione and R may be H or an alkyl moiety. The enzyme is found in all vertebrates and in some invertebrates, but is absent from insects, earthworms, plants, and micro-organisms (56). Glutathione peroxidase appears to be important in removal of both H_2O_2 and lipid peroxides in mammalian cells (14). Reduced glutathione which is abundant in mammalian cells can be regenerated from the oxidized state, thus increasing the effectiveness of reaction 15 as a mechanism of H_2O_2 scavenging.

A group of enzymes classified as non-selenium-containing glutathione peroxidases has also been described (8). These enzymes have a much lower affinity for H_2O_2 than for lipid peroxides. Their chief role is to catalyze the transfer of the sulfur of glutathione and they are more appropriately designated as *S*-transferases (see below).

Another peroxidase, cytochrome *c* peroxidase, found in yeast mitochondria (22) and bacteria (21) appears to protect against H_2O_2 by the reaction:

$$H_2O_2 + 2\text{cyt. } c\ (Fe^{2+}) + 2H^+ \rightarrow 2H_2O + 2\text{cyt. } c\ (Fe^{3+}) \quad \textbf{(16)}$$

Another important scavenger of H_2O_2 in some bacteria is a peroxidase which utilizes NADH to reduce H_2O_2 (66).

A. *Quenchers and Interceptors*

Thus far, we have seen how enzyme catalyzed reactions directly remove the "electron activated" species $O_2^{\overline{\cdot}}$ and H_2O_2. Enzymes which catalyze the removal of $OH^{\cdot}$ or singlet oxygen have not been found and these dangerous species must be removed by reactions using quenching or intercepting molecules. These quenchers and interceptors also can aid in removal of $O_2^{\overline{\cdot}}$ and H_2O_2.

Glutathione. Glutathione (GSH) is widely distributed in biological systems. It is a tripeptide with the structure

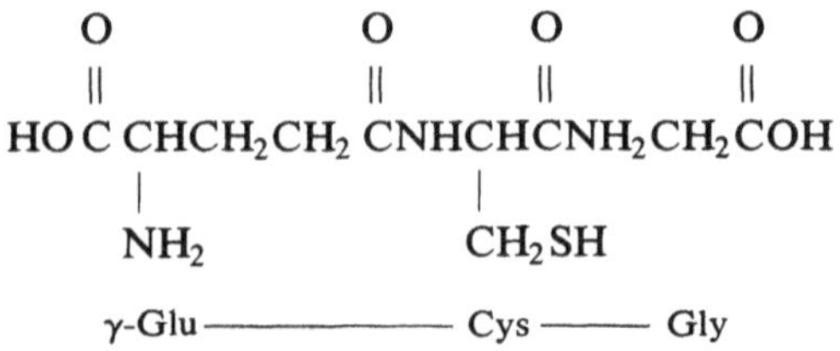

In tissues, glutathione is maintained in the reduced form where it acts as a major source of reducing power in the cell and functions in reduction of oxidized tissue components (2).

Glutathione is a major source of reducing power for the removal of H_2O_2 in the reaction catalyzed by glutathione peroxidase:

$$H_2O_2 + GSH \rightarrow 2H_2O + GSSG \quad \textbf{(17)}$$

This reaction is a specific example of reaction 15.

Glutathione can also interact with $O_2^{\overline{\cdot}}$, $OH^{\cdot}$, and O_2^{*} nonenzymatically, which are specific examples of reactions 5, 7, and 11 (38):

$$2O_2^{\overline{\cdot}} + 2H^+ + 2GSH \rightarrow GSSG + 2H_2O_2 \quad \mathbf{(18)}$$

$$2OH^{\cdot} + 2GSH \rightarrow GSSG + 2H_2O \quad \mathbf{(19)}$$

$$O_2^{*} + GSH \rightarrow GS_{ox}(\text{cysteic acid}) \quad \mathbf{(20)}$$

The chemistry of these reactions is complicated by the probable formation of the thiyl radical ($GS^{\cdot}$) as an intermedate during the reduction of activated oxygen. Since the thiyl radical is potentially toxic, these nonenzymatic reactions may be only partially protective to the cell.

Vitamin E. Vitamin E or α-tocopherol has the structure

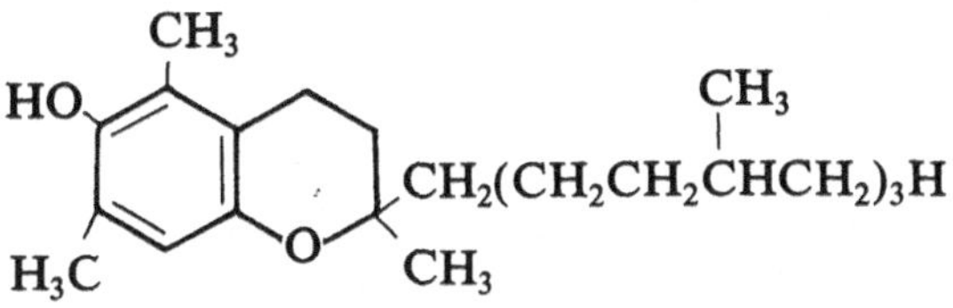

It appears to function in antioxidant defense mainly through its ability to reduce polyunsaturated lipid oxide free radicals by a nonenzymatic reaction:

$$\text{Vitamin E} + RO^{\cdot} \rightarrow \text{vitamin E}^{\cdot} + ROH \quad \mathbf{(21)}$$

The vitamin E radical must then be reduced by interaction with another agent such as vitamin C (see below). In vitro, vitamin E prevents the rapid autooxidation of unsaturated fatty acids which occurs when a thin layer of the lipid is exposed to air (52). In vivo, vitamin E and similar compounds in plants appear to be effective scavengers of singlet O_2 and protect lipids by competition for "energy activated" O_2 (31,48). Although $OH^{\cdot}$ and $O_2^{\overline{\cdot}}$ can react with vitamin E, there is no evidence that α-tocopherol functions in this fashion in vivo. In mammals, signs of vitamin E deficiency seem to occur only if the diet is high in unsaturated fatty acids in which case lipid peroxides become detectable in body fat (18).

Ascorbic acid. Ascorbic acid (vitamin C) is synthesized by most plants and animals. However, in some species, including man, it 's not synthesized and is an essential dietary component.

It has been proposed that ascorbate functions as an antioxidant in several reactions. Nishikimi (47) demonstrated that ascorbate reacts with $O_2^{\overline{\cdot}}$:

$$2O_2^{\overline{\cdot}} + 2H^+ + \text{ascorbate} \rightarrow H_2O_2 + \text{dehydroascorbate} \quad \mathbf{(22)}$$

This is a specific example of reaction 5 with a second-order rate constant of approximately $3 \times 10^5\ M^{-1}\ \text{sec}^{-1}$. Although this rate constant is several orders of magnitude lower than that for the superoxide dismutase reaction, the presence of high concentrations of ascorbate in tissues could allow this reaction to be of physiological importance. Ascorbate can also react rapidly with $OH^{\cdot}$ as an example of reaction 7 (1). It has also been proposed that ascorbate participates in restoration of vitamin E radical to reduced vitamin E (60):

$$2\text{Vitamin E}^{\cdot} + \text{ascorbate} \rightarrow 2\text{vitamin E} + \text{dehydroascorbate} \quad \mathbf{(23)}$$

These reactions actually proceed through the intermediate formation of semidehydroascorbate, a fairly long-lived free radical that can then dismute or react with another free radical:

$$2\text{Semidehydroascorbate} \rightarrow \text{ascorbate} + \text{dehydroascorbate} \quad \mathbf{(24)}$$

$$\text{Semidehydroascorbate} + O_2^{\overline{\cdot}} \rightarrow O_2 + \text{dehydroascorbate} \quad \mathbf{(25)}$$

$$\text{Semidehydroascorbate} + \text{vitamin E}^{\cdot} \rightarrow \text{vitamin E} + \text{dehydroascorbate} \quad \mathbf{(26)}$$

Thus, ascorbate participates in antioxidant defence both as a direct scavenger of "activated oxygen" and indirectly in restoring the pool of reduced vitamin E. As ascorbate radical (semidehydroascorbate) and dehydroascorbate (which will form the radical through reversal of reaction 24) are potentially hazardous to cells (31), their reduction to ascorbate by nonezymatic reaction with GSH (60) or enzymatic reduction by NADH-linked semidehydroascorbate reductase (57) or GSH-dependent dehydroascorbate reductase (30,40) is an important link in antioxidant protection.

Polyunsaturated fatty acids. Much of the damage caused in tissues by hyperoxia appears related to lipid peroxidation so that polyunsaturated fatty acids are a target rather than defense against oxygen toxicity. However, if lipids act competitively in trapping oxygen radicals and singlet oxygen, other molecules such as DNA and proteins which may not be easily regenerated are spared. The mechanism of lipid peroxidation is currently under investigation (36,51), but O_2^* and $O_2^{\bar{}}$ (in combination with H_2O_2 and $OH^{\cdot}$) probably react with polyunsaturated fatty acids (RH) to produce lipid peroxides:

$$RH + O_2^{\bar{}} + OH^{\cdot} + H^+ \rightarrow ROOH + H_2O \qquad \textbf{(27)}$$

Pryor and co-workers (62) have proposed that lipid peroxides further react with $O_2^{\bar{}}$ producing $RO^{\cdot}$ which starts a free radical chain of lipid peroxidation. As described above, vitamin E can reduce $RO^{\cdot}$ to the alcohol (reaction 21) and glutathione peroxidase can reduce ROOH to the alcohol (reaction 15). Thus, both vitamin E and glutathione peroxidase can terminate lipid peroxidation. A small degree of lipid peroxidation might then spare other antioxidants.

β-carotene. β-carotene is an effective quencher of O_2^* by reaction (28) (a specific example of reaction 8). The energy transferred to β-carotene is released by a decay process (reaction 29) or absorbed in a reversible isomerization (reaction 30) (24):

$$O_2^* + \beta\text{-carotene} \rightarrow O_2 + \beta\text{-carotene*} \qquad \textbf{(28)}$$

$$\beta\text{-Carotene*} \rightarrow \rightarrow \beta\text{-carotene} + \text{energy (heat)} \qquad \textbf{(29)}$$

$$\beta\text{-carotene*} \rightarrow \text{all-}\textit{trans}\text{-}\beta\text{-carotene} \qquad \textbf{(30)}$$

In reaction 30, the energy is absorbed by an internal conversion of a cis-double bond to the trans configuration. These reactions are nondestructive to the quencher. One molecule of β-carotene can quench up to 1000 O_2^* molecules consistent with the high second-order rate constant of $3 \times 10^{10}\ M^{-1}\ sec^{-1}$ (24). Since interaction of unsaturated lipids and O_2^* has a second-order rate constant of approximately 10^5 to $10^6\ M^{-1}\ sec^{-1}$ (34), relatively low concentrations of β-carotene are effective in preventing lipid oxidation by O_2^*.

Water and other nonspecific quenchers. Water quenches both the high and low energy states of singlet oxygen by absorbing energy in the form of heat (reactions 8 and 9) (34). Because of the high heat capacity of H_2O and its high intracellular concentration (55 M), it constitutes a very effective agent of defense. H_2O reacts with O_2^{**} with a second-order rate constant of approximately $10^9\ M^{-1}\ sec^{-1}$ and constitutes the primary defense against this oxidant (34). The product of the reaction of O_2^{**} with H_2O is O_2^* which is still a potent oxidizer. The rate constant for quenching of O_2^* to the ground state by H_2O is approximately $10^4\ M^{-1}\ sec^{-1}$ so that other quenchers may be more important for this form of singlet oxygen (34).

Many other normal cellular components are able to function as quenchers of "activated" oxygen species. Histidine, tryptophan, cysteine, tyrosine, methionine, and other amino acids can trap singlet oxygen (23) and are expendable when not part of a protein.

Ethanol reacts rapidly with $OH^{\cdot}$ (45) although it is not found in high enough concentration (in sober organisms) to be a useful defense. In the presence of O_2, formate converts $OH^{\cdot}$ into $O_2^{\bar{}}$ via the intermediate carbonate radical (41):

$$HCOO^{\overline{\cdot}} + OH^{\cdot} \rightarrow CO_2^{\overline{\cdot}} + H_2O \quad (31)$$

$$CO_2^{\overline{\cdot}} + O_2 \rightarrow CO_2 + O_2^{\overline{\cdot}} \quad (32)$$

Many sugars (e.g., mannitol) are also fairly good scavengers of $OH^{\cdot}$ (65). However, $OH^{\cdot}$ will more likely react with an essential component such as the sugar moiety of a nucleic acid unless formate or a free sugar is present at high concentrations. For example, in a model system, 63 m*M* formate is required to inhibit by 50% the interaction of $OH^{\cdot}$ and 8.3 m*M* *p*-cresol (29).

C. *Quenching of "Activated" Oxygen Species*

The various mechanisms for removal of activated oxygen have been presented through a discussion of the quenchers. Table 1 summarizes the major proposed traps and the reaction product for each of the activated oxygen species. $O_2^{\overline{\cdot}}$ and H_2O_2 are removed by both enzymatic and nonenzymatic mechanisms. Cells do not have specific enzymes for coping with singlet oxygen but do have nonenzymatic defenses. On the other hand, the hydroxyl radical has little specificity and reacts with any oxidizable molecule. Therefore, the best defense against any $OH^{\cdot}$ is to prevent its formation. This can be accomplished through the action of superoxide dismutase, catalase, and peroxidases since $OH^{\cdot}$ is formed in biological systems mainly through a series of reactions involving $O_2^{\overline{\cdot}}$ and H_2O_2 with metal ions or complexes.

The subcellular localization of the antioxidant defenses is of some interest. Superoxide dismutase is present in both mitochondria and cytosol as discussed in Chapter 13. In mammalian cells, catalase is localized predominantly in the peroxisomes and is used for their identification (19). Glutathione peroxidase in liver cells can be detected in both the mitochondrial and soluble fractions (8) and α-tocopherol and β-carotene are localized in lipids and are therefore present primarily in membranes. GSH and ascorbic acids are probably distributed throughout the cell. These subcellular localizations of antioxidants help to explain their functional characteristics in terms of radical scavenging.

Table 1. *Quenching of Activated Oxygen*

Active species	Antioxidant	Product
$O_2^{\overline{\cdot}}$	Superoxide dismutase	$H_2O_2 + O_2$
	Ascorbate (vit. C)	H_2O_2
	GSH	H_2O_2
H_2O_2	Catalase	$H_2O + O_2$
	GSH peroxidase	H_2O
	Cyt. *c* peroxidase	H_2O
	GSH	H_2O
$OH^{\cdot}$	Formate	$H_2O + O_2^{\overline{\cdot}}$
	Polyunsaturated fatty acids	Lipid peroxide
	Sugars	Oxidized sugar
O_2^{*}	β-carotene (vit. A)	O_2
	Polyunsaturated fatty acids	Lipid peroxide
	Amino acids	Oxidized amino acids
	GSH	Sulfone or cysteic acid
	α-tocopherol (vit. E)	Oxidized α-tocopherol
O_2^{**}	H_2O	O_2^{*}

The antioxidant protective mechanisms described above involve in part oxidation of expendable compounds. These antioxidant molecules, however, are expendable only provided they can be regenerated through metabolic processes or replaced in the diet. The next section discusses the mechanisms for restoration of normal levels of cellular antioxidants.

D. *Restoration of Oxidized Components*

The glutathione-NADPH system. In addition to direct reactions between GSH and "activated" O_2 species, GSH has a major role in restoring other antioxidants to their reduced states. The oxidized compounds that can be reduced by GSH include lipid peroxides, oxidized protein, sulfhydryls, and vitamin C:

$$2GSH + \begin{Bmatrix} ROOH \rightarrow ROH + H_2O \\ R'S_{ox} \rightarrow R'SH \\ R'SSR' \rightarrow 2R'SH \\ \text{Dehydro-ascorbate} \rightarrow \text{ascorbate} \end{Bmatrix} + GSSG \quad \textbf{(33)}$$

In these reactions, R′ represents a protein containing an oxidized cystine (S_{ox}) or disulfide and R represents a lipid moiety. GSSG is produced by the formation of a disulfide linkage between two molecules of GSH. The reduction of oxidized compounds with GSH is catalyzed by the enzymes glutathione peroxidase for lipid peroxides, glutathione-*S*-transferases for oxidized protein sulfhydryls, and GSH-dehydroascorbate reductase for dehydroascorbate. Although these reactions can be enzyme catalyzed, nonenzymatic reduction by glutathione also probably occurs due to its high intracellular concentration and high reducing power.

The function of GSH as an intracellular reductant is important for protection of cells since GSH can be readily regenerated through the reaction catalyzed by glutathione reductase:

$$GSSG + NADPH + H^+ \rightarrow 2GSH + NADP^+ \quad \textbf{(34)}$$

In this reaction, GSSG is reduced by NADPH. NADPH can be regenerated through the reactions of the pentose phosphate shunt and other pathways of intermediary metabolism as outlined in Fig. 2.

NADH. Although NADPH appears to be the primary nucleotide involved with restoration of oxidized tissue components, NADH also has a role. The major contribution of NADH is in the transhydrogenation reaction (Fig. 2) which forms NADPH by transfer of electrons from NADH to $NADP^+$. A more direct "antioxidant" role for NADH is in the reduction of semidehydroascorbate to ascorbate by the action of semidehydroascorbate reductase (57). In some bacteria, NADH is used to reduce H_2O_2 by a peroxidase (66). NADH can be readily regenerated from

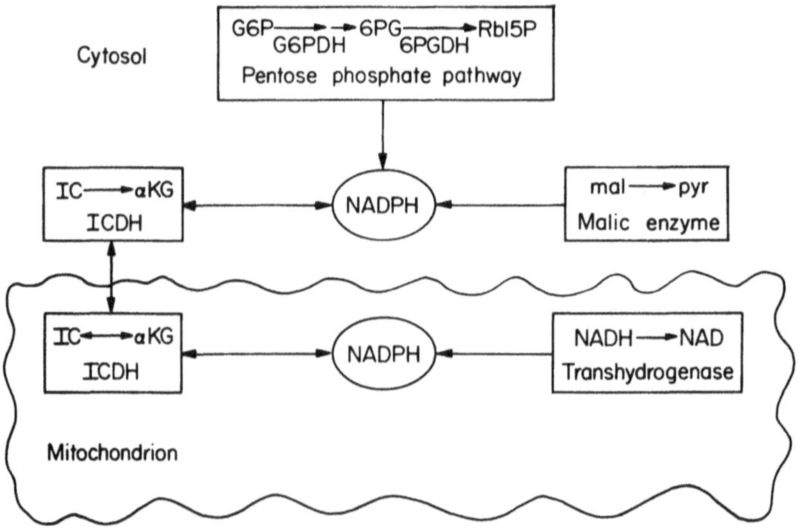

Figure 2. Scheme of NADPH producing reactions in the mitochondrial and cytosolic compartments. NADPH is a product of the reactions catalyzed by the malic enzyme which converts malate (mal) to pyruvate (pyr), cytosolic and mitochondrial isocitrate dehydrogenases (ICDH) which reversibly converts isocitrate to α-ketoglutarate (αKG), transhydrogenase which essentially transfers H from NAD to NADP, and the pentose phosphate shunt enzymes glucose-6-phosphate dehydrogenase (G6PDH) and 6-phospho-gluconate dehydrogenase (6PGDH) which produce ribulose-5-phosphate (Rbl5P). The mitochondrial membrane is impermeable to NADPH but IC and αKG are able to cross this barrier.

NAD$^+$ through the many dehydrogenase reactions involved in catabolism of tissue nutrients and possibly through reversed mitochondrial electron transport (10).

Glutathione-S-transferases. Glutathione-*S*-transferases catalyze a group of reactions where GSH becomes covalently bound to another compound (33). Specific types of these reactions are involved in antioxidant defense, since they serve to reduce oxidized tissue components.

$$R'S^*S^*R' + 2GSH \rightarrow 2R'SH + GS^*S^*G \quad (35)$$

$$R'S^*O_3 + 2GSH \rightarrow R'SH + GSS^*O_3 \quad (36)$$

$$ROOH + 2GSH \rightarrow ROH + H_2O + GSSG \quad (37)$$

In these reactions, R′ is a protein, R is a lipid and the sulfur asterisk is used merely for identification purposes. Oxidized protein sulfhydryls or lipid peroxides are restored to reduced form. Enzymes that catalyze reaction 35 and 36 are also called glutathione thiol transferases (3). Reaction 37 is the same as that catalyzed by the selenium containing glutathione peroxidase (see reaction 15); thus, those glutathione-S-transferases that catalyze reaction 37 have been called nonselenium-dependent glutathione peroxidases (8,54).

Dietary components. Three of the antioxidants discussed, ascorbate (vitamin C), α-tocopherol (vitamin E) and β-carotene (vitamin A) are vitamins for man and other animals and must be obtained through the diet. The unsaturated fatty acids linoleic and linolenic acids are also obtained through the diet rather than through synthesis. Dietary elements with important antioxidant function include copper, manganese and zinc (for SOD) and selenium (for glutathione peroxidase). The essential amino acid cysteine is required for glutathione synthesis. Finally, the nutritional status of the organism may influence availability of intracellular reducing power in the form of NADH and NADPH. Because of these nutritional requirements, antioxidant defenses and their replenishment are subject in part to dietary control. Interconnections of the dietary components are summarized in Fig. 3.

III. Evidence for the Role of Antioxidant Defense Mechanisms

Thus far, we have discussed mechanisms based on either nonenzymatic chemistry or studies of enzymatic mechanisms. In this section, the experimental support for the role of these mechanisms is presented. The models which have been studied in greatest detail are the measurement of metabolic flux on exposure to O_2, the effects of dietary deficiencies, bacterial mutations and metabolic diseases, and the enzyme alterations that accompany prolonged O_2 exposure.

A. Metabolic Flux

According to the models presented, oxidant stress should cause an increased flux through the pathways with an increased rate of GSSG generation. The excess GSSG produced may be partially released from the tissues before it can be reduced by the action of GSH reductase. This explanation has been proposed for the observation that isolated lungs and liver obtained from vitamin E deficient rats release GSSG into perfusate at an increased rate when exposed to hyperbaric oxygen (46). The increased activity of GSH reductase requires an increased supply of NADPH. In the lung, the major source of NADPH appears to be the pentose phosphate shunt pathway (4). Isolated perfused lungs from normal rats show increased glucose utilization through the pentose phosphate pathway with an increase in NADPH production from 72 to 130 mmol/hr per g dry wt, upon changing the atmosphere from 0.2 atm absolute O_2 to 5 atm absolute O_2 (5). Thus, both increased GSSG release and increased pentose phosphate shunt activity on exposure

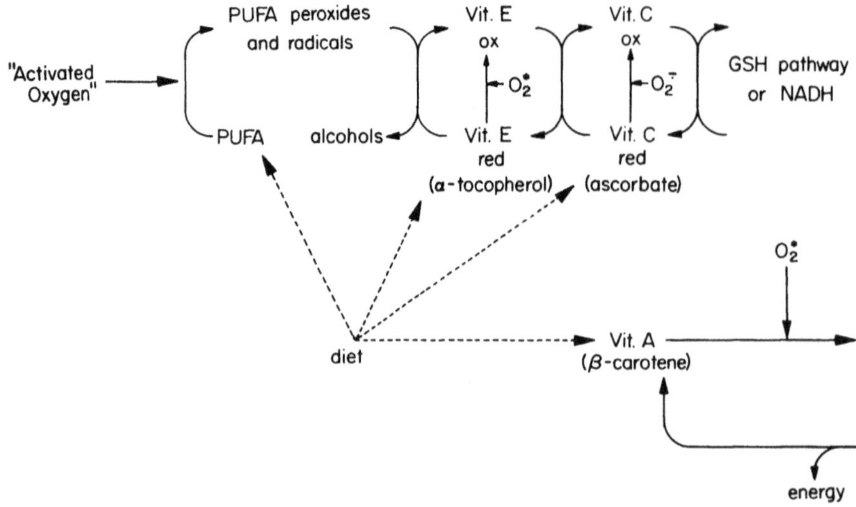

Figure 3. Dietary components of the antioxidant defense. Polyunsaturated fatty acids (PUFA) and vitamins E, C, and A are supplied by the diet. In a series of nonenzymatic reactions, PUFA reacts with several species of "activated oxygen" and is oxidized in the process, vitamin E reduces oxidized PUFA to the alcohol, and vitamin C reduces the oxidized vitamin E. Vitamin E can also trap singlet oxygen (O_2^*) and vitamin C can trap superoxide anion (O_2^-). Reduced ascorbate can be regenerated by enzymatically catalyzed reactions using glutathione (GSH) or NADH. Vitamin A is the most efficient quencher of O_2^* and is regenerated by a nonenzymatic process in which energy is discharged.

to hyperoxia suggest the importance of the NADPH-glutathione system in antioxidant defense.

B. Dietary Deficiencies and Body Composition

Tierney et al. (64) have shown that vitamin E deficient rats have decreasd survival when placed in 0.9 atm absolute O_2. Block and Fisher have shown a decreased ability of isolated perfused lungs from vitamin E deficient rats to clear serotonin after O_2 exposure (7); this observation was felt to indicate damage by O_2 to the pulmonary endothelium. The previously mentioned studies of Nishiki et al. (46) showed that isolated perfused lungs and livers of vitamin E deficient animals released significantly more GSSG upon exposure to hyperbaric O_2 than did normal animals. These vitamin E deficient animals also had increased lung content of glucose-6-phosphate dehydrogenase and glutathione peroxidase (46), possibly to compensate for the dietary deficiency of antioxidant. Vitamin E administration to newborn humans who were given supplemental O_2 for treatment of respiratory distress syndrome appeared to prevent development of bronchopulmonary dysplasia (20), although this preliminary finding has not subsequently been confirmed. While the precise metabolic role of vitamin E has not been established, these studies suggest that it does function in biological systems as an antioxidant.

Animals with higher lung content of unsaturated fatty acid are relatively resistant to O_2 toxicity (13). This may explain in part O_2 resistance of neonatal rats which demonstrate a greater amount of lipid peroxidation than older animals (35). This suggests that the unsaturated fatty acids may serve as a trap for activated O_2 species.

The importance of glutathione peroxidase in antioxidant defenses has been suggested by the effects of selenium deficiency in rats. Selenium is an important cofactor for the enzyme. In liver (8) and lungs (50) of these rats, catalase activity is increased possibly in a partially compensatory role to facilitate H_2O_2 removal in these tissues.

C. *Bacterial Mutations and Metabolic Diseases*

Bacterial mutations through which the organism has lost the ability to synthesize an antioxidant have been useful to study antioxidant requirements. Glutathione has been shown to be important in preventing oxygen enhancement of radiation damage in *E. coli* (44) since a mutant which could not produce GSH was found to be more susceptible than the normal GSH^+ strain. A mutant of the microorganism, *Sarcina lutea*, which can not produce carotenoids has been demonstrated to be less resistant than normal cells to killing by photosensitized toluidine blue (presumably due to O_2^*) (42) or by phagocytizing polymorphonuclear leukocytes which appear to kill bacteria through production of activated oxygen species (39).

Studies of metabolic diseases have also provided insights into antioxidant mechanisms. The hemolysis that occurs in drug-induced hemolytic anemia appears to depend on the production of H_2O_2 (15). This syndrome becomes manifest in the presence of genetic deficiency of glucose-6-phosphate dehydrogenase or glutathione peroxidase (6) indicating that these enzymes have important antioxidant roles. A similar syndrome occurs in individuals with decreased cellular glutathione associated with genetic lack of glutathione synthetase (43). A different syndrome occurs with acatalasia, a genetic deficiency in catalase (59), which results in chronic damage of oral tissues as a result of H_2O_2 secreted by oral bacteria. This suggests an important role for catalase in removal of H_2O_2 in some tissues.

D. *Prolonged O_2 Exposure*

In mammals the lungs are the primary site for oxygen toxicity due to hyperoxic exposure (12). While prolonged exposure of rats to 100% O_2 at 1 atm is usually lethal (average time to death is approximately 72 hr), rats exposed for 5 to 7 days to 80–85% O_2 are then able to survive indefinitely when exposed to 100% O_2 (53). Tierney et al. showed a doubling of glucose-6-phosphate dehydrogenase activity per whole lung in adult rats exposed to 85% oxygen for 7 days (32). A number of subsequent studies, summarized in Table 2, have indicated that glutathione peroxidase, glutathione reductase, and catalase as well as superoxide dismutase increase in the lungs of O_2 tolerant rats. Mice, guinea pigs, and hamsters show little or no increase of enzyme activity in 85% O_2 and do not develop O_2 tolerance (17,63). Exposure of adult rats as well as other species to 95% O_2 for 1 day did not result in increased lung antioxidant enzymes (25). However, neonatal rats, mice and rabbits did show increased lung enzyme content with these levels of O_2 and are resistant to O_2 toxicity, while neonatal hamsters and guinea pigs showed neither enzyme induction nor O_2 resistance (25). These studies indicate that O_2 tolerance and increased lung antioxidant enzyme content can be correlated. Chow and Tappel have shown similar increases in antioxidant

Table 2. *Alterations of "Antioxidant" Enzymes with O_2 Exposure in Lung from Adult Rats*

%O_2	Challenge duration (days)	% Change in activity[a]						Ref.
		Cat.	GSHPx	G6PDH	GSHRed	Mito. SOD	Cyto. SOD	
85	7	—	—	+99	—	—	—	(63)
90	7	—	+217	+313	+75	+145	+45	(37)
85	14	—	+122	+190	+22	—	—	(37)
85	5	+176	—	+87	—	+84	+45	(16)

[a]Cat., catalase, GSHPx, glutathione peroxidase; G6PDH, glucose-6-phosphate dehydrogenase; GSHred, glutathione reductase; Mito., mitochondrial; Cyto., cytosolic; SOD, superoxide dismutase.

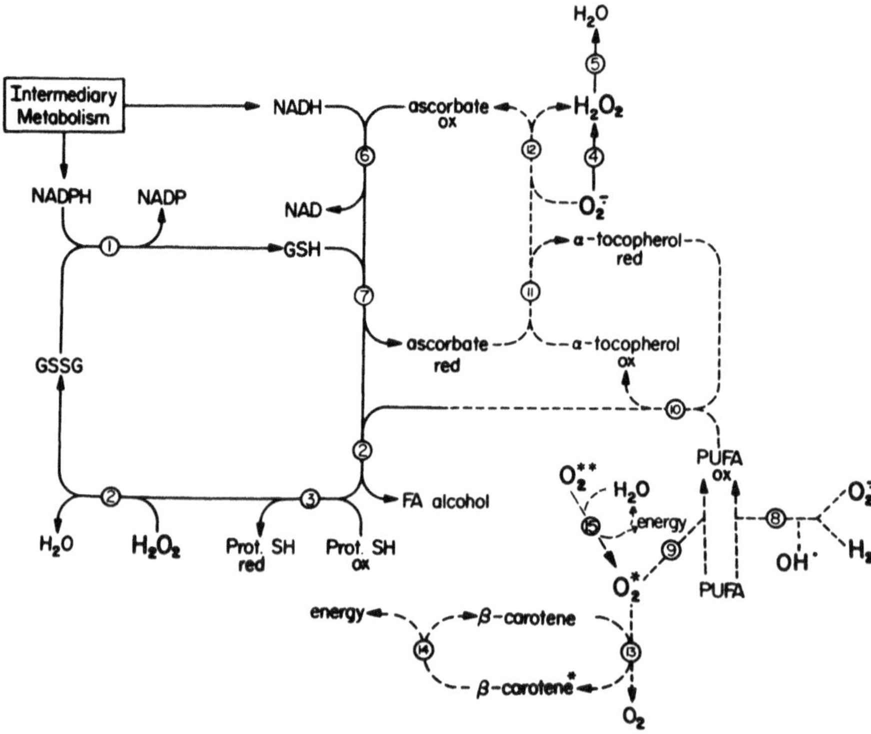

Figure 4. Overall interrelationships of antioxidant defense mechanisms. The scheme indicates the methods for removal of "activated oxygen" species which are indicated by the larger lettering. Solid lines indicate enzymatically catalyzed reactions while broken lines indicate nonenzymatic reactions. Oxidized glutathione (GSSG) is reduced to GSH by the action of an NADPH dependent-glutathione reductase [1]. GSH is used by glutathione peroxidase [2] to convert H_2O_2 to H_2O or oxidized polyunsaturated fatty acids (PUFA) to fatty acid (FA) alcohols. GSH is also used by GSH-*S*-transferases [3] to convert oxidized protein sulfhydryls to cysteinyl residues. Superoxide dismutases [4] convert O_2^- to H_2O_2 (and O_2) while catalase [5] converts H_2O_2 to H_2O (and O_2). Oxidized ascorbate is reduced by an NADH-dependent semidehydroascorbate reductase [6] or by a GSH-dependent dehydroascorbate reductase [7]. Polyunsaturated fatty acids (PUFA) trap $OH^{\cdot}$ and the "activated" products of the metal catalyzed interaction of O_2^- and H_2O_2 (possibly $OH^{\cdot}$) [8]. PUFA also trap singlet oxygen [9]. The oxidized PUFA can be reduced by α-tocopherol [10] and the oxidized α-tocopherol can be reduced by ascorbate [11]. Ascorbate can also trap O_2^- [12]. β-carotene can trap O_2^* [13] and is regenerated when the singlet form (β-carotene*) releases energy [14]. H_2O is the primary scavenger for the higher energy singlet O_2 (O_2^{**}) [15].

enzymes in lungs of rats exposed to low level ozone (11). Catalase and a peroxidase which can use dianisidine as an acceptor are induced in some bacteria by O_2 exposure and correlate with O_2 tolerance (32). These studies have suggested that increased cellular content of antioxidant enzymes may be responsible for O_2 tolerance.

IV. Summary

Protection of biological systems against the toxic effects of O_2 due to oxidation of cellular components involves a series of defenses among which are direct enzymatic conversions of activated oxygen to less toxic forms, trapping of activated oxygen with substances which are oxidized or which quench by absorbing energy, and regeneration of reduced forms of these trapping agents. Figure 4 illustrates the connections of all the major components of these systems. The biological response to oxygen toxicity is also influenced by the replacement of damaged molecules by resynthesis. The latter is not only a problem related to oxidant damage but also relates to recovery of tissues from any type of stress. It is these factors that link the study of oxidant injury to those processes involved in general tissue injury and repair.

Acknowledgments

Aron B. Fisher is an Established Investigator of the American Heart Association.

References

1. Anbar, M., and Neta, P. (1967). A compilation of specific biomolecular rate constants for the reactions of hydrated electrons, hydrogen atoms and hydroxyl radicals with inorganic and organic compounds in aqueous solution. Intern. J. Appl. Radiat. Isotopes 18:495.
2. Arias, I. M., and Jakoby W. B. (1976). Glutathione: Metabolism and Function. New York: Raven Press.
3. Askelof, P., Axelsson, K, Erkisson, S., and Mannervik, B. (1974). Mechanism of action of enzymes catalyzing thiol-disulfide interchange; thioltransferases rather than transhydrogenases. FEBS Lett. 38:263.
4. Bassett, D. J. P., and Fisher, A. B. (1976). Pentose cycle activity of the isolated perfused rat lung. Am. J. Phsyiol. 231:1527.
5. Bassett, D. J. P., and Fisher, A. B. (1979). Glucose metabolism in rat lung during exposure to hyperbaric O_2. J. Appl. Physiol.: Respirat. Environ. Excercise Physiol. 45:943.
6. Beutler, E. (1972). Drug-induced anemia. Fed. Proc. 32:141.
7. Block, E. R., and Fisher, A. B. (1977). Depression of serontonin clearance by rat lungs during oxygen exposure. J. Appl. Physiol.: Respirat. Environ. Exercise. Physiol. 42:33.
8. Burk, R. F., Nishiki, K., Lawrence, R.A., and Chance, B. (1978). Peroxide removal by selenium-dependent and selenium-independent glutatione peroxidases in hemoglobin-free perfused rat liver. J. Biol. Chem. 253:43.
9. Chance, B. (1951). Enzyme-substrate compounds. Adv. Enzymol. 12:153.
10. Chance, B., Jamieson, D., and Coles, H. (1965). Energy-linked pyridine nucleotide reduction: Inhibitory effects of hyperbaric oxygen in vitro and in vivo. Nature 206:257–263.
11. Chow, C. K., and Tappel, A. L. (1972). An enzymatic protective mechanism against lipid peroxidation damage to lungs of ozone-exposed rats. Lipids 7:518.
12. Clark, J. M., and Lambertsen, C. J. (1971). Pulmonary oxygen toxicity: A review. Pharmacol. Rev. 23:37.
13. Clements, J. A. (1971). Comparative lipid chemistry of lungs. Arch. Intern. Med. 127: 387.
14. Cohen, G., and Hochstein, P. (1963). Glutathione peroxidase: The primary agent for the elimination of hydrogen peroxide in erythrocytes. Biochemistry 2:1420.
15. Cohen, G., and Hochstein, P. (1964). Generation of hydrogen peroxide by hemolytic agents. Biochemistry 3:895.
16. Crapo, J. D., Sjostrom, K., and Drew, R. T. (1978). Tolerance and cross-tolerance using NO_2 and O_2. I. Toxicology and biochemistry. J. Appl. Physiol.: Respirat. Environ. Excercise Physiol 44:364.
17. Crapo. J. D., and Tierney, D. F. (1974). Superoxide dismutase and pulmonary oxygen toxicity. Am. J. Physiol. 226:1401.
18. Dam, K. (1957). Influence of antioxidants and redox substances on signs of vitamin E deficiency. Pharmacol. Rev. 9:1.
19. DeDuve, C., and Baudhuin, P. (1966). Peroxisomes (microbodies and related particles). Physiol. Rev. 46:323.
20. Ehrenkranz, R. A., Bonta, B. W., Albow, R. C., and Warshaw, J. B. (1978). Amelioration of bronchopulmonary dysplasia after vitamin E administration. New Engl. J. Med. 299: 564.
21. Ellfolk, N., Ronnberg, M., and Soininen, R. (1973). Pseudomonas cytochrome c peroxidase. Acta Chem. Scand. 27:2171.
22. Erecinska, M., Oshino, N., Loh, P., and Brocklehurst, E.(1973) In vitro studies on yeast cytochrome c peroxidase and its possible function in the electron transfer and energy coupling reactions. Biochim. Biophys. Acta 292:1.
23. Foote, C. S. (1968). Mechanisms of photosensitized oxidation. Science 162:963.
24. Foote, C. S., Denny, R. W., Weaver, L., Chong, Y., and Peters, J (1970). Quenching of singlet oxygen. Ann. N.Y. Acad. Sci. 171:139.
25. Frank, L., Bucher, J. R., and Roberts, R. J. (1978). Oxygen toxicity in neonatal and adult animals of various species. J. Appl. Physiol.: Respirat. Environ. Excercise Physiol. 45: 699.
26. Fridovich, I. (1974). Superoxide dismutase In: Molecular mechanisms of oxygen activation. O. Hayaishi, ed. New York: Academic Press, pp. 453–477.
27. Fridovich, I., and Handler, P. (1962). Xan-

thine oxidase, V. Differential inhibition of the reduction of various electron acceptors. J. Biol. Chem. 237:916–921.
28. Gerschman, R. (1964). Biological effects of oxygen. In: Oxygen in the Animal Organism. F. Dickens and E. Neil, eds. New York: Macmillan, pp. 475–494.
29. Goscin, S. H., and Fridovich, I. (1972). The role of superoxide radical in a nonenzymatic hydroxylation. Arch. Biochim. Biophys. 153: 778.
30. Grimble, R. F., and Hughes, R. E. (1967). A dehydroascorbic acid reductase factor in guinea-pig tissues. Experientia 23:362.
31. Halliwell. B. (1978). Biochemical mechanisms accounting for the toxic action of oxygen on living organisms: the key role of superoxide dismutase. Cell Biol. Intern. Rep. 2.
32. Hassan, H. M., and Fridovich, I. (1977). Enzymatic defenses against the toxicity of oxygen and streptonignin in *Escherichia coli.* J. Bacteriol. 129:1574.
33. Jakoby, W. B., and Keen, J. H. (1977). A triple threat in detoxification: The glutathione-S-transferases. Trends Biochem. Sci. 2:229.
34. Kearns, D. R. (1971). Physical and chemical properties of singlet oxygen. Chem. Rev. 71: 395.
35. Kehrer, J. P., and Autor, A. P. (1977). Changes in the fatty acid composition of rat lung lipids during development and following age-dependent lipid peroxidation. Lipids 12: 569.
36. Kellogg, E. W., III, and Fridovich, I. (1975). Superoxide, hydrogen peroxide and singlet oxygen in lipid peroxidation by a xanthine oxidase system. J. Biol. Chem. 250:8812.
37. Kimball, R. E., Reddy, K., Pierce, T. H., Schwartz, L. W., Mustafa, M. G., and Cross, C. E. (1976). Oxygen toxicity: Augmentation of antioxidant defense mechanisms in rat lung. Am. J. Physiol. 230:1425.
38. Kosower, N. S. and Kosower, E. M. (1976). The glutathione-glutathione disulfide system. In: Free Radicals in Biology. Pryor, W. A., ed, Vol. II. New York: Academic Press.
39. Krinsky, N. I. (1974). Singlet excited oxygen as a mediator of the antibacterial action of leukocytes. Science 186:363.
40. Mapson, L. W. (1958). Metabolism of ascorbic acid in plants: Function. Annu. Rev. Plant Physiol. 9:119.
41. Matheson, M. S., Mulac, W. A., Weeks, J. L., and Rabani, J. (1966). The pulse radiolysis of deaerated aqueous bromide solutions. J. Phys. Chem. 70:2092.
42. Matthews-Roth, M. M., and Krinsky, N. I. (1970). Studies on the protective function of the carotenoid pigments of *Sarcina lutea.* Photochem. Photobiol. 11:419.
43. Mohler, D. N., Majerus, D. W., Minnich, V., Hess, C. E., and Garrick, M. D. (1970). Glutathione synthetase deficiency as a cause of hereditary hemolytic disease. New Engl. J. Med. 283:1253.
44. Morse, M. L., and Dahl, R. H. (1978). Cellular glutathione is a key to the oxygen effect in radiation damage. Nature 271:660.
45. Neta, P., and Dorfman, L. M. (1968). Pulse radiolysis studies. XIII. Rate constants for the reaction of hydroxyl radicals with aromatic compounds in aqueous solutions Adv. Chem. Ser. 81:222.
46. Nishiki, K., Jamieson, D., Oshino, N., and Chance B. (1976). Oxygen toxicity in the perfused rat liver and lung under hyperbaric conditions. Biochem. J. 160:343.
47. Nishikimi, M. (1975). Oxidation of ascorbic acid with superoxide ion generated by the xanthine-oxidase system. Biochem. Biophys. Res. Commun. 63:463-468.
48. Nishikimi, M., and Machlin, L. J. (1975). Oxidation of α-tocopherol model compound by superoxide anion. Arch. Biochem. Biophys. 170:684.
49. Oh, S. H., Ganther, H. E., and Hockstra, W. G. (1974). Selenium as a component of glutathione peroxidase isolated from ovine erythrocytes. Biochemistry 13:1825.
50. Omaye, S. T., Reddy, K. A., and Cross, C. E. (1978). Enhanced lung toxicity of paraquat in selenium-deficient rats. Toxicol. Appl. Pharmacol. 43:237.
51. Pedersen, T. C., and Aust, S. D. (1973). The role of superoxide, hydrogen peroxide and singlet oxygen in lipid peroxidation promoted by xanthine oxidase. Biochem. Biophys. Res. Commun. 52:1071.
52. Porter, W. L., Lavasseur, L. A., Jeffers, J. J., and Henick, A. S. (1971). UV spectrophotometry of autoxidized lipid monolayers while on silica gel. Lipids 6:16.
53. Rosenbaum, R. M., Wittner, M., and Lenger, M. (1969). Mitochondrial and other ultrastructural changes in great alveolar cells of oxygen-adapted and poisoned rats. Lab. Invest. 20:516.
54. Shreve, M. R., Morrissey, P. G., and O'Brien, P. J. (1979). Lipid and steroid hydro-

peroxides as substrates for the non-selenium-dependent glutathione peroxidase. Biochem. J. 177:761.
55. Singh, A., and Petkau, A. (1978). Singlet oxygen and related species in chemistry and biology. Photochem. Photobiol. 28:429–433.
56. Smith, J., and Shrift, A. (1979). Phylogenetic distribution of glutathione peroxidase. Comp. Biochem. Physiol. 63B:39.
57. Staudinger, H., Krisch, K., and Leonhauser, S. (1961). Role of ascorbic acid in microsomal electron transport and the possible relationship to hydroxylation reactions. Ann. N.Y. Acad. Sci. 92:195.
58. Stern, K. G. (1936). On the mechanisms of enzyme action: A study of the decomposition of monoethyl hydrogen peroxide by catalase and of an intermediate enzyme-substrate compound. J. Biol. Chem. 114:473.
59. Takahara, S. (1952). Progressive oral gangrene, probably due to lack of catalase in the blood (actalasaemia): Report on nine cases. Lancet 2:1101.
60. Tappel, A. L. (1969). Vitamin E as the biological lipid antioxidant. Vitam. Horm. (N.Y.) 20:493.
61. Taube, H. (1975). Mechanisms of oxidation with oxygen. J. Gen. Physiol. 49, Part 2:29.
62. Thomas, M. J., Mehl, K. S., and Pryor, W. A. (1978). The role of the superoxide anion in the xanthine oxidase-induced autoxidation of linoleic acid. Biochem. Biophys. Res. Comm. 83:927.
63. Tierney, D., Ayers, L., Herzog, S., and Yang, J. (1973). Pentose pathway and production of reduced nicotinamide adenine dinucleotide phosphate. Am. Rev. Respir. Dis. 108:1348.
64. Tierney, D. F., Ayers, L., and Kasuyama, R. S. (1977). Altered sensitivity to oxygen toxicity. Am. Rev. Respirat. Dis. 115, Part 2:59.
65. Von Sonntag, C., and Dizdaroglu, M. (1971). The reaction of OH radicals with D-ribose in deoxygenated and oxygenated aqueous solution. Carbohyd. Res. 58:21.
66. Walker, G. A., and Kilgour, G. L. (1965). Pyridine nucleotide oxidizing enzymes of *Lactobaccilus casei*. Arch. Biochem. Biophys. 111:534.

13

Superoxide Radical and Superoxide Dismutases

IRWIN FRIDOVICH

The metalloenzymes that have been named superoxide dismutases have several attributes. They are abundant, widely distributed, essential as a defense against oxygen toxicity; they are also unique among enzymes in that their natural substrate is an unstable free radical. They are evolution's answer to the biological production of the superoxide radical (O_2^-). This free radical, previously considered only in connection with the high energy irradiation of oxygenated aqueous media, is also produced under ordinary circumstances within respiring cells. The superoxide radical is intrinsically reactive and can furthermore engender other enormously reactive radicals and excited states which are potentially capable of destroying the delicate chemical architecture of the cell. The intermediates of oxygen reduction, i.e., O_2^-, H_2O_2, and $OH\cdot$, are the primary cause of oxygen toxicity; the enzymes, which catalytically scavenge them or prevent their production, are the foremost defenses which limit that toxicity. H_2O_2 and the catalases and peroxidases, which scavenge it, are discussed in other chapters of this volume. We shall devote ourselves to superoxide and the superoxide dismutases.

I. Some Biological Sources of Superoxide

The electronic structure of dioxygen, in its ground state, and the dictates of quantum mechanics, lead to a spin restriction which actually favors the univalent pathway of oxygen redution (1–5). This most facile route of dioxygen reduction generates, in turn, the superoxide radical (O_2^-), hydrogen peroxide (H_2O_2), the hydroxyl radical ($\cdot OH$) and finally water. The intermediates of dioxygen reduction are too reactive to be well tolerated within living systems, and their production is largely avoided by enzymes which carry out the divalent and even the tetravalent reduction of dioxygen, without the release of intermediates. Cytochrome oxidase, which is responsible for most biological dioxygen consumption, produces water, without the release of intermediates (6,7). The blue, copper-containing oxidases likewise produce H_2O; while the flavin-containing oxidases release H_2O_2. Respiring cells thus avoid exposure to the full potential of oxygen toxicity, by minimizing the univalent pathway of dioxygen reduction. Nevertheless, some production of the intermediates of dioxygen reduction does occur.

Cells are rich in reductants which can react directly with dioxygen, albeit at moderate rates, and oxygen reduction which occurs without the benefit of enzymatic mediation is most likely to proceed by the univalent route. Among the compounds of biological relevance which are known to autoxidize in vitro and in so doing to produce O_2^- are hydroquinones (8), leukoflavins (8,9), catecholamines (10,11), thiols (12–15), tetrahydopterins (16–19), reduced ferredoxins (20–25), ferrocytochrome *c* (26,27), ferrocytochrome b_5 (28), hemoglobins (29–33), and myoglobin (34). There are enzymes that catalyze the univalent reduction of dioxygen. Among these are xanthine oxidase (35–37), aldehyde oxidase (38,39), dihydroorotic dehydrogenase (40,41), indoleamine dioxygenase (42–44), 2-nitropropane dioxygenase (45–47), and undoubtedly many others, yet to be discovered. Fragments of subcellular organelles, such as respiring mitochondria (48–57) and illuminated chloroplasts (58–74), have been shown to produce and liberate substantial amounts of O_2^-. Furthermore, intact cells of the phagocytic type, such as polymorphonuclear leukocytes and macrophages, have been shown to release large amounts of O_2^-, when activated. Since this O_2^- production appears to be an essential component of the bactericidal machinery of these specialized cells, and since a human genetic defect in this ability to make O_2^- results in a serious susceptibility to infection (chronic granulomatous disease), a great deal of work has been done in this area and this extensive literature has recently been reviewed (75).

O_2^- is clearly produced within respiring cells, but attempts to estimate how much is made are usually frustrated by the ubiquity of superoxide dismutases. An upper limit for the O_2^- production inside intact cells can be obtained from the cyanide-resistant respiration. Thus, cyanide is used to suppress the cytochrome *c* oxidase and the residual oxygen reduction is, as a first approximation, assumed to result in O_2^- production. In actively growing *Escherichia coli*, this method suggests that no more than 3% of the total oxygen uptake results in O_2^- production. Compounds capable of subverting intracellular electron flow from the cytochrome pathway and of increasing the production of O_2^-, dramatically increase this cyanide-resistant respiration (76,77). The extent of O_2^- production in homogenates of *Streptococcus faecalis*, respiring with NADH as the substrate, was measured after suppressing endogenous superoxide dismutase with a specific inhibiting antibody. In these broken cell preparations 17% of the total oxygen uptake proceeded by the univalent pathway (78). This is a large number and it seems probable that a smaller relative O_2^- production would occur within *S. faecalis* in vivo. O_2^- is thus a minor product of biological oxygen reduction but, as we shall document in the remainder of this chapter, it is not a trivial product, since its production would have lethal consequences were it not for the protective effects of the superoxide dismutases.

II. The Potential Consequences of Superoxide Production

Fluxes of O_2^- generated enzymatically, photochemically, or radiochemically have been shown to peroxidize lipids (79–87), depolymerize polysaccharides (88,89), cleave DNA (90–97), inactivate viruses (98), inactivate enzymes (98–100), kill bacteria (98,101–104), lyse erythrocytes and vesicles formed from erythrocyte stroma (84, 105–107b), and kill mammalian cells (108). There is, however, uncertainty and controversy concerning the exact mechanism of these effects. Thus, physical chemists and others, who work with free radicals in chemically simple and well-defined reaction mixtures, perceive O_2^- to be rather benign, compared to other free radicals, and they have difficulty conceiving that it could cause widespread biological damage.

A possible explanation for the gulf between the expectations of some traditional free radical chemists and the biochemical realities appeared in 1970. Methional was found to be oxidized and cleaved to yield ethylene, when exposed to the O_2^- and H_2O_2 produced by the enzymatic action of xanthine oxidase on its substrate, xanthine (109). Since superoxide dismutase could suppress ethylene production in these reaction mixtures, O_2^- was taken to be an essential reactant in the attack upon methional. Catalase likewise inhibited ethylene production. H_2O_2 was therefore also taken to be an essential reactant. Superoxide dismutase does not catalytically scavenge H_2O_2 and catalase does not scavenge O_2^-. It follows that neither O_2^- nor H_2O_2 was individually capable of converting methional to ethylene, but that both were required. A reasonable explanation for this dual dependence on O_2^- plus H_2O_2 was found in an earlier paper by Haber and Weiss (110), dealing with the catalytic decomposition of H_2O_2 by iron salts. They had proposed a free radical chain process, one of whose component reactions was

$$O_2^- + H_2O_2 \rightarrow O_2 + OH^- + OH\cdot \quad \textbf{(1)}$$

If O_2^- plus H_2O_2 could indeed give rise to the vastly more reactive $OH\cdot$, then this might have been the species directly responsible for the conversion of methional to ethylene, during the xanthine oxidase reaction. In that case, compounds capable of scavenging $OH\cdot$, but incapable of reacting with O_2^- or with H_2O_2, should prevent ethylene production. Ethanol and benzoate have the desired properties and they were tested (109) and behaved as expected.

O_2^- or H_2O_2 may thus not be terribly life threatening individually but, if present together, they conspire in the production of a product reactive enough to convert methional to ethylene and reactive enough to be efficiently scavenged by such weakly reactive substances as alcohols or benzoate. This reactive product seems likely to be $OH\cdot$. Ethylene is indeed produced when methional is exposed to a known source of $OH\cdot$, such as the Fenton's reagent (111) or pulse radiolysis (112) and methional does not react with O_2^- (112), although it can react with a variety of alkoxy radicals (113). If O_2^- plus H_2O_2 could give rise to something as reactive as $OH\cdot$, that could explain the biological need for superoxide dismutases, catalases, and peroxidases. The conversion of methional, or of the closely related 2-keto-4-thiomethylbutyric acid, to ethylene has since been seen with illuminated chloroplast fragments (62, 114) and activated neutrophils (115,116).

The interaction of O_2^- with H_2O_2 to generate very reactive products such as $OH\cdot$, and possibly also singlet oxygen (81,117), has been invoked many times and on the basis of convincing data (11,62,81,84,85,90, 92,94,95,102,103,109,114,118–133); yet equally careful measurements have shown that O_2^- does not rapidly react with H_2O_2 (124,134–140).

A recent paper by McCord and Day (133) provides one answer to this apparent discrepancy. They exposed tryptophan to the xanthine oxidase reaction and noted changes in its optical spectrum which could be attributed to hydroxylation. They used superoxide dismutase and catalase to demonstrate the role of O_2^- and of H_2O_2, respectively, and they used alcohols to expose the involvement of $OH\cdot$. They found that the presumed hydroxylation of tryptophan was markedly augmented by catalytic levels of FE(III)-EDTA. Iron in the form of transferrin was also active. Iron compounds could catalyze the interaction of O_2^- and H_2O_2 as follows:

$$Fe^{3+} + O_2^- \rightarrow Fe^{2+} + O_2 \quad \textbf{(2)}$$

$$Fe^{2+} + H_2O_2 \rightarrow Fe^{3+} + OH^- + OH\cdot \quad \textbf{(3)}$$

In reaction 2 O_2^- reduces the ferric complex to a ferrous complex and in reaction 3 the ferrous complex reduces H_2O_2 to OH^- + $OH\cdot$, as in the well-known Fenton's reaction (141,142). Fe(II)-EDTA has been shown to efficiently react with H_2O_2, as in reaction 3 (143). Iron compounds were surely present in the many enzymatic reaction mixtures, whose behavior was explained by the Haber-Weiss reaction and were, just as certainly, absent from the well-defined experiments specifically performed to test the reality of the Haber-Weiss reaction. Iron compounds abound within living cells and the production of O_2^- and of H_2O_2 in such a milieu would lead to formation of OH·. It might be useful hereafter to speak of the *biological or the catalyzed Haber-Weiss reaction* to indicate appreciation of the essential role of catalytic amounts of iron or of other transition metal compounds.

Attention to experimental detail is always important, but is especially so in controversial areas where experiments are sometimes done with the explicit goal of disproving the opposing point of view. An illustration of this arose during the studies of McCord and Day (133,144). Thus, *p*-nitroso-dimethylaniline is bleached by OH· and has been used as an indicating scavenger of OH· (145,146). McCord and Day attempted to use it in their studies of OH· production by the xanthine oxidase reaction (144). They found that it was bleached by this enzymatic reaction, but superoxide dismutase did not inhibit this bleaching. Had they been overly ardent in trying to prove that O_2^- was not involved in the production of OH·, they might have stopped there and reported their result. They were, however, trying to find out what was truly occurring, so they prevented the production of O_2^- by removing dioxygen and noted that *p*-nitrosodimethylaniline was bleached by xanthine oxidase plus xanthine even in the absence of oxygen. This could clearly be related to the ability of xanthine oxidase to directly reduce aromatic nitro compounds all the way to the corresponding amines (147–150). Nitro and nitroso compound were therefore judged useless for detecting OH· production in the xanthine oxidase system and they then selected and used tryptophan for further studies (133).

III. Assays for Superoxide Dismutases

Great ingenuity has been exercised in devising assays for superoxide dismutases, with the result that a great variety of methods are available. This allows flexibility, but the basis of the methods used must be understood, because the possibility of being misled by artifact is ever present.

A. Direct Assays

Direct assays, in which the effect of the enzyme on the rate of decay of O_2^- is observed, are possible, but are technically difficult, because of the intrinsic instability of this free radical substrate. This instability can best be expressed in terms of the rates of the spontaneous dismutation reactions. Since O_2^- is the conjugate base of the weak acid $HO_2\cdot$, whose pK_a is 4.8, three dismutation reactions must be considered (151). Thus:

$$HO_2\cdot + HO_2\cdot \rightarrow H_2O_2 + O_2 \qquad k = 8.6 \times 10^5 M^{-1}\text{sec}^{-1} \qquad \textbf{(4)}$$

$$HO_2\cdot + O_2^- + H^+ \rightarrow H_2O_2 + O_2 \qquad k = 1.0 \times 10^8 M^{-1}\text{sec}^{-1} \qquad \textbf{(5)}$$

$$O_2^- + O_2^- + 2H^+ \rightarrow H_2O_2 + O_2 \qquad k = <0.35 M^{-1}\text{sec}^{-1} \qquad \textbf{(6)}$$

The spontaneous dismutation would thus be most rapid at pH 4.8 and would fall an order of magnitude for each unit change in pH in the neutral and alkaline range. Since the salts of metals, such as Cu(II), can catalyze the dismutation of O_2^-, the noticeable effect of trace impurities becomes increasingly difficult to avoid as the pH is raised and the noncatalyzed rate falls. The spontaneous dismutation of O_2^- [reaction (6)] is very slow, partly because electrostatic repulsions would tend to decrease collisions between the like charged anion radicals, but mostly because, during the existence of the collisional complex, repulsion between electronic charges would prevent the asymmetric distribution of the two negative charges to yield O_2 and O_2^{2-} (152).

EDTA complexes with trace metal impurities and largely eliminates their ability to catalyze the dismutation of O_2^-, while having no effect on the activity of superoxide dismutases. This greatly facilitates direct assays, in which the rate of decay of O_2^- is obsserved by ultraviolet spectrophotometry. Most often, the O_2^- is introduced very rapidly into the reaction mixtures by pulse radiolysis. These pulse methods have been very useful in exploring the mechanism of action of superoxide dismutases (153–161), but are not suitable for routine assays. An attempt to adapt direct observation of the decay of O_2^-, for routine assays of the copper-zinc superoxide dismutases, was marginally successful and was possible only at elevated pH (pH $\geq$ 9.5) (162).

B. Indirect Assays—Negative

Assays of great sensitivity and convenience can be achieved by introducing a steady flux of O_2^- into a solution containing an indicating scavenger for this radical. Superoxide dismutase is then measured in terms of its ability to intercept O_2^- and thus to prevent it from reacting with the indicator. The first assay devised for superoxide dismutase activity, and the one which guided its isolation (163), was of this type. The xanthine oxidase reaction was used to generate O_2^- and cytochrome *c* was used as the indicating scavenger. Superoxide dismutase competed with the cytochrome *c* for the O_2^- and thus inhibited its reduction. One unit of the enzyme was defined as that amount which caused 50% inhibition. Under the conditions specified (163), 0.3 μg of the enzyme per 3.0 ml reaction mixture equaled one unit. This type of assay can be considered a negative assay because the superoxide dismutase is assayed in terms of its ability to decrease the rate of the process actually measured, in this case, the reduction of cytochrome *c*.

This method is reliable, provided that the possible artifacts are kept in mind. Thus, any inhibitor of the xanthine oxidase reaction could falsely give the appearance of acting like a superoxide dismutase. The critical difference is that superoxide dismutase prevents cytochrome *c* reduction by scavenging O_2^-, without affecting the turnover of xanthine oxidase, and thus without modifying the rate of *production* of O_2^-. In contrast, inhibitors of xanthine oxidase prevent the production of O_2^-. A somewhat more subtle artifact would involve the oxidation of ferrocytochrome *c*. Thus, cytochrome oxidase or a cytochrome *c* peroxidase could decrease the rate of accumulation of reduced cytochrome *c*, in the xanthine oxidase system, and could also be mistaken for superoxide dismutases. True superoxide dismutase is, of course, without effect on reduced cytochrome *c*. Another possible diversion is provided by compounds which can mediate electron transfer between the xanthine oxidase and cytochrome *c*. Menadione is such a compound and in its presence much of the cytochrome *c* reduction is caused directly by reduced menadione, rather than by O_2^-, and is thus not inhibitable by superoxide dismutase. Menadione will, indeed, enable xanthine oxidase to reduce cytochrome *c* even anaerobically. All of the these potential problems can easily be controlled by appropriate experimental design.

This assay can be modified in several ways. Thus, the sensitivity of the assay depends on competition for O_2^-, between superoxide dismutase and the indicating scavenger. If the scavenger reacts slowly with O_2^-, then less superoxide dismutase will be required to compete equally with it. The rate of reaction of O_2^- with cytochrome *c* at pH 7.2 and at 21°C is $1.1 \times 10^6\ M^{-1}\ sec^{-1}$ (164). Under these conditions the rate of reaction of O_2^- with superoxide dismutase is approximately $2 \times 10^9\ M^{-1}\ sec^{-1}$, so that one part of the enzyme will compete equally with 2000 parts of cytochrome *c*. At pH 8.5 the rate for cytochrome *c* is $1.1 \times 10^5\ M^{-1}\ sec^{-1}$ (165), whereas the rate for the superoxide dismutase is still $\simeq 2 \times 19^9\ M^{-1}\ sec^{-1}$. Hence at pH 8.5 the reduction of cytochrome *c* by O_2^- will be ten times more sensitive to inhibition by superoxide dismutase than at pH 7.2; at more alkaline pH, the sensitivity becomes even greater, because the rate of the O_2^-–cytochrome *c* reaction becomes even slower (164,166). Such competitive kinetics have been used to measure rate constants for a variety of superoxide dismutases (167) and to provide an ultrasensitive assay for these enzymes (168,169). Acetylated cytochrome *c* has been used as the indicating scavenger, because it can still be reduced by O_2^-, while having lost much of its abilities to interact with cytochrome *c* reductases and oxidases (50,170). Acetylated cytochrome *c* thus largely eliminates interferences by these activities and is useful in assaying crude extracts.

Indirect negative assays can be based on indicating scavengers other than cytochrome *c* and on sources of O_2^- other than the xanthine oxidase reaction. Thus, O_2^- reduces tetranitromethane to the nitroformate anion (171) and oxidizes epinephrine to adrenochrome (163) and both of these indicating scavengers have been used (163). Tetranitromethane reacts with O_2^- at neutral pH four orders of magnitude faster than does cytochrome *c* (171,172) and so gives an assay which is proportionately less responsive to inhibition by superoxide dismutase. In contrast, epinephrine reacts with O_2^- more slowly than does cytochrome *c* (173) and so provides greater sensitivity to superoxide dismutase. Nitroblue tetrazolium (NBT) is a useful indicating scavenger for O_2^- in that its rate of reaction with O_2^- is approximately 6×10^4 M^{-1} sec^{-1} (174), which is fast enough to compete against the spontaneous dismutation, yet slow enough to be competed against by low levels of superoxide dismutase. Furthermore, the reduced compound (formazan) is deeply blue-purple and insoluble in water and so lends itself to activity staining of polyacrylamide gel electropherograms (175). There are potential artifacts peculiar to nitroblue tetrazolium. Thus there are many enzymes capable of the direct reduction of this dye. Indeed xanthine oxidase, which does produce O_2^- when acting aerobically, will reduce NBT as well in the absence as in the presence of oxygen. Under aerobic conditions most, but not all, of the NBT reduction is mediated by O_2^- and so is inhibited by superoxide dismutase, whereas anaerobically O_2^- cannot be involved and superoxide dismutase has no effect. Another artifact involves the production of O_2^- by autoxidation of an NBT radical. NBT can thus mediate the production of O_2^- in systems which do not ordinarily produce it. In one example of this effect Auclair et al. (176) noted an NBT-dependent NADPH oxidation by the NADPH-cytochrome P_{-450} reductase, which was greatly augmented by superoxide dismutase. Their explanation involved the production of NBT radicals which could react with each other, or with oxygen. The reaction with oxygen was a mobile equilibrium with O_2^- as a product, and superoxide dismutase displaced this equilibrium [reaction (10)] and so increased oxygen consumption. Thus,

$$\mathrm{Enz} + \mathrm{NADPH} + \mathrm{H}^+ \leftrightarrow \mathrm{NADP}^+ + \mathrm{Enz\,H_2} \quad \textbf{(7)}$$

$$\mathrm{Enz\,H_2} + 2\mathrm{NBT} \leftrightarrow \mathrm{Enz} + 2\mathrm{NBTH}\cdot \quad \textbf{(8)}$$

$$2\mathrm{NBTH}\cdot \leftrightarrow \mathrm{NBT} + \mathrm{NBTH_2} \quad \textbf{(9)}$$

$$\mathrm{NBTH}\cdot + \mathrm{O_2} \leftrightarrow \mathrm{NBT} + \mathrm{O_2^-} \quad \textbf{(10)}$$

Nishikimi (18) reported another unusual property of the interaction of NBT with O_2^-. Triton X-100, above its critical micellar concentration, was noted to increase the rate of reduction of NBT, by autoxidizing tetrahydropterins or by the xanthine oxidase reaction. Nishikimi proposed that the effect was due to stabilization of a colloid of the insoluble reduced formazan. Probably several effects were operating, at least one of which was simply an increase in the effective light path, due to tortuosity of that light path in turbid suspensions. The extent of light path lengthening by tortuosity can be dramatic. Thus, Butler and Norris (177) increased the intensity of the absorption bands of cytochrome *c* 70-fold, merely by suspending powdered $CaCO_3$ in the solution. Other indicating scavengers for O_2^-, which have been used as the basis of assays for superoxide dismutase, include sulfite (35,178), hydroxyl amine (179), luminol (180,181), and crocin (182).

Chemical and photochemical sources of O_2^- have been used in providing simple and convenient assays for superoxide dismutase. Nishikimi et al. (183) used NADH, phenazinemethosulfate, and NBT. They supposed that the phenazine was reduced by NADH and then, in turn, generated O_2^- by reaction with oxygen. However, phenazine methosulfate, reduced by NADPH plus the ferredoxin-NADP$^+$ reductase, failed to cause a superoxide dismutase-inhibitable reduction of cytochrome *c* (184). It thus appears likely that the autoxidation of phenazine methosulfate is not a very good source of O_2^-. NBT was probably mediating the production of O_2^-, in the system of Nishikimi et al. (183), much as it did in the system of Auclair et al. (176). The autoxidation of phenyl hydrazine also produces O_2^- and this was coupled with NBT to provide an assay useful at elevated pH (185). Ballou et al. (9) illuminated a mixture of tetraacetyl riboflavin plus EDTA to generate the reduced flavin which, upon admixture with oxygenated buffer, generated O_2^-. The radical was measured by EPR, after freeze-quenching, and superoxide dismutase was assayed by its ability to decrease the EPR signal. Mixtures of riboflavin, EDTA and NBT darken upon illumination because the O_2^-, generated by autoxidation of the photoreduced flavin, reduces NBT to the formazan. This is the basis of a widely used solution assay and activity stain for superoxide dismutase (175).

C. Indirect Assays—Positive

The negative assays, described above, are entirely satisfactory, in practice, but many enzymologists dislike measuring an enzyme activity by its ability to *inhibit* the process being observed. It is certainly more usual to detect enzymes by their acceleration of the parameter being measured. This can be achieved in illuminated mixtures of riboflavin and dianisidine (186,187). The riboflavin photooxidizes the dianisidine to a quinone diimine, which self-couples to a colored bisazobiphenyl compound. Superoxide dismutase augments the rate of accumulation of this product. The mechanism of this augmentation involves the production of O_2^-, by autoxidation of the photoreduced flavin, and reduction of a partially oxidized intermediate of dianisidine, back to the parent compound by this O_2^-. This back reduction by O_2^- thus decreased the net accumulation of the colored bisazobiphenyl and superoxide dismutase eliminates this effect of O_2^-. In these augmentation assays, superoxide dismutase is thus doing what it usually does, i.e., scavenge O_2^-.

D. Autoxidation Assays

O_2^- frequently functions as a chain-propagating intermediate in autoxidations. Superoxide dismutase therefore is often seen to inhibit autoxidations and the extent of such inhibitions can serve as a measure of the superoxide dismutase. In essence, compounds which autoxidize, by an O_2^--propagated chain reaction, act both as the source of O_2^- and as the indicating scavenger. When, upon oxidation, they yield a colored product, they provide the basis of exceedingly convenient assays for superoxide dismutase. Of course, free radical chain reactions always involve at least two radical intermediates; substances that intercepted the organic radical, in an autoxidation process, could also inhibit and might then wrongly be credited with scavenging O_2^-. Epinephrine autoxidizes to adrenochrome and, at pH ≤ 8.5, O_2^- is an important chain carrier. Inhibition of this autoxidation was used to assay superoxide dismutase (10). A recent improvement of this assay (188) involves following the autoxidation of epinephrine in the near ultraviolet, rather than in the visible. Pyrogallol is less expensive than epinephrine and moreover autoxidizes, at more nearly neutral pH, by an O_2^--propagated chain mechanism. It provides the basis for a very useful assay of superoxide dismutase (189). Other autoxidations which have been used for assaying superoxide dismutase include that of 6-hydroxydopamine (190,191) and of hydroxyl amine (192).

E. Miscellaneous Assays

The cathodic reduction of oxygen involves the transient production of O_2^- (193). When the cathode is coated with a hydrophobic film of triphenylphosphine oxide, the likelihood that O_2^- will diffuse away from the vicinity of the cathode is increased, because H^+ cannot then diffuse to the O_2^-, as it is produced at the cathode surface. If the O_2^- diffuses far from the cathode surface (it is being electrostatically repelled from that surface) then the P_{O_2} in the vicinity of that surface decreases. In contrast, if the O_2^- dismutes close to the cathode, it produces oxygen which tends to prevent the drop in P_{O_2}. Superoxide dismutase causes this dismutation to occur close to the cathode surface and thus increases the oxygen supply and thus the height of the polarographic reduction wave. This method has been adapted to the assay of superoxide dismutase (194,195).

When O_2^- reduces an indicating scavenger, such as NBT or cytochrome *c*, O_2 is produced. If oxygen uptake is being monitored in a reaction producing O_2^-, addition of such reducible scavengers of O_2^- will reduce the rate of oxygen uptake. Superoxide dismutase will prevent the effect of the scavengers and will thus increase oxygen uptake. This method was used in studies of the production of O_2^- by the xanthine oxidase reaction (37) and it has been applied as the basis of an oxygen electrode assay for superoxide dismutase (196).

All of the foregoing assays measured the catalytic activity of superoxide dismutase. There are assays that are directed toward specific antigenic determinants on the enzymes. These are radioimmunoassays and are capable of great sensitivity, but must be devised and used with a specific superoxide dismutase, since the enzyme from another source is likely to be antigenically distinct. Such assays have been reported for the bovine erythrocyte copper-zinc enzyme (197) and for the corresponding human enzyme (198,199).

F. Spin Trapping

Nitrones react with free radicals to give relatively stable nitroxide radical adducts (199a). These can be detected by EPR, which often also allows identification of the original radical. This method has been applied to the detection of O_2^- and OH· production by illuminated chloroplasts (65), liver microsomes (199b,c), and the mixed function amine oxidase (199d). Some confusion has developed because workers in this field did not at first recognize that the most widely used spin trap 5,5'-dimethyl-1-pyrroline-*N*-oxide (DMPO) gave an unstable O_2^- adduct. This was noted by Buettner and Oberley (199d). Finkelstein et al. (199e) subsequently showed that the O_2^- adduct of DMPO rapidly decays to the same product as is given by the reaction of DMPO with OH·. They further noted that 2,5,5-trimethyl-1-pyrroline-*N*-oxide (TMPO) gives a stable O_2^- adduct and is thus suitable for distinguishing between O_2^- and OH·. The rate of reaction of TMPO with O_2^- is only 7 M^{-1} sec^{-1} at 25°C and pH 7.8. This is to be compared with 6×10^5 M^{-1} sec^{-1} for the reaction of O_2^- with cytochrome *c* and with 2×10^9 M^{-1} sec^{-1} for its reaction with SOD. It is clear that TMPO is not an efficient scavenger of O_2^- and must be used at very high concentrations to be at all useful. Finkelstein et al. (199e) also noted that iron-EDTA could itself catalyze the oxidation of DMPO to a nitroxide radical. Since some workers have added iron-EDTA to their reaction mixtures (199f) their results must be reinterpreted. Given the hard won recognition of the limitations and potential pitfalls of spin trapping, it now seems certain that this technique will be increasingly useful in studies of the biochemistry of O_2^- and of OH·.

IV. Varieties of Superoxide Dismutases

A stringent selection pressure persistently applied to a varied biota is likely to result in parallel evolutionary adaptations. Oxygenation of the biosphere applied such a selection pressure to what must have been a very varied anaerobic biota. It is not surprising, therefore, that the superoxide dismutases, already discovered, fall into three classes when their prosthetic groups are considered, and into two classes when evolutionary relatedness is considered. There are superoxide dismutases based on manganese (MnSOD), iron (FeSOD), or copper plus zinc (CuZnSOD). The Mn and the Fe enzymes are related, as shown by extensive amino acid sequence homology; while the Cu–Zn enzymes represent an independently evolved group of proteins. All of these enzymes have achieved comparable catalytic efficiency in catalyzing the dismutation of O_2^-.

Respiring organisms, including bacteria (200,201), fungi (202), algae (203), higher plants (204), molluscs and arthropods (205), fish (206), birds (207), and mammals (163), all contain superoxide dismutases. The only independently living organisms thus far found free of this activity are certain oxygen-sensitive obligate anaerobes (208). Only organisms which never produce O_2^- can safely dispense with this enzyme activity. Given the ubiquity of oxygen, it is perhaps not surprising that some anaerobes do contain superoxide dismutase (209–214) since occasional exposure to oxygen must be expected. In a variety of anaerobic bacteria, obtained as clinical isolates, SOD content paralleled oxygen resistance (215). At the other extreme *Mycobacterium lepraemurium,* which can proliferate within macrophages, contain an unusually high level of superoxide dismutase, i.e., 7% of the extractable protein (216). This may provide a defense against the superoxide radical elaborated by activated neutrophils and macrophages. The multiplicity of superoxide dismutases and their distribution among living things is what would be expected of a family of enzymes evolved to provide a defense against a major aspect of oxygen toxicity.

A. CuZnSOD

The copper and zinc containing superoxide dismutases are characteristically found in the cytosols of eukaryotic cells and not in prokaryotes. One exception is known and that is the symbiotic bacterium *Photobacterium leiognathi*, which does contain a CuZnSOD (217). Since the host pony fish has evolved a special gland to house this photobacterium, the symbiosis is one of long standing and we can speculate that the bacterium obtained the CuZnSOD by gene transfer from the host fish. Copper- and zinc-containing superoxide dismutases have been isolated from many sources including bovine erythrocytes (163), chicken liver (207), swordfish liver (206), wheat germ (218), yeast (219), *Neurospora crassa* (202), and spinach (204). Great evolutionary conservatism is evident in the gross structural and functional similarity of these enzymes.

The enzyme from bovine erythrocytes has been most thoroughly studied and will be described as the prototype of this class of superoxide dismutases. It is a homodimer with a subunit weight of 16,000 daltons (220,221). Its basic structure and the arrangement at the active site have been elucidated by X-ray crystallography (222, 223). The major structural feature of the subunit is a cylinder composed of β structure. A pair of nonhelical loops project from the top and bottom of one side of this cylinder, enclosing and creating the active site. The two subunits are joined by noncovalent, predominantly hydrophobic interactions between the β barrels such that the active sites are on opposite sides of the dimer. This is an unusually stable enzyme and was unaffected by an isolation procedure which made liberal use of organic solvents (163). Moreover, it retains its catalytic activity and its native dimeric structure in 8.0 *M* urea or in 1% sodium dodecyl sulfate; although it can be denatured by heating or by 6.0 *M* guanidinium chloride (224–226). The copper is held within the active site by ligation to the nitrogens of the imidazole rings of histidines 44, 46, 61, and 118. These ligands are in a distorted square planar arrangement. The zinc is bridged to the copper by ligation to the imidazole of histidine 61 and is also ligated to histidines 67 and 78 and to aspartate 81. The ligands to the zinc are in a tetrahedral array.

B. FeSOD

Iron-containing superoxide dismutases have been found only in prokaryotes and algae. This family of enzymes is composed of subunits of 23,000 daltons and the individual enzymes are usually dimeric, although one tetrameric FeSOD has been noted in *Mycobacterium tuberculosis* (227). Iron-containing superoxide dismutases have been isolated from *Escherichia coli* (201), *Photobacterium leiognathi* (228), *Bacillus megaterium* (229), *Pseudomonas ovalis* (230), *Chromatium vinosum* (213), *Chlorobium thiosulfatophilum* (231), *Plectonema boryanum* (203,232), and *Euglena gracilis* (233). If we assume that these enzymes are composed of identical subunits, we should expect 1.0 Fe/subunit. The reported values range between 0.5–1.0, probably reflecting varying degrees of loss of Fe during the isolation procedure. The iron in the resting enzyme is Fe(III) (234).

C. MnSOD

Manganese-containing superoxide dismutases have been found in both prokaryotes and eukaryotes. In the latter case it is usually localized in the mitochondrial matrix, supporting speculations concerning the symbiotic origin of these organelles. In prokaryotes, the manganese enzyme may be the only superoxide dismutase present in the cell, or it may occur in combination with the corresponding iron enzyme. The iron and the manganese-containing enzymes are related and exhibit the same subunit weight and oligomeric structure and occur in dimeric or tetrameric forms in different organisms. Amino acid sequence analysis confirms the close evolutionary relationship between MnSOD and FeSOD (235–239). Manganese-containing superoxide dismutases have been isolated from several species of bacteria including *E. coli* (200), *M. lepraemurium* (216), *M. phlei* (240), *M. tuberculosis* (227), *R. spheroides* (241), *S. faecalis* (78), *S. mutans* (242), and *T. aquaticus* (243). Most of these bacterial enzymes are dimeric, but the *M. phlei* and *T. aquaticus* enzymes are tetrameric. MnSOD has also been isolated from the red alga *P. cruentum* (244), the yeast, *S. cerevisiae* (245), and the fungus, *P. olearius* (246,247). It has also been isolated from tissues of higher organisms including chicken liver (207), rat liver (169), and human liver (248). The chicken (207) and rat (169,178) enzymes are mitochondrial, while the human (248) enzyme is found in both mitochondrial and cytosol fractions. All of the MnSODs from higher organisms were found to be tetrameric and contained between 0.5–1.0 atoms of manganese per subunit.

V. Mechanisms of Enzymatic Superoxide Dismutation

The superoxide dismutases are metalloenzymes and all operate by a redox cycle in which the active site metal is reduced by one O_2^- then reoxidized by another O_2^-. Thus,

$$Me^n + O_2^- \leftrightarrow Me^{n-1} + O_2 \qquad (11)$$

$$Me^{n-1} + O_2^- + 2H^+ \leftrightarrow Me^n + H_2O_2 \qquad (12)$$

In the case of the CuZnSOD, the copper oscillates between the divalent and monovalent states during the catalytic cycle, while in the cases of the MnSOD and FeSOD the corresponding valence states are trivalent and divalent. Pulse radiolysis permits the very rapid introduction of relatively high concentrations of O_2^- into aqueous solutions and it has been very useful in studying the mechanisms of the superoxide dismutases. The second-order rate constant for the reaction of O_2^- with the CuZnSOD is approximately $2 \times 10^9\ M^{-1}\ sec^{-1}$ (153–156) and no saturation of the enzyme was evident. Rotilio et al. (154) estimated that the K_m, if at all measurable, must exceed 0.5 mM. The enormous rate constant suggests diffusion limitation, so that saturation by substrate would not be expected. The enzymatic rate was unaffected by pH in the range 5.5–9.5. The rates of reactions 11 and 12 are essentially equal and the expected partial reduction of the oxidized enzyme and conversely oxidation of the reduced enzyme, by pulses of O_2^-, was noted.

The possibility that proton transfer might be rate limiting in the catalytic turnover of carbonic anhydrase was considered (249) and an augmentation of rate by buffer assisted proton transfer was noted (250). The extremely high rate of the superoxide dismutase reaction and the need for protons in forming the products ($O_2^- + O_2^- + 2H^+ \leftrightarrow H_2O_2 + O_2$) suggests that proton conduction might be rate limiting here too. Yet, as already stated, the catalytic rate is unaffected by a change in $[H^+]$ of four orders of magnitude. The existence of an imidazolate-bridged copper–zinc complex at the active site of CuZnSOD led to the proposal that the bridging ligand was released from the copper during its reduction by O_2^- and then served in proton conduction during its reoxidation by a second O_2^- (251). There is some support for this proposal. Thus, a proton binding group with an apparent pK_a greater than 9.0 was generated by reduction of the enzyme (252). However, the MnSOD and FeSOD do not contain ligand-bridged bimetal complexes, yet they are equally effective as the CuZnSOD in catalyzing the dismutation of O_2^-. At this time such details of mechanism as the need for proton conduction, or outer sphere versus inner sphere electron transport between O_2^- and the active site metal, remain to be elucidated.

A. Inhibitors

The CuZnSOD (156,218,251,253) and FeSOD (232) are irreversibly inhibited by H_2O_2. This is probably without physiological significance, since the effect becomes apparent only at concentrations of H_2O_2 orders of magnitude higher than those possible in vivo and only at alkaline pH (251). The process is nevertheless interesting in that it exposes some of the properties of the active site and can be used to distinguish CuZnSOD and FeSOD, which are inactivated by H_2O_2, from MnSOD, which is not (232). H_2O_2 rapidly reduces the copper of CuZnSOD (251) and the cuprous enzyme does slowly reduce dioxygen to O_2^-. Indeed, the enzyme will catalyze the production of O_2^- from $H_2O_2 + O_2$, although the equilibrium lies far in the other direction (254). Cu(I) at the active site of the reduced enzyme can react with a second H_2O_2, in what is an analogue of the Fenton's reaction, to generate a powerful oxidant, which can attack adjacent amino acid residues and thus inactivate the enzyme. A variety of exogenous electron donors, including xanthine, urate, or formate, could prevent inactivation of the enzyme by H_2O_2, presumably by competing with the critical endogenous donor for the oxidant generated at the active site (251). The production of a powerful oxidant by the interaction of CuZnSOD with H_2O_2, followed by the discharge of that oxidant by exogenous electron donors is the basis for the weak, nonspecific peroxidase activity of this enzyme (255). Since it is the Cu(I) enzyme which reacts with H_2O_2 to generate the self-destructive oxidant and since the Cu(I) enzyme can be oxidized by dioxygen, then dioxygen should protect CuZnSOD against inactivation by H_2O_2 and it does so (251). Inactivation of CuZnSOD by H_2O_2 seems due to oxidative attack on one of the histidines in ligand field of the active site copper. Thus, inactivation was associated with changes in both EPR and optical spectra and a loss of 1 histidine residue per subunit (256). Furthermore, the loss of activity and of histidine were mutually proportional and the residue modified was shown to be either histidines 44, 46, or 61 (257).

Cyanide inhibits CuZnSOD (154) reversibly and does so by complexing to Cu(II) via its carbon atom (258). The human enzyme was 50% inhibited at pH 8.2 by 45–50 μM cyanide (259). Since cyanide does not inhibit either MnSOD or FeSOD, it is very useful in distinguishing these enzymes from the CuZnSOD in crude extracts of cells or of subcellular organelles. It was used in this way to demonstrate a cyanide-insensitive SOD in chicken liver, which was then found to be localized in the mitochondria and when purified was shown to be a MnSOD (207). Azide also inhibits superoxide dismutase in a reversible fashion and it inhibits all classes of these enzymes, albeit to different degrees. FeSOD, from a blue green alga, was inhibited 92% by 10 m*M* N_3^- while MnSOD from this organism was inhibited only 55% (232). N_3^- does bind onto the copper of CuZnSOD (260). Inhibition by azide was found to be characteristic of the type of SOD and independent of its biological source. Thus, under specific assay conditions 50% inhibition for a variety of FeSODs, MnSODs or CuZnSODs was noted at 4 m*M*, 20 m*M*, and 32 m*M* N_3^-, respectively (261). Susceptibility to N_3^- inhibition was thus useful as a means of distinguishing between these enzymes in cell extracts.

Diethyldithiocarbamate removes the copper from CuZnSOD and thus inactivates it (190). This chelating agent has been applied to intact erythrocytes and shown to decrease the SOD content of these cells, rendering them more susceptible to an oxidative stress imposed with 1,4-naphthoquinone-2-sulfonate (105). Diethyldithiocarbamate was similarly used in depleting the SOD content of cells from rabbit intestine (44). Treatment of young rats with this chelating agent decreased lung SOD and made them more susceptible to the lethality of hyperoxia (262). It is clear that inhibitors of SODs can be used to probe their physiological function and that more specific and effective inhibitors would be extremely useful experimental tools.

VI. The Physiological Role of Superoxide Dismutases

The discovery of superoxide dismutase (35, 163) led almost immediately to the proposal that O_2^- was an important agent of oxygen toxicity and that SOD was the defense against this threat. The first test of this theory involved surveying numerous microorganisms for their contents of SOD, catalase, and peroxidase, in the expectation that actively respiring cells should have more SOD than microaerophiles and that only sensitive obligate anaerobes could do without SOD. The results were in accord with these expectations (263). One apparently aberrant organism was *Lactobacillus plantarum*, which was aerotolerant yet devoid of SOD. It was then found not to consume dioxygen (263,101), so it could not produce O_2^- and did not need SOD.

Induction of the biosynthesis of an enzyme by a particular substrate is probably the surest indication that its in vivo function is to act on that substrate. Exposure to dioxygen has been seen to cause increased levels of SOD in numerous cells and tissues including *Streptococcus faecalis* and *Escherichia coli* (264–266), rat lung (267–270), *Saccharomyces cerevisiae* (271), *Photobacter leiognathi* (217), leukocytes (272), rat mammary carcinoma (273), *Propionibacterium shermanii* (274), neonatal rat lung (275), alveolar macrophages (276), and potato slices (277). Increased intracellular levels of SOD, achieved through oxygen induction, correlated with enhanced resistance of *S. faecalis* and of *E. coli* toward the lethality of hyperbaric oxygen (264–266) and toward the oxygen enhancement of the lethality of streptonigrin (265). These results are suggestive but not conclusive, since there is a definite circularity in using dioxygen to induce an enzyme such as SOD and then noting increased resistance toward hyperbaric oxygen (HPO). We chose to correlate the increased resistance toward HPO with the elevated levels of SOD, but we cannot exclude the possibility that exposure to dioxygen induced some other enzyme which was actually responsible for imparting the observed resistance to HPO. Much more convincing correlations could be made if we could manipulate the level of SOD without changing P_{O_2}. Several strategies have been successfully used to achieve this goal.

1. *E. coli* K12 was found to be exquisitely sensitive to oxygen induction of SOD. When grown under truly oxygen-free conditions this organism contained only one SOD and that was FeSOD. The immeasurably low levels of

oxygen present in deep still liquid cultures, open to the air, sufficed to cause induction of MnSOD; while exposure to increased levels of dioxygen, by agitation in air, induced catalase and peroxidase. Induction of MnSOD correlated with increased resistance toward HPO and toward the oxygen enhancement of streptonigrin lethality. Moreover, anaerobically grown cells, prevented from inducing MnSOD by the presence of puromycin, were killed and visibly damaged (electron microscopy) by exposure to air (278).

2. *E. coli* were grown in a vigorously aerated chemostat and their growth rate was increased by augmenting the rate of inflow of fresh medium. Increased growth rate caused increased respiration and a parallel increase in SOD, but not of catalase or peroxidase. Increased SOD correlated, in turn, with enhanced resistance toward the lethality of hyperoxia (279). This correlation has special significance since the SOD level was manipulated by changing the growth rate, while oxygenation was abundant and constant.

3. Paraquat (methyl viologen) caused a profound increase in the rate of biosynthesis of MnSOD in *E. coli* (280). Since paraquat is readily reduced to a radical which reacts very rapidly with dioxygen, to yield O_2^-, it appeared likely that paraquat was subverting electron flow within the cells from the normal water-producing cytochrome pathway to an O_2^--producing pathway. In accord with this supposition was an increase in cyanide-resistant respiration, imposed by paraquat (76). Cells whose content of SOD had been increased by exposure to paraquat were then found to be markedly resistant toward the lethality of hyperoxia and toward the oxygen enhancement of the lethality of streptonigrin. Paraquat can cause increased production of O_2^- in *E. coli* only if dioxygen is present to oxidize the paraquat radical and only if a source of electrons is present to reduce paraquat to its radical. Furthermore, the induction of SOD should be an adaptive response and preventing it should markedly increase the susceptibility of the cells to the toxicity of this compound. These effects have been clearly demonstrated (77). Thus, when *E. coli* can respond to paraquat by increased synthesis of MnSOD, they easily tolerate millimolar levels. In contrast, when prevented from achieving rapid synthesis of MnSOD, either because of the paucity of the medium or because of the presence of inhibitiors of protein synthesis, their growth is strongly suppressed by micromolar levels. This toxic effect is seen only in the presence of both dioxygen and a metabolizable souce of electrons. Furthermore, cells whose level of SOD was initially high, due to prior induction of SOD biosynthesis by means other than paraquat, were found to be resistant to the lethality of paraquat.

4. Paraquat is not the only compound able to increase the rate of intracellular production to O_2^-. thus pyocyanine, phenazine methosulfate, streptonigrin, juglone, menadione, plumbagin, methylene blue, and azure c were all effective. They all caused an increase in cyanide resistant respiration and an induction of MnSOD in *E. coli* (281). In the case of induction by paraquat it was shown that the observed increase in MnSOD was due to *de novo* protein synthesis, since it was prevented by inhibitors of transcription or of translation, but not by an inhibitor of DNA replication. Any increase in O_2^- production, of necessity, leads to an increase in H_2O_2 production. The redox active compounds, listed above, might therefore also be expected to cause induction of catalase and this was observed (281). If one or more components of the normal electron transport chain were autoxidizable and able to generate O_2^-, then blocking the cytochrome *c* oxidase should increase the steady-state concentration of the reduced form of such carriers and should thus lead to increased O_2^- production and to induction of SOD. This too was observed (281).

In toto, these data lead inexorably to the conclusion that O_2^- can be generated inside respiring cells and would damage and kill those cells, were it not for superoxide dismutases which, by scavenging this radical, keep its steady-state concentration vanishingly small. Increasing the rate of O_2^- production, whether by raising the P_{O_2}, increasing the growth rate, replacing glucose by nonfermentable substrates, adding redox active compounds or by adding inhibitors of cytochrome oxidase, leads to an adaptive increase in the biosynthesis of SOD. When this adaptive response is blocked, the cells succumb. The several types of superoxide dismutases must be presumed to serve the same function, i.e., the scavenging of O_2^-, although they may function in different subcellular locations. In the case of facultative prokaryotes, such as *E. coli*, the constitutive FeSOD appears to be a backup defense, maintained even during anaerobic growth, so that sudden exposure to air will not kill the cells before induction of SOD synthesis can occur. In contrast, the MnSOD, which is under stringent repression control, allows an efficient fine tuning of the net level of SOD to the net rate of production of O_2^-. This same end could be achieved by having a single type of SOD which was under a less stringent repression control. Organisms that control oxygen toxicity in this way are now being studied in our laboratory. In the fungus *Dactylium dendroides* there are two superoxide dismutases, one a MnSOD and the other a CuZnSOD. When the level of CuZnSOD is decreased, because of growth on a copper-deficient medium, the level of MnSOD increases such that net SOD remains constant (282). This is a clear indication of the ability of one SOD to substitute for another. Since there have been numerous indications that O_2^- and H_2O_2 cooperate in producing extremely potent oxidants, such as $OH\cdot$ (281a,b), we should also anticipate that very high intracellular levels of SOD might compensate for a relative lack of catalase or peroxidase and vice versa.

A. Superoxide and Oxygen-Enhanced Lethalities

Paraquat, which increases the rate of production of O_2^- within cells, is vastly more toxic in the presence than in the absence of oxygen. In effect, paraquat enhances the toxicity of oxygen and oxygen enhances the toxicity of paraquat. We have already noted several compounds that act like paraquat (281). A number of antibiotics also exhibit oxygen enhancements of toxicity; increased production of O_2^- is probably the root cause in most if not all cases. This is certainly the case with streptonigrin which has been discussed above.

Adriamycin, also called doxorubicin, is an important anticancer drug whose applicability is limited by severe cardiotoxicity. It is a polycyclic paraquinone and should be capable of cyclical reduction and reoxidation, with the production of O_2^-. That free radical production was involved in its cardiotoxicity was suggested by the cardiac lipid peroxidation it caused and by the protective effect of pretreatment with the antioxidant α-tocopherol (283). Anthracyclines such as adriamycin have been seen to cause an NADPH-dependent production of O_2^- by microsomes (284) and by the NADPH-cyt P-450 reductase (285). Adriamycin caused a sixfold increase in O_2^- production by heart submitochondrial particles (286). Finally, aerobic solutions of adriamycin have been reported to cleave DNA, but only in the presence of reductants (96).

Bleomycin is a peptide antibiotic useful in treating certain neoplastic diseases. In vitro, bleomycin will cause degradation of DNA and this action depends on both dioxygen and reductants (287). Bleomycin is a powerful chelating agent and, as used, contains adventitious iron. Fe(II) was shown to be essential for the attack upon DNA by bleomycin + O_2 and Fe(III) would not replace Fe(II) (288). SOD protects DNA against the attack by bleomycin (95), but was seen to be effective only when low concentrations of bleomycin were used (289). The reason for this was not clear but we must consider the possibility that bleomycin might inhibit SOD. Bleomycin, as the iron complex, thus seems to bind to DNA and then to undergo a cycle of reduction by any of a variety of electron donors and reoxidation by dioxygen, with the generation of O_2^-, H_2O_2 and $OH^\cdot$, which attack the DNA (290). Free radical mechanisms have been proposed for a great number of quinone anticancer drugs, all of which have been shown to augment the NADPH oxidase activity of microsomes and probably do so by an O_2^--producing redox cycle (291).

Oxygen enhancement of radiation lethality also appears to involve O_2^-. Foetal calf myoblasts in culture were protected against X-ray damage by SOD added to the medium (108) and SOD injected into mice provided some protection against the lethality of whole body irradiation (292) and also protected their bone marrow hematopoietic stem cells (293). Solutions of DNA were protected against the effects of ^{60}Co irradiation by SOD or catalase added to the medium (91). Artificial phospholipid vesicles were peroxidized by exposure to ionizing radiation and SOD prevented this damage (294). Lymphocytes in culture exhibited chromosomal abnormalities after exposure to ionizing radiation and SOD or catalase provided significant protection, while both enzymes, in concert, provided marked protection (295). Mice, given 680 rad of whole body radiation, suffered 70–86% lethality and this could be reduced to 7.4% by SOD injected 1 hr prior to and 1 hr following irradiation (296). The ability of SOD to exert a postirradiation protection suggests that irradiation sets in motion a continuing deleterious physiological process, similar to inflammation, in which O_2^- plays a key role.

Dilute suspensions of *E. coli* exhibited an oxygen enhancement of radiation lethality (OER) of 2.4-fold. This was decreased to 1.4 by SOD or by catalase and to 1.2 by SOD plus catalase. Heat inactivated SOD was without effect (103). Very similar results were obtained with *E. coli* in several laboratories (104,297), but apparently cannot be repeated routinely (298). Since these effects deal with extracellular events, it may be that strain differences in the external bacterial membrane are the cause of the differences noted in different laboratories. Indeed, OER and radiosensitivites of *E. coli* have been reported to vary from strain to strain and the diminution of OER by a singlet oxygen scavenger also showed such differences among strains (299). Bone marrow stem cells lost proliferative capacity when irradiated in vitro. They were protected when SOD was added to the medium (300).

The multiplicity of reports of protections against oxygen-dependent radiation lethality, by SOD, argues that O_2^- is often an important agent of the OER.

B. Oxygen Radicals, Phagocytosis, and Inflammation

Phagocytic leuckocytes, when activated, exhibit a dramatic increase in respiration which is due to activation of a membrane-associated NADPH oxidase. This oxidase, which is cyanide insensitive, produces O_2^- which in turn gives rise to H_2O_2. Both the O_2^- and the H_2O_2 are important components of the machinery these cells use in killing ingested microorganisms. Indeed, phagocytes from persons afflicted with chronic granulomatous disease, which do not exhibit the respiratory burst upon activation, are markedly defective in the ability to kill certain bacterial species. All of this has recently been reviewed (75).

Because the NADPH oxidase of phagocytes is a membrane-associated enzyme, a goodly proportion of the O_2^- produced during the respiratory burst is released into the suspending medium (301). Aggregations of phagocytes, such as are attracted to sites of injury by chemotaxis, will release substantial amounts of O_2^- and of H_2O_2 into the extracellular fluid and these fluids contain very little SOD, i.e., less than 1% of the concentration found within cells (88). The released O_2^- could therefore exert its effects unopposed. These effects appear to be of at least two sorts. Thus, O_2^- can kill cells and attack the polysaccharide ground substance of connective tissue. A suspension of granulocytes, when activated, suffer a self-inflicted mortality, which can be diminished by SOD added to the medium (302). Hyaluronic acid, which lends viscosity and lubricating properties to synovial fluid, is depolymerized by an enzymatically generated flux of O_2^- and SOD protects (88). Starch, pectin, and methyl cellulose are similarly degraded by O_2^- (89). Quite independently, O_2^- can act on components of plasma to generate potent chemotaxins. Thus, McCord (303) noted the production of a macromolecular chemotaxin, when normal human plasma was exposed to a flux of O_2^-, while Perez and Goldstein (304) noted that O_2^- converted arachidonic acid to a product which was chemotactic at only 3.0 ng/ml. It seems possible that the macromolecular serum component which McCord (303) studied was serum albumin with arachidonic acid bound to it.

It is clear from these results that O_2^- probably plays an important role in inflammation. Thus, activation of the first phagocytes which arrive at a site of injury or infection will result in O_2^- production, which will generate chemotaxins capable of attracting increasing hordes of phagocytes. The massed phagocytes will then release enough O_2^- and H_2O_2 to damage themselves and surrounding tissues (305). Finally, lytic enzymes released from damaged phagocytes will increase the extent of tissue damage. Given this scenario and the paucity of SOD in extracellular fluids one might expect that SOD introduced into the extracellular space might have an anti-inflammatory effect. Extensive clinical trials affirm this expectation (306). Moreover, a variety of anti-inflammatory drugs have been reported to diminish O_2^- production by activated macrophages (307,308). Activated phagocytes have been seen to damage endothelial cells and this was prevented by SOD and catalase (305). If a localized source of O_2^- could indeed initiate inflammation, then injection of an enzymatic source of O_2^- should suffice. In the rat paw edema test, xanthine oxidase plus hypoxanthine was active. Allopurinol, which inhibits xanthine oxidase, diminished this edema as did SOD and catalase (309). An appreciation of the biological production and enzymatic scavenging of O_2^- clearly adds a new dimension to our understanding of the metabolism of oxygen. We can already see practical advantages flowing from this new knowledge and we can anticipate even more.

References

1. Taube, H. (1965). J. Gen. Physiol. 49: 29–52.
2. Collman, J. P. (1968). Acc. Chem. Res. 1, 136–143.
3. Hamilton, G. A. (1969). Adv. Enzymol. 32:55–96.
4. Hamilton, G. A. (1971). Prog. Bioorg. Chem. 1:83–157.
5. Hamilton, G. A. (1974). In: Molecular Mechanisms of Oxygen Activation, Hayaishi, O., ed. New York: Academic Press, pp. 405–451.
6. Chance, B. (1952). Nature 169:215–221.
7. Antonini, E., Brunori, M., Greenwood, C., and Malmstrom, B. G. (1970). Nature 228:936–937.
8. Misra, H. P., and Fridovich, I. (1972). J. Biol. Chem. 247:188–192.
9. Ballou, D., Palmer, G., and Massey, V. (1969). Biochem. Biophys. Res. Commun. 36:898–904.
10. Misra, H. P., and Fridovich, I. (1972). J. Biol. Chem. 247:3170–3175.
11. Cohen, G., and Heikkila, R. E. (1974). J. Biol. Chem. 249:2447–2452.
12. Misra, H. P. (1974). J. Biol. Chem. 249: 2151–2155.
13. Al-Thannon, A. A., Barton, J. P., Packer, J. E., Sims, R. J., Trumbore, C. N., and Winchester, R. V. (1974). Int. J. Radiat. Phys. Chem. 6:233–250.
14. Lai, M. (1973). Radiat. Effects 22:237–242.
15. Baccanari, D. P. (1978). Arch. Biochem. Biophys. 191:351–357.
16. Fisher, D. B., and Kaufman, S. (1973). J. Biol. Chem. 248:4300–4304.
17. Heikkila, R. E., and Cohen, G. (1975). Experientia 31:169–170.
18. Nishikimi, M. (1975). Arch. Biochem. Biophys. 166:273–279.
19. Hasegawa, H., Nakanishi, N., and Akino, M. (1978). J. Biochem. 84:499–506.
20. Orme-Johnson, W. H., and Beinert, H. (1969). Biochem. Biophys. Res. Commun. 36:905–911.
21. Nilsson, R., Pick, F. M., and Bray, R. C. (1969). Biochim. Biophys. Acta 192:145–148.
22. Nakamura, S. (1970). Biochem. Biophys. Res. Commun. 41:177–183.

23. Misra, H. P., and Fridovich, I. (1971). J. Biol. Chem. 246:6886–6890.
24. Nakamura, S., and Kimura, T. (1972). J. Biol. Chem. 247:6462–6468.
25. Allen, J. F. (1975). Biochem. Biophys. Res. Commun. 66:36–43.
26. Cassell, R. H., and Fridovich, I. (1975). Biochemistry 14:1866–1868.
27. Markossian, K. A., and Nalbandyan, R. M. (1975). Biochem. Biophys. Res. Commun. 67:870–876.
28. Berman, M. C., Abrams, C. M., Ivanetich, K. M., and Kench, J. E. (1976). Biochem. J. 157:237–246.
29. Misra, H. P., and Fridovich, I. (1972). J. Biol. Chem. 247:6960–6962.
30. Brunori, M., Falcioni, G., Fioretti, E., Giardina, B., and Rotilio, G. (1975). Eur. J. Biochem. 53:99–104.
31. Winterbourn, C. C., McGrath, B. M., and Carrell, R. W. (1976). Biochem. J. 155: 493–502.
32. Lynch, R. E., Lee, G. R., and Cartwright, G. E. (1976). J. Biol. Chem. 251:1015–1019.
33. Lynch, R. E., Thomas, J. E., and Lee, G. R. (1977). Biochemistry 16:4563–4567.
34. Gotoh, T., and Shikama, K. (1976). J. Biochem. (Tokyo) 80:397–399.
35. McCord, J. M., and Fridovich, I. (1968) J. Biol. Chem. 243:5753–5760.
36. Knowles, P. F., Gibson, J. F., Pick, F. M., and Bray, R. C. (1969). Biochem. J. 111: 53–58.
37. Fridovich, I. (1970). J. Biol. Chem. 245: 4053–4057.
38. Rajagopalan, K. V., Fridovich, I., and Handler, P. (1962). J. Biol. Chem. 237: 922–928.
39. Branzoli, U., and Massey, V. (1974). J. Biol. Chem. 249:4339–4345.
40. Forman, H. J., and Kennedy, J. (1975). J. Biol. Chem. 250:4322–4326.
41. Miller, R. W. (1975). Can. J. Biochem. 53:1288–1300.
42. Hirata, F., and Hayaishi, O. (1975). J. Biol. Chem. 250:5960–5966.
43. Hirata, F., Ohnishi, T., and Hayaishi, O. (1977). J. Biol. Chem. 252:4637–4642.
44. Taniguchi, T., Hirata, F., and Hayaishi, O. (1977). J. Biol. Chem. 252:2774–2776.
45. Kido, T., Soda, K., Suzuki, T., and Asada, K. (1976). J. Biol. Chem. 251:6994–7000.
46. Kido, T., Soda, K., and Asada, K. (1978). J. Biol. Chem. 253:226–232.
47. Kido, T., Hashizume, K., Soda, K., and Asada, K. (1978). Photochem. Photobiol. 28:729–732.
48. Loschen, G., Azzi, A., Richler, C., and Flohé, L. (1974). FEBS Lett. 42:68–72.
49. Tyler, D. D. (1975). Biochim. Biophys. Acta 396:335–346.
50. Azzi, A., Montecucco, C., and Richter, C. (1975). Biochem. Biophys. Res. Commun. 65:597–603.
51. Boveris, A., and Cadenas, E. (1975). FEBS Lett. 54:311–314.
52. Boveris, A., Cadenas, E., and Stoppani, A. O. M. (1976). Biochem. J. 156:435–444.
53. Boveris, A. (1977). Adv. Exp. Med. Biol. 78:67–82.
54. Cadenas, E., Boveris, A., Ragan, C. I., and Stoppani, A. O. M. (1977). Arch. Biochem. Biophys. 180:248–257.
55. Nohl, H., and Hegner, D. (1978). Eur. J. Biochem. 82:563–567.
56. Rich. P. R., and Bonner, W. D., Jr. (1978). Arch. Biochem. Biophys. 188:206–213.
57. Boveris, A., Sanchez, R. A., and Beconi, M. T. (1978). FEBS Lett. 92:333–338.
58. Asada, K., and Kiso, K. (1973). Eur. J. Biochem. 33:253–257.
59. Epel, B. L., and Neuman, J. (1973). Biochim. Biophys. Acta 325:520–529.
60. Asada, K., Kiso, K., and Yoshikawa, K. (1974). J. Biol. Chem. 249:2175–2181.
61. Elstner, E. F., and Konze, J. R. (1974). FEBS Lett. 45:18–21.
62. Allen,. J. F., and Hall, D. O. (1974). Biochem. Biophys. Res. Commun. 58: 579–585.
63. Greenstock, C. L., and Miller, R. W. (1975). Biochim. Biophys. Acta 396:11–16.
64. Harbour, J. R., and Bolton, J. R. (1975). Biochem. Biophys. Res. Commun. 64: 803–807.
65. Schmid, R. (1975). FEBS Lett. 60:98–102.
66. Miller, R. W., and MacDowell, F. D. H. (1975). Biochim. Biophys. Acta 387:176–187.
67. Elstner, E. F., and Heupel, A. (1975). Planta 123:145–154.
68. Halliwell, B. (1975). Eur. J. Biochem. 55:355–360.
69 Elstner, E. F., Stoffer, C., and Heupel, A. (1975). Z. Naturforsch., Teil C 30:53–57.
70. Asada, K., Kanematsu, S., Takahashi, M., and Kona, Y. (1976). Adv. Exp. Med. Biol. 74:551–564.

71. Kono, Y., Takahashi, M., and Asada, K. (1976). Arch. Biochem. Biophys. 174: 454–462.
72. Elstner, E. F., Wildner, G. F., and Heupel, A. (1976). Arch. Biochem. Biophys. 173:623–630.
73. Asada, K., Takahashi, M., Tanaka, K., and Nakano, Y. (1977). In: Biochemical and Medical Aspects of Active Oxygen, Hayaishi, O., and Asada, K., eds. Tokyo: University of Tokyo Press, pp. 45–63.
74. Halliwell, B. (1978). Mol. Biol. 33:1–54.
75. Babior, B. M. (1978). New Engl.J. Med. 298:659–668, 721–725.
76. Hassan, H. M., and Fridovich, I. (1977). J. Biol. Chem. 252:7667–7672.
77. Hassan, H. M., and Fridovich, I. (1978). J. Biol. Chem. 253:8143–8148.
78. Britton, L., Malinowski, D. P., and Fridovich, I. (1978). J. Bacteriol. 134: 229–236.
79. Goda, K., Chu, T.-W., Kimura, T., and Schaap, A. P. (1973). Biochem. Biophys. Res. Commun. 52:1300–1306.
80. Petkau, A., and Chelack, W. S. (1974). Fed. Proc. 33:1505.
81. Kellogg, E. W., III, and Fridovich, I. (1975). J. Biol. Chem. 250:8812–8817.
82. Takahama, U., and Nishimura, M. (1976). Plant Cell Physiol. 17:111–118.
83. Goldstein, I. M., and Weissman, G. (1977). Biochem. Biophys. Res. Commun. 75:604–609.
84. Kellogg, E. W., III, and Fridovich, I. (1977). J. Biol. Chem. 252:6721–6728.
85. Gutteridge, J. M. C. (1977). Biochem. Biophys. Res. Commun. 77:379–386.
86. Svingen, B. A., O'Neal, F. O., and Aust, S. D. (1978). Photochem. Photobiol. 28:803–809.
87. Thomas, M. J., Mehl, K. S., and Pryor, W. A. (1978). Biochem. Biophys. Res. Commun. 83:927–932.
88. McCord, J. M. (1974). Science 185:529–531.
89. Kon, S., and Schwimmer, S. (1976). Fed. Proc. 35:1553.
90. Wong, K., Morgan, A. R., and Paranchych, W. (1974). Can. J. Biochem. 52:950–958.
91. Van Hemmen, J. J., and Meuling, W. J. A. (1975). Biochim. Biophys. Acta 402:133–141.
92. Morgan, A. R., Cone, R. L., and Elgert, T. M. (1976). Nucleic Acids Res. 3:1139–1149.
93. Lown. J. W. Begleiter. A., Johnson, D., and Morgan, A. R. (1976). Can. J. Biochem. 54:110–119.
94. Cone, R., Hasan, S. K., Lown, J. W., and Morgan, A. R. (1976). Can. J. Biochem. 54:219–223.
95. Lown, J. W., and Sim, S. K. (1977). Biochem. Biophys. Res. Commun. 77: 1150–1157.
96. Lown, J. W., Sim, S. K., Majundar, K. C., and Chang, R. Y. (1977). Biochem. Biophys. Res. Commun. 76:705–710.
97. Lown, J. W., and Weir, G. (1978). Can. J. Biochem. 56:296–304.
98. Lavelle, F., Michelson, A. M., and Dimitrejevic, L. (1973). Biochem. Biophys. Res. Commun. 55:350–357.
99. Lin, W. S., Armstrong, D. A., and Lal, M. (1978). Int. J. Radiat. Biol. 33:231–243.
100. Armstrong, D. A., Buchanan, J. D. (1978). Photochem. Photobiol. 28:743–755.
101. Gregory, E. M., and Fridovich, I. (1974). J. Bacteriol. 117:166–169.
102. Babior, B. M., Curnutte, J. T., and Kipnes, R. S. (1975). J. Lab. Clin. Med. 85:235–244.
103. Misra, H. P., and Fridovich, I. (1976). Arch. Biochem. Biophys. 176:577–581.
104. Oberley, L. W., Lindgren, A. L., Baker, S. A., and Stevens, R. H. (1976). Radiat. 68:320–328.
105. Goldberg, B., and Stern, A. (1976). J. Biol. Chem. 251:6468–6470.
106. Bartosz, G., Fried, R., Grzelinska, E., and Leyko, W. (1977). Eur. J. Biochem. 73: 261–264.
107a. Stone, D., Lin, P. S., and Kwock, L. (1978). Int. J. Radiat. Biol. 33:393–396.
107b. Lynch, R. E., and Fridovich, I. (1978). J. Biol. Chem. 253:1838–1845.
108. Michelson, A. M., and Buckingham, M. E. (1974). Biochem. Biophys. Res. Commun. 58:1079–1086.
109. Beauchamp, C., and Fridovich, I. (1970). J. Biol. Chem. 245:4641–4646.
110. Haber, F., and Weiss, J. (1934). Proc. Roy. Soc. London A147:332–351.
111. Lieberman, M., Kunishi, A. T., Mapson, L. W., and Wardale, B. A. (1965). Biochem. J. 97:449–459.
112. Bors, W., Lengfelder, E., Saran, M., Fuchs, C., and Michel, C. (1976). Biochem. Biophys. Res. Commun. 70: 81–87.
113. Pryor, W. A., and Tang, R. H. (1978).

Biochem. Biophys. Res. Commun. 81: 498–503.
114. Konze, J. R., and Elstner, E. F. (1976). FEBS Lett. 66:8–11.
115. Weiss, S. J. (1976). Fed. Proc. 35:1396.
116. Tauber, A. I., and Babior, B. M. (1977). J. Clin. Invest. 60:374–379.
117. Rosen, H., and Klebanoff, S. J. (1978). Clin. Res. 26:404A.
118. Goscin, S. A., and Fridovich, I. (1972). Arch. Biochem. Biophys. 153:778–783.
119. Hodgson, E. K., and Fridovich, I. (1976). Arch. Biochem. Biophys. 172:202–205.
120. Mieyal, J. J., Ackerman, R. S., Blumer, B. L., and Freeman, L. S. (1976). J. Biol. Chem. 251:3436–3441.
121. Fong, K.-L., McCay, P. B., Poyer, J. L., Misra, H. P., and Keele, B. B. (1976). Chem.-Biol. Interact. 15:77–89.
122. McCord, J. M., and Salin, M. L. (1976). Miami Winter Symp. 12:289–303.
123. Halliwell, B. (1977). Biochem. J. 163: 441–448.
124. Van Hemmen, J. J., and Meuling, W. J. A. (1977). Arch. Biochem. Biophys. 182: 743–748.
125. Heikkila, R. E., and Cabbat, F. S. (1977). Res. Commun. Chem. Pathol. Pharmacol. 17:649–662.
126. Heikkila, R. E., and Cabbat, F. S. (1978). Photochem. Photobiol. 28:677–680.
127. Buettner, G. R., Oberley, L. W., and Chan-Leuthauser, S. W. H. (1978). Photochem. Photobiol. 28:693–695.
128. Halliwell, B. (1978). FEBS Lett. 96:238–242.
129. Weiss, S. J., Rustagi, P. K., and Lo Buglio, A. F. (1978). J. Exp. Med. 147:316–323.
130. Klebanoff, S. J., and Rosen, H. (1978). J. Exp. Med. 148:490–506.
131. Miles, P. R., Castranova, V., and Lee, P. (1978). Am. J. Physiol. 235:C103–108.
132. Halliwell, B. (1978). FEBS Lett. 92:321–326.
133. McCord, J. M., and Day, E. D., Jr. (1978). FEBS Lett. 86:139–142.
134. Ferradini, C., and Seide, C. (1969). Int. J. Radiat. Phys. Chem. 1:219–228.
135. McClune, G. J., and Fee, J. A. (1976). FEBS Lett. 67:294–298.
136. Halliwell, B. (1976). FEBS Lett. 72:8–10.
137. Rigo, A., Stevenato, R., Finazzi-Agro, A., and Rotilio, G. (1977). FEBS Lett. 80: 130–132.
138. Gibian, M. J., and Ungermann, T. (1979). J. Am. Chem. Soc. 101:1291–1293.
139. Czapski, G., and Ilan, Y. A. (1978). Photochem. Photobiol. 28:651–653.
140. Weinstein, J., and Bielski, B. H. J. (1979). J. Am. Chem. Soc. 101:58–62.
141. Fenton, H. J. (1894). J. Chem. Soc. 65: 899–910.
142. Walling, C. (1975). Acct. Chem. Res. 8:125–131.
143. Walling, C., Partch, R. E., and Weil, T. (1975). Proc. Nat. Acad. Sci. U.S.A. 72: 140–142.
144. McCord, J. M., and Day, E. D., Jr. (1978) Personal communication.
145. Baxendale, J. H., and Khan, A. A. (1969). Int. J. Radiat. Res. 1:11–24.
146. Kraljic, E., and Trumbore, C. N. (1965). J. Am. Chem. Soc. 87:2547–2550.
147. Bueding, F., and Jolliffe, N. (1946). J. Pharmacol. Exp. Ther. 88:300–312.
148. Taylor, J. D., Paul, H. E., and Paul, M. F. (1951). J. Biol. Chem. 191:223–231.
149. Murray, J. N., and Chaykin, S. (1966). J. Biol. Chem. 241:3468–3473.
150. Stöhrer, G., and Brown, G. B. (1969). J. Biol. Chem. 244:2498–2502.
151. Bielski, B. H. J. (1978). Photochem. Photobiol. 28:645–649.
152. Koppenol, W. H. (1978). Personal communication.
153. Klug, D., Rabani, J., and Fridovich, I. (1972). J. Biol. Chem. 247:4839–4842.
154. Rotilio, G., Bray, R. C., and Fielden, E. M. (1972). Biochim. Biophys. Acta 268: 605–609.
155. Klug, D., Fridovich, I., and Rabani, J. (1973). J. Am. Chem. Soc. 95:2786–2790.
156. Fielden, E. M., Roberts, P. B., Bray, R. C., and Rotilio, G. (1973). Biochem. Soc. Trans. 1:52–53.
157. Pick, M., Rabani, J., Yost, F., and Fridovich, I., (1974). J. Am. Chem. Soc. 96:7329–7333.
158. Fielden, E. M., Roberts, P. B., Bray, R. C., Lowe, D. J., Mautner, G. N., Rotilio, G., and Calabrese, L. (1974). Biochem. J. 139:49–60.
159. Lavelle, F., McAdam. M. E., Fielden, E. M., Roberts, P. B., Puget, K., and Michelson, A. M. (1977). Biochem. J. 161:3–12.
160. McAdam, M. E., Fox, R. A., Lavelle, F.,

and Fielden, E. M. (1977). Biochem. J. 165:71–79.
161. McAdam, M. E., Lavelle, F., Fox, R. A., and Fielden, E. M. (1977). Biochem. J. 165:81–87.
162. Marklund, S. (1976). J. Biol. Chem. 251: 7504–7507.
163. McCord, J. M., and Fridovich, I. (1969). J. Biol. Chem. 244:6049–6055.
164. Koppenol, W. H., Van Buuren, K. J. H., Butler, J., and Braams, R. (1976). Biochim. Biophys. Acta 449:157–168.
165. Land, E. J., and Swallow, A. J. (1971). Arch. Biochem. Biophys. 145:365–372.
166. Simic, M. G., Taube, T.A., Tacci, J., and Hurwitz, P. A. (1975). Biochem. Biophys. Res. Commun. 62:161–167.
167. Forman, H. J., and Fridovich, I. (1973). Arch. Biochem. Biophys. 158:396–400.
168. Salin, M., and McCord, J. M. (1974). J. Clin. Invest. 54:1005–1009.
169. Salin, M., Day, E. D., and Crapo, J. D. (1978). Arch. Biochem. Biophys. 187: 223–228.
170. Buchanan, A. G., and Lees, H. (1976). J. Microbiol. 22:1643–1646.
171. Rabani, J., Mulac, W. A., and Matheson, M. S. (1965). J. Phys. Chem. 69:53–70.
172. Czapski, G. H., and Bielski, B. H. J. (1963). J. Phys. Chem. 67:2180–2184.
173. Asada, K., and Kanematsu, S. (1976). Agr. Biol. Chem. 40:1891–1892.
174. Bielski, B. H. J., and Richter, H. W. (1977). J. Am. Chem. Soc. 99:3019–3023.
175. Beauchamp, C., and Fridovich, I. (1971). Anal. Biochem. 44:276–287.
176. Auclair, C., Torres, M., and Hakim, J. (1978). FEBS Lett. 89:26–28.
177. Butler, W. L., and Norris, K. H. (1960). Arch. Biochem. Biophys. 87:31–40.
178. Tyler, D. D. (1975). Biochem. J. 147: 493–504.
179. Elstner, E. F., and Heupel, A. (1976). Anal. Biochem. 70:616–620.
180. Michelson, A. M. (1973). Biochemistry 55:465–479.
181. Hodgson, E. K., and Fridovich, I. (1973). Photochem. Photobiol. 18:451–455.
182. Montalbini, P., Koch, F., Burba, M., and Elstner, E. F. (1978). Physiol. Plant Pathol. 12:211–223.
183. Nishikimi, N., Rao, N. A., and Yagi, K. (1972). Biochem. Biophys. Res. Commun. 46:849–853.
184. McCord, J. M., and Fridovich, I. (1970). J. Biol. Chem. 245:1374–1377.
185. Misra, H. P., and Fridovich, I. (1976). Biochemistry 15:681–687.
186. Misra, H. P., and Fridovich, I. (1977). Arch. Biochem. Biophys. 181:308–312.
187. Misra, H. P., and Fridovich, I. (1977). Arch. Biochem. Biophys. 183:511–515.
188. Sun, M., and Zigman, S. (1978). Anal. Biochem. 90:81–89.
189. Marklund, S., and Marklund, G. (1974). Eur. J. Biochem. 47:469–474.
190. Heikkila, R. E., Cabbat, F. S., and Cohen, G. (1976). J. Biol. Chem. 251:2182–2185.
191. Heikkila, R. E., and Cabbat, F. S. (1976). Anal. Biochem. 75:356–362.
192. Kono, Y. (1978). Arch. Biochem. Biophys. 186:189–195.
193. Forman, H. J., and Fridovich, I. (1972). Science 175:339.
194. Rigo, A., Viglino, P., and Rotilio, G. (1975). Anal. Biochem. 68:1–8.
195. Rigo, A., and Rotilio, G. (1977). Anal. Biochem. 81:157–166.
196. Marshall, M. J., and Worsfold, M. (1978). Anal. Biochem. 86:561–563.
197. Kelly, K., Barefoot, C., Seton, A., and Petkau, A. (1978). Arch. Biochem. Biophys. 190:531–538.
198. Del Villano, B. C., and Tischfield, J. A. (1978). Fed. Proc. 37:1749.
199. Del Villano, B. C., and Tischfield (1979). J. Immunol. Meth. 29:253–262.
199a. Janzen, E. G. (1971) Acct. Chem. Res. 4:31–40.
199b. Saprin, A. N., and Piette, L. H. (1977). Arch. Biochem.Biophys. 180:480–492.
199c. Rauckman, E. J., Rosen, G. M., and Kitchell, B. B. (1979). Mol. Pharmacol. 15:131–137.
199d. Buettner, G. R., and Oberley, L. M. (1978). Biochem. Biophys. Res. Commun. 83:69–74.
199e. Finkelstein, E., Rosen, G. M., Rauckman, E. J., and Paxton, J. (1979). Biochem. Pharmacol. 16:676–685.
199f. Lai, C. -S., Grover, T. A., and Piette, L. H. (1979). Arch. Biochem. Biophys. 193: 373–398.
200. Keele, B. B., Jr., McCord, J. M., and Fridovich, I. (1970). J. Biol. Chem. 245:6176–6181.
201. Yost, F. J., Jr., and Fridovich, I. (1973). J. Biol. Chem. 248:4905–4908.

202. Misra, H. P., and Fridovich, I. (1972). J. Biol. Chem. 247:3410–3414.
203. Misra, H. P., and Keele, B. B., Jr. (1975). Biochim. Biophys. Acta 379:418–425.
204. Asada, K., Urano, M., and Takahashi, M. (1973). Eur. J. Biochem. 36:257–266.
205. Tegelstrom, H. (1975). Hereditas 81:185–198.
206. Bannister, J. V., Anastasi, A., and Bannister, W. H. (1976). Comp. Biochem. Physiol. 53B:235–238.
207. Weisiger, R. A., and Fridovich, I. (1973). J. Biol. Chem. 248:3582–3592.
208. McCord, J. M., Keele, B. B., Jr., and Fridovich, I. (1971). Proc. Nat. Acad. Sci. U.S.A. 68:1024–1027.
209. Lindmark, D. G., and Muller, M. (1974). J. Biol. Chem. 249:4634–4637.
210. Hewitt, J., and Morris, J. G. (1975). FEBS Lett. 50:315–318.
211. Gregory, E. M., Kowalski, J. B., and Holdeman, L. V. (1977). J. Bacteriol. 129: 298–302.
212. Gregory, E. M., Moore, W. E., and Holdeman, L. V. (1978). Appl. Environ. Microbiol. 35:988–991.
213. Kanematsu, S., and Asada, K. (1978). Arch. Biochem. Biophys. 185:473–482.
214. Wimpenny, J. W. T., and Samah, O. A. (1978). J. Gen. Microbiol. 108:329–332.
215. Tally, F. P., Goldin, B. R., Jacobus, N. V., and Gorbach, S. L. (1977). Infect. Immunol. 16:20–25.
216. Ichihara, K., Kusunose, E., Kusunose, M., and Mori, T. (1977). J. Biochem. (Tokyo) 81:1427–1433.
217. Puget, K., and Michelson, A. M. (1974). Biochem. Biophys. Res. Commun. 58: 830–838.
218. Beauchamp, C. O., and Fridovich, I. (1973). Biochim. Biophys. Acta 317:50–64.
219. Goscin, S. A., and Fridovich, I. (1972). Biochim. Biophys. Acta 289:276–283.
220. Evans, H. J., Steinman, H. M., and Hill, R. L. (1974). J. Biol. Chem. 249:7315–7325.
221. Steinman, H. M., Naik, V. R., Abernethy, J. L., and Hill, R. L. (1974). J. Biol. Chem. 249:7326–7338.
222. Richardson, J. S., Thomas, K. A., Rubin, B. H., and Richardson, D. C. (1975). Proc. Nat. Acad. Sci. U.S.A. 72:1349–1353.
223. Richardson, J. S., Thomas, K. A., and Richardson, D. C. (1975). Biochem. Biophys. Res. Commun. 63:986–992.
224. Forman, H. J., and Fridovich, I. (1973). J. Biol. Chem. 248:2645–2649.
225. Abernethy, J. L., Steinman, H. M., and Hill, R. L. (1974). J. Biol. Chem. 249: 7339–7347.
226. Malinowski, D. P., and Fridovich, I. (1979). Biochemistry. 18:5055–5060.
227. Kusunose, E., Ichihara, K., Noda, Y., and Kusunose, M. (1976). J. Biochem. (Tokyo) 80:1343–1352.
228. Puget, K., and Michelson, A. M. (1974). Biochimie 56:1255–1267.
229. Anastasi, A., Bannister, J. V., and Bannister, W. H. (1976). Int. J. Biochem. 7:541–546.
230. Yamakura, F. (1976). Biochim. Biophys. Acta 422:280–294.
231. Kanematsu, S., and Asada, K. (1978). FEBS Lett. 91:94–98.
232. Asada, K., Yoshikawa, K., Takahashi, M., Maeda, Y., and Enmanji, K. (1975). J. Biol. Chem. 250:2801–2807.
233. Kanematsu, S., and Asada, K. (1979). Arch. Biochem. Biophys. 195:535–545.
234. Villafranca, J. J. (1976). FEBS Lett. 62: 230–232.
235. Steinman, H. M., and Hill, R. L. (1973). Proc. Nat.Acad. Sci. U.S.A. 70:3725–3729.
236. Bridgen, J., Harris, J. I., and Northrop, F. (1975). FEBS Lett. 49:392–395.
237. Harris, J. I., and Steinman, H. M. (1977). In: Superoxide and Superoxide Dismutases, Michelson, A. M., McCord, J. M., and Fridovich, I., eds. London: Academic Press, pp. 225–230.
238. Bruschi, M., Hatchikian, E. D., Bonicel, J., Bovier-Lapierre, G., and Couchoud, P. (1977). FEBS Lett. 76:121–124.
239. Steinman, H. (1978). J. Biol. Chem. 253: 8708–8720.
240. Chikata, Y., Kusunose, E., Ichihara, K., and Kusunose, M. (1975). Osaka City Med. J. 21:127–136.
241. Lumsden, J., Cammack, R., and Hall, D. O. (1976). Biochim. Biophys. Acta 438: 380–392.
242. Vance, P. G., Keele, B. B., Jr., and Rajagopalan, K. V. (1972). J. Biol. Chem. 247:4782–4786.
243. Sato, S., and Harris, J. I. (1977). Eur. J. Biochem. 73:373–381.
244. Misra, H. P., and Fridovich, I. (1977). J. Biol. Chem. 252:6421–6423.
245. Ravindranath, S. D., and Fridovich, I.

(1975). J. Biol. Chem. 250:6107–6112.
246. Lavelle, F., Durosay, P., and Michelson, A. M. (1974). Biochimie 56:451–458.
247. Lavelle, F., and Michelson, A. M. (1975). Biochimie 57:375–381.
248. McCord, J. M., Boyle, J. A., Day, E. D., Rizzolo, L. J., and Salin, M. (1977). In: Superoxide and Superoxide Dismutases, Michelson, A. M., McCord, M. J., and Fridovich, I., eds. London: Academic Press, pp. 129–138.
249. Lindskog, S., and Coleman, J. E. (1973). Proc. Nat. Acad. Sci. U.S.A. 70:2505–2508.
250. Silverman, D. N., and Tu, C. K. (1975). J. Am. Chem.Soc. 97:2263–2269.
251. Hodgson, E. K., and Fridovich, I. (1975). Biochemistry 14:5294–5299.
252. Fee, J. A., and DiCorleto, P. E. (1973). Biochemistry 12:4893–4899.
253. Symonyan, M. A., and Nalbandyan, R. M. (1972). FEBS Lett. 28:22–24.
254. Hodgson, E. K., and Fridovich, I. (1973). Biochem. Biophys. Res. Commun. 54: 270–274.
255. Hodgson, E. K., and Fridovich, I. (1975). Biochemistry 14:5299–5303.
256. Bray, R. C., Cockle, S. H., Fielden, E. M., Roberts, P. B., Rotilio, G., and Calabrese, L. (1974). Biochem. J. 139:43–48.
257. Malinowski, D. P., and Fridovich, I. (1977) 174th Am. Chem. Soc. Nat. Meet. Biol. Chem. Abstr. 20.
258. Haffner, P. H., and Coleman, J. E. (1973). J. Biol. Chem. 248:6626–6629.
259. Marklund, S., Beckman, G., and Stigbrand, T. (1976). Eur. J. Biochem. 65:415–422.
260. Beem, K. M., Richardson, D. C., and Rajagopalan, K. V. (1977). Biochemistry 16:1930–1936.
261. Misra, H. P., and Fridovich, I. (1978). Arch. Biochem. Biophys. 189:317–322.
262. Frank, L., Wood, D. L., and Roberts, R. J. (1978). Biochem. Pharmacol. 27:251–254.
263. McCord, J. M., Keele, B. B., Jr., and Fridovich, I. (1971). Proc. Nat. Acad. Sci. U.S.A. 68:1024–1027.
264. Gregory, E. M., and Fridovich, I. (1973) J. Bacteriol. 114:543–548.
265. Gregory, E. M., and Fridovich, I. (1973). J. Bacteriol. 114:1193–1197.
266. Gregory, E. M., Yost, F. J., and Fridovich, I. (1973). J. Bacteriol. 115:987–991.
267. Crapo, J. D., and Tierney, D. (1973). Clin. Res. 21:222.
268. Crapo, J. D., and Tierney, D. L. (1974). Am. J. Physiol. 226:1401–1407.
269. Kimball, R. E., Reddy, K., Pierce, T. H., Schwartz, L. W., Mustafa, M. G., and Cross, C. E. (1976). Am. J. Physiol. 230:1425–1431.
270. Autor, A. P., Frank, L., and Roberts, R. J. (1976). Pediatr. Res. 10:154–158.
271. Gregory, E. M., Goscin, S. A., and Fridovich, I. (1974). J. Bacteriol. 117: 456–460.
272. Rister, M., and Baehner, R. L. (1975). Blood 46:1016.
273. Petkau, A., Monasterski, L. G., Kelly, K., and Friesen, H. G. (1977). Res. Commun. Chem. Pathol. Pharmacol. 17:125–132.
274. Pritchard, G. G., Wimpenny, J. W. T., Morris, H. A., Lewis, M. W. A., and Hughes, D. E. (1977). J. Gen. Microbiol. 102:223–233.
275. Autor, A. P., and Stevens, J. B. (1978). Photochem. Photobiol. 28:775–780.
276. Nerurkar, L. S., Zeligs, B. J., and Bellanti, J. A. (1978). Photochem. Photobiol. 28: 781–786.
277. Boveris, A., Sanchez, R. A., and Beconi, M. T. (1978). FEBS Lett. 92:333–338.
278. Hassan, H. M., and Fridovich, I. (1977). J. Bacteriol. 129:1574–1583.
279. Hassan, H. M., and Fridovich, I. (1977). J. Bacteriol. 130:805–811.
280. Hassan, H. M., and Fridovich, I. (1977). J. Bacteriol. 132:505–510.
281. Hassan, H. M., and Fridovich, I. (1979). Arch. Biochem. Biophys. 196:385–395.
281a. Cohen, G., and Cederbaum, A. I. (1979). Science 204:66–68.
281b. Repine, J. E., Eaton, J. W., Anders, M. W., Hoidal, J. R., and Fox, R. B. (1979). J. Clin. Invest. 64:1642–1651.
282. Schatzman, A. R., and Kosman, D. (1978). Biochim. Biophys. Acta 544:163–179.
283. Myers, C. E. M., McQuire, W. P., Liss, R. H., Infrim, I., Grotzinger, K., and Young, R. C. (1977). Science 197:165–167.
284. Handa, K., and Sato, S. (1975). Gann 66:43–47.
285. Goodman, J., and Hochstein, P. (1977). Biochem. Biophys. Res. Commun. 77: 797–803.
286. Thayer, W. S. (1977). Chem. Biol. Interact. 19:265–278.
287. Oshida, R., and Takehashi, T. (1975).

Biochem. Biophys. Res. Commun. 66: 1432–1438.

288. Sausville, E. A., Peisach, J., and Horwitz, S. (1976). Biochem. Biophys. Res. Commun. 73:814–822.
289. Sausville, E. A., Stein, R. W., Peisach, J., and Horwitz, S. B. (1978). Biochemistry 17:2746–2754.
290. Sausville, E. A., Peisach, J., and Horwitz, S. B. (1978). Biochemistry 17:2740–2745.
291. Bachur, N. R., Gordon, S. L., and Gee, M. V. (1978). Cancer Res. 38:1745–1750.
292. Petkau, A., Chelack, W. S., Pleskach, S. D., Meeker, B. E., and Brady, C. M. (1975). Biochem. Biophys. Res. Commun. 65:886–893.
293. Petkau, A., Kelly, K., Chelack, W. S., Pleskach, S. D., Barefoot, C., and Meeker, B. E. (1975). Biochem. Biophys. Res. Commun. 67:1167–1174.
294. Petkau, A., and Chelack, W. S. (1976). Biochim. Biophys. Acta 433:445–456.
295. Nordenson, I., Beckman, G., and Beckman, L. (1976). Hereditas 82:125–126.
296. Petkau, A., Chelack, W. S., and Pleskach, S. D. (1976). Int. J. Radiat. Biol. 29:297–299.
297. Niwa, T., Yamaguchi, H., and Yano, K. (1978). Agr. Biol. Chem. 42:689–695.
298. Samuni, A., and Czapski, G. (1978). Radiat. Res. 76:624–632.
299. Anderson, R. F., and Patel, K. B. (1978). Photochem. Photobiol. 28:881–885.
300. Petkau, A. (1978). Photochem. Photobiol. 28:765–774.
301. Drath, D. B., and Karnovsky, M. L. (1975). J. Exp. Med. 141:257–262.
302. Salin, M. L., and McCord, J. M. (1975). J. Clin. Invest. 56:1319–1323.
303. Petrone, W. F., English, D. K., Wong, K., and McCord, J. M. (1980). Proc. Nat. Acad. Sci. U.S.A. 77:1159–1163.
304. Perez, H. D., and Goldstein, I. M. (1979). Fed. Proc. 38:1170 Abstr.
305. Sacks, T., Moldow, C. F., Craddock, P. R., Bowers, T. K., and Jacob, H. S. (1978). J. Clin. Invest. 61:1161–1167.
306. Menander-Huber, K. B., and Huber, W. (1977). In: Superoxide and Superoxide Dismutases. Michelson, A. M., McCord, J. M., and Fridovich, I., eds. London: Academic Press, pp. 536–549.
307. Ōyanagui, Y. (1976). Biochem. Pharmacol. 25:1473–1481.
308. Nelson, D. H., Meikle, A. W., Benowitz, B., and Ruhmann-Wennhold, A. (1978). Clin. Res. 26:550A.
309. Ohmori, H., Komariya, K., Azuma, A., Hashimoto, Y., and Korozumi (1978). Biochem. Pharmacol. 27:1397–1400.

14

Pulmonary Oxygen Toxicity

GARY L. HUBER AND DAVID B. DRATH

Oxygen is potentially toxic to all living cells. The nature and the degree of this toxicity are dependent on a number of factors, including the oxygen tension delivered, the duration of exposure, specific tissue variables (blood supply, metabolic rate, immunological defenses), and the susceptibility of the exposed cells to oxygen poisoning (129,249). Because at atmospheric pressures oxygen is a gas and its incorporation into the host is usually by inhalation, the lung, as the portal of entry, is exposed to higher relative oxygen tensions than any other body organ. Thus, at exposure tensions of 1 atm or less, the lung is usually the first vital organ to respond adversely to an increased delivery of oxygen; at exposures of greater than 1 atm, central nervous system damage occurs concurrently with (or may even precede) lung damage (30,64,73,76,83,117,307,342). The ultimate end-point of toxicity, therefore, most commonly is a pulmonary-related death with oxygen administration of 1 atm or less, and a mortality due to combined pulmonary and central nervous system failure caused at higher exposure tensions. Because supplemental oxygen is seldom administered to man at greater than 1 atm pressure, the response of the lungs usually limits the safe use of higher than normal oxygen exposures.

I. Terminology and Relative Concepts

A. Oxygen Concentration and Oxygen Tension

The literature on pulmonary oxygen toxicity is sometimes complicated by misuse of the concepts of oxygen concentration and oxygen tension. In addition, the effects of oxygen on the lung are also influenced by other sometimes poorly appreciated factors, such as humidity, temperature, and related variables. A unified understanding of these terms and concepts is essential to reviewing the relevant published contributions on pulmonary oxygen toxicity; therefore, a brief preliminary clarification of these and other terms may be helpful.

Atmospheric gases consist of specific molecules of individual weight, determined by the pull of gravity with a finite force. At sea level, this downward force is termed *atmospheric pressure*, and will support a column of mercury 760 mm high; atmospheric pressure is the sum of individual pressures of the component gases, generally termed the *partial pressures* (*P*). Gas *tension* and gas *partial pressure* are basically equivalent terms. One atmosphere of pressure at sea level is represented by the term *atm*, or by 760 *torr* on the Toricelli scale. Gas tensions or partial pressures, as well as atm (or torr), are expressed usually in units of millimeters of mercury (mm Hg).

B. Gas Laws and Oxygen Behavior

Exposure of the lung to varying oxygen tensions can be quantified by specific gas relationships. The molecules of gas in the atmosphere, including oxygen, are constantly in motion and randomly collide, and as such their behavior is governed by several predictable relationships or gas laws, described in detail elsewhere (187). One such relationship (Avogadro's law) states that equivalent volumes of different gases at the same temperature and pressure contain the same number of molecules. Under ideal circumstances, Avogadro's number (6.02×10^{23}) is the number of molecules in a mole of gas. As a consequence of this, if two or more gases interact to form a gaseous product (such as oxygen with nitrogen and other gases of atmospheric air), the volume of the gaseous product and the reacting gases are related to each other as simple proportions (Gay-Lussac's law of combining volumes), which can be expressed in small whole numbers. Furthermore, the total pressure exerted by a mixture of gases is equal to the summation of the separate partial pressures exerted if the individual gases occupied alone the given volume (Dalton's law of partial pressures). As an example, the barometric or atmospheric pressure (P_B) is equal to the sum of the combined atmospheric gas pressures of nitrogen (P_{N_2}), oxygen (P_{O_2}), water (P_{H_2O}), carbon dioxide (P_{CO_2}), and so on. This can be expressed by the equation,

$$P_B = P_{N_2} + P_{O_2} + P_{H_2O} + P_{CO_2} + \ldots \quad \textbf{(1)}$$

The gas to which the lung is exposed is not dry, as it contains certain amounts of water in the gaseous state (water vapor). This water exerts its own gas tension, and has a finite partial pressure (P_{H_2O}). Thus, it reduces proportionately the effective pressure or tension of the other gases present. The amount of water in any gas as a whole is a function, primarily of temperature. The higher the temperature of a gas, for example, the more water vapor it can contain. The relationship between temperature and water vapor tension is termed the *absolute humidity* and is defined as the mass of water vapor (mm Hg or torr) present in unit volume of the gas. The ratio of the amount of water vapor actually present in the gas relative to the quantity of water vapor that would saturate the gas at an existing temperature is termed the *relative humidity* (%). As a practical application, at normal body temperature (37°C) the wet linings of the upper respiratory tract and the airways normally would humidify any inspired gas to a constant water vapor tension of 47 mm Hg.

Applying these principles, we can express the partial pressure (P) of any gas (mm Hg) in the atmosphere as directly proportionate to its fractional concentration (F), corrected for any water vapor (P_{H_2O}) potentially present. Thus, the partial pressure of a gas, such as oxygen, can be determined by multiplying the total gaseous pressure by the fractional concentration of the gas in the mixture. For example, the fractional concentration of oxygen in our atmosphere is 20.95% of the total gas content present. If the barometric pressure of the total gas at sea level is 760 mm Hg, the partial pressure of normally inspired oxygen, then, can be calculated, corrected for the expected water vapor present (usually 47 mm Hg), as follows:

$$P_{O_2} = F_{O_2} \times (P_B - P_{H_2O}) \quad \textbf{(2a)}$$

$$P_{O_2} = 0.2095 \times (760\text{ mm Hg} - 47\text{ mm Hg}) \quad \textbf{(2b)}$$

$$P_{O_2} = 149\text{ mm Hg} \quad \textbf{(2c)}$$

The fractional content (% by volume) and the atmospheric partial pressures of some of the other important gases, as well as oxygen, in our environment are summarized in Table 1.

Table 1. Partitional Composition of Atmospheric Air

Gas	Fractional content (% by volume)	Atmospheric partial pressure (mm Hg)
Nitrogen (N_2)	78.084	593.44
Oxygen (O_2)	20.948	159.20
Argon (Ar)	0.934	7.10
Carbon dioxide (CO_2)	0.031	0.24
All others	0.003	0.02
	100%	760.00

Molecules are exchanged between gases and liquids (including those in our lungs and in other tissues) until an equilibrium is reached. The quantity of a gas, therefore, that will dissolve into a liquid (or into a tissue) at a given temperature, therefore, is nearly proportional to the tension or partial pressure of that gas in the gas phase (Henry's law). The *absolute concentration* of a dissolved gas in a liquid or a tissue is determined by the *solubility* of the gas, which is the ratio of the amount of gas dissolved to the amount in the gas phase.

C. Oxygen Tension, Oxygen Concentration, and Oxygen Toxicity

It appears that over the course of universal evolution most mammalian organisms have reached a state of adaptation to an optimum oxygen tension (148). It must be emphasized that it is the *oxygen tension*, however, not the *oxygen concentration* that is the significant consideration. Exposure of the host to supplemental concentrations of oxygen (including exposures to 100% oxygen) produces no toxicity when the ambient pressure is such that the alveolar oxygen tension is not greater than normal. For example, at a constant ambient oxygen concentration the maximum oxygen partial pressure or tension exerted falls proportionately to any reduction in total barometric pressure. This can occur normally with increasing altitude (as shown in Table 2), or may be artificially regulated or set (as in hypobaric chambers or spacecraft). If the ambient total barometric pressure were reduced to 187 mm Hg, for instance, an oxygen concentration of 100% in the inspired gas would deliver a normal alveolar oxygen gas tension of 100 mm Hg, corrected for water vapor ($P_{H_2O} = 47$ mm Hg) and the metabolically produced carbon dioxide normally released in the alveoli ($P_{CO_2} = 40$ mm Hg). Therefore, any discussion of pulmonary "oxygen toxicity" must be undertaken in the perspective of the oxygen tension (and not the oxygen concentration) delivered to the lungs. Whereas under appropriately reduced pressures high concentrations of oxygen are not considered toxic, Clark and Lambertsen (75), as well as many others (7,13,14,17–19,30,87, 94,95,97,130,131,152,169,242,243,357, 384), have emphasized that prolonged ex-

Table 2. Oxygen Pressure Relationships and Altitude

Altitude (ft)	Barometric pressure (mm Hg)	Oxygen tension[a] (mm Hg)
Sea level	760	149
2,000	707	138
4,000	656	127
6,000	609	118
8,000	564	108
10,000	523	100
12,000	483	91
14,000	446	83
16,000	412	76
18,000	379	69
20,000	349	63
30,000	226	37

[a]Oxygen concentration of 20.9%, with barometric pressure corrected for water vapor (47 mm Hg).

posure to oxygen at a tension or partial pressure greater than that normally encountered in ambient air induces toxic effects which are progressively more severe as the duration of exposure or the inspired oxygen tension is increased.

D. *Physiological Effects versus Toxic Effects of Oxygen*

The host reaches a state of equilibrium with "normal" oxygen tension present in its environment. Increasing the inspired oxygen tension above this tension will have either a physiological or a toxic effect on the lung, as well as on other body organs (241,242,244). The *physiological effects* of oxygen can be defined as those responses to increased oxygen tensions that are not lethal and do not normally endanger life. Thus, the physiological effects of oxygen do not impose any significant or serious limitations on its use therapeutically (or on its administration for whatever reasons) at increased tensions. These physiological effects of oxygen are usually rapidly demonstrable and measurable, promptly and completely reverse on cessation of administration of the supplemental oxygen delivery, and do not injure the host.

The *toxic effects* of oxygen, on the other hand, do injure the host and potentially can be lethal. The exposure–response relationships of these toxic effects of oxygen (as well as the development of tolerance to these toxic effects) are shown schematically in Fig. 1 (73,74,133,213,395). The concept of *exposure* in oxygen toxicity is based on both the tension of oxygen inspired and the duration of administration of the increased tension; the duration from the initiation of exposure to the manifestation of detectable toxicity is inversely proportional to the degree of increased tension. The toxic effects of oxygen (or the degrees of the toxic responses) are usually delayed in onset or difficult to measure by their earliest manifestations. The reversal of the toxic effects of oxygen requires time, whereas the reversal of the physiological effects are immediate or nearly immediate.

E. *Relative Nature of Oxygen Toxicity*

Pulmonary oxygen toxicity is, in many ways, a *relative* concept that appears in the literature sometimes masked in different disguises and described in different or divergent terms. No endeavor here will adequately clarify this problem. Part of the dilemma is that what has been described as so-called pulmonary "oxygen toxicity" in man usually appears as a complication in the course of therapy for the otherwise severely injured lung or as a complication of support of a failing respiratory system. Under these conditions, it is difficult if not generally impossible to separate clearly the oxygen-induced pathology from the underlying disease process.

One would think this difficulty could have been resolved readily by precisely controlled experimental studies, but such has not been the case. First, most experimental studies have utilized animals with normal, not injured, lungs exposed for prolonged periods to very high oxygen tensions, conditions unlikely to be encountered in humans except under very rare circumstances. Second, there has been very little standardization of exposure conditions in experiments designed to clarify the nature of pulmonary oxygen toxicity. This is unfortunate, because the idealized exposure–response curve (for the intact animal, for individual organs or organ components, and for cellular and subcellular systems) in oxygen administration takes the form of a rectangular hyperbola (Fig. 1), with relatively limited linear components. In addition, exposure and dosage measurements have not always been defined in terms of both *duration* and *level* of oxygen delivery. There is not only a threshold of detectable response in oxygen toxicity, but a variable response relationship that is, as well, highly dependent on the exact exposure of oxygen tension. Therefore, what is "oxygen toxicity" at 0.85 atm oxygen delivery may be very different in nature from what is "oxygen toxicity" at 1.0 atm exposure tension, and the two responses may not necessarily be an extension of the same continuing spectrum. These considera-

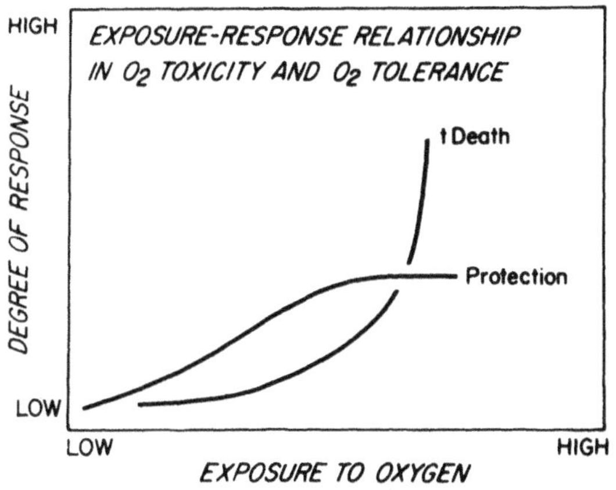

Figure 1. Conceptual presentation of the exposure–response relationship in oxygen toxicity and oxygen tolerance. The degree of oxygen toxic response is depicted along the ordinate in relative terms. Exposure to oxygen is a function of both the delivered oxygen partial pressure or tension and the duration of oxygen administration. The oxygen toxic response in most biological assays can be characterized as a rectangular hyperbola, leading ultimately to death. With the development of adaptive tolerance to the toxic effects of oxygen, this characteristic response is attenuated.

tions become even more vastly complicated with hyperbaric (>1.0 atm) exposure conditions.

A number of terms have been employed to describe some of these concepts. *Oxygen poisoning* is defined as the administration of oxygen at a tension that is injurious to the health or integrity of the organism or dangerous to the life of the host. *Oxygen toxicity* is a poisonous state, induced by breathing high partial pressures of oxygen. Pulmonary oxygen toxicity frequently has been divided into a so-called *exudative phase* and a *proliferative phase*. The term exudative phase is usually used to describe the edemagenic or hemorrhagic condition of the lung, encountered within the first few days of exposure to high tensions (generally >0.8 atm) of oxygen, that frequently leads to death. The term proliferative phase, in contrast, refers generally to the repairative adaptation of the lung parenchyma to oxygen tensions greater than normal ambient levels (>0.21 atm), and is characterized by replication or differentiation of cuboidal-like alveolar lining cells, the appearance of interstitial fibrosis, and other organ reparative responses. Probably, during the course of oxygen toxicity, both the exudative and proliferative phases overlap, if indeed they do not occur concurrently. The term tolerance is used in the field of toxicology to describe the power of resisting the action of a poison. Pulmonary *oxygen tolerance*, therefore, is a state of reduced pulmonary oxygen toxicity; how this term in fact has been used to differ from the proliferative phase of oxygen toxicity is not always clear. *Oxygen cross-tolerance* is used to describe the protective adaptation of the lung to oxygen that is induced by something other than oxygen. The term *oxygen resistance* suggests the presence within the host of an opposing or retarding force against oxygen toxicity; oxygen resistance may be described as native or *inherent* (derived on the basis of species or strain evolution) or it can be *acquired* (achieved by manipulation of some external factor or force).

II. The Lung as a Target Organ of Oxygen Toxicity

The lung is the host organ with the largest surface area in continuous contact with the gases of the external environment. Using first light microscopic and then later ultrastructural morphometric techniques, the internal pulmonary surface area has been estimated to be approximately 160 m^2 to 200 m^2 in size in the average human (192,411). Indeed, from the standpoint of molecular interaction, this surface area is probably even considerably larger. The primary function of the lung is to provide an interface by which oxygen is internalized and carbon dioxide, as the end product of oxidative metabolism, is released from the organism. In addition to being extensive in surface area, the membrane separating the pulmonary capillary blood from the gas in the alveolar spaces is also very thin (about 0.2 μm) and delicate. If indeed oxygen is potentially toxic to all living cells, it is not surprising, then, that this extremely large structure, with its very thin blood–air barrier, is highly vulnerable to the harmful effects of inspiring oxygen at increased tensions. Although these toxic effects of oxygen on the lung have been studied for the better part of a century, they are not as yet well understood. There are several reasons for this.

It is important to emphasize again that except in extremely uncommon circumstances (such as exist in hyperbaric diving chambers, under certain rare environmental conditions, and associated with some types of air and space flights), the lungs of the normal man and other animals are not generally exposed to increased tensions of oxygen. In contrast, supplemental oxygen at higher than normal partial pressures is frequently administered to patients with underlying lung disease or with hypoxemia of nonpulmonary origin. In this clinical setting, the toxic effects of oxygen on the already damaged airways and injured lung parenchyma are extremely difficult (and at times nearly impossible!) to study. In contrast, extensive literature on the toxic effects of exposure to high tensions of oxygen on the *normal* animal lung has been accumulated. How much of this information is directly applicable to man, and especially to humans with abnormal or acutely injured lungs, is conjectural; the effect of oxygen adminstration on the experimentally injured lung in animal models has not been investigated adequately (262). In presenting a summary of these topics, we will first review the clinical manifestation of pulmonary oxygen toxicity, including a discussion of pulmonary physiological function in the presence of hyperoxia. This will be followed by a discussion of the pathology of pulmonary oxygen toxicity, the development of adaptive tolerance to oxygen, and related matters.

III. Clinical Manifestations of Pulmonary Oxygen Toxicity

A. Clinical Symptoms

The clinical manifestations of exposure to high oxygen tensions have been studied experimentally in man and in animals, and to a lesser extent in the acutely ill patient (30,36, 64,74,76,83,96,307,426). Early clinical symptoms include a subjective sensation of restlessness, lethargy, anorexia, and in some instances, nausea and vomiting. Dyspnea, or a subjective awareness of difficulty or distress in breathing, occurs as a later symptom with sustained exposure; this is frequently associated with an increased respiratory rate and occurs both at rest and upon exertion. Late symptoms of exposure include those manifestations characteristic of severe respiratory distress.

The early and intermediate clinical symptoms of oxygen toxicity are related to several factors. High tensions of oxygen acutely induce tracheobronchitis, with substernal burning, cough, and pain on inspiration. Irritant receptors, with their impulses conducted over myelinated afferent nerve fibers, appear to be associated with these responses (197). Nociceptive responses, conducted over both myelinated and nonmyelinated pathways throughout the respiratory system also are important to the early clinical manifestations of oxygen toxicity, and appear to respond to edema formation, at a microscopic or submicroscopic level, in the interstitial spaces in the pulmonary tissue (197). The resultant edema-induced mechanical distortion, especially when it involves the distal pulmonary parenchymal architecture, causes frequent and variable nociceptive discharges, giving rise to the sensation of dyspnea and respiratory discomfort.

B. Clinical Signs

The physical signs of pulmonary oxygen toxicity parallel the clinical symptoms, and usually share a common origin (30,64,83, 96,207,307). Early in exposure, there is a hyperemia, edema, and swelling of the mucous membranes of the upper respiratory tract and the airway epithelium and mucosa. Tenacious tracheal secretions frequently accumulate. Bubbling rales develop first in the lower or basal posterior lung fields, primarily on deep inspiration; they eventually may spread to involve the entire lung. Labored, gasping breathing is a late manifestation of pulmonary oxygen toxicity, and is often accompanied by the development of frothy, bloody sputum. Utilization of extrapulmonary muscles of respiration is clinically evident, and with the development of respiratory distress intercostal retractions on inspiration can be seen. Cyanosis is another late manifestation, even in the presence of the administration of supplemental oxygen. Convulsions can occur, but usually are seen either only terminally or with exposure to oxygen tensions in excess of 1.0 atm.

C. Radiological Manifestations of Oxygen Toxicity

The early and late radiological manifestations of pulmonary oxygen toxicity are depicted in Fig. 2A, 2B, and 2C, and have been described at length in the literature (52, 64,68,83,95,207,288,419). The radiologically detectable signs of oxygen toxicity are usually absent early in the course of exposure (Fig. 2A). Small (0.2 cm) to larger (2.0 cm), diffuse, bilateral pulmonary densities, which are irregular in shape, develop with prolonged exposure (Fig. 2B). These may extend and coalesce to opacify entire lung fields (Fig. 2C). Pleural effusions also are frequently observed as both intermediate and late radiological manifestations of pulmonary oxygen toxicity. The eventual resolution of these radiological manifestations is variable (207). Some signs may remain evident for considerable time, persisting perhaps longer in infants than they do in adults (304,358). If significant interstitial parenchymal fibrosis develops, there may be correlative radiological manifestations that do not resolve with time.

IV. Pulmonary Function in Oxygen Toxicity

A. Respiration

Demonstrable abnormalities in pulmonary function during the development of oxygen toxicity are insidious in onset, progress with the increasing severity of the developing lesions, and result ultimately in hypoxemia and death. In the normal individual and, depending on the underlying cause of hypoxemia, in the patient requiring the clinical administration of supplemental oxygen, an increase in the arterial oxygen tension will usually depress both respiratory frequency and respiratory depth (241,242). As oxygen toxicity becomes manifest, however, net respiration will increase. This increase occurs primarily through an increase in respiratory

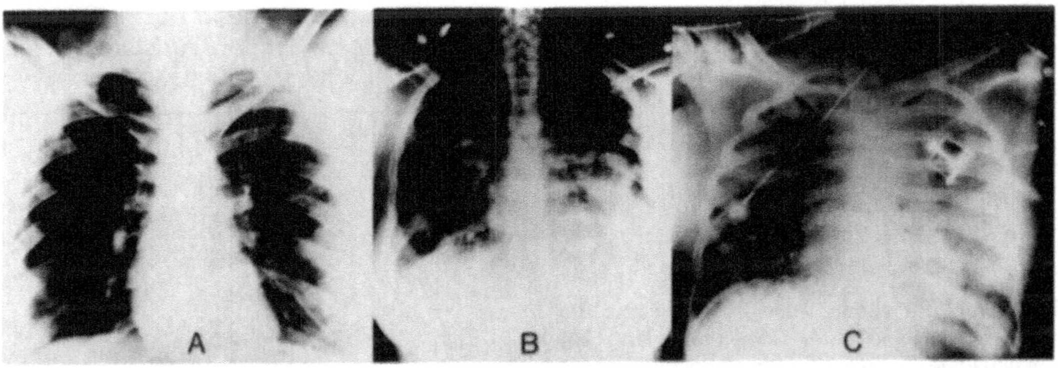

Figure 2. **A.** Initial chest film of a patient with mild respiratory discomfort following administration of high tensions of supplemental oxygen for nonpulmonary reasons. Although there may be suggestions of blunting of the costodiaphragmatic angles from early fluid accumulation, this chest radiograph was interpreted by the radiologists as "normal." **B.** Intermediate radiological manifestations of pulmonary oxygen toxicity with diffuse, bilateral, irregular pulmonary densities of variable size (0.2–2.0 cm). The bases of the lung are involved more than the apices. **C.** Late radiological manifestations of pulmonary oxygen toxicity, with extension and coalescence of the infiltrates. Radiological manifestations are due to interstitial and intraalveolar edema, atelectasis, and the accumulation of cellular debris and hyaline membranes in alveolar spaces and in the terminal airways.

rate, even if tidal volume is diminished (76, 96). If systemic acidosis (↓ arterial pH) and hypercapnia (↑ arterial P_{CO_2}) also develop, the increase in respiratory frequency will be even more significant (246).

B. Lung Volumes

The lung volumes for the normal and for the oxygen toxic lung are shown in Fig. 3. There is both an acute and a progressive reduction in vital capacity following the administration of high tensions of oxygen and the development of oxygen toxicity (30,64,73,74,76,83, 96,307). This reduction is potentially most severe following exposure at 2 atm of oxygen or more, with a net decrease in the vital capacity beyond probably that which can be attributed to oxygen administration alone (74,76). The loss of vital capacity in oxygen toxicity appears to be due in part to a reduction in inspiratory capacity (or inspiratory reserve volume) and an increase in functional residual capacity (or residual volume), as depicted in Fig. 3 (74,76). The reasons for these changes are severalfold. Early in the course of oxygen toxicity, chest pain and substernal discomfort may limit inspiration. Atelectasis, which develops early as the inert gases are washed out of the alveolar spaces and is accentuated later as further alveolar collapse and pulmonary edema occur secondary to increased alveolar surface tension forces and structural alterations in the blood–air barrier, may physically reduce tidal volume and the total inspiratory capacity. The increases in surface and tissue elastic recoil forces induced by prolonged exposure to high oxygen tensions serve to reduce tidal volume per given unit of change in transpulmonary pressure, to increase the functional residual capacity, and, with the collapse of small airways and terminal bronchioles, to increase the residual volume of gas trapped in the lung. These alterations in lung volumes and lung capacities can be demonstrated in the absence of clinical symptoms, and usually have an eventual but slow return towards normal in the recovery phase of oxygen toxicity.

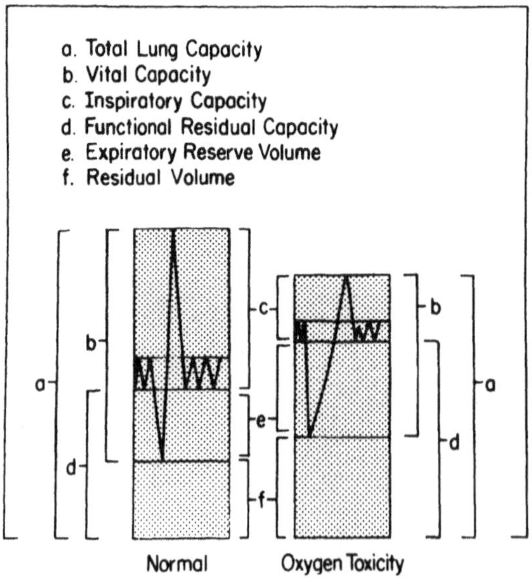

Figure 3. Lung volumes and lung capacities for the normal lung and for the oxygen toxic lung. The lung capacities include total lung capacity (TLC), vital capacity (VC), inspiratory capacity (IC), and functional residual capacity (FRC). The lung volumes are the (IRV), expiratory reserve volume (ERV), and the residual volume (RV). With oxygen toxicity, there is a reduction in TLC (and VC), primarily as a result of a diminished IC and IRV and in increase FRC (and RV).

C. Airflow Resistance

Normal and oxygen-toxic expiratory flow relationships are shown in Fig. 4. Results of measurements of airflow resistance in man

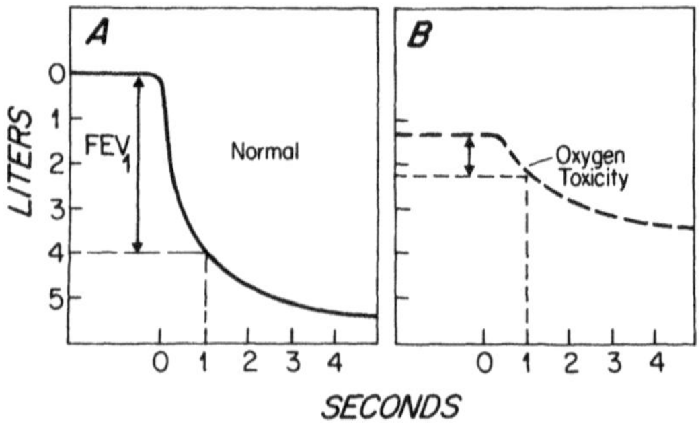

Figure 4. Expiratory airflow relationships in the normal (A) and in the oxygen-toxic lung (B). As oxygen toxicity becomes severe, there is a reduction in the forced expiratory volume at 1 sec (FEV_1).

and in experimental animals have been inconsistent in studies on pulmonary oxygen toxicity (64,74,76,117). Some investigators have reported that there is no significant change in airway resistance and expiratory flow rates (64,117), whereas others have indicated that there is a reduction in the forced expiratory volume at 1 sec (FEV_1) and in the maximum mid-expiratory flow rate (74,76). Alterations in expiratory flow rates are more prominent late in the course of oxygen toxicity, as airway obstruction develops from edema and lumenal blockage by airway secretions and debris.

D. *Diffusion Capacity*

The gas-diffusion capacity (*DL*) across a normal diffusion membrane (such as the air-blood barrier of the lung) is depicted in Fig. 5. The diffusion capacity for any gas across the membrane is related to the amount of gas ($\dot{V}_G$) that diffuses across the barrier in response to a pressure gradient ($P_A - P_a$) from the alveolar space to the alveolar capillary, as determined by the Bohr equation (Fig. 5). The gas-diffusion capacity of the air–blood barrier can be altered either by increasing the thickness of the diffusion membrane (as theoretically occurs with pulmonary edema or with the accumulation of hyaline membranes and cellular debris on the alveolar surfaces) or, more commonly, by reducing the area of the total membrane available for diffusion. Generally, a change in the normal alveolar–arterial oxygen differences, implying an alteration in the gas-diffusion capacity of the lung, is not considered an early manifestation of oxygen toxicty (64,76). Alterations in the diffusion capacity after several hours of exposure to

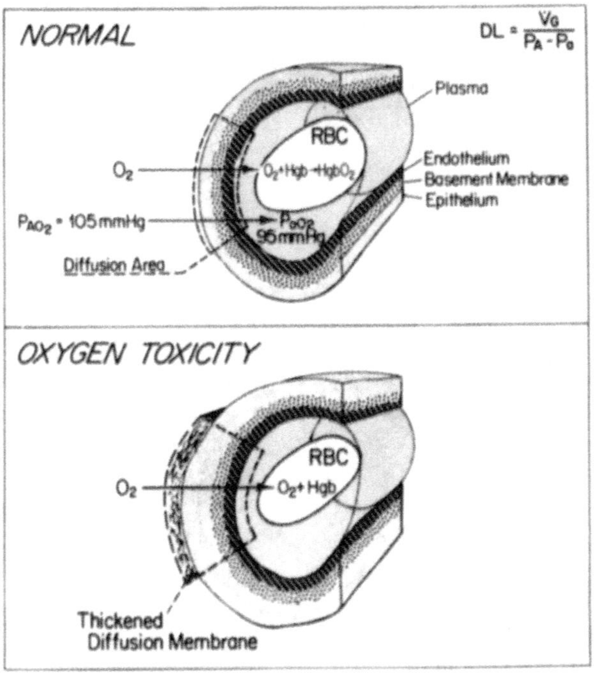

Figure 5. The influence of the diffusion membrane on the diffusion capacity of oxygen. The gas-diffusion capacity (*DL*) of the air–blood barrier is determined by the relationship: $DL = (V_a)/(P_A - P_a)$, where V_a is the volume of gas (O_2), P_A is the alveolar gas tension, and P_a is the alveolar capillary tension. The diffusion membrane is comprised of the alveolar epithelium (EPI), the capillary endothelium (ENDO), and the interposed basement membrane (BM). A thickened diffusion membrane can develop when edema forms within the air–blood barrier and by the accumulation of cellular debris and hyaline membranes in the alveolar space.

high oxygen tensions, however, have been considered significant (64,342). These alterations may be due to edema alone or to endothelial damage and alveolar capillary destruction, whereas more prolonged manifestations of an altered diffusion capacity may be due to an increase in alveolar lining epithelial cell thickness and interstitial fibrosis.

The early changes (after just hours of exposure to high oxygen tensions) that have been reported demonstrable in the pulmonary diffusion capacity are questionable for several reasons (64,342). First, there are no corresponding early morphological changes on which to base a physiological change in the diffusing capacity. Second, there are considerations in the diffusion capacity testing procedures that may contribute to erroneous interpretations when it is measured under conditions of hyperoxia. For example, the diffusion capacity is usually measured by a procedure that quantified the uptake of small amounts of carbon monoxide, and in the presence of very high oxygen tensions oxygen itself may alter the displacement of carbon monoxide from hemoglobin and affect the physiological measurements.

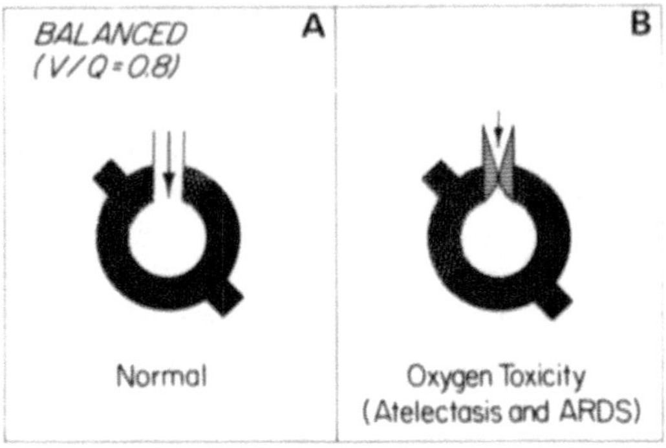

Figure 6. **A.** Schematic representation of the normal ventilation–perfusion ratio for an "ideal" gas-exchanging unit and for an oxygen-toxic lung. On a practical basis, the ratio of these two variables cannot be determined for individual alveoli, as each gas-exchanging unit does not have a perfect "matchup." The ratio of ventilation to blood flow for all units of the lung as a whole, however, can be averaged. This physiological ratio can be maintained if ventilation and perfusion are reduced proportionately. **B.** The normal ratio of ventilation to perfusion is 0.8 (balanced $\dot{V}/\dot{Q} = 0.8$). In oxygen toxicity, ventilation may be either diminished or increased, but on a relative basis perfusion generally significantly exceeds ventilation, especially as intraalveolar edema and atelectasis develop, reducing the net $\dot{V}/\dot{Q}$ ratio below 0.8.

E. Ventilation–Perfusion Abnormalities

Although it is unclear just how much of it is utilized at any one time in gas exchange, the very large internal surface of the lung (approximately 160–200 m^2, with over 100,000 terminal respiratory units and 300 or more million alveoli) must be in balance with its blood supply. The total pulmonary capillary blood volume, however, is relatively small, or no more than about 100 ml or less, at any one time. Thus, the exchange of gases across the air–blood barrier is dependent on the near aposition of an extremely thin film of mixed venous blood over a very large gas-exchanging surface. The relationship of ventilation to this surface and the supply of blood for gas exchange can be expressed as the *ventilation–perfusion ratio*, or $\dot{V}/\dot{Q}$. The normal ventilation–perfusion relationship ($\dot{V}/\dot{Q} = 0.8$), and an imbalance of this relationship in oxygen toxicity are shown schematically in Fig. 6. In oxygen toxicity, both ventilation and perfusion change, but not necessarily in direct proportion to each other or even in the same direction (64,342).

A falling arterial oxygen tension in the presence of the administration of progressively increasing tensions of inspired oxygen is the *sine qua non* component to diagnosing pulmonary oxygen poisoning. Although a reduced diffusion capacity and other abnormalities of respiration attributed to oxygen toxicity may be important contributors to this hypoxemia, it is primarily due not to these factors but to imbalances in ventilation and perfusion. The severity of the hypoxemia has two primary bases. The first of these is the nature of oxygen binding to and dissociation from hemoglobin, and the shape of the oxyhemoglobin dissociation curve. The second

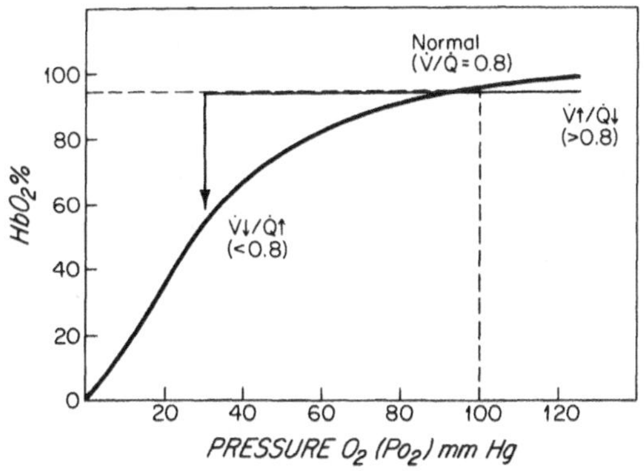

Figure 7. The relationship of ventilation and perfusion to the saturation of hemoglobin. In the oxygen toxic lung there are areas with alveolar units of both increased and decreased ventilation perfusion ratios. Increasing the $\dot{V}/\dot{Q}$ ratio, however, does not enhance the oxygen content of blood in areas of relative over ventilation, or the net oxygen content of blood leaving the lung, as the carrying capacity of the red blood cell for oxygen is near saturation on the plateau phase of the oxyhemoglobin dissociation curve. Decreasing the $\dot{V}/\dot{Q}$ ratio, however, markedly reduced oxygen loading on the steep phase of the oxyhemoglobin dissociation curve.

involves the development of intrapulmonary physiological shunts. As depicted in Fig. 7, there will be areas within the lung that have both increased and decreased ventilation–perfusion ratios. Increasing the ventilation–perfusion ratio ($\dot{V}/\dot{Q} > 0.8$) can occur in oxygen toxicity where ventilation and respiratory frequency increase in some gas-exchanging units; this does not effectively improve by very much the oxygen content of the blood in these areas of relative over-ventilation, however, as the carrying capacity of red blood cell hemoglobin is near saturation at the plateau phase of the oxyhemoglobin dissociation curve. When net ventilation is reduced by airway obstruction, diminished transpulmonary compliance, a reduction in effective lung volumes, and the accumulation of fluid and debris within alveolar spaces, the ventilation–perfusion ratio falls ($\dot{V}/\dot{Q} < 0.8$), and in so doing markedly reduces the oxygen loading-capacity of red blood cell hemoglobin in the steep phase of the oxyhemoglobin dissociation curve.

F. Shunt

An anatomic shunt is a vascular communication that morphologically bypasses the usual circulatory channels; small anatomic shunts exist normally in the lungs of all healthy people, as well as in many other organs and tissues. Physiological or functional shunts also occur normally in the lung as a generally more severe ventilation-to-perfusion mismatch, and when accentuated become significant as the major contributing factor to hypoxemia in any pulmonary disease process that has atelectatic or consolidated alveolar spaces that continue to be perfused. If, for example (Fig. 8), mixed venous blood, which is low in oxygen, perfuses nonventilated alveolar spaces (because of atelectasis or consolidation), significant hypoxemia will develop. Shunt-related hypoxemia cannot be corrected by an overall increase in ventilation, as the nonshunted areas are already nearly fully saturated with oxygen. In other words, any further increase in the oxygen tension to already well-perfused hyperventilated areas will not add quantitatively significant amounts of oxygen to the

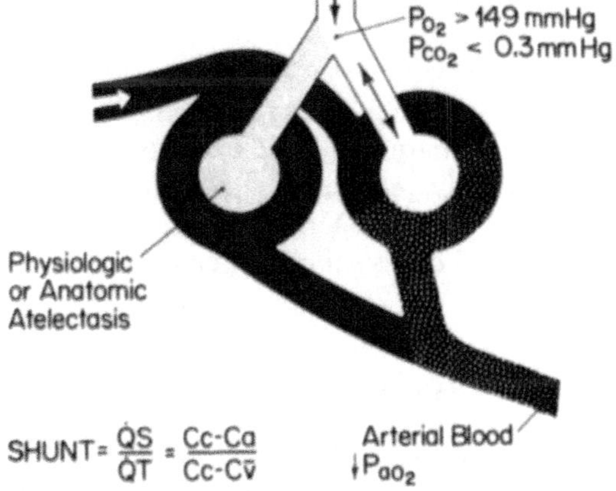

Figure 8. Physiological shunt ($\dot{Q}_s/\dot{Q}_v$). The shunting of nonoxygenated mixed venous blood around atelectatic alveoli or consolidated airways, in the presence of normal or even increased ventilation to the rest of the lung, dramatically reduced the oxygenation of arterial blood ($P_{a_{O_2}}$) leaving the lung for the systemic circulation.

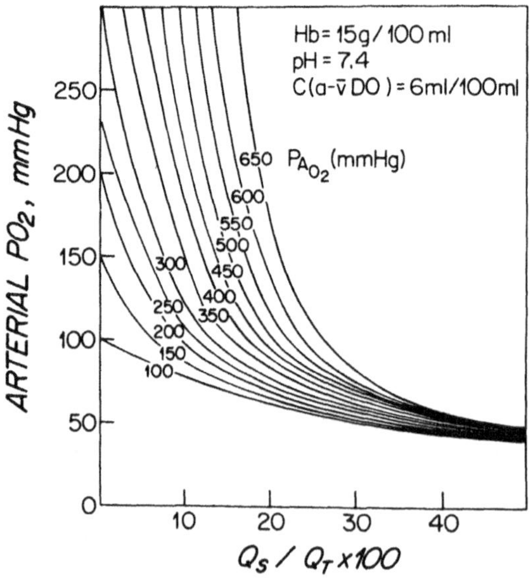

Figure 9. The effect of shunt ($\dot{Q}_s/\dot{Q}_v$) on arterial oxygen tension ($P_{a_{O_2}}$) at varying alveolar oxygen tensions ($P_{a_{O_2}}$). A normal hemoglobin (Hb: 15 mg%), arterial acid–base balance (pH = 7.4), and cardiac output [$C(a - \bar{v}DO_2)$] are assumed. In severe imbalances of ventilation and perfusion (shunt), as occurs in the oxygen toxic lung with atelectasis and intraalveolar accumulation of hyaline membranes and cellular debris, arterial oxygenation is dramatically reduced as the degree of ventilation–perfusion imbalance or shunt increases. This affect is accentuated, as well, with any reduction in cardiac output.

mixed aggregate of post-shunt blood from underoxygenated areas (101). In the patient with pulmonary oxygen toxicity, this shunting dramatically reduces the oxygen content of arterial blood (249), regardless of the tension of inspired or alveolar oxygen (Fig. 9).

G. Respiratory Mechanics

Representative inflation–deflation volume–pressure relationships for the normal lung and for the oxygen-toxic lung are shown in Fig. 10. Even when corrected for percent maximum lung inflation, there is in oxygen toxicity a major "shift-to-the-right" of the volume–pressure characteristics of the lung (64,76, 117,342). Surface elastic recoil increases early following exposure to high oxygen tension, probably as a result of increased alveolar surface tensions. *Dynamic lung compliance*, or the change in lung volume per change in transpulmonary pressure ($\Delta V/\Delta P$), falls within hours of exposure to high oxygen tensions in both man (64,117,155) and the experimental animal (33,56,155,210,328, 356,380). Early in the course of pulmonary oxygen toxicity, some of this shift in the volume–pressure relationships may be overcome with deep inspirations or hyperinflation of the lung, presumably reversing microatelectasis or reactivating the alveolar-lining surface active materials. Late in the course of oxygen toxicity, an increase in surface elastic forces, as well as interstitial fibrosis, develops and tissue rigidity ensues (76,117,342).

V. The Pulmonary Pathology of Oxygen Toxicity

The pulmonary pathology of oxygen toxicity has been reported extensively. Selected references are included (13,71,158,175,221, 224, 233, 300, 301, 307, 312, 333, 335, 351–353,380,405,418); these include descriptions of alterations in both the airways of the lung and in the pulmonary parenchyma. In general, the pulmonary pathological changes appear to be most significant at exposures of 0.5 atm to 1 atm and less severe at higher exposure pressures, as in exposure conditions above 1 atm of oxygen central nervous system damage becomes a more important factor for the organism than lung injury. For perspective, these nonpulmonary manifestations of oxygen toxicity include central nervous system convulsions, paralysis, and brain death (7,8, 13, 30, 42, 92, 98, 167, 211, 214, 242, 261, 374,384,404,407). Additional nonpulmonary factors contributing to the morbidity and mortality of the host involve occular toxicity, visual field contraction (35,427,428), retinal damage (34,141,278,303), blindness (5,6, 327), testicular damage (131,308,400), and red blood cell hemolysis and intravascular damage (174,225,286,357).

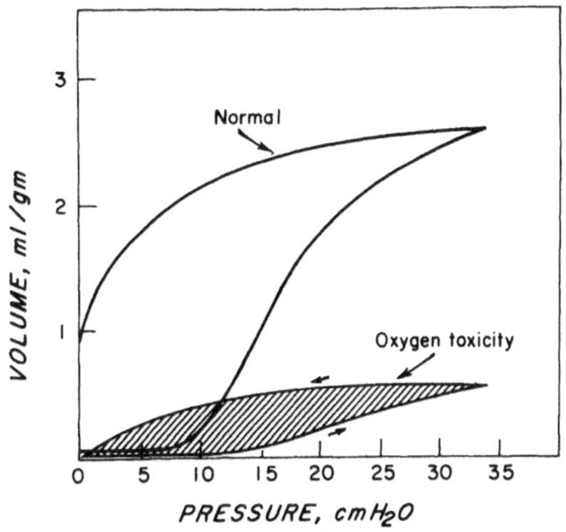

Figure 10. The static pressure–volume relationships of the normal lung and of the oxygen-toxic lung. Transpulmonary dynamic compliance is, in effect, the slope of the volume–pressure curve at resting lung volume (a careful volume history, therefore, should be ascertained in reporting any measurements of lung compliance). With the development of oxygen toxicity, the pressure–volume relationship of the lung is shifted to the right (and downward), reflecting increased surface-elastic and tissue-elastic recoil forces, and a decrease in transpulmonary compliance.

A. Pulmonary Pathomorphological Changes

A tissue section from an oxygen toxic human lung is shown in Fig. 11. The pulmonary pathological changes of oxygen toxicity include atelectasis, edema, hemorrhage, consolidation, congestion of the pulmonary capillaries, inflammation, fibrin formation, alveolar thickening and intra-alveolar hyalinization, alveolar cell desquamation and degeneration, and a bronchitis characterized by an accumulation of secretions and cellular debris in the terminal airways. Evidence of attempted tissue repair within the oxygen-toxic lung is often present concurrently with tissue destruction, with type II alveolar epithelial cell and airway-lining cell hypertrophy and hyperplasia the most evident manifestation of attempted recovery.

The sequence of the pathological cellular response within the lung depends on the susceptibility of the individual cell type to enhanced tensions of oxygen. Indeed, the specific sequences of response may be different for alternate cell lines at different oxygen exposure tensions. The absence of detectable or obvious pathological changes at any one point in a hyperoxic exposure sequence, even in the presence of morphological or metabolic injury to other cell types, does not mean exclusion of damage. Cells may appear normal to microscopic examination, even by ultrastructural techniques, through given durations or periods of exposure, and then very rapidly or abruptly deteriorate within the course of just a few hours of additional exposure to high oxygen tensions.

The literature describing the pulmonary pathology of oxygen toxicity is not inherently consistent, however, and there probably are multiple explanations for this. The primary reason is that much literature is derived from animal research, where there are both species and strain differences in the response of normal animals to oxygen toxicity, and differences as well in individual animal susceptibility within a given species or strain. The age of the animals under study, as well as other factors of potential variability, including nutrition, also are important to the pathological response. Younger or more immature animals, for example, appear to be more resistant to the morphopathological manifestations of pulmonary oxygen toxicity, and this may be due to a delay in the onset of the toxic effects of oxygen rather than to any inherent net total resistance to toxicity (13, 55,107,135,143,237,327). Low protein diets (161) and partial fasting (13,66,151,308, 384) protect against oxygen toxicity. These inconsistencies and variations in the pathology of oxygen toxicity become less apparent when they are evaluated after comparatively equitable dosages.

B. Atelectasis

Atelectasis is a state of airlessness of the lung, due to failure of expansion or to reabsorption of air from the alveoli. Atelec-

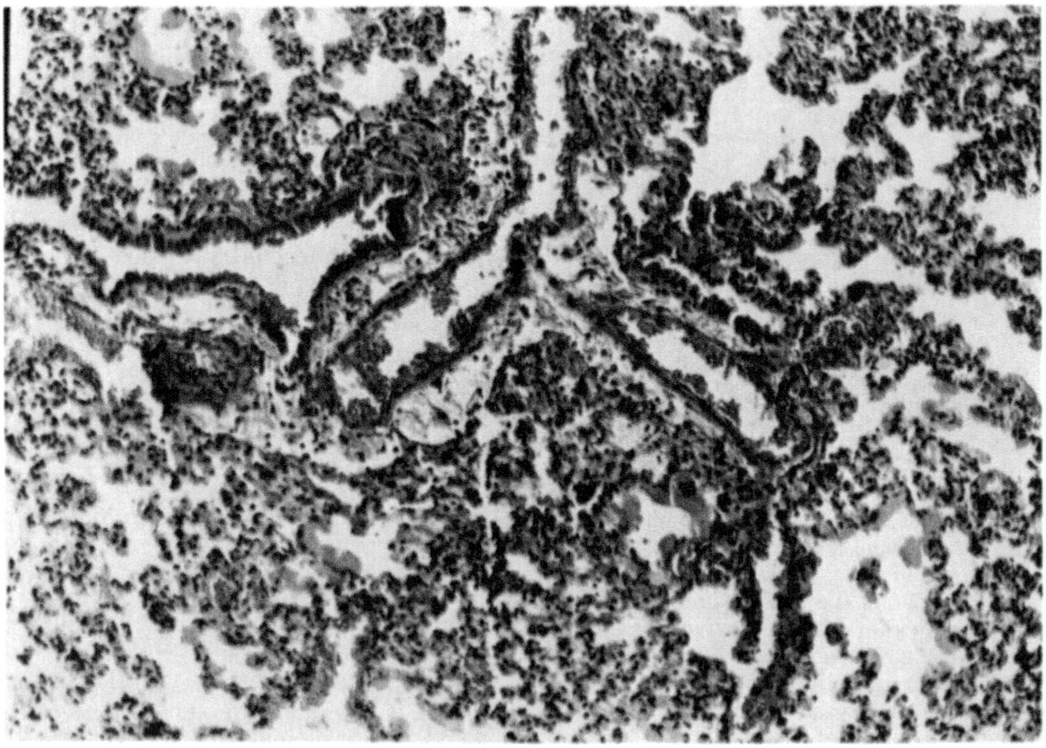

Figure 11. General histological presentation of pulmonary oxygen toxicity, with congestion, atelectasis, hypercellularity, edema, and hyaline membranes. The lining epithelium of the terminal bronchioles and alveolar ducts is partially denuded, with the intralumenal accumulation of cellular debris, fluid, and in some instances hyaline membranes.

tasis is an almost always present component of pulmonary oxygen toxicity, and is probably the result of several combining factors (13,30, 71,93,227,228,301,382,388,405,418). The demonstration of atelectasis in tissue sections, however, may be missed if the lungs are fixed for postmortem preservation by intratracheal insufflation. There is some controversy as to whether atelectasis is primarily an early or a late response in the course of oxygen toxicity; it most surely is both. It can be demonstrated clinically, directly by radiological techniques and indirectly by abnormalities in the ventilation–perfusion relationships of the lung (405). This can be overcome to a considerable degree in the intact host by increasing, through assisted ventilation, airway pressure, and especially by employing a positive end-expiratory pressure with mechanical assisted ventilation.

The pathophysiological mechanism by which atelectasis develops in oxygen toxicity has received considerable attention; edema is thought to be important in its pathogenesis (93,224,226,232,233). Transudation of fluid from the capillary bed and from the lymphatics of the lung into the alveolar spaces and parenchymal interstitium occurs very early in oxygen toxicity; fluid accumulates, in addition, in the pleural spaces. Simply by its physical presence, either within the lung or in the pleural compartments, the accumulated fluid can directly compress alveolar spaces and contribute to atelectasis. In addition, the edema fluid, by an indirect effect, can also alter alveolar surface forces in a way that further enhances alveolar collapse.

A second important factor in the pathogenesis of atelectasis appears to be airway obstruction (330). Inhalation of high tensions of oxygen will eventually wash out proportionately any inert gases (such as nitrogen) normally in the lung. If any obstruction to ventilation or airflow then develops (as with the plugging by debris of an airway or a terminal bronchiole), continued vascular perfusion distal to that obstruction will result in the absorbance of all oxygen, carbon dioxide, and water vapor present in the alveolar spaces, leading to collapse (86,91, 344,350). This may occur in the normal, noninjured lung. Aviators, for example, who at times inhale pure oxygen at decreased tensions, end up with fewer gas molecules present in their alveolar spaces, which may lead under some circumstances to the rapid absorption of oxygen and alveolar collapse (152,350). As a second example, patients with severe ventilation–perfusion imbalances (especially as occurs with emphysema and other forms of chronic obstructive airflow disease) are predisposed to gas trapping. When the oxygen concentration is increased in the gas ventilated by these individuals, absorption of oxygen distal to the obstruction occurs, with a resultant atelectasis. As a final example, in the acutely injured lung, accumulation of edema fluid (transudates as well as exudates) and cellular debris in the terminal airways leads to obstruction, with absorption atelectasis.

A third important factor in the pathogenesis of atelectasis is the pulmonary surfactant system. Regardless of the importance of the role of other factors, alveolar surface forces are crucial in stabilizing alveoli and preventing their collapse (252). The important nature and significance of alterations in pulmonary surface active materials as a function of exposure to high tensions of oxygen will be discussed in detail later.

C. Edema

Edema is the most common manifestation of pulmonary oxygen toxicity, and has been studied extensively (1,25,93,224,226,228, 299,311,312,335,361,378,380,382,414, 418). It may be present only in microscopically detectable amounts (demonstrable ultrastructurally by intracellular vacuolization and widening of the interstitial spaces) or in more florid presentations of pleural effusions and frothy tracheal discharges.

In animals, accumulation of fluid in the lungs during oxygen toxicity has been studied by measuring lung weights; a very limited amount of data of comparable nature have been accumulated from investigations in man. In animals, the lung-to-body weight ratio may as much as triple with fluid accumulation in oxygen toxicity (124,175). The increase in wet lung weight may be either on the basis of fluid accumulation (2,351,405), as water follows sodium and plasma proteins in transudates, or on the basis of a net gain in lung mass by nonsoluble proteins (2,175). The fluid that can be recovered for analysis is similar to diluted serum, with some additional proteinaceous contributions being derived from injured cells (4,93,354).

The mechanisms of edema formation are probably multiple. Capillary damage is one of the earliest structural alterations demonstrable in oxygen toxicity (1,49,62,71,224, 232,233,266,277,299,300,307,314,333, 355,363,405,408,425). Increases in systemic blood pressure and increases in pulmonary blood pressure also have been reported (39, 53,54,77,246,420–422), but these are inconsistent findings (13,35,36,373); significant alterations in blood pressure, especially in the systemic circulation, may very well be factors only at oxygen tension deliveries in excess of 1 atm pressure (246, 420–422). Increases in retained carbon dioxide tensions also have been suggested as being important to pathogenic mechanisms of edema formation in oxygen toxicity, but increases in lung-to-body weight ratios have been reported to occur before carbon dioxide retention becomes physiologically or clinically significant (275).

D. Congestion

As shown in Fig. 12, significant changes occur early in the pulmonary vasculature in oxygen toxicity. In addition to the intra-alveolar edema and hemorrhage that develops, fluid accumulates in the interstitial and perivascular spaces (13,39,49,54,71,93,206, 226,307,379,405,414). A number of studies, including those employing morphometric techniques, have suggested that the first morphologically detectable injury in the lung involves the capillary endothelium (224,232, 233), which becomes vacuolated and swollen with what may be described as pinocytotic vesicles; additional studies indicate that the endothelial replication rate may be reduced transiently (103). The walls of the pulmonary arterioles and arteries (Fig. 12) become thickened and their lumens narrowed, with eventual thrombosis ensuing (39,235,379, 414). Polymerization of fibrin clotting factors within the vessel walls may result in a hyalinization of the vasculature. Fibrinous tissue forms outside of the media of arterial walls (Fig. 12), and the arterial elastic tissue and smooth muscles thicken (39); vasoconstriction of the arteries, arterioles, and pulmonary venules accentuates these findings (418). The veins, in fact, may necrose, accelerating the potential susceptibility to intravascular thrombosis (414); erosion of the arterial media results in the media merging with the adventia of the vessel (39,235).

If the lung recovers from these insults, or if adaptive oxygen tolerance develops at a rate greater than oxygen toxicity, the initial vascular erosion and damage will be followed by a state of endothelial hypertrophy (235). With or without repair, pulmonary hypertension (39,53,54,77), increased pulmonary vascular resistance (53,54), right heart hypertrophy (53,54), and eventual right heart dilation (39,227,228) develop.

E. Hyaline Membranes

Hyaline membranes, consisting of homogenous strands of fibrin and cellular debris lining the walls of alveoli and terminal bronchioles, have been reported as a late component of pulmonary oxygen toxicity (1,3, 40, 58, 60, 71, 88, 254, 259, 299, 335, 336, 356,370,372,402). These are similar (but not necessarily identical) to the hyaline membranes demonstrable in the lungs of infants dying with the respiratory distress syndrome (40,58). The development of hyaline membranes in pulmonary oxygen toxicity, however, is variable (57,88,206), with the reported responses differing among animal species and even within the same species or strain (122,254). When all other variables can be controlled under experimental conditions, their appearance seems to be, in part, a function of the duration of hyperoxic exposure (254,336). Age is also an important variable in the potential development of hyaline membranes following oxygen exposure, with younger animals apparently more susceptible to their formation (1,122,173, 254,306,336).

The pathogenesis of alveolar hyaline membranes most likely involves a stepwise process. Increased capillary permeability and cell destruction of the blood–air barrier result in a transudation and exudation of plasma clotting factors, leading to fibrin polymerization and the entrapment of cell debris within the fibrin meshwork (1,3,40,58,60, 198,253,322,323,325,405). Concurrently, the fibrinolytic activity of the lung appears to be reduced with progressive exposure to high tensions of oxygen (253,334). Under normal conditions, the fine capillary network of the pulmonary vasculature acts as a giant "sieve" or "filter," so to speak, for the removal of small microemboli from the circulation; as such it must be rich in fibrinolytic activity. Oxygen diminishes profibrinolysin activity and fibrinolytic inhibitors (334), reduces on a relative basis the levels of plasminogen activator (254), and appears to prolong the clotting time of systemic blood (297). Because of these responses, a role for heparin or epsilon amino caproic acid (EACA) administration in the control of hyaline membrane formation has been suggested (123). In that hyaline membranes are not demonstrable in the lungs of stillborn humans (88,123) or stillborn animals (56,88), it seems that extrauterine respiration is essential to all of these pathogenic mechanisms.

Several factors affect the rate of oxygen-induced hyaline membrane formation in the

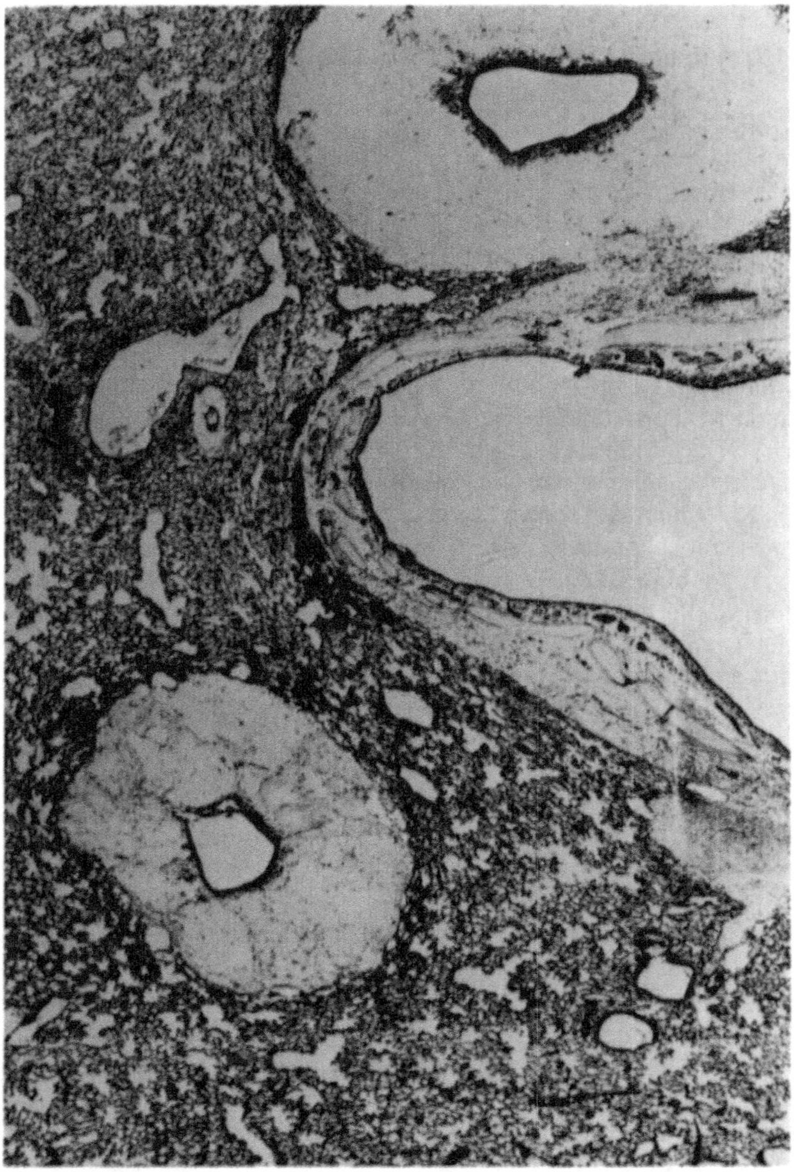

Figure 12. Low power light micrograph of an oxygen toxic lung. Perivascular and peribronchial edematous "cuffing" are dominant, with parenchymal edema, vascular congestion, intraalveolar hemorrhage, and generalized atelectasis. A hypercellularity also is evident throughout, especially within and around the alveolar spaces.

susceptible species. Lowered body temperature reduces the rate of formation (254). Administration of adrenocorticotropic hormone (ACTH), cortisone, atropine, heparin, pilocarpine, and aerosols of hyaluronidase, trypsin, or amniotic fluid experimentally accentuate hyaline membrane formation (3, 58,60,123,258,321). Certain thiol compounds [such as 5-(2-aminothiol)-isothiouronium bromide hydrobromide], antifibrinolytic agents (EACA), and allegedly aerosolization of surface active agents (ethanol, alteraine) reduce hyaline membrane formation in some experimental studies (3,58,123). These findings are less than well substantiated, however, and need to be investigated further.

VI. Oxygen Toxicity and the Pulmonary Surfactant System

A. Surfactant

A surface active agent, or surfactant, is any substance that can reduce or lower the surface tension (usually measured in dynes/cm) of a liquid at a liquid–gas interface. Surfactants include substances commonly referred to as wetting agents, surface tension depressants, detergents (338), or substances described in related terms. Surface activity, by definition, is a relative concept, and surfactants, therefore, are also relative. One of the most potent surfactants known to man appears to line, probably when exerting its function as a monomolecular layer, the alveolar surfaces. The exact nature of this material is not yet fully understood, but it appears to be a lipoprotein complex, the surface active components of which are highly saturated phospholipids (the most predominant of which is dipalmitoyl lecithin) (113,282).

B. Physiological Effects of a Pulmonary Surfactant System

It is important to grasp fully the physiologic functions of the surfactant system and their alterations in oxygen toxicity in order to understand the pathophysiological mechanisms of injury; therefore, these functions will be reviewed briefly here (79). The pulmonary surfactant system has several important physiological functions, the foremost of which is to stabilize alveolar spaces at low (at approximately 20% total vital capacity) lung volumes. Inspiration is a muscularly active process, comprised of two normally integrated but independent components, the diaphragm and the thoracic cage musculature. Expiration, on the other hand, normally is a passive process generated by both tissue elastic forces and surface elastic forces created by the spherical shape of the alveoli and the interface between alveolar gas and a wet alveolar lining. These surface forces are governed by the law of LaPlace:

$$\text{Alveolar collapsing pressure} = \frac{2 \times \text{surface tension}}{\text{alveolar radius}} \tag{3}$$

As shown diagramatically in Fig. 13, the surface tension forces of collapsing alveoli are not constant but variable, probably from a dilutional effect of any surface active material present on the alveolar surface at high lung volumes (large alveolar surface areas). That is to say, at full lung inflation or at relatively high alveolar volumes, the surface active material on the alveolar surface is probably not contiguous and the relative alveolar surface tension is raised (for example, to 40

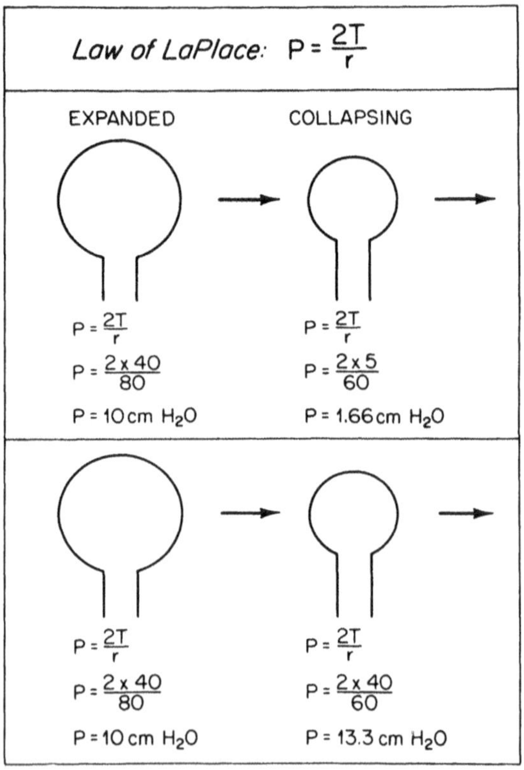

Figure 13. The effect of constant versus variable surface tension forces (T) on the collapsing pressure (P) of spheres with an changing radius (r), as might occur in the alveoli of the lung. The collapsing forces are governed by the law of LaPlace: $P = 2T/r$.

dynes/cm or the approximate surface tension of plasma), creating high alveolar collapsing pressures (10 cm H_2O or more, for example, for an alveolar space with radius of 80 μm). As the alveolar diameter is reduced by these high collapsing pressures, the surface active material becomes concentrated enough to eventually form along the alveolar surface a monolayer that exerts a very low surface tension force (0–5 dynes/cm), and by application of the law of LaPlace thus reduces collapsing pressures (less than 2 cm H_2O, for example, for an alveolar radius of 50 μm and a surface tension of 5 dynes/cm). The net effect of this variable alveolar surface tension force is to enhance alveolar collapse (and expiration) at high lung volumes and to prevent alveolar collapse (to stabilize expiration) at low lung volumes. The alveolar lining material probably has other functions, as well, which will be discussed later.

C. *Oxygen Toxicity and Alveolar Surface Forces*

A knowledge of the effect of high oxygen tensions on the pulmonary surfactant system is important to understanding the pathogenesis of oxygen toxicity. Although this subject has received considerable attention, some of the reported results are still inconsistent. Most investigators have reported that hyperoxia induces a reduction in pulmonary surface activity (1,33,48,82,146,164,209, 217, 229, 248, 268, 290, 292, 295, 356, 410, 418), whereas others have observed no change in the pulmonary surfactant (48,124, 146,209,210,267,302,324). Most of these inconsistencies, perhaps, can be explained by experimental differences in the approach to measurement of pulmonary surface tension forces, and these issues will be addressed separately below.

A lowering of alveolar surface tension forces has several important normal physiological functions. The first and foremost of these is to prevent collapse of the lung on expiration (thus reducing the energy needed for reexpansion on inspiration and optimizing the work of respiration). This function is lost in oxygen toxicity. If surface tension forces in the lung are increased by exposure to high tensions of oxygen (that is to say, the activity of the pulmonary surfactant is diminished), alveolar surface tension forces will remain high and atelectasis will occur. This is schematically depicted in Fig. 13B, where the presence of a constantly high surface tension force (40 dynes/cm) will result in an increase (rather than the decrease observed when surface tension is lowered) in the alveolar collapsing pressures as the radius is reduced on lung deflation. Instead of stabilizing at low lung volumes, then, there will be further strong forces generated for continued collapse of the lung to an airless state.

A second normal important physiological function of reducing alveolar surface forces, and thus alveolar collapsing pressures, is to stabilize the flow of gases between alveolar spheres of different sizes. This function also is lost in oxygen toxicity. The collapse of alveoli connected to a common airway, or interconnected to adjacent alveoli by pores of Kohn, also will be governed by the law of LaPlace (Fig. 14). If, for example, alveolar surface tension forces were high and constant (50 dynes/cm), small alveoli (50 μm or less in diameter) would, because of their smaller radius (as governed by the law of LaPlace) have higher collapsing pressures (20 cm H_2O) than the collapsing pressure (6–7 cm H_2O) of larger alveoli (150 μm or more in diameter). The net effect would be to have extensive areas of parenchymal atelectasis surrounding a few overinflated alveoli or expanded alveolar ducts, as indeed does occur in the oxygen toxic lung (Fig. 15).

D. *Measurement of Pulmonary Surface Activity in Oxygen Toxicity*

Measurements of pulmonary surface activity in studies of pulmonary oxygen toxicity generally have utilized three approaches:

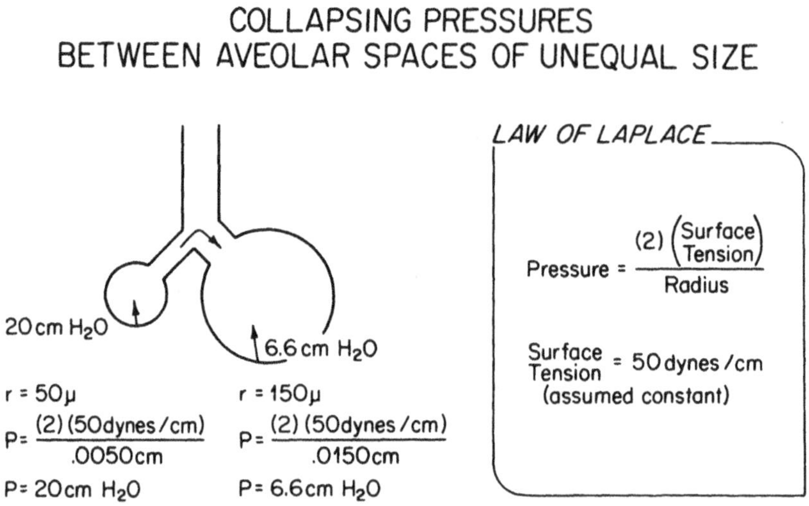

Figure 14. The law of LaPlace applied to two spheres (alveoli) of different diameter connected to a common conduit (airway). At a constant high surface tension force, there will be a tendency for spheres of small diameter to empty into spheres of large diameter (note Fig. 15).

surface tension balance studies, alveolar bubble stability studies, and evaluations of the surface contributions to the volume–pressure relationships in the lung. Figure 16 depicts the use of a modified Wilhelmy surface tension balance to measure surface tension forces. With this technique, surface active materials can be studied best after recovery from the lung by bronchoalveolar lavage. Generally, this approach is most

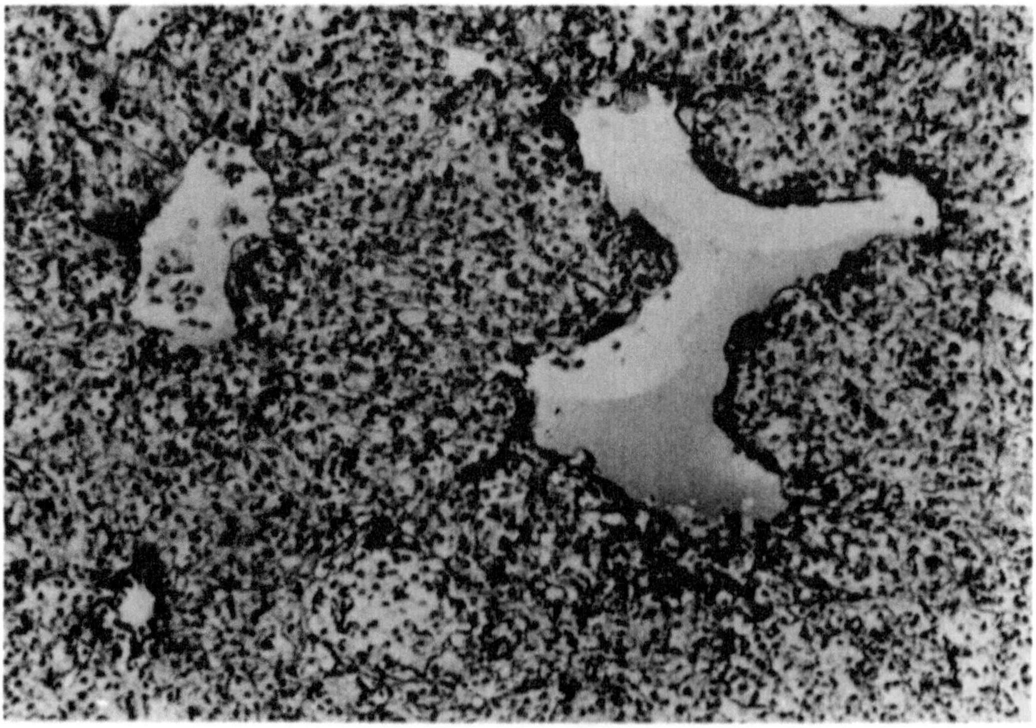

Figure 15. Pulmonary atelectasis adjacent to areas of alveolar duct overdistension in an oxygen toxic lung, presumably due to alterations in alveolar surfactant materials (note Fig. 14).

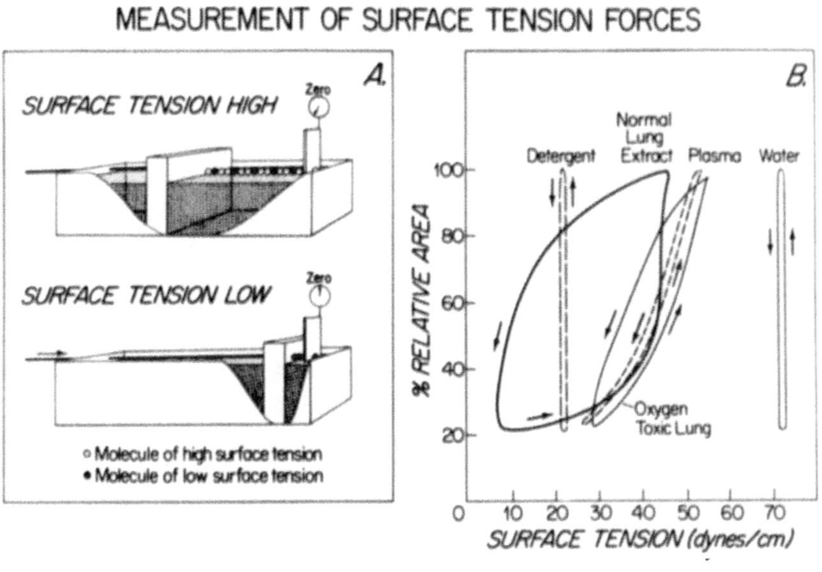

Figure 16. **A.** Schematic representation of a modified Wilhelmy surface tension balance. The balance consists of a trough (T), a movable barrier (B) for compressing the surface film, a platinum float (P) against which surface molecules can adhere, and a strain gauge (F) to record the surface forces. As a surface comprised of both molecules of low surface tension (a surfactant or surface active agent) and molecules of high surface tension is compressed, the molecules of high surface force will leave the surface for the interior of the liquid and the most surface active molecules will be selectively retained at the liquid–gas interface, thus lowering the measured surface tension. ***B.*** Representative surface tension balance tracings, plotting compression–expansion loops of the measured surface tension (in dynes/cm) against relative area (%). A normal lung extract and an extract from an oxygen-toxic lung are included. The normal pulmonary surfactant demonstrates significant hysteresis; the extract from the oxygen-toxic lung has diminished hysteresis and does not lower surface tension to the same degree as extracts from normal lungs.

feasible in large animals, where a progressive loss of recoverable surfactants appears to accompany prolonged exposure to oxygen at very high tensions. Alternatively, the whole lung can be minced or homogenized, and the filtered "extract" studied directly on the surface tension balance (251); data generated by this technique have been much less consistent in the demonstration of altered surface activity following exposure to high tensions of oxygen. It would appear, from a comparison of these two approaches, that recovery by lavage provides a more sensitive assay of a alveolar-lining surface activity, whereas mincing pieces of whole lung provides a mixture of alveolar-lining and tissue-derived surface active forces, as well as the release of a number of potential surfactant inhibitors.

A second approach, depicted in Fig. 17, employs the suspending of small bubbles expressed from the lung (through a cut pleural surface) or recovered from endotracheal lavage fluid in a hanging drop on a hollow ground slide with an overlying coverslip. The bubbles then can be viewed and measured through a light microscope, and their stability quantified by measuring the change in bubble diameter over the course of time, according to the following relationship (317–323,325, 326)

$$\text{Bubble stability index (BSI)} = \frac{d_t^2}{d_i^1} \qquad \textbf{(4)}$$

where d_i is the initial bubble diameter and d_t is the final bubble diameter at some later time, t (usually 20 min). A bubble stability index of 0.80 or greater usually indicates the presence of highly surface active materials (normal mammalian lung often has a sustained BSI of 0.95–0.98 for a prolonged period); a bubble stability index of less than

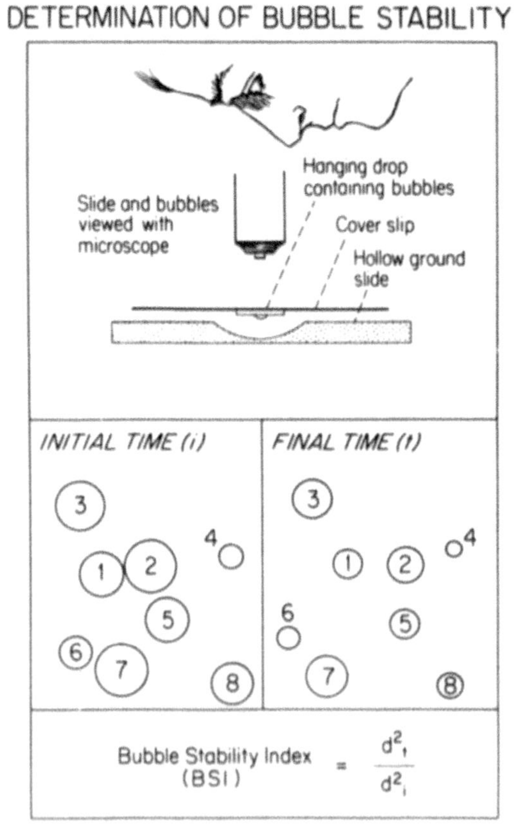

Figure 17. Determination of bubble stability index ($BSI = d_f^2/d_i^2$). Bubbles (usually squeezed from a cut lung surface) are suspended in a hanging drop on a hollow-ground slide. The bubbles can be photographed or measured by eye through a light microscope at an arbitrary initial time (t_i) and at a later time (t_f), usually 20 min.

0.80 generally indicates a loss of pulmonary surfactant or the presence of a pulmonary surfactant inhibitor (291). The lower the bubble stability index, the higher are the alveolar surface tension forces (the oxygen toxic lung may have a BSI of 0.10–0.20, for example). This approach has the advantage of measuring pulmonary surface activity in bubbles approximately equivalent in size to alveolar spaces. The results in such evaluations in oxygen toxicity, however, have been inconsistent, for several different reasons. Perhaps, the sometimes focal nature of the response in the oxygen toxic lung delivers bubbles from both involved and noninvolved areas, increasing the variation in measured results. Furthermore, the technique, by its very nature, may selectively deliver bubbles from more aerated than nonaerated alveoli, less so from partially atelectatic areas, and not at all from those collapsed areas where surface tension forces are highest.

Finally, a third approach that has been used to quantify alveolar surface forces indirectly is the determination of the volume–pressure relationships of either excised lungs or the lungs in situ (Fig. 18) (219). Normally, there is a relatively predictable relationship between the inflation pressure (or deflation pressure) and the resultant lung volume of the lung. When the lung is filled with saline or another insufflating liquid, the usually present air–surface interface is eliminated and the resultant volume–pressure relationship reflects only the tissue elastic recoil forces of the lung. In constrast, when the lung is filled with a gas, the resultant volume–pressure relationships reflect a combination of both the tissue elastic recoil forces and the tissue–air surface elastic recoil forces. The difference between the two curves (Fig. 18), then, represents the effects of surface active forces alone. With high surface tension forces, as occurs in oxygen toxicity, there is a shift of the gas-filled pressure–volume relationship to the right and downward (Fig. 18). Since the internal surface of the lung is comprised of greater than 99% alveolar surface and well less than 1% airway surface, this approach provides a reliable evaluation of alveolar surface tension forces. A careful "volume history" must be obtained however, and as the lung becomes progressively consolidated in the course of oxygen toxicity the volume–pressure relationship must be normalized for percent relative maximum inflation.

E. The Biochemistry of the Surfactant in Oxygen Toxicity

A number of studies have attempted to characterize directly or indirectly the lipid composition of the pulmonary surfactant (33, 63,82,113,115,146,164,217,268,292,295, 399, 425); several evaluations of the effect of prolonged exposure to high tensions of oxy-

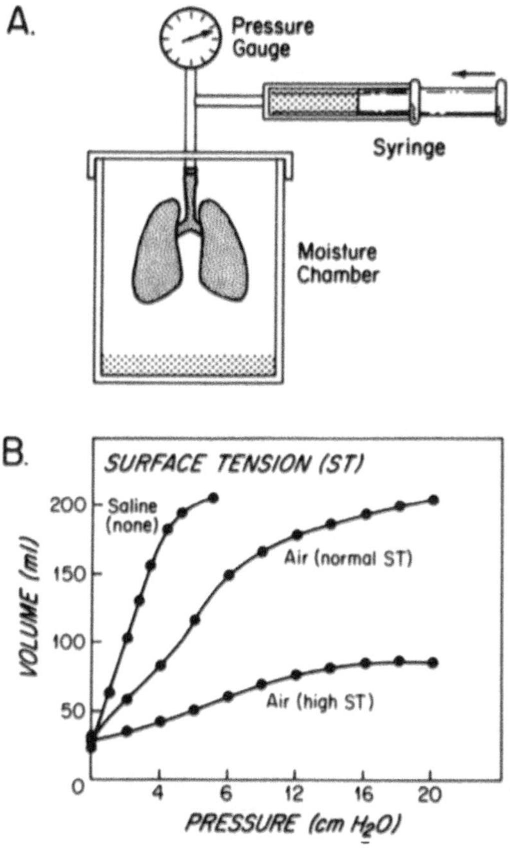

Figure 18. A. The volume pressure relationships of the lung can be determined by inflating the lung with a measured or fixed volume (ml) and recording pressure (cm H_2O). Usually the lung is inflated from a collapsed state (the "volume history" is is determined) with a gas of known composition, temperature, and relative humidification. **B.** Volume–pressure relationships for a saline-filled lung and for air-filled lungs of normal and of high surface tension forces. By eliminating the tissue–gas interface with saline inflation–deflation only tissue-elastic forces, and not surface tension forces, are measured. The difference (arrows) between the air-filled and the saline-filled volume–pressure relationships is due to alveolar surface forces; with high surface tension forces, as occurs in pulmonary oxygen toxicity, this difference is increased.

gen on the surfactant system, however, have been inconclusive (1,124,146,217,267,292, 348). Some of these inconsistencies appear related to the method of phospholipid recovery from the lung, with those studies employing the harvest of surfactant by endobronchial lavage having the highest reproducibility and those approaches which utilize minced lung extracts having the greatest variability (191,292–294). The most surface active chemical species recoverable from the lung appear to be the highly saturated phospholipids, particularly dipalmitoyl lecithin (282). At prolonged exposure to 0.7–1.0 atm oxygen tension, there appears with exposure duration to be a progressive reduction in the percent of total lung lipid recoverable as phospholipid or, specifically, as lecithin. In addition, in the lecithin fraction of the recovered lipids, the relative proportion of palmitic acid is reduced with increasing severity of exposure, with a concomitant reduction of total saturated fatty acids (Fig. 19). The biochemical reduction in fatty acid saturation and in the total lecithin recoverable is accompanied clinically by respiratory distress. With further exposure to high tensions of oxygen, and with the development of pulmonary edema and the entry of plasma protein components into the alveolar spaces, there is a progressive further decrease in total saturated fatty acids, and specifically in palmitic acid.

These changes in phospholipid composition have significant physiological significance. It appears, for example, that the surface activity of phospholipids is determined, in part, by the degree of their side-chain fatty acid saturation (185,293). This is schematically depicted in Fig. 20, along with relevant tracings of modified Wilhelmy surface tension balance measurements. Phosphatidyl choline with a high proportion of unsaturated fatty acids, harvested from an oxygen toxic lung, exerts high surface tension forces (23–25 dynes/cm). When the side-chain fatty acids of this isolated lipid from the oxygen-toxic lung are hydrogenated experimentally, producing a fully saturated phospholipid, surface activity is restored. Thus, reduction of total relative phospholipid saturation as a function of prolonged exposure to high tensions of oxygen appears to be the key to the pathogenesis, induced by elevated surface tension forces, of the physiological abnormalities demonstrable in the oxygen toxic lung.

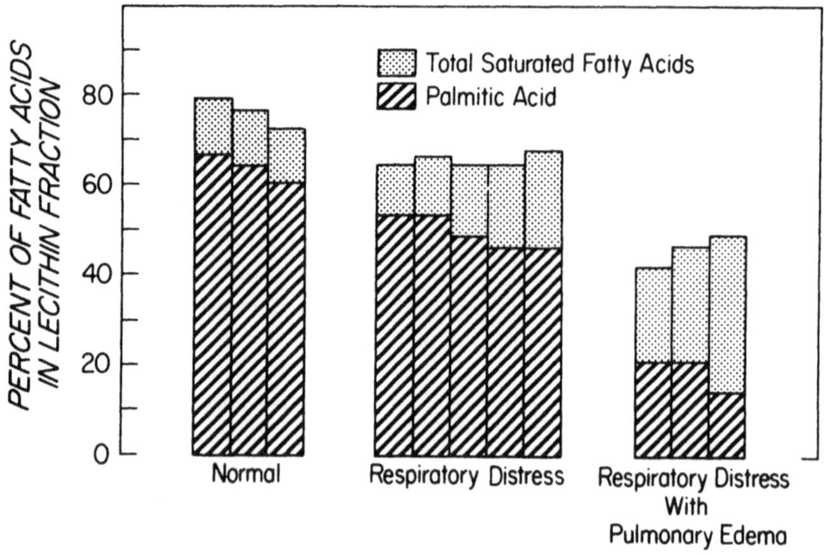

Figure 19. The effect of exposure to oxygen on fatty acid saturation in the lecithin fraction of phospholipids recovered by endobronchial lavage from the normal and from the oxygen toxic lung. With progressive exposure to increased tensions of oxygen, and with the development of respiratory distress and pulmonary edema, there is a progressive reduction in total saturated fatty acids.

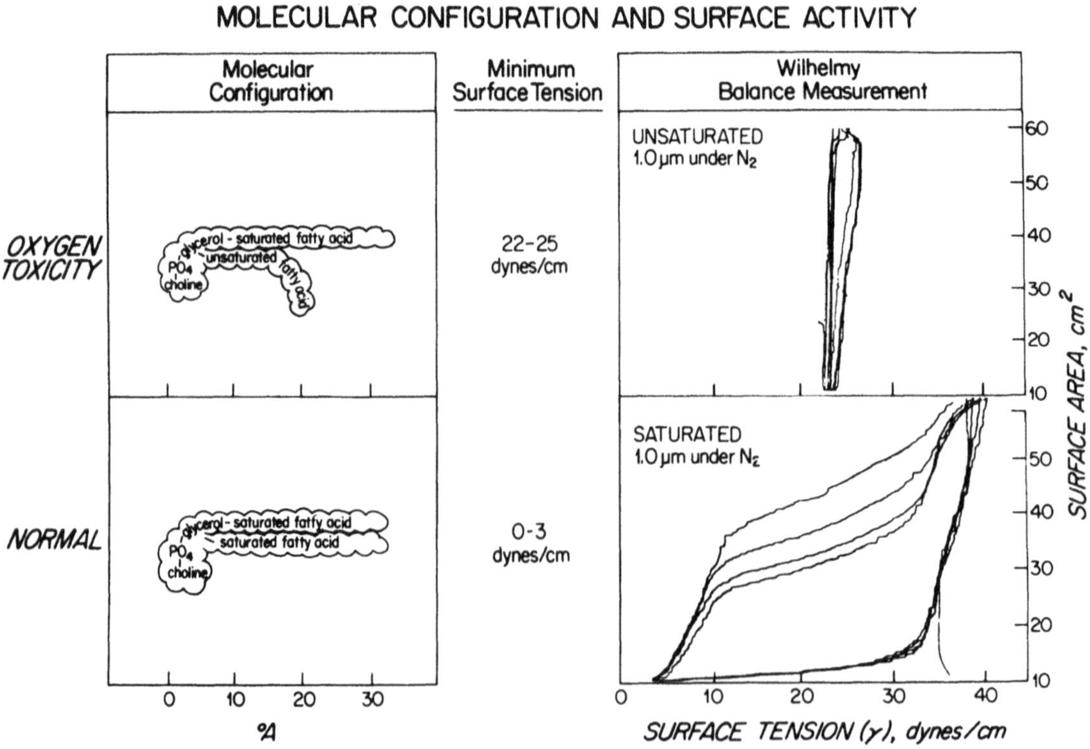

Figure 20. The effect of molecular configuration on surface activity. With an increase proportion of unsaturated side-chain fatty acids, lecithins recovered from the oxygen toxic lung lose their surface activity and exert high surface tension forces (with minimal hysteresis) when measured on a modified Wilhelmy surface tension balance. When the unsaturated molecule is experimentally hydrogenated, and the side-chain fatty acids saturated, surface activity is restored, with low surface tension values and normal hysteresis.

F. The Surfactant and Pulmonary Edema

As depicted in Fig. 21, alterations in the surface activity of phospholipids lining alveolar spaces also may be very important to the pathogenesis of pulmonary edema in the oxygen-damaged lung (78). In the normal lung, where alveolar surface tension forces are on balance very low, the plasma oncotic pressure is balanced by, for the most part, a capillary perfusing blood pressure and a comparably significant tissue oncotic pressure, with small additional contribution from the alveolar-lining surface pressure. Assuming all other variables are constant, an increase of alveolar surface tension forces (to 25–30 dynes/cm or even more) would disrupt this balance, with a net force transmitted across the blood–air barrier that would favor, on the basis of the increased alveolar forces alone, transudation of plasma proteins and fluid from the parenchymal capillaries into the alveolar spaces.

VII. Oxygen Toxicity and Pulmonary Infection

Surprisingly, the effects of exposure to high tensions of oxygen on pulmonary host defenses and the direct effects of oxygen on the development of pulmonary infection have not been extensively studied. That is primarily in that pulmonary bacterial infection is an extremely frequent and probably the most significant complication of clinical oxygen toxicity in man, as well as in other forms of the adult respiratory distress syndrome in humans. A limited number of studies indicate that oxygen, as well as other oxidants, depress intrapulmonary antibacterial defenses in a dose-dependent manner (189,193–196,200,201,250). In that the antimicrobial defenses of the lung in animals are comparable, in many respects, to similar defenses in man, oxygen administration at high tension may have similar effects and important clinical implications in humans.

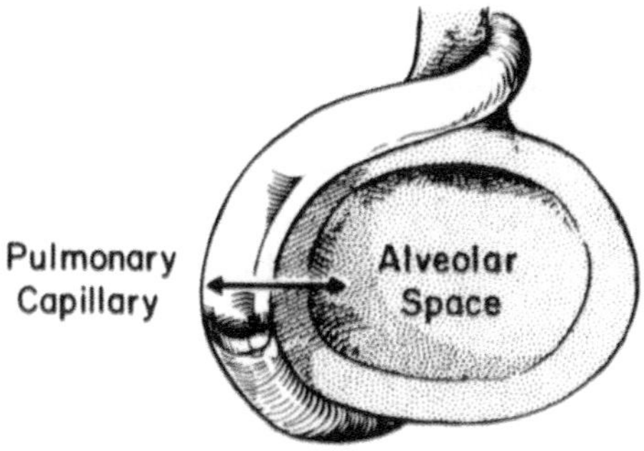

	←		→		→		→
	PLASMA ONCOTIC PRESSURE	=	CAPILLARY BLOOD PRESSURE	+	TISSUE ONCOTIC PRESSURE	+	SURFACE PRESSURE
NORMAL VALUES	37	=	15	+	18	+	4
OXYGEN TOXICITY	37	<	15	+	18	+	25 to 30

Figure 21. The normal balance across the blood–air barrier between plasma oncotic pressures within the pulmonary capillaries and capillary blood pressure and tissue oncotic pressures. Unlike other organs, surface forces exerted within the alveoli are important to this balance. With an increase in surface tension forces, as occurs in pulmonary oxygen toxicity, the balance is disrupted in favor of a net transudative force from the pulmonary vasculature into the alveolar spaces. After Clements (78).

A. Airway Defenses

Although the collective internal surface of the airways is relatively small (less than 1% of the entire internal lung surface), the airways, as conduits to the external environment, are essential to normal pulmonary function (186). The airway defense mechanisms are integrally related to the defense network of the respiratory system as a whole, and consist basically of two main components: mucus and cilia (383) (Fig. 22).

The noncellular secretory materials that line the airways include more than simply the mucous and serous products derived from submucosal cells and airway-lining epithelial secretory structures (171). Serum and tissue transudates, as well as inflammatory exudates in certain circumstances, combine with mucous and serous secretions to form what is commonly termed *sputum*. Serum-derived constituents comprise, by dry weight, a five times greater macromolecular contribution to the total sputum weight than any other airway secretory product. Glycoproteins, formed as conjugated protein–carbohydrate compounds, are produced by mucous cells, serous cells, and the Clara cells of the distal airway epithelium. In the airway, the accumulated noncellular materials layer into a gel and a sol phase. The *gel phase* primarily contains the relatively insoluble epithelial-derived glycoproteins and the *sol phase* is comprised of the more soluble glycoproteins. A number of specific immunoglobins also are secreted into or retained in this airway-lining noncellular layer.

Cilia are cellular projections of specialized airway-lining epithelial cells, and have a motility of their own. Cilia basically are "naked" projections and must be bathed in a overlying serous, watery blanket. Ciliary movement is complex and beyond the scope of this review (171). In that the specialized cells to which the cilia are fixed are structurally anchored to the airway itself, any movement of the cilia will be transmitted to a propulsion of the closely apposed overlying seromucous layer. The net effect of ciliary movements, then, is to transport the seromucous layer up and out of the lung.

B. Oxygen Toxicity and Airway Defenses

Exposure to oxygen at high tensions reduces ciliary function, reduces intra-airway mucous flow rates, and, with the poisoning of secretory cells in the airway-lining epithelium and submucosal glands, reduces the effective secretion of antimicrobial products (41,247).

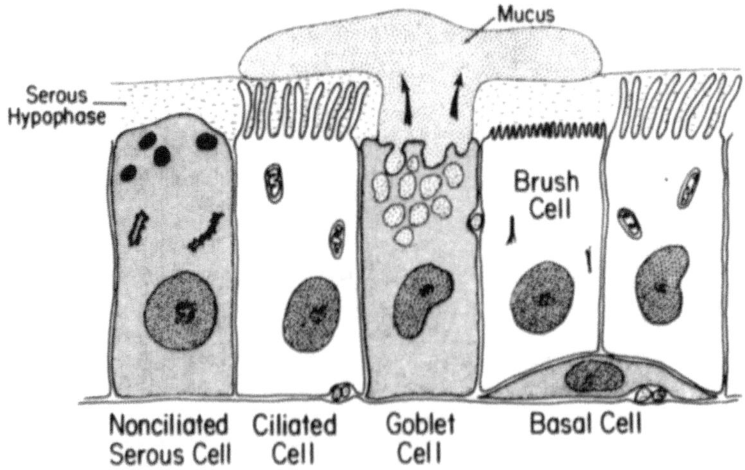

Figure 22. Schematic representation of the structure of an airway epithelial surface. Airway defenses are comprised of ciliated cells with projections that move accumulated secretions out of the lung, and of an overlying seromucous layers. Microorganisms landing on the seromucous blanket can be inactivated *in situ* by antimicrobial enzymes and immunoglobulins or can be physically transported out of the lung.

In addition, airway obstruction in oxygen toxicity enhances conditions for microbial growth in the distal lung (41,239,247). Although most infective organisms are inherently of such small size, or are carried on droplets of such small size, that few will deposit on airway surfaces during inhalation, these alterations in airway defenses render the lung more susceptible to infection that potentially do so.

C. *Alveolar Defenses*

Greater than 99% of the internal surface of the lung is made up of alveolar walls, alveolar ducts, and terminal bronchioles. Most potentially infective microorganisms are carried on droplets of 2–5 μm or less in diameter, and as such they will impact on inhalation not on the airways but primarily on this extensive alveolar surface. The cell most responsible for the defense of this distal lung is the alveolar macrophage (365) (Fig. 23); in addition, the distal lung surfaces are coated in part with noncellular substances, including immunoglobulins and the pulmonary surfactant, that aid the alveolar macrophage in its two key functions, phagocytosis and intracellular inactivation of microbes and potentially harmful nonviable particles (188,190). If the challenge to the distal lung is greater than this integrated surface-lining alveolar-macrophage defense network can handle, additional defenses can be recruited from nonpulmonary sources by a variety of processes (188,190).

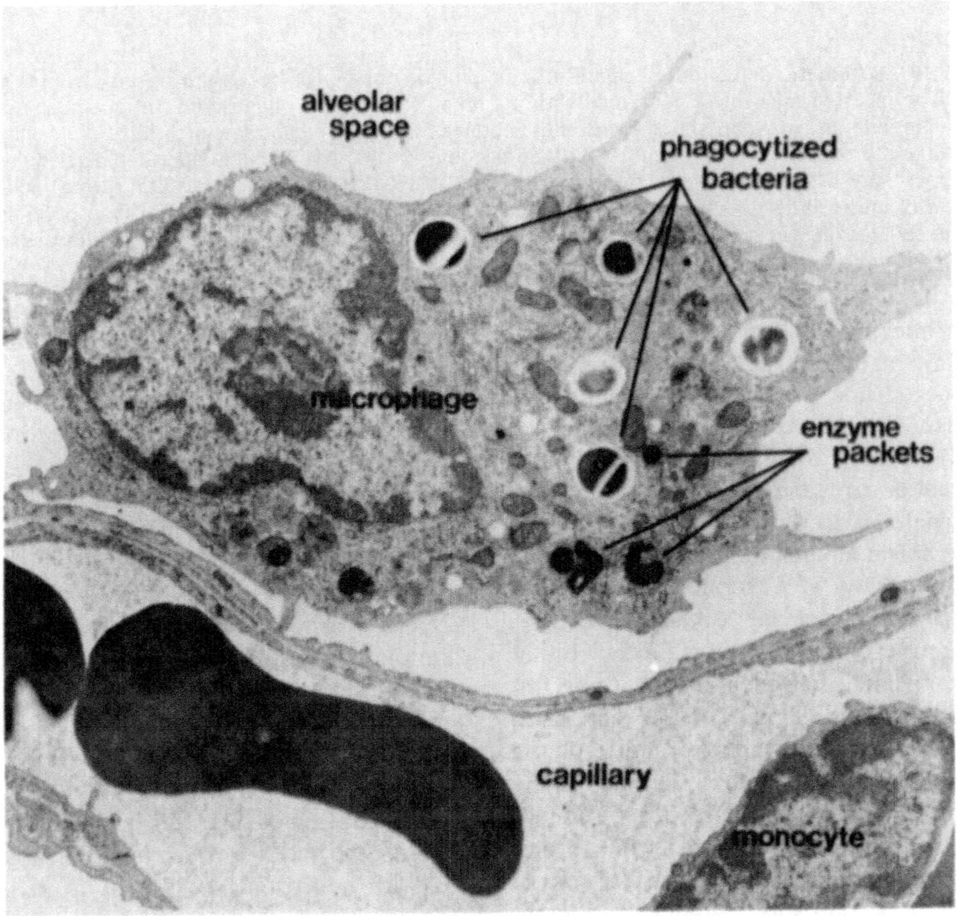

Figure 23. A pulmonary alveolar macrophage on the alveolar surface, containing recently ingested *S. aureus*, Radioisotopic, electron microscopic, and immunofluorescent studies demonstrate that inhaled organisms are engulfed and inactivated at the alveolar surface by macrophages, the key pulmonary host defense cell.

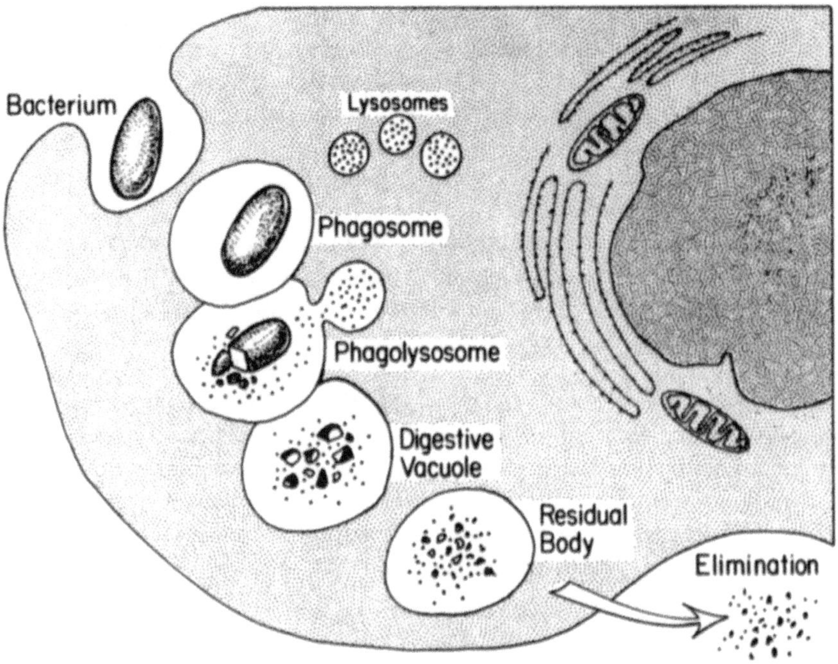

Figure 24. Schematic depiction of engulfment or phagocytosis of a bacterium by an alveolar macrophage. The ingested microbe is contained within a membrane-bound phagosome. Intracellular fusion of phagosome with enzyme-containing lysosomes produces a phagolysosome, which in turn by enzymatic digestion and processing of the engulfed particle becomes a digestive vacuole and then a residual body. Lysosomes may be primary (generated *de novo* by the cell) or secondary (resulting from incomplete digestion of intracellular or extracellular products). Some of the processed material may be eliminated from the cell, leaving (in some instances) an enzymatically active residual body, the secondary lysosome.

The mechanisms by which pulmonary alveolar macrophages inactivate inhaled microorganisms are not fully understood. Because the ultralow surface tension of alveolar lining materials imparts unique detergent or spreading properties on anything with which it comes in contact, it is likely that any organism landing on the alveolar surface is rapidly coated with a monolayer or more of a lipoprotein surfactant (80,81). Coating of an organism or a noninfective particle by the surfactant lining has several potentially important functions. First, it would appear to facilitate transport or approximation of the microorganism or the inert particle to the alveolar macrophage, where it can be engulfed and physically removed from further contact with the alveolar surface (Fig. 24) (269,274). Second, this coating by surfactant will carry with it other macromolecules (including antimicrobially active immunoglobulins opsonins, and potentially antimicrobial enzymes) that may enhance phagocytosis and intracellular destruction and digestion (188,193,269–279); the role of the surfactant per se in facilitating phagocytosis of microorganisms or killing of microorganisms has not been adequately characterized. Finally, as depicted in Fig. 25, in some species (including, based on limited data, in man) coating of certain bacteria by alveolar lining materials prior to engulfment of alveolar macrophages appears essential to normal intracellular killing (190, 240,296); the process by which this is achieved is not known, but may involve coating the bacteria with a peroxidizable lipid or with a substance essential to activating the macrophage intracellular bactericidal systems.

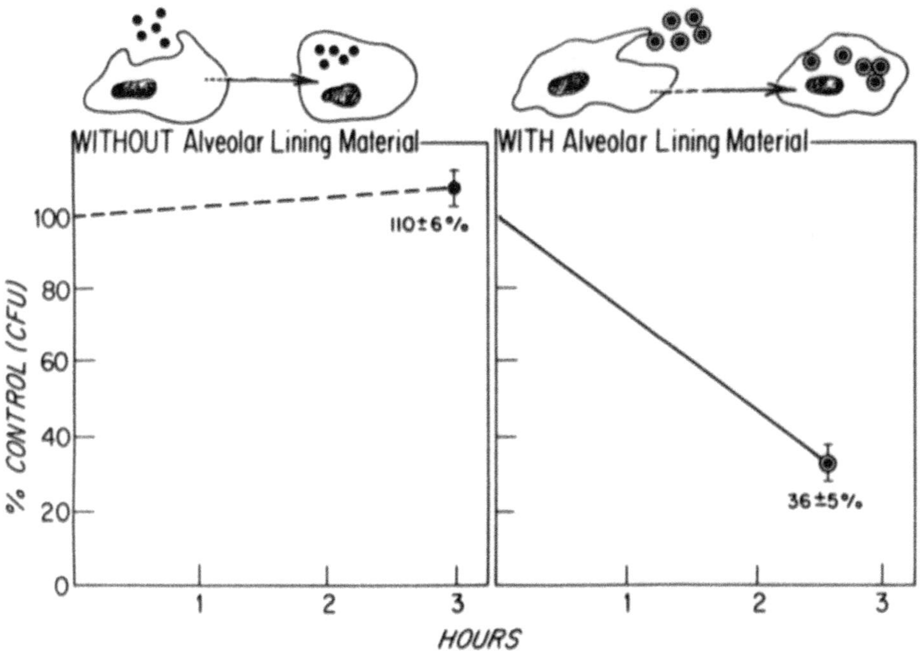

Figure 25. Phagocytosis and intracellular killing of *S. aureus* with and without coating of the bacteria by a cell-free, alveolar-lining surface active material prior to engulfment. Coating of the microbes by the surfactant fraction of alveolar-lining material does not appear to affect the rate of phagocytic uptake. Only those organisms that are coated with alveolar-lining material, however, are inactivated or killed within the macrophage. Intramacrophage staphylococcal viability is expressed relative to macrophage-free control colony forming units (CFU) of the bacteria in culture over a comparable incubation period.

D. *Oxygen Toxicity and Alveolar Antibacterial Defenses*

Exposure to higher than normal ambient oxygen tensions impairs the antibacterial defenses of the lung in a dose-dependent manner, relative to either duration or level of exposure (111,193,194,199,263). This is shown in Fig. 26 for data derived from the pathogen-free rat challenged with an aerosolized inoculum of *Staphylococcus aureus* (193). Experiments with other bacteria and other animal species show comparable results (189,193–196,200,201,204,205); comparable data for humans and subhuman primates are sparse, however.

Over the first 48 hrs or so of exposure of the experimental animal to 1 atm of oxygen (Fig. 26), the dose-dependent depression of intrapulmonary antibacterial defenses is relatively linear. This is a period of response in this animal model wherein the lung is relatively edema free and void of the structural pathological changes characterisitic of end-stage oxygen toxicity. It appears that this early impairment of alveolar antibacterial defenses may be related to as yet undefined changes in the alveolar lining material (200), and not to any significant direct effect on the pulmonary alveolar macrophage. With sustained exposure, the bactericidal capacities of both the alveolar lining material and the alveolar macrophage are affected and intrapulmonary bacterial replication of an aerosolized challenge exceeds intrapulmonary bacterial inactivation (Fig. 26).

Alveolar macrophages harvested during the early phases of impaired bactericidal functions of the lung are metabolically normal and have subtle structural alterations demonstrable only by morphometric techniques (202), whereas during this early phase

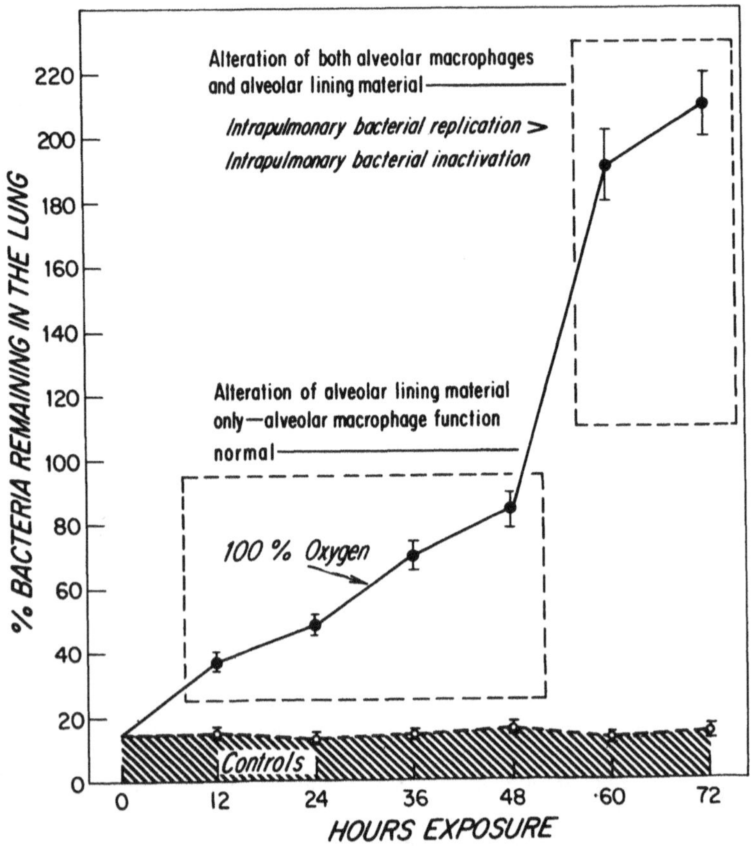

Figure 26. The effect of continuous exposure to 100% oxygen at 1 atm tension on intrapulmonary antibacterial activity. Control animals inactivated all but approximately 14–16% of the test microbe (*S. aureus*) over 6 hr of intrapulmonary incubation after inhalation of an aerosolized bacterial challenge. Exposure to 1 atm of oxygen depressed intrapulmonary antistaphylococcal defenses in a direct dose-dependent relationship over the first 48 hr of exposure, with intrapulmonary bacterial replication exceeding intrapulmoanry bacterial inactivation thereafter. The experimental animal in these studies is the pathogen-free rat.

the role of alveolar lining material in stimulating intracellular bacterial killing is diminished. With longer exposures to oxygen at high tensions, the alveolar machrophage undergoes structural and functional alterations that diminish its capacity to phagocytize and to inactivate organisms once they are engulfed (Fig. 27). Furthermore, as edema develops, and as cellular debris and hyaline membranes accumulate in alveolar spaces, surface phagocytosis is impaired (205), and the bactericidal capacity of the distal lung is reduced by multiple influences to the point where intrapulmonary bacterial replication of a microbial challenge exceeds intrapulmonary bacterial inactivation.

VIII. Biochemical Mechanisms of Pulmonary Oxygen Toxicity

It is almost impossible to identify precisely and to investigate adequately the biochemical target sites of oxygen toxicity in the lung. The primary reason for this difficulty is that the lung is a cellular heterogeneous organ, the components of which are difficult to separate

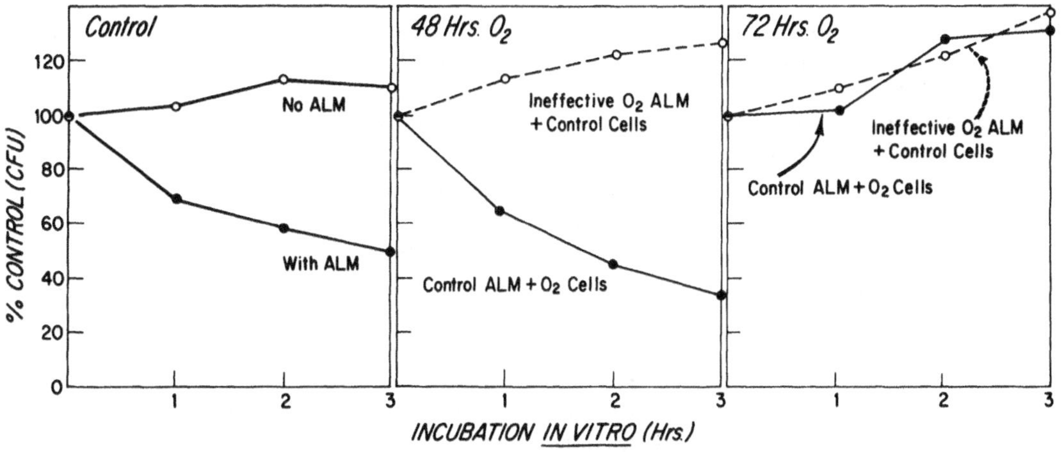

Figure 27. The effect of exposure of the intact experimental animal (pathogen-free rat) to 1 atm of oxygen on the *in vitro* intracellular inactivation of *S. aureus*. Alveolar macrophages harvested by bronchopulmonary lavage from the lungs of control animals intracellulary inactivated bacteria precoated with surface active alveolar lining materials (ALM), but not uncoated bacteria. After 48 hr exposure to 1 atm oxygen, macrophages from oxygen-exposed animals could still inactivate bacteria coated with normal alveolar lining material from controls (but not the abnormal alveolar lining material from oxygen-exposed animals). The alveolar lining material from oxygen exposed animals was not effective, even with bacteria phagocytized by control cells. After 72 hr or more of exposure to oxygen, neither component (macrophages or alveolar lining material) from oxygen treated animals was bactericidally effective.

and to study independently. Whereas the sequence of the pulmonary morphological adaptations to hyperoxic conditions now have been carefully mapped by several investigators at varying levels of exposure in different species, it is not equally easy to separate individual cell types for the generation of comparable biochemical information. The morphometric characterizations indicate a sequential differential by cell type in the development of pulmonary oxygen toxicity, complicating even further the interpretation of biochemical assays performed on cellular heterogenous lung preparations.

Despite a lack of clear identification of specific target sites, however, it is generally held that hyperoxic conditions inactivate enzyme systems essential to cellular metabolism and both directly and indirectly affect cell membranes. The best available information on these reactions is derived from experiments on isolated cellular or subcellular systems, reviewed extensively elsewhere in this series of contributions (118, 121). In general, the biochemical mechanisms of oxygen toxicity have been grouped into four broad categories: (1) sulfhydryl enzymatic reactions, (2) release of cellular mediators, (3) lipid peroxidations, and (4) oxidizing radicals of molecular oxygen.

A. Sulfhydryl Enzymatic Reactions

Evidence from in vivo and in vitro research efforts indicate that exposure of the lung (or lung components) to increased oxygen tensions can adversely alter subcellular metabolism by the oxidation of enzymes containing sulfhydryl groups. At very high exposure levels (above 1 atm), sulfhydryl and dehydrogenase activity in the lung as a whole is diminished well within 1 hr of oxygen treatment (215,216), during which time there are few (if any) demonstrable morphological changes. With longer exposure durations at very high oxygen tension, total pulmonary disulfide activity apparently increases (215, 216). Other enzymes containing sulfhydryl reaction sites (alkaline phosphatase, 5′-

nucleotidase, adenosine triphosphatase, certain lipases), as well as the net nonprotein sulfhydryl content of the lung, are reduced at exposure to very high oxygen tensions (3–5 atm), as reported by several investigators (168,215,216,305). These effects can be blocked, to some degree, by the administration of exogenous seratonin (168), pentobarbital and cysteamine (214,215), dibenzyline and reserpine (305), and by adrenal medullectomy (305).

Whether or not these responses are demonstrable or important at exposure tensions of 1 atm or less is unclear. There is a slight reduction in dehydrogenase activity at exposure conditions of 0.9 atm for 72 hr in one species (26), for example, but there is a general suggestion that susceptible sulfhydryl-containing enzymes are protected from oxidation at exposure tensions of much less than 1 atm by soluble reducing substances demonstrable in lung tissue.

B. Cellular Mediators

Several cellular mediators, particularly serotonin and histamine, have been invoked as biochemical determinants of pulmonary oxygen toxicity. Histamine, a depressor amine derived from histidine by decarboxylation, is released from mast cells and other tissue sources in the lung (188). It can cause constriction of both airway and vasculator smooth muscle (188). The histamine content in the lung appears to increase early in the course of oxygen exposure (119), and the exogenous administration of histamine enhances the severity of the pulmonary pathology (157). It has been implied that histamine, perhaps through a postcapillary venoconstrictive mechanism, promotes pulmonary edema and transcapillary movement of plasma proteins. In support of this, the edemagenic response of the lung to high oxygen tension is diminished by pre-exposure adminstration of promethazine, metyramine, and thiazinanium (128). Contradictory to this, however, pulmonary histamine depletion and the administration of promethazine do not extend the average survival time of animals exposed to 1.0 atm oxygen (157). The differences in these findings may be due to species variation or to as yet undefined other factors (128, 157).

The lung is also rich in stores of serotonin (5-hydroxytryptamine), a smooth muscle constrictor stored in platelets and in mast cells. It is inactive in the bound state, but once released becomes bioactive in very small concentrations and as such can be an important mediator of vasoconstriction and pulmonary edema (222). At exposure to very high oxygen tensions (6 atm), its content in the lung is reduced dramatically (61,62). In addition, prevention of mast cell degranulation by phenobarbital or by the administration of certain proteinase inhibitors (trasylol) alters the severity of pulmonary oxygen toxicity (61,62). Exogenous administration of serotonin, on the other hand, prolongs survival time and the duration of exposure to toxicity at high (>1.0 atm) exposure tensions (43,238,406). Reserpine-induced depletion of serotonin does not prolong the survival times of guinea pigs and rats in high exposure tensions, however (418). Thus, on review of the existing literature, one is left with a less than clear understanding of the role of these mediators in the pathogenesis of oxygen toxicity.

C. Lipid Peroxidation

Lipid peroxidation, involving primarily lipids in cell membranes, by high oxygen tensions may be important to the pathogenesis of oxygen toxicity. Peroxidation may occur by oxygen or through oxidizing radicals, as described below.

D. Oxidizing Free Radicals

There is considerable indirect evidence that suggests that the inactivation of enzymes and enzyme cofactors by molecular oxygen involves the intermediate formation of *oxi-*

dizing free radicals (87,120,130,131,147–149,169,287). Similarities have been observed, for example, between the toxic effects of oxygen and the biological effects of irradiation, where free radicals are also involved (38,72,110,125,134,215,255,394,403). It was proposed as early as 1933 that oxygen toxicity might be related to the formation of oxygen free radicals, chain reactions, and destructive oxidations (30,150), with the suggestion that aerobic animals developed defenses to enable them to survive in a normal atmospheric oxygen. These mechanisms of protection from harmful oxidation may be overwhelmed, however, by exposure to higher than normal ambient tensions of oxygen.

The generation of reactive species of oxygen, their effects on biological systems, and the evolution of antioxidant defenses against oxygen radicals is discussed at length in separate chapters contributed to this monograph (118,121). Superoxide anions may be particularly important in the genesis of oxygen induced cell damage. Superoxide is produced by a wide variety of enzymatic and nonenzymatic oxidations, and is important to such divergent biological phenomena as granulocyte phagocytosis and mitochondrial respiratory chain activity (99,264,289). The rate of superoxide production for many biological reactions has been shown to be directly proportional to available oxygen tensions (289). Although the precise role of superoxide anions in the pulmonary cytotoxicity of hyperoxia is yet to be fully elucidated, superoxide anions can react with hydrogen peroxide to produce the hydroxyl free radical, one of the most reactive free radicals known (100). Free radical-initiated peroxidation of unsaturated lipids may destroy essential cellular components and form lipid peroxides which are themselves cytotoxic. Aerobic biological systems possess a number of potentially protective antioxidant cellular defense mechanisms, discussed in greater detail elsewhere in this monograph (118,121). Two of these systems involve mechanisms for detoxifying superoxide and the lipid peroxides.

IX. Sequence of Injury in Pulmonary Oxygen Toxicity

Most surely, there are reactions at a molecular level within the lung that occur immediately upon exposure of the host to high oxygen tensions. How these evolve into detectable biochemical or morphological alterations is essentially conjectural with our current state of knowledge. In that it is impossible to investigate sequential responses of exposure to high oxygen tensions in humans in any degree of significance, most of our information must of necessity be derived from animal studies, where the sequential response of the normal lung has been characterized in considerable detail (71,224,227,232,233,405). Reaction rates in potential cytological or subcellular target sites will be governed by biological time constants that differ, however, as a function of the species (or even the strain) of animal studied, the cell type under consideration, the exposure tension, and many other factors. In addition, the sequence of cellular response may be quite different under different oxygen exposure conditions.

A. Response of the Vasculature

Whereas our understanding of the biochemical changes that occur in the lung in hyperoxia are at present inadequately quantified, the progression of structural alterations in pulmonary oxygen toxicity (Fig. 28) have been more extensively quantified, especially by a series of carefully performed morphometric studies by Weibel, Kistler, Kapanci, and co-workers (224,232,233,411). In addition to an early impairment in the function of the pulmonary alveolar lining material, it would appear that the earliest demonstrable structural changes involve alterations in the gas-exchanging surfaces of the lungs (Fig. 29), and especially in the pulmonary capillary endothelium (1,49,71,224,232,233,266,277,299,300,314,333,355,363,408,425).

Initially, the clear cytoplasm of the capillary endothelium (Fig. 29) appears on

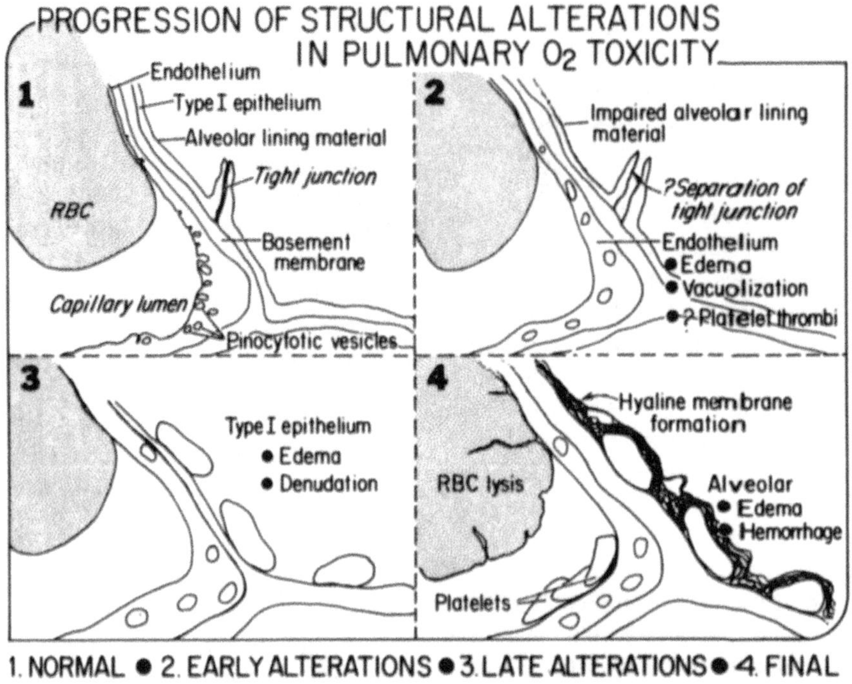

Figure 28. The sequential progression of structural alterations in pulmonary oxygen toxicity. In stage 1 (note Fig. 29), the normal air–blood barrier is presented in cross section. The early morphological alterations of the oxygen toxic lung are summarized in stage 2, and the late alterations in stage 3. The final, and usually fatal manifestations, for pulmonary oxygen toxicity include alveolar hyaline membrane formation, edema, hemorrhage, intravascular hemolysis, platelet thrombi, and destruction of the air–blood barrier.

ultrastructural examination to be swollen or edematous, with morphometric analyses demonstrating a relative increase in the capillary-to-parenchymal volume density (224). The endothelium, which normally contains numerous small pinocytotic vacuoles in its cytoplasm, becomes more extensively vacuolated (both the number and size of the cytoplasmic vacuoles increase). Relatively inconspicuous changes in the intracellular junctions of endothelial cells (and of epithelial cells, as well) develop, with eventual complete separation of these junctions as edema becomes more paramount. The pathophysiological mechanism responsible for these morphological changes are not known. As soon as endothelial surface alterations occur, platelet thrombi may form and release serotonin and other potentially harmful mediators. Eventually, the endothelial replication rate falls, there is a 30 to 50% loss in endothelial cell volume density and capillary endothelial surface area on morphometric analysis, and intravascular thrombosis becomes significant (224).

B. Alveolar Epithelium

The alveoli are lined by two types of cells: a squamous or type I alveolar lining epithelium and a cuboidal or type II epithelium. Both types of epithelial cells rest closely apposed to an underlying basement membrane. The type I epithelial cell (Fig. 29) is a flat and very morphologically advanced cell in that it can extend to frequently cover, in part, three or more alveolar spaces. For the most part, its cell cytoplasm is very sparse and the cell, comprising the alveolar component of the air–blood barrier, is thin and tenuous. Early in the course of oxygen toxicity, the type I epithelium becomes swollen and denuded, contributing to the accumulating intra-

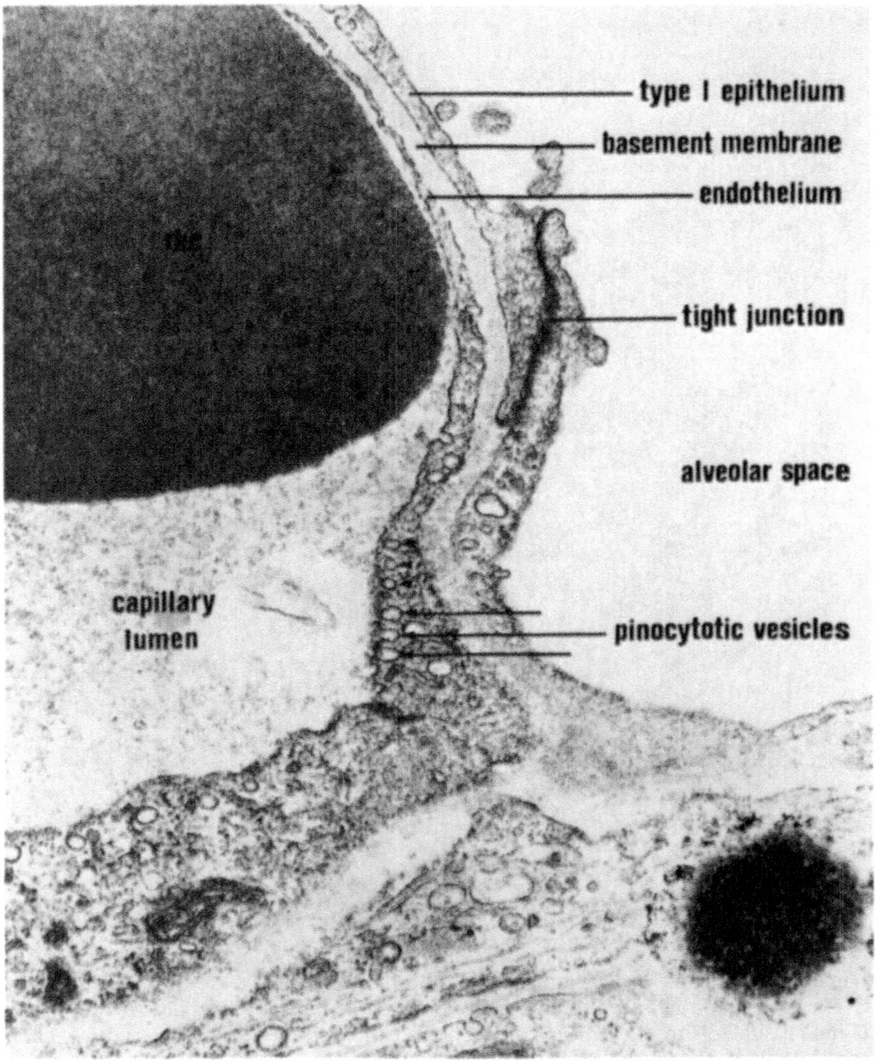

Figure 29. An electron micrograph of the normal air–blood barrier (usually 0.5 μm or less in thickness, comprised of an alveolar-lining endothelium, and interposed basement membranes. Normally, the epithelial cells are connected by tight junctions and fluid transfer occurs via pinocytotic vesicles. An alveolar-lining layer of the pulmonary surfactant, at times only a molecule in thickness is not visible in this preparation. A part of an interstitial fibroblast, important in both the pathogenesis and in the repair of the oxygen-toxic lung, is shown in the lower portion of the micrograph. These components are altered in a predictable sequence in the course of pulmonary oxygen toxicity.

alveolar cellular debris. In severe oxygen toxicity, up to 90% or more of the alveolar epithelium is destroyed, leaving a bare basement membrane covered sparsely with fibrin strands. Because the type I cell is morphologically advanced, with its extended spatial coverage, it cannot readily replicate when injured.

The type II alveolar-lining epithelium (Fig. 30), also closely apposed to an underlying basement membrane, is cuboidal in shape, does not have extended projections, and covers only a limited portion of one alveolus. This cell, also called the granular pneumocyte, is metabolically active, is the presumed origin of the pulmonary surfactant (59), and because of its contained displacement can readily replicate. As such, it is key to repair

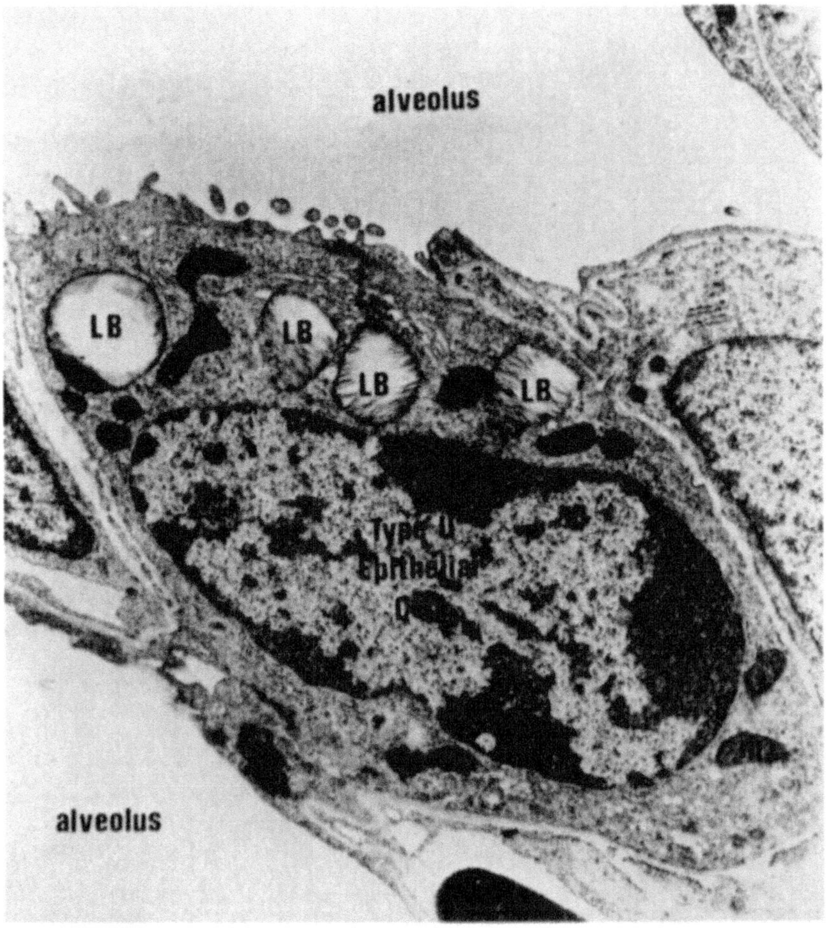

Figure 30. A representative type II epithelial cell, or granular pneumocyte. These cells are cuboidal in shape and generally contribute to a portion of only one alveolar space and rest on an underlying basement membrane. The type II epithelial cell, rich in mitochondria (M) and endoplasmic reticula, is metabolically active. It is the presumed site of synthesis of the pulmonary surface active material, which appears to be stored intracellularly as lamellated bodies (LB) in the granular pneumocyte.

of the injured alveolar cellular lining and thus to the recovery phase of oxygen toxicity. In recovery, or the so-called proliferation phase of oxygen toxicity, type II alveolar-lining epithelial cells may increase two- to threefold or more in number, and result in a severalfold increase in the thickness of the air–blood barrier. With time, the type II cell can "dedifferentiate" into a squamoid type I alveolar-lining cell.

C. Fibrinous Exudation

In the majority of instances in oxygen toxicity, the later alterations of alveolar epithelial denudation and endothelial cell destruction proceed to a final phase of fibrinous exudation. Plasma proteins, including clotting factors, reach the alveolar surface, as pulmonary edema and intra-alveolar hemorrhage predominate. Alveolar hyaline membranes form as fibrin polymerization traps cellular debris. Platelets, thrombocytes, leukocytes, and other cells and cellular fragments accumulate intravascularly in the pulmonary capillary bed. The capillary endothelium, like the type I alveolar epithelium may denude and be lost. Red blood cells fragment and lyse. Intrastitial edema, as well as perivascular and peribronchial edema and hemorrhage, become profound.

Alveolar macrophages become loaded with phagocytized cellular debris, fragmenting red blood cells, hyaline membranes, and other materials not normally present in the alveolar spaces; their bactericidal functions are impaired. Surface phagocytosis by macrophages is inhibited or restricted by edema fluid and cellular remains. The contribution of the alveolar lining material to host antibacterial defenses is lost, as are other noncellular components of alveolar antimicrobial defenses (immunoglobulins, macrophage mediators released from lymphocytes, other factors), and the lung becomes susceptible to bacterial infection.

Airways become inflammed, filled with exudates and debris, and obstructed. As the alveoli become progressively atelectatic, edema filled, and congested with red blood cells and various cellular debris, and as the perfusion capacity of the pulmonary capillary bed is reduced with progressive endothelial cell injury and destruction, the prime function of the lung, that of gas exchange, is lost. Death of the host usually ensues.

D. *Intravascular Mediation of Oxygen Toxicity*

There has been evidence generated in recent years to implicate an important role for various blood factors to the genesis of oxygen toxicity (233). The finding that endothelial cells were the first to be damaged by exposure to high oxygen tension (rather than epithelial cells despite the fact that the latter were directly exposed to the external toxic environment) prompted several studies. These studies led to the hypothesis that other secondary factors were necessary for oxygen to become toxic, and that these factors were supplied by the blood. The fact that erythrocytes were hemolyzed by increased oxygen tensions further supported the proposition that a close correlation might exist between hematological alterations and pulmonary responses in oxygen poisoning.

In a series of experiments performed by Weibel and co-workers, this correlation was studied in vitamin E-deficient rats, as a lack of this vitamin is known to greatly increase the sensitivity of erythrocytes to injury by oxygen. It was demonstrated that the osmotic fragility of red blood cells increased dramatically after only 12 hr of pure oxygen breathing. Thickening of the air–blood barrier by interstitial edema formation was accelerated occurring after 24 hr of exposure, whereas it requires normally at least 48 hr or more for the first signs of pulmonary morphological changes to occur in rats with adequate vitamin E levels.

As further support for a potential role of blood factors in pulmonary oxygen toxicity, Sacks et al. demonstrated, in an in vitro system consisting of cultured human endothelial cells, that damage to these cells resulted when they were incubated in the presence of granulocytes which were exposed to activated complement components (360). This damage was found to be primarily mediated by toxic oxygen metabolites released from stimulated granulocytes. It was proposed by Sacks et al. that C5a in the complement cascade was a critical effector of endothelial injury under these conditions, and that a close physical proximity between granulocytes and endothelium was necessary for the damage to occur. This was demonstrated by pretreating the granulocyte with cytocholasin B, which to some extent (perhaps via an inhibition of spreading or presenting adherence to the endothelial cells) resulted in a diminution in effector–target cell interaction, and thereby prevented endothelial damage. From this model, it was hypothesized that the pulmonary fibrosis/calcinosis syndrome seen in some chronically hemodialized patients may be caused by repeated pulmonary endothelial damage as a consequence of a complement-directed granulocyte-mediated inflammatory reaction, and that similar factors may be important in the pathogenesis of pulmonary oxygen toxicity. Whatever the direct mechanism or cause of oxygen toxicity, it is clear that adaptations must be made by the specific cells involved initially and then by the tissue as a whole, if the organism is to survive under hyperoxic conditions for any length of time.

E. *Perspectives on Morphological Interpretations*

Most of our understanding of the sequential morphological progressions that occur in the course of pulmonary oxygen toxicity have been derived from studies on animals, with a very limited number of correlations made in humans. That in itself not only poses a significant problem, but is complicated even more by a widely neglected consideration of experimental studies on oxygen toxicity. All investigators who have worked in this field are (or should be) aware that experimental animals may be quite comfortable and in no apparent distress while in a high oxygen tension environment, often even of prolonged periods. But once removed from any sustained exposure to the high oxygen tension, however, most animals become acutely dyspneic, gasping, and cyanotic in room air (232,233). This should not be surprising, and one must keep in focus the consideration that the sudden change in lowering ambient oxygen tensions may directly, or indirectly through acute physiological adaptations, introduce a variable that by itself can alter the morphology and biochemistry of the lung. Without adequate control of this variable, convulsions and death may occur, and the lungs appear grossly mottled when the thoraxes are opened. The effect of this transit exposure to room air on the morphological and biochemical results generated by subsequent analyses is seldom controlled for, as could easily be achieved by sacrificing the experimental animals within the high oxygen environment; because of the apparent inconvenience involved, however, it would appear most investigators have not done so. Although it is not easy, or sometimes not feasible, to recover tissue samples or harvest materials from the lungs or other organs of oxygen-exposed animals while they remain in a high oxygen environment, it is reasonable to sacrifice such animals without an abrupt exposure to air and normal oxygen tensions. Additional, carefully controlled research is needed to clarify this variable and important consideration. Until such controls are routinely included or are well clarified by other means, considerable caution must be exercised in interpreting results, especially those related to the early changes attributed to oxygen toxicity.

X. Pathogenesis of Oxygen Toxicity

A. *The Acute Respiratory Distress Syndromes*

Pulmonary oxygen toxicity is considered but one in a family of acute respiratory distress syndromes (347); other members of the acute respiratory distress syndrome family are listed in Table 3, although this list is not intended to be all-inclusive. The acute respiratory distress syndrome is defined by the presentation of the relatively rapid onset of respiratory failure, usually attributable to a specific cause; potential causes of underlying chronic respiratory disease or respiratory insufficiency are excluded, by definition. It is not the intent to review these syndromes here, except to emphasize that pulmonary oxygen toxicity is by itself one of the components of this general classification, that pulmonary oxygen toxicity is often superimposed on other causes of acute lung injury, and that all members of the so-called acute respiratory distress syndrome family share (beyond the initial lung insult) common or

Table 3. *Members of the Acute Respiratory Distress Syndrome Family*

Fat embolism	Postperfusion lung
Aspiration pneumonitis	Viral penumonia
Septicemia	Acute renal failure
Burns	Drug overdose
Traumatic wet lung	Shock lung
Smoke or gas inhalation	
Infatile respiratory distress syndrome	
Disseminated intravascular coagulation	

closely related sequences or pathways in their pathophysiological development and evolution.

B. The Respiratory Distress Syndrome of the Newborn

Generally, the respiratory distress syndrome of the newborn (or hyaline membrane disease of the newborn) is excluded from the acute respiratory distress syndromes (170,184, 185,304,358). By definition, it presents clinically at birth or shortly thereafter (usually well within 48 hrs), is characterized by severe respiratory distress and respiratory failure, and, like its adult counterpart, has a high mortality rate. The morphological presentation of this disorder is similar, but not identical, to the pulmonary oxygen toxic lung, as shown in Fig. 31. Although perinatal or postnatal pulmonary insults may initiate the eventual presentation of this syndrome, it is in most instances due to an immature lung that cannot generate adequate amounts of normally functional pulmonary surfactant. Surfactant synthesis, especially when potential deficiencies are suspected in high risk mothers prenatally and are diagnosed by amniocentesis before birth, can be induced by steroids; the role of steroid-induced synthesis of the surfactant in the adult lung is less clearly defined. Pulmonary oxygen toxicity can be and frequently is superimposed on the lungs of neonates with the respiratory distress syndrome of the newborn (372), but the development of hyaline membranes and other pathognomonic morphological and biochemical manifestations of this disorder can occur without oxygen administration.

C. A Unified Concept of Pathophysiological Mechanisms

A unified concept of the pathophysiological mechanisms of the injury sequence in pulmonary oxygen toxicity, as well as in other forms of the acute respiratory distress syndrome, is presented in Fig. 32. The initial

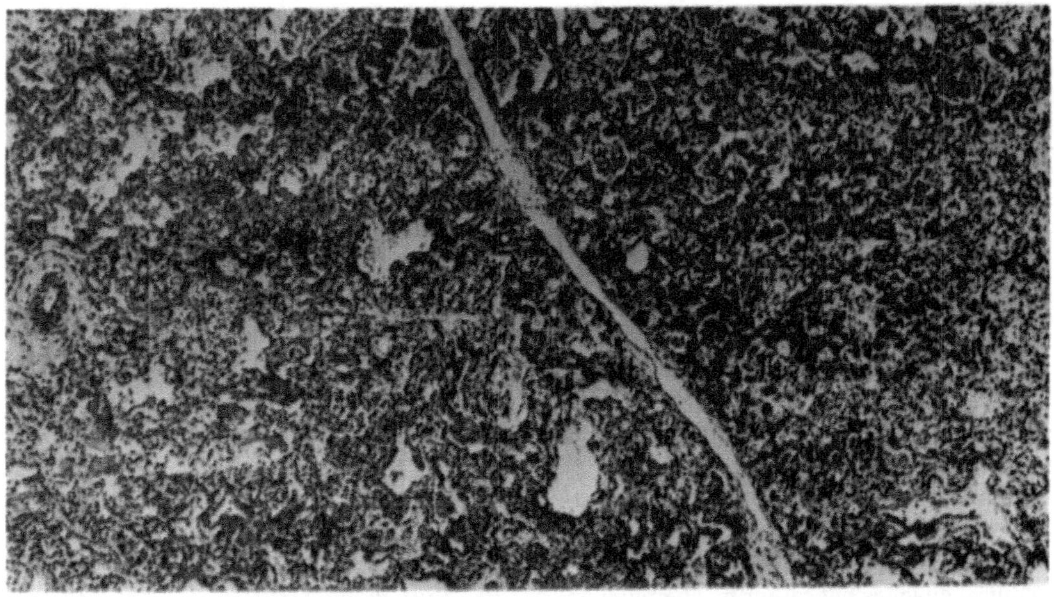

Figure 31. The histomorphological presentation of the lung of an infant with the respiratory distress syndrome of the newborn. There are several similarities to the oxygen-toxic lung. The lung parenchyma is atelectatic, and hyaline membranes have formed in alveolar spaces in terminal bronchioles, and alveolar ducts. Although high tensions of oxygen are frequently administered to neonates with this disorder, the morphological abnormalities demonstrable in this section can develop in the absence of supplemental oxygen.

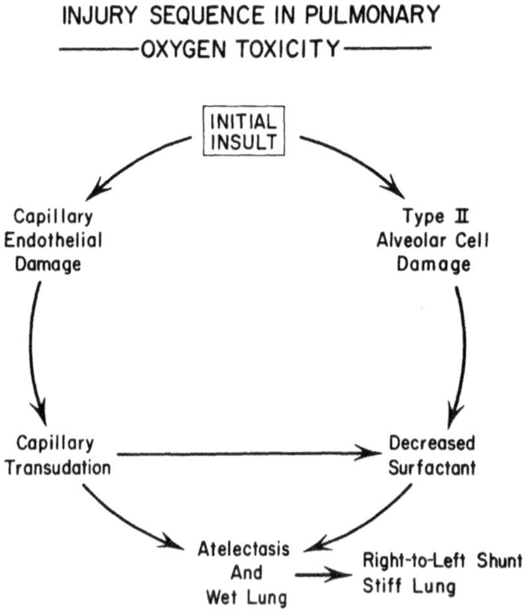

Figure 32. The injury sequence in the pathogenesis of pulmonary oxygen toxicity. The initial insult, induced by as yet not adequately defined biochemical mechanisms, injures both the capillary endothelium and the type II alveolar-lining epithelial cell. Endothelial damage leads to capillary transudation and type II alveolar cell damage results in decreased surfactant production. Increased alveolar surface forces further favor capillary transudation (producing the so-called "wet lung syndrome"); some of the proteinaceous materials in the capillary transudate inhibit surface activity, which enhances atelectasis and results in a right-to-left shunt of blood through the lung.

insult, presumably of a molecular or biochemical nature, is not as yet well understood. The capillary lining endothelial cell appears to show the first morphological evidence of injury, becoming edematous and vacuolated with a protein-rich fluid. The type II alveolar lining epithelium, or granular pneumocyte, also apparently is damaged very early in the course of oxygen toxicity, manifest not as much by morphologically demonstrable structural changes but rather by chemical and functional alterations in its primary secretory product, the pulmonary surface active material. Damage to the capillary endothelium, as well as to the type I alveolar-lining epithelium and interposed basement membranes, results in a transudation of proteinacious materials across the blood–air barrier from the capillary lumen to the alveolar spaces. Some of these proteins appear to be able to inhibit the already decreased pulmonary surfactant (114,116), further enhancing increased alveolar surface forces. The combination of increased alveolar surface tension forces from decreased pulmonary surfactant activity and the transcapillary leakage of proteins and edema fluid lead to atelectasis and the so-called "wet lung syndrome." Continued perfusion of mixed venous blood low in oxygen through atelectatic pulmonary parenchyma presents clinically as a right-to-left circulatory shunt. The combination of increased surface tension forces, atelectasis, alveoli lined with hyaline membranes, and airways partially obstructed by exudates and cellular debris gives rise to what has been termed the "stiff lung syndrome," which is further enhanced by the proliferation of type II alveolar epithelial cells, fibroblast infiltration, and the development of interstitial collagen deposition.

XI. Oxygen Tolerance

Oxygen tolerance is an adaptive state of the host whereby the otherwise toxic and potentially lethal manfestations of exposure to high oxygen tensions are reduced in severity. The classic rectangular hyperbolic shape of the oxygen exposure–response curve is attenuated (Fig. 1). This attenuation has been described in terms of the reversibility of oxygen toxicity,the development of adaptive tolerance to oxygen toxicity, and the resistance of the host to oxygen toxicity.

A. The Reversibility of Oxygen Toxicity

Once the host is exposed to high tensions of oxygen, two closely related but divergent pathways of response are initiated. In the first instance, there is a potentially lethal reaction, characterized by the edemagenic, hemorrhagic, and exudative phase of oxygen toxicity. At the same time that alveolar-lining and capillary-lining cells are being destroyed by the high oxygen tensions, however, there is an additional protective effort of the host to reverse the toxic effects of oxygen (1,3,58,173, 224,226,339,340,351,355,379). This protective adaptation has several components.

The granular pneumocyte is stimulated to replicate, and in so doing to aid in the reparative mechanism by recovering the bare basement membrane that has been denuded of its usual alveolar lining, the type I cell. In addition, there is an influx of interstitial fibroblasts and an increased deposition of interstitial collagen fibers. Edema may still be prevalent. On the basis of these reactions, the relative proportion of the interstitial volume to the lung parenchyma as a whole may increase severalfold, as may the thickness of the blood–air barrier as the cuboidal-shaped type II cells replace the squamoid-shaped type I cells (112). Whereas there eventually may be a net reduction in the total capillary bed, an apparent capillarization is evident as endothelial cells attempt replication.

If these reparative phenomena occur at a rate that exceeds the continued destruction induced by exposure to the high oxygen tension, a state of *adaptive oxygen tolerance* is induced. The definition of oxygen tolerance is expressed as a relative concept in the perspective of oxygen toxicity, which in itself varies as a function of animal species and strain. Thus, the definition of oxygen tolerance depends, to a considerable degree, on a knowledge and characterization of the dose-dependent oxygen-poisoning relationship for any species (Fig. 1). Usually, this dose-dependent relationship of oxygen toxicity has the form of a rectangular hyperbola for most cytologic or subcellular systems, with a progressively increasing rate of toxicity as exposure tension and duration are increased. A concept of oxygen tolerance, therefore, must define as well the lowest level of oxygen tension and duration of exposure that will produce demonstrable pulmonary damage.

B. Morphological Adaptation of the Type II Cell

The pulmonary structural changes are the key to understanding the development of oxygen tolerance. In contrast to the exudative phase of oxygen toxicity, (where interstitial and alveolar edema, hemorrhage, fibrinous exudation, hyaline membrane formation, and the destruction and desquamation of type I alveolar-lining epithelial cells and the capillary-lining endothelium predominate), the development of oxygen tolerance is characterized by a proliferative response, expecially of the type II epithelial cell at the alveolar level (112).

Because it is so advanced topographically in being spread out to cover parts of one to three or more alveolar spaces, the type I or squamoid epithelial cell has minimum potential for the "rounding-up" process necessary to cell division; therefore, upon injury it cannot readily replicate as part of a reparative process. The type II cell, or granular pneumocyte, on the other hand, is cuboidal in shape and as such has greater potential to divide. It also probably has the metabolic capacity to generate better antioxidative defenses.

As depicted in Fig. 33, repair of the initial structural damage induced by exposure to high tensions of oxygen may result in alveolar spaces being lined predominantly by type II epithelial cells, or granular pneumocytes. In the normal lung, 15% or less of the alveolar space is lined by these cuboidal-shaped cells. With fully developed oxygen tolerance, usually greater than 80% (and up to 100%) of the alveolar air surface is comprised of type II cells. This can thicken anatomically, of course, the air–blood barrier by as much as 300% (Fig. 34) (46,256,257,333,379,381, 408). This increased thickness of the air–blood barrier will reduce the diffusion capacity accordingly (Fig. 5), and may result in death to the host from hypoxemia upon abrupt return to normal air (46,379,381,408). The morphological adaptations to oxygen toxicity and oxygen tolerance are summarized in Fig. 24, based on morphometric data derived from subhuman primates (224).

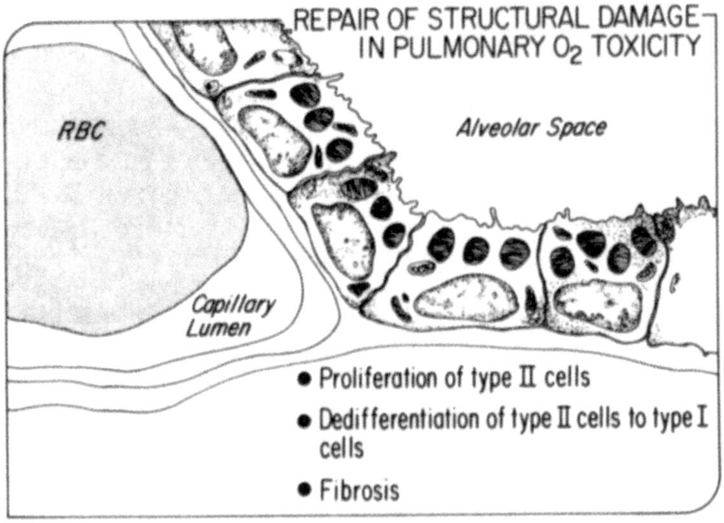

Figure 33. The repair of structural damage in pulmonary oxygen toxicity. Alveolar type II cells (granular pneumocytes) proliferate to line from 80 to 100% of the alveolar surface, increasing the thickness of the air–blood barrier by threefold or greater. With recovery, these cells may dedifferentiate into a type I alveolar-lining epithelium, reducing the thickness of the alveolar-capillary air–blood barrier toward normal. An influx of fibroblasts into the interstitium occurs, where they replicate and collagenize the pulmonary parenchyma; permanent fibrosis may result.

C. Mitochondrial and Other Ultrastructural Changes

In addition to increasing in number, there are other changes in the internal structure of granular pneumocytes in the development of oxygen tolerance (355). The cells become somewhat smaller in size as they replicate, with a relative increase in the ratio of their cell surface area to cell volume. Free ribosomes increase in number and density and the cisternae of the endoplasmic reticula dilate.

The effect of exposure to high tensions of oxygen and the development of oxygen tolerance on the mitochondria of type II cells have been the subject of several studies (1,71, 232,233,300,355,364,399,425). Mitochondria appear to increase in size and number, in part secondary to an apparent pleomorphic elongation (355). Some studies have reported that they become swollen (1,71,232,233), whereas others have not observed this change (399). It has been suggested that mitochondria may become vacuolated (364,399), lose their internal matrix (300,355), and degenerate (425). These alterations are sometimes very subtle, however, and can be quantified only by morphometric analyses; occasionally the changes are nonspecific and inconsistent.

D. Other Parenchymal Adaptations in Tolerance

The type I alveolar lining epithelium can be totally denuded and lost in oxygen toxicity, and may be absent in the oxygen tolerant lung as well. When the type I cell initially is lost during the exudative phase of oxygen toxicity, the underlying or remaining basement membrane may remain bare or be covered partially by a fibrinous hyaline membrane, proteinaceous materials, and cellular debris. With recovery from oxygen toxicity, or with the development of oxygen tolerance this intra-alveolar exudate and accumulation of materials is partly or completely resolved. Although some of these debris may be removed physically via the airways, most of the accumulated exudate is resolved through phagocytosis by pulmonary alveolar macrophages. With time and further recovery,

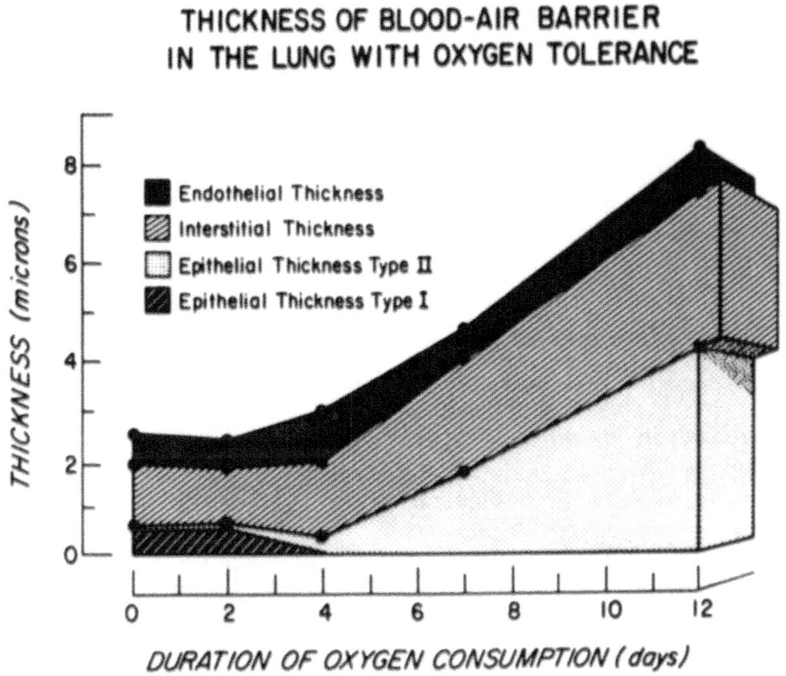

Figure 34. The effect of oxygen poisoning and the development of adaptive oxygen tolerance on the alveolar-capillary air–blood barrier (258). The type I epithelial cell is destroyed and denuded, leaving the alveolar space lined entirely by type II granular pneumocytes. The average endothelial thickness increases slightly during the early phases of toxicity, presumably from edema and vacuolization, but generally remains relatively constant in its contribution to the air–blood barrier, in spite of a net loss of total alveolar capillaries. The interstitial space also increases in thickness as a result of fibroblast influxation and proliferation, as well as on the basis of fibroblast deposition of collagen. Edema accumulation may also contribute to the increase in the interstitial compartment. All data derived on the basis of morphometric analyses (224).

the type II alveolar lining may dedifferentiate in part into the more topographically advanced type I cell and in so doing reduce the thickened air–blood diffusion membrane toward more normal dimensions.

Morphometric analyses demonstrate that the interstitial space may increase by 80% or in volume density relative to the total parenchyma in the adaptive reparative process. Some of this relative increase in the interstitium may be due to retained edema, but most is probably due to an influx in fibroblasts and the deposition of interstitial collagen. Leukocytes may also migrate into the interstitium from the pulmonary capillary bed. Generally, there will be some residual interstitial fibrosis if the host survives, persisting for months to years radiologically and probably forever by histological examinations in most instances. The proliferative and fibroblastic responses are prominant only with survival of the host from oxygen toxicity.

E. The Capillaries and Pulmonary Vasculature

The capillary endothelium is the first cell in which injury can be demonstrated morphologically upon exposure to high oxygen tensions for prolonged periods. The course of capillary destruction and revascularization in the course of toxicity and in the development of tolerance has been difficult to define. The total capillary endothelium may decrease by as much as 50%. Normally, however, only a relatively small proportion of the capillary bed is perfused at any one time under the usual physiological demands. As oxygen toxicity progresses, and as some of the capillary bed is destroyed, the remainder of the pulmonary capillaries dilate to compensate for the otherwise lost perfusion. Revascularization occurs, but whether or not the pulmonary capillary bed can be fully restored on recovery is questionable.

F. Mechanisms of Tolerance Induction

The precise mechanisms by which oxygen tolerance develops are not fully understood, and most surely are multifactorial. Several general principles are self-evident. The means to neutralize the toxic effects of oxygen, especially if these toxic effects are mediated via reactive oxygen radicals, must evolve in the presence of a continued high oxygen tension. The alveolar macrophage defense system, as well as other means of alveolar clearance, must be sustained at a level sufficient to clear debris. Finally, repair at a structural level must return the lung toward normal function.

These principles are illustrated in Fig. 35. Based much more on the knowledge that has been derived from the study of the development of oxidant tolerance in animals exposed to environmental air pollutants (a review of which is beyond the scope of this contribution), the formation of altered proteins, either involving the lung tissue or lipoprotein alveolar lining materials, and the subsequent stimulation of immunological responses, have been considered key to mediating the cellular proliferative response in oxygen tolerance. By these mechanisms, it is proposed that the initial insult induced by exposure to high levels of oxidants causes an alveolar leak, primarily through a direct insult to the integrity of cells comprising the air–blood barrier. An immunological response to the altered cellular proteins (or to the altered acellular alveolar-lining lipoproteins) then serves as a stimulus to the proliferation of type II alveolar cells and the development of oxygen tolerance. The latter pathway can be

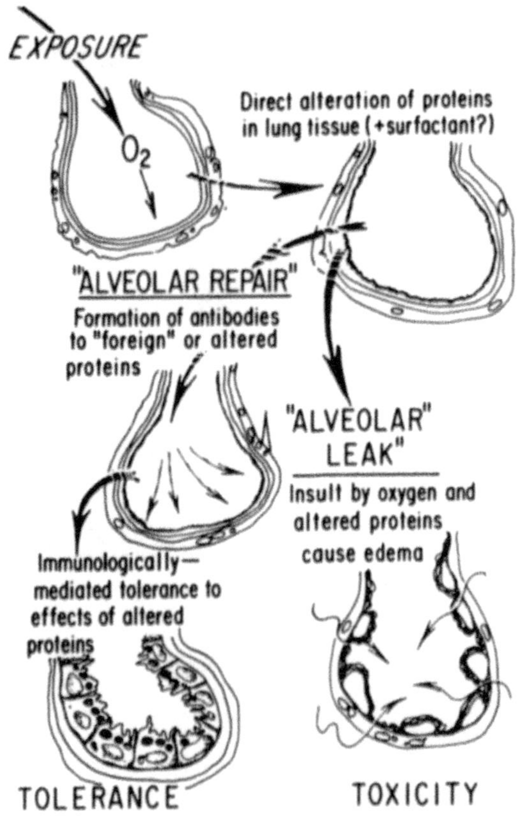

Figure 35. Schematic summary of pathways of response in the development of oxygen toxicity versus oxygen tolerance. In both instances, an early cellular lesion involves the alteration of proteins in the lung tissue and in the lipoprotein alveolar lining surfactant.

inhibited by immunosuppression, either by systemic steroids (332) or by other more specific immunosuppressive agents (313, 314). Additional numerous studies employing measurements of and experimental alterations of adrenocortical–hypophyseal function before, during, and after the development of oxygen toxicity indicate that any increase in adrenal cortical activity enhances toxicity and depresses the development of tolerance.

The development of tolerance by these mechanisms is contingent on the evolution during the course of oxygen toxicity of a means to remove intra-alveolar fluid as fast as it is formed (231), presumably by the pulmonary lymphatic drainage. The initial toxicity is accompanied by a decreased rate of synthesis of net pulmonary DNA, RNA, and total protein synthesis (283), proportionate to the accumulative exposure level (exposure tension plus exposure duration). If the tolerance pathway predominates, DNA turnover increases dramatically in the type II alveolar epithelial cells (283), and in the total pulmonary structural (but not secretory) proteins (283).

G. *Superoxide Dismutase and Oxygen Tolerance*

The development of potential defenses against oxidant injury, including the neutralization of reactive species of oxygen by superoxide dismutase and other antioxidant substrates (218), is discussed in greater detail elsewhere in this monograph (118); some of the more important pulmonary considerations will be summarized briefly here. The development of oxygen tolerance employing antioxidant metabolic adaptations has several important considerations (105,230,382,396). The tolerant cell must maintain reservoirs of oxidizable substrate above that amount needed for the normal cellular processes. They must maintain these in a manner that provides acceptable levels of the normal oxidative–reductive systems. Finally, the oxygen tolerant cells must induce new or augment existing antioxidant or antioxidative free radical defense systems.

Early studies on the effect of a high oxygen tension environment on bacteria indicated that their survival was proportional to their ability to generate superoxide dismutase (265). More recently, Crapo and a number of co-workers (84,85) reported that prolonged sublethal exposure to oxygen (approximately 0.6 atm for 7 days) induces oxygen tolerance, and that the development of this tolerance is accompanied by a 50% or more increase in total lung superoxide dismutase and a significant extension of host survival time (from less than 4 days survival in nontolerant animals to greater than 7 days in oxygen-tolerant animals). The development of tolerance in these animals (rats) paralleled the increase in superoxide dismutase (85). There appears to be species variability in this response, however, with guinea pigs, mice, and hamsters being less prone to develop tolerance than rats and other species.

It has been indicated that type II alveolar epithelial cells may have increased levels of superoxide dismutase activity with progressive exposure to sublethal oxygen tensions (343), coincident with their replication in the reparative process of oxygen tolerance. There are also relatively increased numbers of type II cells topographically in the immature lung of some species, as there are increased levels of superoxide dismutase and an enhanced resistance to oxygen toxicity (343).

One of the prime difficulties in quantifying changes in superoxide dismutase activity in the development of oxygen tolerance is the impracticability of recovering isolated cell types from the lung as a whole. The alveolar macrophage is one cell type that can be recovered as a relatively pure population, however, or probably at least those that are harvested by bronchopulmonary lavage (203). The results of studies on superoxide dismutase activity in alveolar macrophages have been inconsistent, however (84,90, 385). There appears to be no increase in superoxide dismutase activity in pulmonary alveolar macrophages recovered from adult rats during the development of oxygen tolerance (90), whereas there is an apparent increase in superoxide dismutase activity in the mitochondria of alveolar macrophages harvested from young rats (385). In addition, there is an increase in superoxide dismutase in alveolar macrophages from the mouse and guinea pig lungs with prolonged exposure to high tensions of oxygen (349,375), which may be related to changes in the recoverable cell populations.

H. *Additional Biochemical Mechanisms of Tolerance*

Other biochemical changes following sublethal exposure to elevated oxygen tensions, particularly in young animals, include an initial decrease in DNA, RNA, and protein synthesis, followed by a marked increase in their synthesis upon prolonged exposure to oxygen, with this level of increase dependent upon the increase above ambient levels of oxygen (283). Radioautographic labeling studies have indicated that it is the type II cells which are mainly responsible for the increase in DNA synthesis, especially as oxygen tolerance develops (104). The increase in protein synthesis during the development of tolerance appears to be specifically related to an increase in cell structural proteins, as opposed to secretory protein. It might be considered, therefore, that biochemical tolerance to oxygen toxicity would entail several processes. These would include maintaining reservoirs of oxidizable substrate (above that needed for normal metabolic process), of oxidative–reductive systems, and the ability to quickly augment or induce antioxidant and anti-free radical systems so as to limit damage to an absolute minimum.

In addition to considerations of the superoxide system, the antioxidant systems most studied in oxygen toxicity and oxygen tolerance are the NADPH-generating systems and thiol-generating enzymes (230,360,396). The NADPH-generating systems, including enzymes of the pentose pathway, malate dehydrogenase and isocitric dehydrogenase, have been shown to be present in lung tissue. In experiments performed by Tierney and coworkers, it was demonstrated that rats (but not mice) exposed for 7 days to 85% oxygen increased their glucose-6-phosphate dehydrogenase activity twofold (396). Upon exposure to 99% O_2 for 7 days, less than 25% of control or air-exposed rats survived, while 100% of the rats previously exposed to elevated oxygen survived. It was only in those rats that developed tolerance that an increase in glucose-6-phosphate dehydrogenase activity also developed. The authors suggested that the increased levels of glucose-6-phosphate dehydrogenase activity contributed to the development of tolerance by providing reducing equivalents to reduce oxidants, with a concurrent increase in the capacity to repair damaged cells, as well as an increased capacity of type II cells to proliferate and replace injured cells. The hexose–monophosphate shunt pathway as reflected by this enzyme activity, would be expected to supply more ribose 5-phosphate for nucleic acid repair and synthesis. The increased NADPH production provides reducing equivalents for reductive biosynthesis and is required for synthesis of fatty acids, lipoprotein, and other cell constituents, and may therefore be necessary for cell division (230,396). It has been postulated that an increase in glucose-6-phosphate dehydrogenase with the development of oxygen tolerance is related to an increase in type II cells and that is is possible that the resistance of the type II cell to high tensions of oxygen is related to this high glucose-6-phosphate dehydrogenase activity, which enables these cells to reduce oxidants rapidly and replace injured cell components by active synthesis requiring the pentose pathway.

Glutathione peroxidase, a ubiquitous intracellular enzyme that utilizes lipid peroxide substrates, is relatively active in lung tissue. It reduces toxic lipid peroxides to the correspondingly less toxic hydroxy fatty acids, utilizing GSH as a cofactor (230). The levels of reduced glutathione and total nonprotein sulfhydryl compounds are known to be elevated in rats exposed to sublethal amounts of oxygen (169,215). Glutathione reductase and NADPH-generating enzyme reactions, such as those catalyzed by glucose-6-phosphate dehydrogenase, malate dehydrogenase, isocitrate dehydrogenase, and 6-phosphogluconate dehydrogenase also are important in this detoxification reaction. The reaction of glutathione peroxidase in reducing lipid peroxides oxidizes GSH. Glutathione reductase, in turn, replenishes GSH by reducing oxidized glutathione at the expense of NADPH. NADPH-generating pathways are thus required to provide reducing equivalents to glutathione reductase.

It can be concluded that sublethal exposures above ambient oxygen tensions result in simultaneous increases in lung GSH levels and in activities of two potentially important protective antioxidant enzyme systems, superoxide dismutase and the glutathione peroxidase system linked to the pentose pathway. The observed elevations in enzymatic activities and in GSH and nonprotein sulfhydryl levels may be a reflection of multiple cytodynamic events occurring in the lung following oxygen exposure. The superoxide dismutase and glutathione peroxidase systems, if augmented, would represent a potent synergistic mechanism for diminishing lung susceptibility to oxidant injury, as perhaps mediated by superoxide or lipid peroxides. This adaptive response would be important in decreasing the susceptibility of lung tissue to continued oxygen toxicity (396).

I. Oxygen Tolerance in Man

Concepts of oxygen tolerance have important applications to several considerations in man, including the therapeutic administration of supplemental oxygen to patients in respiratory failure, aerospace medicine and conditions of hyperbaric diving. Unfortunately, most information on oxygen tolerance has been derived from studies of experimental animals, and these data cannot be extrapolated fully to man. Because there commonly is an associated pulmonary abnormality, many descriptions of oxygen toxicity in humans may not be specific for the effects of oxygen, but rather reflect a combination of the influence of oxygen administration and the evolution of the underlying pathology of the disease process in the lung.

Nash has addressed this problem specifically (298), as have others (11,52,68–70, 126,127,154,162,197,260,288,339–341). Results of descriptions of oxygen toxicity and tolerance in man appear similar to those in experimental animals, with both exudative and proliferative phases reported following exposures to tension of 0.4–1.0 atm. As in animals, the exudative phase in humans is characterized by congestion, edema, intraalveolar hemorrhage, fibrinous exudates, and hyaline membrane formation, whereas the proliferative phase of oxygen toxicity and oxygen tolerance are characterized by thickening of alveolar and interlobar septa, residual edema, fibroblast proliferation, fibrosis, and alveolar hyperplasia. Experimental comparisons indicate that the response of subhuman primates appears to resemble that of man better than any other experimental animal model (224,226,351,353,412).

Generation of acceptable oxygen tolerance prediction curves in man will not be feasible because of an inability to quantify any but the most crude of alterations in the lungs of humans on an ongoing basis. Quantitation of changes in vital capacity has been used as an index of projected oxygen tolerance (73,74), with values generated over a wide range of exposure tensions. Predictions of this nature (Fig. 36) are based on the theory that oxygen tolerance curves comparing exposure levels with a biological response, have the form of a family of rectangular hyperbolas with *a* asymptotes at time zero and an assumption

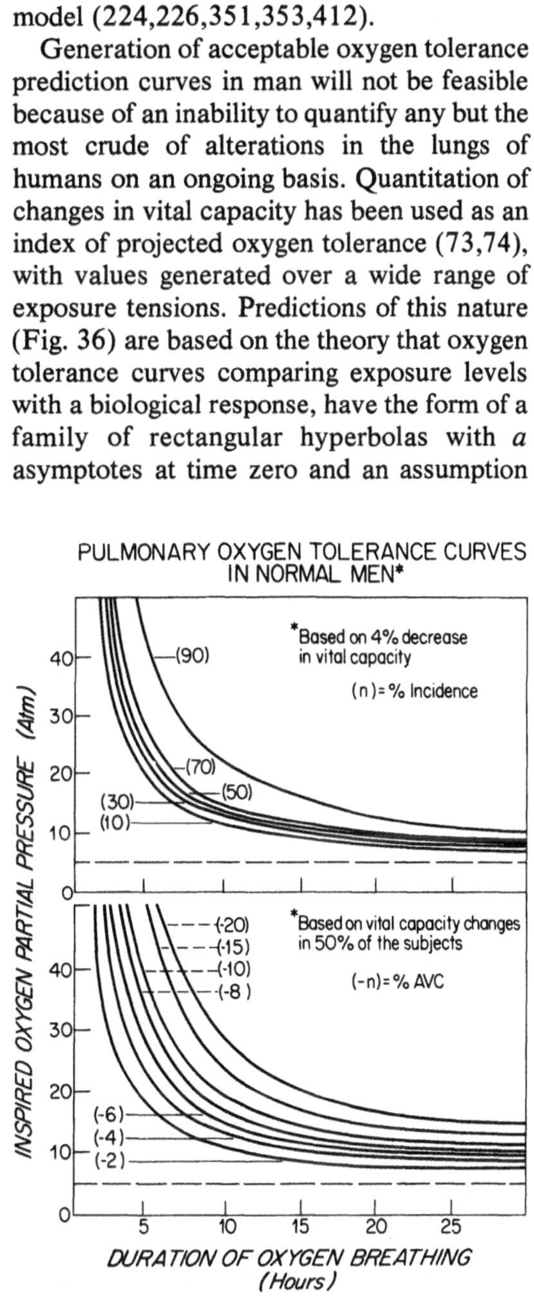

Figure 36. Oxygen tolerance in man, as estimated by pulmonary function. Modification from Clark and Lambertsen (75).

that pulmonary oxygen toxicity or tolerance responses will develop in 50% of individuals exposed to increased partial pressures of oxygen (73). Most such projections are based, as well, on the assumption that normal man can breathe an oxygen delivery of 0.5 atm for a prolonged time without severe poisoning. These may not be valid assumptions (165, 234,413,429), as data derived from some animal exposures indicate that toxicity can occur at these exposure conditions and that so-called "normal man" is not the best focus for such projections.

J. Resistance to Oxygen Toxicity

Generally, humans appear to be able to recover from exposures to 1 atm of oxygen for periods of 50–100 hr or so, with the longest reported exposure and survival at 110 hr of exposure (95,96). As depicted in Figure 37, some animal species, on the other hand, are inherently resistant to oxygen toxicity (415–417). Cold blooded animals appear especially resistant, with the turtle, for example, tolerating 1.0 atm of oxygen for approximately 6 months at an ambient normal temperature (23°–26°C; when warmed to human body temperatures (37°C), however, turtles become very toxic on exposure to increased oxygen tensions. Birds also appear inherently resistant to oxygen toxicity, and whether that is due to the differences in avian metabolism or to the presence of a semirigid pulmonary apparatus that is less dependent on the presence of a surface active agent is not known. Age (107,327,379,381,417), genetic factors (423), and other variables (47, 88,107,261,327,352,379,381,416,417) are also important.

How inherent resistance to oxygen toxicity is related to the potential development of oxygen tolerance is unclear. Resistance, perhaps manifest as oxygen tolerance, to oxygen toxicity can be induced by intermittent subtoxic exposures to high tensions of oxygen (9,44,45,46,408), and the overt pulmonary manifestations of toxicity avoided with subsequent prolonged exposures at high tensions

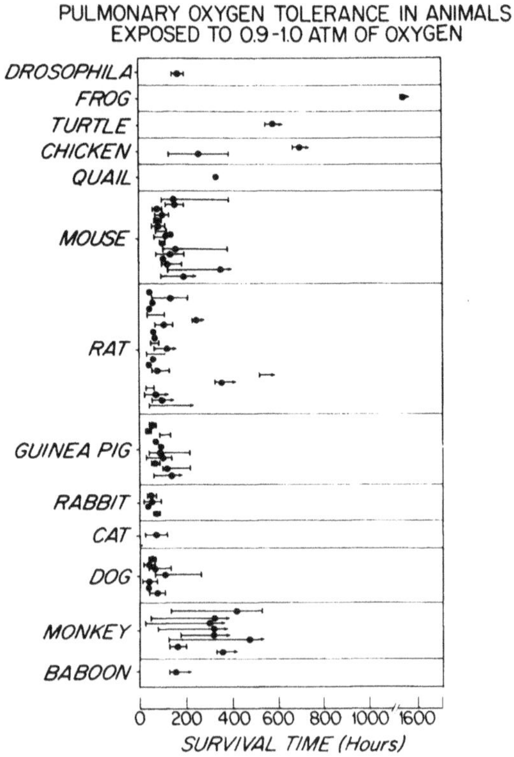

Figure 37. Relative resistance of animal species to oxygen toxicity. Modification from Clark and Lambertsen (75).

(236,355). Hyperoxia of variable duration also appears to induce resistance or tolerance to oxygen toxicity, perhaps mediated through adaptation in the pulmonary circulation or the lymphatic drainage of the lung (51). Tolerance, once induced, can persist for several weeks (9,379,381,391), although tolerance may not protect the host to exposures much greater than 1.0 atm. There is considerable interspecies and intraspecies variations, as well, with tolerance readily difficult to induce in those animals with minimal resistance to oxygen toxicity (such as the dog, rabbit, guinea pig, and mouse).

The development of selective resistance or tolerance within one population of a specific strain or species uniformly exposed to high oxygen tensions (especially in the rat) has been frequently described (50,93,256,333, 345,379,381,414). Under these conditions, a variable proportion of a group of animals exposed will develop pulmonary oxygen poisoning and expire within 3 to 5 days. The survivors, however, may tolerate continued exposure to 1.0 atm of oxygen for weeks, although they too will eventually die from at times undetermined causes; generally, such survivors have diminished food intake and a relative weight loss (379,381).

XII. Modification of Oxygen Toxicity

Although the lung is the prime target organ of oxygen toxicity, especially at exposure tensions of 1.0 atm or less, the influences of other nonpulmonary organs can significantly alter the response of the lung to oxygen. Some of these factors are summarized in Fig. 38. Although not fully understood, the nonpulmonary modifiers of pulmonary oxygen toxicity most likely are interdependent. Some of the more important considerations will be summarized here by component parts; a more detailed review has been published previously (75).

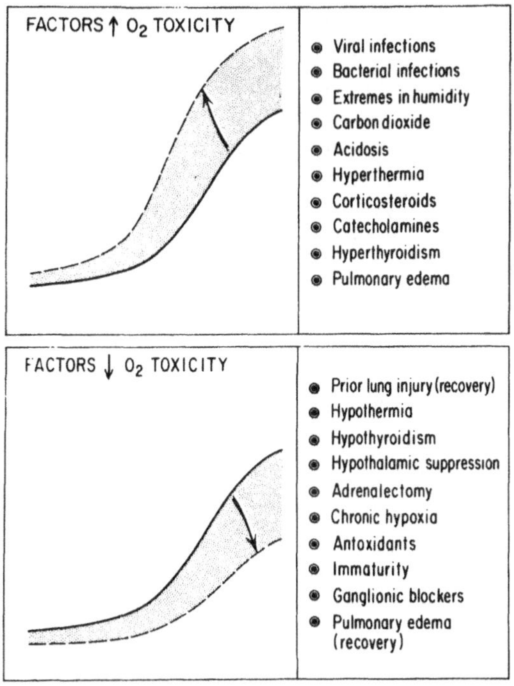

Figure 38. Factors that alter the course of oxygen toxicity.

A. Central Nervous System

The role of the central nervous system in the pathogenesis of pulmonary oxygen toxicity has been appreciated for a long time, but remains far less than fully understood. Both at exposure tensions of 1.0 atm, and more commonly at exposures of greater than 1.0 atm, central nervous system convulsions occur in the course of oxygen toxicity (23,26, 28,29,168,221,315,316,346,393,424). The explanations for the seizures and the interrelationship to the pulmonary manifestations are unclear, however. Convulsions can occur without evidence of pulmonary poisoning (26,29,221,374,395,424), and pulmonary toxicity can be severe without convulsions (212,214,395). Cerebral trauma, without any oxygen administration, can induce pulmonary pathologic alterations indistinguishable from some stages of oxygen toxicity, whether or not convulsion occurs (22,31,32). The pulmonary manifestations of oxygen poisoning can be prevented, in some animal models, with prevention of the oxygen-induced seizures, and the pulmonary response can be enhanced if seizures are induced. General anesthetics can protect the lung from the changes attributed to oxygen poisoning (12,13,26,27,37,42,61,62,143, 156, 163, 167, 172, 212, 213, 215, 216, 332, 374,405), presumably due to their effect on the central nervous system; the protective effect of general anesthesia is not secondary to general decreases in metabolism. Additional research efforts are needed to clarify these relationships.

B. Adrenocortical and Sympathomedullary Activity

Adrenalectomy protects the lung from the toxic effects of high oxygen tensions (24,138, 139,158,376,389,390), whereas systemic administration of epinephrine enhances pulmonary oxygen toxicity and removes the protective effects of adrenalectomy (24,136, 138,145,279,359,390). In like manner, hypophysectomy (presumably through reduced output of adrenocorticotropic hormone) protects the lung, and administration of exogenous adrenocorticotropic hormone (ACTH) enhances pulmonary oxygen toxicity (58,277). The protective effects of hypophysectomy are delayed in onset (presumably associated with the time course of ACTH depletion) and reduce as well the incidence and severity of central nervous system convulsions (12,23,25,67).

Attempts have been made to quantify endogenous adrenocortical activity in the course of oxygen toxicity. Attempted measurements of adrenocortical hormones with exposure to oxygen have provided equivocal results (182,280), presumably in part due to stress factors inherent to the exposure protocols. The effects of exposure to high tensions of oxygen on adrenocortical activity are increased at greater than 1.0 atm (208). Intermittent exposures to high oxygen tensions result in adrenal hypertrophy (10,20, 182,401), with an apparent depletion of adrenal gland ascorbic acid (132,362,369).

Efforts to alter the severity and nature of pulmonary oxygen toxicity by exogenous administration of adrenal cortical steroids and adrenal extracts have also produced nonuniform results and equivocal findings (138–140,389,390). Cortisone, hydrocortisone, adrenal cortical extracts, and related synthetic hormones, when administered in low dosages, appear to provide a protective function and reverse the protection afforded by adrenalectomy (138,389,390); these effects do not appear related to or mediated by fluid or electrolyte alterations (138,389,390).

Exposure to high tensions of oxygen stimulates the sympathetic autonomic system (62,106,144,166), and catecholamine depletion and adrenal medullectomy reduce the pulmonary damage and associated mortality of oxygen toxicity (102,139,141,158,305, 309,359,376). Administration of ganglionic blockers (hexamethonium, tetraethylammonium, chlorpromazine, reserpine), in part presumably due to hypothalamic-mediated suppression of both the adrenal cortex and the adrenal medulla (15,16,128,136,277, 285,305,406), for example, reduce the toxic manifestations of exposure to high oxygen tensions. Adrenergic blocking agents (dibenzyline, dibenamine, SKF501, dehydrobenzoperidol) also provide a protective effect in oxygen toxicity (22,28,32,128,136,139,140, 221,223,276,279,305,310). The effects of vagotomy on pulmonary oxygen toxicity have been controversial, with both protection (153, 158) and enhancement (158,366–368,371) implied. In general, there appears to be agreement that, in addition to the direct toxic effects of oxygen on the lung, oxidative cyclization of epinephrine to highly reactive indoles (adrenochrome, adrenolutin, soluble malanin) may accentuate the course of pulmonary oxygen toxicity (144,178–181, 183,238,305). In that both experimental results from animal studies are less than clearly defined and good information from clinical trials is less than satisfactory, the employment of supplemental steroid administration in the patient receiving high tensions of supplemental oxygen must be undertaken with considerable caution (331).

C. Thyroid Function and Other Hormonal Factors

Whereas the role of adrenal medullary and adrenocortical influences on pulmonary oxygen toxicity are less than precisely defined, the influence of the thyroid gland is far less equivocal. Thyroidectomy decreases the pulmonary manifestations of oxygen toxicity, and the administration of thyroid extract, thyroxine, or propthiouracil dramatically enhances oxygen toxicity (65,67,143,160,387, 389). Increased lung damage and increased mortality have been demonstrated in multiple species. The role of decreased thyroid function in protection from oxygen toxicity following hypophasectomy, a manifestation more commonly attributed to decreased ACTH release and associated reductions in adrenal cortical activity, must be considered, as hypophysectomy also reduces levels of circulating thyrotropin (21,378). Exposure to high tensions of oxygen leads (perhaps following an initial transient involution) to increased thyroid activity, as evidenced by thyroid gland hypertrophy (increased weight), increased rate of radioactive iodide uptake, and histologically demonstrable hyperplasia (102,142,397,398), although some of these responses have been questioned (108,109).

The susceptibility of the host to oxygen poisoning is influenced significantly by the rate of its cellular metabolism, with hypermetabolic animals more vulnerable to oxygen toxicity (137,138,141,159,337,384); hypothermia and hybernation, for example, reduce oxygen toxicity. The effect of increasing thyroid function, therefore, or of administrating thyroid-active substances, is to increase the cellular metabolism of the host and in so doing render it more reactive to high tensions of oxygen (143,160).

Other hormones have been shown to alter the course of oxygen toxicity. Chorionic gonadotrophin and progesterone administered to male animals, for example, increases the poisoning of oxygen at high exposure tensions, and administration of estrogen to females increases oxygen toxicity, especially in large doses (58).

D. Carbon Dioxide and Acid-Base Balance

Studies on the effects of increasing inspired carbon dioxide tension on oxygen toxicity have had conflicting conclusions. At 1.0 atm exposure tension, it has been reported that carbon dioxide supplementation increases pulmonary oxygen toxicity (27,213,220,275, 307,329,386,405,420), has no effect on animal survival time in oxygen toxicity (9), or both decreases the pulmonary manifestations and lengthens survival times (89,377). Above 3.0 atm exposure tensions, it is clear that even small increases in inhaled carbon dioxide increase pulmonary toxicity and enhance central nervous system convulsions (13,27,145,166,172,176,177,212–214,220, 245, 275, 281, 284, 329, 332, 373, 384, 392, 405,407,409,420). Carbon dioxide decreases intracellular pH, and this variable has not always been controlled or monitored appropriately in the relevant studies. Whether or not carbon dioxide neurogenically potentiates oxygen-induced central nervous system intoxication or, in contrast, suppresses convulsions through its narcoleptic effect has not been fully clarified (27,213,220,275,329, 386,405,420), and may be a function of the concurrent level of oxygen delivery (27,281).

XIII. Summary

The structural and functional effects of high levels of oxygen on the lungs of humans and experimental animals are reviewed. Oxygen is potentially injurious and even lethal to all living cells, although its dose-dependent deleterious effects vary depending on the oxygen tension, duration of exposure, metabolic milieu, and relevant susceptibility of the exposed cells. Since oxygen is gaseous at atmospheric pressures, it enters the host organism through inhalation, exposing the lung to higher oxygen tensions than other organs of the body. Thus, the lung is the primary target organ in oxygen toxicity and the signficant adverse responses are exhibited here first.

Humans and animals respond to hyperoxia in much the same way clinically, with signs and symptoms manifested as lethargy, restlessness, anorexia, nausea, vomiting, dyspnea, edema, hyperemia, and swelling of exposed mucous membranes. As exposure continues, increased difficulty in breathing develops, with increased respiratory rate, frothy, bloody sputum, cyanosis, and convulsions. Radiologically, diffuse, bilateral, irregular pulmonary densities occur that may coalesce to opacify entire lung fields. Pleural effusions also are common.

Lung function is obviously affected in oxygen toxicity. Vital capacity is reduced, airway obstruction develops, edema and blockage by secretions and debris occur, expiratory flow rates are impaired, the diffusion capacity is reduced, imbalances in ventilation and perfusion are evidenced, and dynamic lung compliance falls.

The pulmonary surfactant is of paramount importance to the pathogenesis of several important physiologic functions which develop following exposure to high-oxygen tensions. Specific research results pertaining to alterations in the surfactant have been inconsistent, however, perhaps due to differences in measurement of surface tension and other technical approaches. As the activity of pulmonary surfactant is reduced, alveolar surface tension is increased and atelectasis results.

Airway defenses against bacterial infection appear to be adversely affected in a dose-dependent manner by hyperoxia. Ciliary function, mucous flow rate, and secretion of antimicrobial products are reduced. Clearance is impaired. The antimicrobial alveolar lining material function is reduced, as is alveolar macrophage phagocytosis of foreign matter. These responses have been demonstrated in reports utilizing experimental inhalation of bacteria by experimental animals. Data on the effect of oxygen on host defenses in primates and humans, however, are sparse.

The morphologic changes in the lung due to oxygen have been the focus of extensive research efforts. Atelectasis, edema, hemorrhage, capillary congestion, endothelial injury, inflammation, fibrin formation, and alveolar thickening have been reported. The sequence of injury appears to begin relatively simultaneously with an impairment of alveolar lining material function and structural alterations in the capillary bed endothelium proceeding to more significant structural changes in gas-exchanging surfaces, swollen and denuded alveolar-lining epithelium, formation of platelet thrombi associated with the release of serotonin and other mediators, reduction in endothelial cell replication, capillary-alveolar transudation, and fibrinous exudation.

The host can adapt to the toxic effects of high-oxygen tensions. Granular pneumocytes replicate to cover the denuded alveolar lining, fibroblasts and interstitial collagen are increased quantitatively, and capillarization occurs as the endothelium attempts replication. If these changes predominate over the destructive effects of oxygen, so-called "adaptive tolerance" is produced.

Most work on oxygen toxicity has been done on animals and these data cannot be extrapolated to humans without reservations; however, studies show the responses of most mammalian species to be similar to those in man. A review of the literature is complicated by misuse of terms pertaining to the concepts of "concentration" and "tension," along with a lack of consideration of several environmental factors that might influence research results.

Acknowledgments

The authors gratefully express their appreciation to Ms. Margaret Turner Cooney for her assistance in the development of the manuscript, to Mr. Jerome Reicher of ARRCO Medical Illustration for his creative contributions, and to Mr. Val Pochay for his administrative assistance. This work was supported in part by funding to the Smoking and Health Research Program of Harvard University by the tobacco industry of the United States.

References

1. Adamson, I. Y. R., Bowden, D. H., and Wyatt, J. P. Oxygen poisoning in mice. Ultrastructural and surfactant studies during exposure and recovery. Arch. Pathol. 90: 463–472, 1970.
2. Aikawa, J. K., and Bruns, P. D. Pulmonary lesions in experimental oxygen poisoning. Am. J. Dis. Child 91:614–620, 1956.
3. Ambrus, C. M., Pickren, J. W., Weintraub, D. H., Niswander, K. R., Ambrus, J. L., Rodbard, D., and Levy, J. C. Studies on hyaline membrane disease v. oxygen induced by hyaline membrane disease in guinea pigs. Biol. Neonatorum 12:246–260, 1968.

4. Aschan, G., and Wallenius, G. Electrophoretic studies of transudates caused by experimental oxygen poisoning and oxygen deficiency. Acta. Soc. Med. Up. 58:315–320, 1953.
5. Ashton, N. B., Ward, B., and Serpell, G. Role of oxygen in the genesis of retrolental fibroplasia. Br. J. Ophthalmol. 37:513–520, 1953.
6. Ashton, N., Ward, B., and Serpell, G. Effect of oxygen on developing retinal vessels with particular reference to the problem of retrolental fibroplasia. Br. J. Ophthalmol. 38:397–432, 1954.
7. Balentine, J. D. Pathologic effects of exposure to high oxygen tensions. N. Engl. J. Med. 275:1038–1040, 1966.
8. Balentine, J. D., and Gutsche, B. B. Influence of pentobarbital anesthesia on the distribution of central nervous system lesions in rats exposed to oxygen at high pressure. In: Proceedings of the Third International Conference on Hyperbaric Medicine (Eds.) Brown, I. W., Jr. and Cox, B. G., pp. 145–150, NAS, NRC Publ. No. 1404, Washington, D.C., 1966.
9. Barach, A. L., Eckman, M., Oppenheimer, E. T., Rumsey, C., Jr., and Soroka, M. Observations on methods of increasing resistance to oxygen poisoning and studies of accompanying physiological effect. Am. J. Physiol. 142:462–475, 1944.
10. Bartek, M. J., Daniels, M. B., and Ulvedal, F. Growth of the young male rat in a hyperoxic environment. U. S. Air Force School of Aerospace Medicine, Aerospace Medical Division SAM-TR-67-82, Brooks Air Force Base, Texas, September, 1967.
11. Barter, R. A., Finlay-Jones, L. R., and Walters, M. N.-I. Pulmonary hyaline membrane: sites of formation in adult lungs after assisted respiration and inhalation of oxygen. J. Pathol. Bacteriol. 95:481–488, 1968.
12. Bean, J. W. Effect of high oxygen pressure on carbon dioxide transport, on blood and tissue acidity, on O_2 consumption and pulmonary ventilation. J. Physiol (London) 72:27–48, 1931.
13. Bean, J. W. Effects of oxygen at high pressure. Physiol. Rev. 25:1–147, 1945.
14. Bean, J. W. Hormonal aspects of oxygen toxicity. In: Proceedings of the Underwater Physiology Symposium. (Ed.) Lambertsen, C. J., pp. 13–19, NAS/NRC Publ. No. 377, Washington, D.C., 1955.
15. Bean, J. W. Reserpine and reaction to O_2 at high pressure. Fed. Proc. 15:11–12, 1956.
16. Bean, J. W. Reserpine, chlorpromazine and the hypothalamus in reactions to oxygen at high pressure. Am. J. Physiol. 187:389–391, 1956.
17. Bean, J. W. General effects of oxygen at high tension. In: Oxygen in the Animal Organism. (Eds.) Dickens, F., and Neil, E., pp. 455–474, Macmillan, New York, 1964.
18. Bean, J. W. Problems of oxygen toxicity. In: Clinical Application of Hyperbaric Oxygen. (Eds.) Boerema, I., Brummelkamp, W. H., and Meijne, N. G., pp. 267–276, Elsevier Publ. Co., Amsterdam, 1964.
19. Bean, J. W. Factors influencing clinical oxygen toxicity. Ann. N.Y. Acad. Sci. 117: 745–755, 1965.
20. Bean, J. W., Baker, B. L., and Johnson, P. Cytological alterations of adrenal cortex induced by oxygen of high pressure. Fed. Proc. 12:11–12, 1953.
21. Bean, J. W., and Bauer, R. Thyroid in pulmonary injury induced by O_2 in high concentration at atmospheric pressure. Proc. Soc. Exp. Bio. Med. 81:693–694, 1952.
22. Bean, J. W., and Beckman, D. L. Centrogenic pulmonary pathology in mechanical head injury. J. Appl. Physiol. 27:807–812, 1969.
23. Bean, J. W., and Johnson, P. C. Influence of hypophysis on pulmonary injury induced by exposure to oxygen at high pressure and by pneumococcus. Am. J. Physiol. 171: 451–458, 1952.
24. Bean, J. W., and Johnson, P. C. Epinephrine and neurogenic factors in the pulmonary edema and CNS reactions induced by O_2 at high pressure. Am. J. Physiol. 180:438–444, 1955.
25. Bean, J. W., and Smith, C. W. Hypophyseal and adrenocortical factors in pulmonary damage induced by oxygen at atmospheric pressure. Am. J. Physiol. 172: 169–174, 1953.
26. Bean, J. W., and Zee, D. Metabolism and the protection by anesthesia against toxicity of O_2 at high pressure. J. Appl. Physiol. 20:525–530, 1965.
27. Bean, J. W., and Zee, D. Influence of anesthesia and CO_2 on CNS and pulmonary effects of O_2 at high pressure. J. Appl. Physiol. 21:521–526, 1966.

28. Bean, J. W., Zee, D., and Thom, B. Pulmonary changes with convulsions induced by drugs and oxygen at high pressure. J. Appl. Physiol. 21:865–872, 1966.
29. Bean, J. W., Zee, D., and Thom, B. Pulmonary involvement in oxygen toxicity. Jpn. Heart J. 8:734–736, 1967.
30. Becker-Freyseng, H. Physiological and patho-physiological effects of increased oxygen tension. In: German Aviation Medicine in World War II, vol. I, pp. 493–514, United States Air Force School of Aviation Medicine, 1950.
31. Beckman, D. L., and Bean, J. W. Pulmonary damage and head injury. Proc. Soc. Exp. Biol. Med. 130:5–9, 1969.
32. Beckman, D. L., and Bean, J. W. Pulmonary pressure-volume changes attending head injury. J. Appl. Physiol. 29:631–636, 1970.
33. Beckman, D. L., and Weiss, H. S. Hyperoxia compared to surfactant washout on pulmonary compliance in rats. J. Appl. Physiol. 26:700–709, 1969.
34. Beehler, C. C., Newton, N. L., Culver, J. E., and Treppi, T. Ocular hyperoxia. Aerosp. Med. 34:1017–1020, 1963.
35. Behnke, A. R., Forbes, H. S., and Motley, E. P. Circulatory and visual effects of oxygen at 3 atmospheres pressure. Am. J. Physiol. 113:436–442, 1935.
36. Behnke, A. R., Johnson, F. S., Poppen, J. R., and Motley, E. P. The effect of oxygen on man at pressures from 1 to 4 atmospheres. Am. J. Physiol. 110:565–572, 1934-35.
37. Behnke, A. R., Shaw, I. A., Shilling, C. W., Thomson, R. M., and Messer, A. C. Studies on the effects of high pressure O_2. 1. Effect of high oxygen pressure upon the carbon dioxide and oxygen content, the acidity and carbon dioxide combining power of the blood. Am. J. Physiol. 107: 13–28, 1934.
38. Benjamin, F. B., and Peyser, L. Effect of oxygen on radiation resistance of mice. Aerosp. Med. 35:1147–1149, 1964.
39. Bennett, G. A., and Smith, F. I. C. Pulmonary hypertension in rats living under compressed air conditions. J. Exp. Med. 59:181–196, 1934.
40. Berfenstem, R., Edlund, T., and Zettergren, L. Hyaline membrane disease. Acta Paediat. 47:82–100, 1958.
41. Berfenstam, R., Edlund, T., and Zettergren, L., Hyaline membrane disease: Influence of high oxygen concentration on ciliary activity in respiratory tract: Experimental study on rabbits. Acta. Paediat. 47:527–533, 1958.
42. Bert, P. Barometric Pressure; Researches in Experimental Physiology, translated by Hitchcock, M. A. and Hitchcock, F. A., College Book Co., Columbus, Ohio, 1943.
43. Bertharion, G., and Barthelemy, I. Acute effect of hyperbaric oxygen. Neurophysiological study. Rev. Agressol. 5:583–594, 1964.
44. Binet, L., and Bochet, M. L'adaptation a la vie dans l'oxygene pur. J. Physiol. Pathol. Gen. 38:341–349, 1944–1945.
45. Binet, L., and Bochet, M. De l'adaptation a l'hyperoxie. C. R. Acad. Sci. Paris 221: 161–164, 1945.
46. Binet, L., and Bochet, M. Prolonged exposure to pure oxygen. Biol. Med. (Paris) 40:85–89, 1951.
47. Binet, L., and Bochet, M. Resistance to hyperoxia of the frog *Rana esculenta* maintained in 100 per cent oxygen for fifty-two days. J. Physiol. (Paris) 55:405–412, 1963.
48. Bondurant, S., and Smith, C. Effect of oxygen intoxication on the surface characteristics of lung extracts. Physiologist 5:111, 1962.
49. Bowden, D. H., Adamson, I. Y. R., and Wyatt, J. P. Reaction of the lung cells to high concentration of oxygen. Arch. Pathol. 86:671–675, 1968.
50. Boycott, A. E., and Oakley, C. L. Oxygen poisoning in rats. J. Pathol. Bacteriol. 35: 468–469, 1932.
51. Brauer, R. W., Parrish, D. E., Way, R. D., Pratt, P. C., and Pessotti, R. L. Protection by altitude acclimatization against lung damage from exposure to oxygen at 825 mm Hg. J. Appl. Physiol. 28:471–481, 1970.
52. Brewis, R. A. Oxygen toxicity during artificial ventilation. Thorax 24:656–666, 1969.
53. Brooksby, G. A., Datnow, B., and Menzel, D. B. Pulmonary hypertension induced by high partial pressures of oxygen. Physiologist 10:134, 1967.
54. Brooksby, G. A., Dennis, R. L., Datnow, B., and Clark, D. Experimental emphysema. Histologic changes and alterations in pulmonary circulation. Calif. Med. 107:391–395, 1967.
55. Brooksby, G. A., Dennis, R. L., and Staley,

R. W. Effects of prolonged exposures of rats to increased oxygen pressures. In: Proceedings of the Third International Conference on Hyperbaric Medicine. (Eds.) Brown, I. W., Jr. and Cox, B. G., pp. 208–216, NAS/NRC Publ. No. 1404, Washington, D. C., 1966.
56. Brooksby, G. A., and Staley, R. W. Static volume-pressure relations in lungs of rats exposed to 100 percent oxygen. Physiologist 9:144, 1966.
57. Bruns, P. D., and Shields, L. V. The pathogenesis and relationship of the hyaline-like pulmonary membrane to premature neonatal mortality. Am. J. Obstet. Gynecol. 61:953–965, 1951.
58. Bruns, P. d., and Shields, L. V. High O_2 and hyaline-like membranes. Am. J. Obstet Gynecol. 67:1224–1236, 1954.
59. Buckingham, S., McNary, W., and Sommers, S. Pulmonary alveolar cell inclusions: Their development in the rat. Science 145: 1192, 1964.
60. Buckingham, S., and Sommers, S. C. Pulmonary hyaline membranes. Am. J. Dis. Child. 99:216–227, 1960.
61. Buckingham, S., Sommers, S. C., and McNary, W. F. Sympathetic activation and serotonin release as factors in pulmonary edema after hyperbaric oxygen. Fed. Proc. 25:566, 1966.
62. Buckingham, S., Sommers, S. C., and McNary, W. F. Experimental respiratory distress syndrome. I. Central autonomic and humoral pathogenetic factors in pulmonary injury of rats induced with hyperbaric oxygen and the protective effects of barbiturates and trasylol. Biol. Neonatorum 12:261–281, 1968.
63. Caldwell, P. R. B., Giammona, S. T., Lee, W. L., Jr., and Bondurant, S. Effect of oxygen breathing at one atmosphere on the surface activity of lung extracts in dogs. Ann. N.Y. Acad. Sci. 121:823–828, 1965.
64. Caldwell, P. R. B., Lee, W. L., Jr., Schildkraut, H. S., and Archibald, E. R. Changes in lung volume, diffusing capacity, and blood gases in men breathing oxygen. J. Appl. Physiol. 21:1477–1483, 1966.
65. Campbell, J. A. Oxygen poisoning and the thyroid gland. J. Physiol. (London) 90:91–92P, 1937.
66. Campbell, J. A. Oxygen poisoning and tumor growth. Br. J. Exp. Pathol. 18:191–197, 1937.
67. Cambell, J. A. Effects of oxygen pressure as influenced by external temperature, hormones and drugs. J. Physiol. (London) 92: 29–31P, 1938.
68. Castleman, B., and McNeely, B. V. (Eds.) Case records of the Massachusetts General Hospital, Case 7-1967. N. Engl. J. Med. 276:401–411, 1967.
69. Castleman, B., and McNeeley, B. V. (Eds.) Case records of the Massachusetts General Hospital, Case 20-1970. N. Engl. J. Med. 282:1087–1096, 1970.
70. Cederberg, A., Hellsten, S., and Miorner, G. Oxygen treatment and hyaline pulmonary membranes in adults. Acta Pathol. Microbiol. Scand. 64:450–458, 1965.
71. Cedergren, B., Gyllensten, L., and Wersall, J. Pulmonary damage caused by oxygen poisoning: Electron microscopic study in mice. Acta. Paediat. Scand. 48:477–494, 1959.
72. Christensen, G. M., Dahlke, L. W., Griffin, J. T., Moutvic, J. C., and Jackson, K. L. The effect of high pressure oxygen on acute radiation mortality in mice. Radiat. Res. 37:283–286, 1969.
73. Clark, J. M. Derivation of pulmonary oxygen tolerance curves describing the rate of development of pulmonary oxygen toxicity in man. Ph.D. Thesis, Graduate School of Arts and Sciences, University of Pennsylvania, 1970.
74. Clark, J. M., and Lambertsen, C. J. Pulmonary oxygen tolerance and the rate of development of pulmonary oxygen toxicity in man at two atmospheres inspired oxygen tension. In: Underwater Physiology. (Ed.) Lambertsen, C. J., pp. 439–451, Williams & Wilkins, Baltimore, 1967.
75. Clark, J. M., and Lambertsen, C. J. Pulmonary oxygen toxicity. Pharmacol. Rev. 23:37–133, 1971.
76. Clark, J. M., and Lambertsen, C. J. Rate of development of pulmonary O_2 toxicity in man during O_2 breathing at 2.0 atm abs. J. Appl. Physiol. 30:739–752, 1971.
77. Clarke, G. M., Sandison, A. T., and Ledingham, I. McA. Acute pulmonary oxygen toxicity: A pathophysiological study at 1 and 2 atmospheres absolute (ATA) in spontaneously breathing anesthetized dogs. Br. J. Anaesthesiol. 41:558–559, 1969.
78. Clements, J. Pulmonary edema and permeability of alveolar membranes. Arch. Environ. Health 2:280, 1961.

79. Clements, J. Surface phenomena in relation to pulmonary function. Physiologist 5:11–28, 1962.
80. Clements, J. A. Surface tension of lung extracts. Proc. Soc. Exp. Biol. Med. 95: 170, 1957.
81. Clements, J., Brown, E., and Johnson, R. Pulmonary surface tension and the mucus lining of the lungs: some theoretical considerations. J. Appl. Physiol. 12:262, 1958.
82. Collier, C. R., Hackney, J. D., and Rounds, D. E. Alterations of surfactant in oxygen poisoning. Dis. Chest. 48:233–238, 1965.
83. Comroe, J. H., Jr., Dripps, R. D., Dumke, P. R., and Deming, M. Oxygen toxicity. The effect of inhalation of high concentrations of oxygen for twenty-four hours on normal men at sea level and at a simulated altitude of 18,000 feet. J. Am. Med. Assoc. 128:710–717, 1945.
84. Crapo, J. D., and McCord, J. M. Oxygen induced changes in pulmonary superoxide dismutase assayed by antibody titrations. Am. J. Physiol. 231:1196–1203, 1976.
85. Crapo, J. D., and Tierney, D. F. Superoxide dismutase and pulmonary oxygen toxicity. Am. J. Physiol. 226:1401–1407, 1974.
86. Dale, W. A., and Rahn, H. Rate of gas absoprtion during atelectasis. Am. J. Physiol. 170:606–615, 1952.
87. Davies, H. C., and Davies, R. E. Biochemical aspects of oxygen poisoning. In: Handbook of Physiology, Section 3: Respiration, vol. II. (Eds.) Fenn, W. O., and Rahn, H., pp. 1047–1058, American Physiological Society, Washington, D.C., 1965.
88. De, T. D., and Anderson, G. W. The experimental production of pulmonary hyaline-like membranes with atelectasis. Am. J. Obstet. Gynecol. 68:1557–1567, 1954.
89. DeClement, F. A., and Smith, C. W. Alteration of toxicity of oxygen at atmospheric pressure by addition of CO_2 and of N_2. Fed. Proc. 21:444, 1962.
90. Deneke, S. M., Bernstein, S., and Fanburg, B. L. Absence of inductive effect of hyperoxia on superoxide dismutase activity in rat alveolar macrophages. Am. Rev. Respir. Dis. 118:105–111, 1978.
91. Dery, R., Pelletier, J., Jacques, A., Clavet, M., and Houde, J. Alveolar collapse induced by denitrogenation. Can. Anaesthiol. Soc. J. 12:531–544, 1965.
92. Dickens, F. The toxic effect of oxygen on nervous tissue. In: Neurochemistry. (Eds.) Elliott, K. A., C., Page, I. H., and Quastel, J. H., pp. 851–869, Thomas, Springfield, Ill., 1962.
93. Dickerson, K. H. Pathophysiology of pulmonic toxicity in rats exposed to 100 percent oxygen at reduced pressures. Aviation Medical Acceleration Laboratory, NADC-ML-6403, U.S. Naval Air Development Center, Johnsville, Pennsylvania, May, 1961.
94. Dolezal, V. Effects on the body of prolonged inhalation of oxygen during normal barometric pressure. Cesk. Fysiol. 11:326–355, 1962.
95. Dolezal, V. Some humoral changes in man produced by continuous oxygen inhalation at normal barometric pressure. Rev. Med. Aeronaut. 25:219–233, 1962.
96. Dolezal, V. The effect of longlasting oxygen inhalation upon respiratory parameters in man. Physiol. Behemoslov. 11:149–158, 1962.
97. Dolezal, V. Effect on organism of pure oxygen applied by excessive pressure. Cesk. Fysiol. 17:347–363, 1968.
98. Donald, K. W. Oxygen poisoning in man. I and II. Brit. Med. J. 1:667–672, 712–717, 1947.
99. Drath, D. B., and Karnovsky, M. L. Superoxide production by phagocytic leukocytes. J. Exp. Med. 141:257–262, 1975.
100. Drath, D. B., Karnvosky, M. L., and Huber, G. L. Hydroxyl radical formation in phagocytic cells of the rat. J. App. Physiol. 46:136–140, 1979.
101. Edmunds, L. H., and Austen, W. G. Effect of cardiopulmonary bypass on pulmonary volume-pressure relationships and vascular resistance. J. Appl. Physiol. 21:209–216, 1966.
102. Edstrom, J. E., and Rockert, H. The effect of oxygen at high pressure on the histology of the central nervous system and sympathetic and endocrine cells. Acta Physiol. Scand. 55:255–263, 1962.
103. Evans, M. J., Bils, R. F., and Hackney, J. D. An electron microscopic study of cellular renewal in pulmonary alveolar walls of normal and oxygen poisoned mice. Aspen Emphysema Conf. 11:33–40, 1968.
104. Evans, M. J., Cabral, L. J., Stephens, R. J.,

and Freeman, G. Cell division of alveolar macrophages in rat lung following exposure to NO_2. Am. J. Pathol. 70:199–208, 1973.
105. Evans, M. J., Cabral, L. J., Stephens, R. J., and Freeman, G. Renewal of alveolar epithelium in the rat following exposure to NO_2. Am. J. Pathol. 70:175–198, 1973.
106. Faiman, M. D., and Heble, A. R. The effect of hyperbaric oxygenation on cerebral amines. Life Sci. 5:2225–2234, 1966.
107. Faulkner, J. M., and Binger, C. A. L. Oxygen poisoning in cold blooded animals. J. Exp. Med. 45:865–871, 1927.
108. Felig, P. Oxygen toxicity: Ultrastructural and metabolic aspects. Aerosp. Med. 36: 658–662, 1965.
109. Felig, P., Goldman, J. K., and Lee, W. L., Jr. Protein-bound iodine in serum of rats breathing 99 percent oxygen. Science 145: 601–602, 1964.
110. Fenn, W. O., Gerschman, R., Gilbert, D. L., Terwilliger, D., and Cothran, F. V. Mutagenic effects of high oxygen tensions on *Escherichia coli*. Proc. Nat. Acad. Sci. U.S.A. 43:1027–1032, 1957.
111. Finder, E., Simmons, G., and Huber, G. Pulmonary adaptive tolerance. Fed. Proc. 32:341, 1973.
112. Finder, E., Simmons, G., LaForce, M., and Huber, G. The role of edema in the development of adaptive tolerance to pulmonary oxygen toxicity. Clin. Res. 21:985, 1973.
113. Finley, T., Morgan, T., Fialkow, H., and Huber, G. Surface activity of phospholipid components of dog pulmonary surfactant collected in vivo. Fed. Proc. 23:156, 1964.
114. Finley, T. N., Swenson, E., Clements, J., Gardner, R., Wright, R., and Severinghaus, J. Changes in mechanical properties, appearance, and surface activity of one lung following occlusion of its pulmonary in the dog. Am. Physiol. Soc. Meet. Aug. 23–26, 1960.
115. Finley, T. N., Swenson, E. W., Curran, W. S., Huber, G. L., and Ladman, A. J. Bronchopulmonary lavage in normal subjects and patients with obstructive lung disease. Ann. Intern. Med. 66:651–658, 1967.
116. Finley, T., Tooley, W., Swenson, E., Gardner, R., and Clements, J. Pulmonary surface tension in experimental atelectasis. Am. Rev. Respir. Dis. 89:372, 1964.
117. Fisher, A. B., Hyde, R. W., Puy, R. J. M., Clark J. M., and Lambertsen, C. J. Effect of oxygen at 2 atmospheres on the pulmonary mechanics of normal man. J. Appl. Physiol. 24:529–536, 1968.
118. Forman, H. J., and Fisher, A. B. Antioxidant defenses. This volume, 1981.
119. Franck, C., Lamarche, M., Arnould, P., and Demange, J. M. Action de l'hyperoxie sur le taux d'histamine dans le sang et le tissu pulmonaire du cobaye non anesthesie. J. Physiol. (Paris) 49:176–177, 1957.
120. Frank, L., and Massaro, D. The lung and oxygen toxicity. Arch. Intern. Med. 139: 347–350, 1979.
121. Fridovich, I. Superoxide radical and superoxide dismutases. This volume, 1981.
122. Fujikura, T. Pulmonary hyaline membranes in various strains of mice. Am. J. Obstet. Gynecol. 87:1081–1085, 1963.
123. Fujikura, T. Effect of anticoagulant, fibrinolytic and antifibrinolytic agents on experimental hyaline membranes. Am. J. Obstet. Gynecol. 90:850–853, 1964.
124. Fujiwara, T., Adams, F. H., and Seto, K. Lipids and surface tension of extracts of normal and O_2 treated guinea pig lungs. J. Pediat. 65:45–52, 1964.
125. Furry, D. E. Tolerance of mice X-irradiated in an oxygen-rich environment to explosive decompression. Aerosp. Med. 35:459–461, 1964.
126. Fuson, R. L., Saltzman, H. A., Smith, W. W., Whalen, R. E., Osterhout, S., and Parker, R. F. Clinical hyperbaric oxygenation with severe oxygen toxicity. Report of a case. N. Engl. J. Med. 273: 415–419, 1965.
127. Gable, W. D., and Townsend, F. M. Lung morphology of individuals exposed to prolonged intermittent suplemental oxygen. Aerosp. Med. 33:1344–1348, 1962.
128. Garwacki, J. Effect of neuroplegic drugs on changes in the respiratory organs in rats breathing pure oxygen under pressure (1 atm). Acta. Physiol. Pol. 11:73–85, 1960.
129. Gerschman, R. Oxygen effects in biological systems. In: Proceedings of the International Congress on Physiological Science, 21st, Buenos Aires, pp. 222–226, 1959.
130. Gerschman, R. The biological effects of increased oxygen tension. In: Mans Dependence on the Earthly Atmosphere. (Ed.) Schaefer, K. E., pp. 170–179, Macmillan, New York, 1962.
131. Gerschman, R. Biological effects of oxygen.

In: Oxygen in the Animal Organism (Eds.) Dickens, F. and Neil, E., pp. 475–494, Macmillan, New York, 1964.
132. Gerschman, R., and Fenn, W. O. Ascorbic acid content of adrenal glands of rat in oxygen poisoning. Am. J. Physiol. 176:6–8, 1954.
133. Gerschman, R., Gilbert, D. L., and Caccamise, D. Effects of various substances on survival times of mice exposed to different high oxygen tensions. Am. J. Physiol. 192:563–571, 1958.
134. Gerschman, R., Gibert, D. L., and Frost, J. N. Sensitivity of *Paramecium caudatum* to high oxygen tensions and its modification by cobalt and manganese ions. Am. J. Physiol. 192:572–576, 1958.
135. Gerschman, R., Gilbert, D. L., and Nye, S. W. Survival time of newborn rats and mice in high oxygen pressure (HOP). Studies on oxygen poisoning. USAF School of Aerospace Medicine Report No. 10, pp. 35–36, Randolph Field, Texas, 1955.
136. Gerschman, R., Gilbert, D. L., and Nye, S. W. The effects of autonomic drugs on oxygen poisoning. USAF School of Aviation Medicine Report NO. 56-42, Randolph Air Force Base, Texas, 1956.
137. Gerschman, R., Gilbert, D. L., Nye, S. W., Dwyer, P., and Fenn, W. O. Oxygen poisoning and X-irradiation: A mechanism in common. Science 119:623–626, 1954.
138. Gerschman, R., Gilbert, D. L., Nye, S. W., Nadig, P., and Fenn, W. O. Role of adrenalectomy and adrenal cortical hormones in oxygen poisoning. Am. J. Physiol. 178:346–350, 1954.
139. Gerschman, R., Gilbert, D. L., Nye, S. W., Price, W. E., Jr., and Fenn, W. O. Effects of autonomic drugs and of adrenal glands on oxygen poisoning. Proc. Soc. Exp. Biol. Med. 88:617–621, 1955.
140. Gerschman, R., Gilbert, D. L., Nye, S. W., Price, W. E., Jr., and Fenn, W. O. The role of the adrenal cortical hormones and of dibenzyline in normal and demedullated mice submitted to high oxygen pressure. USAF School of Aviation Medicine Report No. 56-44, Randolph Air Force Base, Texas, 1956.
141. Gerschman, R., Nadig, P. W., Snell, A. D., Jr., and Nye, S. W. Effect of high oxygen concentrations on eyes of newborn mice. Am. J. Physiol. 179:115–118, 1954.
142. Gersh, I. Syndrome of oxygen poisoning in cats. War Med. 8:221–228, 1945.
143. Gersh, I., and Wagner, C. E. Metabolic factors in oxygen poisoning. Am. J. Physiol. 144:270–277, 1945.
144. Gershenovich, Z. S., Krichevskaya, A. A., and Alekseenko, L. P. Adrenaline substance of the brain and suprarenals under increased oxygen pressure. Ukr. Biokhem. Zh. 27:3–11, 1955.
145. Gesell, R. On the chemical regulation of respiration: Regulation of respiration with special reference to the metabolism of the respiratory center and the coordination of the dual function of hemoglobin. Am. J. Physiol. 66:5–49, 1923.
146. Giammona, S. T., Kerner, D., and Bondurant, S. Effect of oxygen breathing at atmospheric pressure on pulmonary surfactant. J. Appl. Physiol. 20:855–858, 1965.
147. Gilbert, D. L. The role of pro-oxidants and anti-oxidants in oxygen toxicity. Radiat. Res. Suppl. 3:44–53, 1963.
148. Gilbert, D. L. Atmosphere and evolution. In: Oxygen in the Animal Organism. (Eds.) Dickens, F., and Neil, E., pp. 641–654. Macmillan, New York, 1964.
149. Gilbert, D. Oxygen and life. Anesthesiology 37:100–111, 1972.
150. Gilbert, D., Gerschman, R., Cohen, J., and Sherwood, W. The influence of high oxygen pressures on the viscosity of solutions of sodium desoxyribonucleic acid and of sodium alginate. J. Am. Chem. Sci. 79: 5677–5680, 1957.
151. Gilbert, D. L., Gerschman, R., and Fenn, W. O. Effects of fasting and X-irradiation on oxygen poisoning in mice. Am. J. Physiol. 181:272–274, 1955.
152. Gottlieb, S. F. Hyperbaric oxygenation. Adv. Clin. Chem. 8:69–139, 1965.
153. Grandpierre, R., Grognot, P., and Senelar, R. Modifications des lesions pulmonaires provoques par l'inhalation d'oxygene apres section d'un nerf pneumogastrique. J. Physiol. (Paris) 48:564–565, 1956.
154. Greenberg, S. D., Gyorkey, F., and O'Neal, R. M. Pulmonary hyaline membranes in adults receiving oxygen therapy. Texas Rep. Biol. Med. 27:1005–1012, 1969.
155. Griffo, Z., and Roos, A. Effect of oxygen breathing on pulmonary compliance. J. Appl. Physiol. 17:233–238, 1962.

156. Grognot, P., and Chome, J. Action de la chlorpromazine et du tetrylammonium sur les reactions pulmonaires provoquees par inhalation d'oxygene pur. C. R. Soc. Biol. 148:1474–1475, 1954.
157. Grognot, P., and Chome, J. Action de l'histamine et de antihistaminiques sur les lesions pulmonaires dues a l'inhalation de l'oxygene pur chez le rat et le cobaye. C. R. Soc. Biol. 149:562–564, 1955.
158. Groshikov, M. A., and Sorokin, P. A. Pathological changes in the lungs of animals under the influence of high oxygen pressures. In: The Effect of the Gas Medium and Pressure on Body Functions, Collection No. 3. (Ed.) Brestkin, M. P., pp. 112–121, published for NASA, USA and NSF, Washington, D.C. by Israel Program for Scientific Translations, NASA TTF-358, TT 65-50136, 1965.
159. Grossman, M. S., and Penrod, K. E. Relationship of hypothermia to high oxygen poisoning. Am. J. Physiol. 156:177–181, 1949.
160. Grossman, M. S., and Penrod, K. E. The thyroid and high oxygen poisoning in rats. Am. J. Physiol. 156:182–184, 1949.
161. Gupta, S. R., and Abraham, S. Some factors that affect susceptibility to toxic effects of oxygen. Indian J. Med. Res. 57: 739–746, 1969.
162. Gutierrez, V. S., Berman, I. R., Soloway, H. B., and Hamit, H. F. Relationship of hypoproteinemia and prolonged mechanical ventilation to the development of pulmonary insufficiency in shock. Ann. Surg. 171:385–393, 1970.
163. Gutsche, B. B., Harp, J. R., and Stephen, C. R. Physiologic responses of the anesthetized dog to oxygen at five atmospheres absolute. Anesthesiology 27:615–623, 1966.
164. Hackney, J. D., Collier, C. R., and Rounds, D. E. Mechanism of pulmonary damage in oxygen poisoning. In: Drugs and Respiration. (Eds.) Aviado, D., and Palecek, I., vol. XI, pp. 93–103, Macmillan, New York, 1964.
165. Hagebusch, O. E. Pathology of animals exposed for 235 days to a 5 psia 100% oxygen atmosphere. In: Proceedings of the Second Annual Conference on Atmospheric Contamination in Confined Spaces. Aerospace Medical Research Laboratories Report AMRL-TR-66-120, pp. 103–107, Wright-Patterson Air Force Base, Ohio, 1966.
166. Haggendal, J. The effect of high pressure air or oxygen with and without carbon dioxide added on the catacholamine levels of rat brain. Acta Physiol. Scand. 69:147–152, 1967.
167. Harp, J. R., Gutsche, B. B., and Stephen, C. R. Effect of anesthetics on central nervous system toxicity of hyperbaric oxygen. Anesthesiology 27:608–614, 1966.
168. Harris, J. W., and van den Brenk, H. A. S. Comparative effects of hyperbaric oxygen and pentylenetetrazol on lung weight and non-protein sulfhydryl content of experimental animals. Biochem. Pharmacol. 17: 1181–1188, 1968.
169. Haugaard, N. Cellular mechanisms of oxygen toxicity. Physiol. Rev. 48:311–373, 1968.
170. Hawker, J. M., Reynolds, E. D. R., and Taghizadeh, A. Pulmonary surface tension and pathological changes in infants dying after respirator treatment for severe hyaline membrane disease. Lancet 2:75–77, 1967.
171. Hayashi, M., and Huber, G. L. Airway defenses, Seminars Respir. Med. 1:233–239, 1980.
172. Hederer, C., and Andre, L. De l'intoxication par les hautes pressions d'oxygene. Bull. Acad. Nat. Med. (Paris) 123:294–307, 1940.
173. Hellstrom, B. and Nergardh, A. The effect of high oxygen concentrations and hypothermia on the lung of the newborn mouse. Acta Paediat. Scand. 54:457–466, 1965.
174. Helvey, W. M., Albright, C. A., Benjamin, F. B., Call, L. S., Peters, J. M., and Rind, H. Effects of prolonged exposure to pure oxygen on human performance. Republic Aviation Corporation, Report 393-1, NASA Contr. NASr-92, 1962.
175. Hemingway, A., and Williams, W. L. Pulmonary edema in oxygen poisoning. Proc. Soc. Exp. Biol. Med. 89:331–334, 1952.
176. Hempleman, H. V. Effect of pre-exposure to carbon dioxide upon resistance to acute oxygen poisoning in the rat. Great Britain MRC-RNPRC, R.N.P. 56/860, U.P.S. 156, March, 1956.
177. Hill, L. Influence of CO_2 in production of oxygen poisoning. J. Exp. Physiol. Cog. Med. Sci. 23:49–50, 1933.

178. Houlihan, R. T. Adaptation to chronic hyperbaric oxygen pressures. Final Scientific Report Contr. Nonr. 656-36. USAF School of Aerospace Medicine, September, 1969.
179. Houlihan, R. T. Rheomelanin accumulation in the blood and lungs and hemolysis in rats poisoned by hyperbaric oxygen. J. Am. Osteopath. Assoc. 69:1040, 1970.
180. Houlihan, R. T., Altschule, M. D., and Hegedus, Z. L. Indole metabolism of catecholamines during exposure to hyperbaric oxygen. Preprints of Annual Scientific Meeting of Aerospace Medical Association, pp. 192–193, 1969.
181. Houlihan, R. T., Altschule, M. D., Hegedus, Z. L., and Cross, M. H. Oxidative cyclization of catecholamines following exposure to hyperbaric oxygen. Fourth International Congress on Hyperbaric Medicine (Program), Sapporo, Japan, pp. 14–15, 1969.
182. Houlihan, R. T., Zavodni, J. J., and Cross, M. H. Adaptation to increased oxygen tension at ambient pressure. Aerosp. Med. 38:995–997, 1967.
183. Houlihan, R. T., Zavodni, J., and Cross, M. Effects of increased oxygen pressure on adrenal steroid and catecholamine release. In: Aviation and Space Medicine. (Eds.) Hannisdahl, B., and Sem-Jacobsen, C. W., pp. 68-73, Universitetsforlaget, Oslo, 1969.
184. Huber, G. L. Congestive atelectasis and respiratory distress. Doctor of Medicine Thesis, University of Washington, School of Medicine, Seattle, Washington, 1966.
185. Huber, G. L. The respiratory distress syndrome. Master of Science Thesis, Department of Biological Structure, University of Washington, Seattle, Washington, 1970.
186. Huber, G. L. Pulmonary function tests and blood gases in the asthmatic. In: The Asthmatic Patient in Trouble. (Ed.) Petty, T. L., pp. 21–29, CPC Communications, Inc., Greenwich, Connecticut, 1975.
187. Huber, G. L. Clinical Application of Arterial Blood Gas and Acid-Base Physiology. Current Concepts Monograph, Upjohn, Kalamazoo, Michigan, 1978.
188. Huber, G. L. Immunologic lung reactions. Seminars Respir. Med. 1:251–273, 1980.
189. Huber, G. L., Burley, S. W., Porter, L., Mason, R. J., and LaForce, F. M. Pulmonary oxygen toxicity. J. Clin. Invest. 50:46, 1971.
190. Huber, G. L., and Davies, P. Alveolar defenses. Seminars Respir. Med. 1:240–250, 1980.
191. Huber, G. L., and Finley, T. N. Alteration in pulmonary morphology and surface activity during experimental oxygen toxicity. Ann. Intern. Med. 70:1097, 1969.
192. Huber, G. L., and First, M. W. Perspectives: Pulmonary host defenses, the host and the development of lung disease. Seminars Resp. Med. 1:187–196, 1980.
193. Huber, G. L., and LaForce, F. M. Comparative effects of ozone and oxygen on pulmonary antibacterial defense mechanisms. Antimicrob. Agents Chemother. 10: 129–136, 1970.
194. Huber, G. L., and LaForce, F. M. Progressive impairment of pulmonary antibacterial defense mechanism associated with prolonged oxygen administration. Ann. Intern. Med. 72:808, 1970.
195. Huber, G. L., LaForce, F. M., and Johanson, W. G., Jr. Experimental models and pulmonary antimicrobial defenses. In: Respiratory Defense Mechanisms, (Eds.) Brain, J. D., Proctor, D., and Reid, L. In: Lung Biology in Health and Disease. (Exec. Eds.) Lenfant, C. and Dekker, M., pp. 983–1022, Marcel Dekker, New York, 1977.
196. Huber, G. L., Laforce, F. M., and Mason, R. J. Impairment and recovery of pulmonary antibacterial defense mechanisms after oxygen administration. J. Clin. Invest. 49:47, 1970.
197. Huber, G. L., and Mahajan, V. K. Cough, respiratory reflexes, and patterns of breathing. Seminars Respir. Med. 1:223–232, 1980.
198. Huber, G. L., Mason, R. J., Boyd, A. E., and Norman, J. C. Experimental pulmonary hyaline membrane disease following disseminated intravascular coagulation. Curr. Top. Surg. Res. 411–430, 1969.
199. Huber, G. L,., Mason, R. J., LaForce, F. M., Spencer, N. J., Gardner, D. E., and Coffin, D. L. Alterations in the lung following the administration of ozone. Arch. Intern. Med. 128:81–87, 1971.
200. Huber, G., O'Connel, D., and LaForce, M. The role of pulmonary surfactant in the bactericidal activity of alveolar macrophages during oxygen toxicity. J. Clin. Invest. 52:42, 1973.
201. Huber, G. L., Redding, R., and Ellenbogen, M. Oxygen therapy and oxygen

toxicity. American College of Chest Physicians, Audiographic Series, Vol. 7, Chicago, 1975.

202. Huber, G., Schauffler, H., Levens, D., Pollack, C., Taffel, W., Korman, G., Hayashi, M., Gillis, B., O'Connell, D., Laguarda, R., and Pereira, W. Stereologic quantification of alterations in alveolar macrophages (AM) recovered from oxygen toxic animals. Clin. Res. 23:347, 1975.

203. Huber, G., Schauffler, H., McCarthy, C., and Hayashi, M. Comparative morphometric analysis of recoverable and *in situ* pulmonary alveolar macrophages. J. Cell Biol. 63:148, 1974.

204. Huber, G. L., and Spencer, N. J. Alterations in pulmonary antibacterial defense mechanisms following exposure to ozone. Am. Rev. Respir. Dis. 101:1016–1017, 1970.

205. Huber, G. L., Vater, C. A., Huber, A. J., Burley, S. W., and LaForce, F. M. An experimental model for correlative quantification of bacterial inactivation with consolidation and fluid accumulation in the lung. Clin. Res. 19:741, 1971.

206. Hudson, L. H., and Erdmann R. R. Pulmonary vascular changes in newborn mice following exposure to increased oxygen tensions under moderate hyperbaric conditions. Angiology 17:819–824, 1966.

207. Hyde, R. W., and Rawson, A. J. Unintentional iatrogenic oxygen pneumonitis—response to therapy. Ann. Intern. Med. 71: 517–531, 1969.

208. Igarashi, S., Hasegawa, K., Kawakami, F., Sugase, Z., and Nishikawa, M. Effect of hyperbaric oxygen on adrenocortical function. Folia Endocrinol. Jpn. 44:358, 1968.

209. Ishizuka, R. A. Surface tension study of the animal lungs: influence of O_2 toxicity due to hyperbaric oxygenation. Jpn. J. Thorac. Surg. 19:578–585, 1966.

210. Ishizuka, R., Miyakawa, K., Maekawa, T., Imamura, B., Akashi, T., Tanaka, N., Kasai, Y., Jockin, H. C., and Bernhard, W. F. Pulmonary surface characteristics in oxygen toxicity. In: Proceedings of the Fourth International Congress on Hyperbaric Medicine. (Eds.) Wada, J., and Iwa, T., pp. 16–21, Williams & Wilkins, Baltimore, 1970.

211. Jamieson, D. Potentiation of HPO paralysis in rats. In: Clinical Application of Hyperbaric Oxygen. (Eds.) Boerema, I., et al, pp. 319–323, Elsevier Publ. Co., Amsterdam, 1964.

212. Jamieson, D. Role of central nervous system and pulmonary damage as cause of respiratory failure in rats exposed to hyperbaric oxygen. In: Proceedings of the Third International Conference on Hyperbaric Medicine. (Eds.) Brown, I. W., Jr., and Cox, B. G., pp. 89–96, NAS/NRC Publ. 1404, Washington, D.C., 1966.

213. Jamieson, D. The effect of anaesthesia and CO_2 on survival time and lung damage in rats exposed to high pressure oxygen. Biochem. Pharmacol. 15:2120–2122, 1966.

214. Jamieson, D., and Cass, N. CNS and pulmonary damage in anesthetized rats exposed to hyperbaric oxygen. J. Appl. Physiol. 23:235–242, 1967.

215. Jamieson, D., Ladner, K., and van den Brenk, H. A. S. Pulmonary damage due to high pressure oxygen breathing in rats. 4. Quantitative analysis of sulfhydryl and disulfide groups in rat lungs. Aust. J. Exp. Biol. Med. Sci. 41:491–497, 1963.

216. Jamieson, D., and van den Brenk, H. A. S. Pulmonary damage due to high pressure oxygen breathing in rats. 2. Changes in dehydrogenase activity of rat lung. Aust. J. Exp. Biol. Med. Sci. 40:51–56, 1962.

217. Jamieson, D., and van den Brenk, H. A. S. Pulmonary damage due to high pressure O_2 breathing in rats. 5. Changes in the surface active lung alveolar lining. Aust. J. Exp. Biol. Med. Sci. 42:483–490, 1964.

218. Jamieson, D., and van den Brenk, H. A. S. The effect of antioxidants on high pressure oxygen toxicity. Biochem. Pharmacol. 13: 159–164, 1964.

219. Johnson, J. W., Permutt, S., Sipple, J. H., and Salem, E. S. Effect of intra-alveolar fluid on pulmonary surface tension properties. J. Appl. Physiol. 19:769–777, 1964.

220. Johnson, P. C., and Bean, J. W. Carbon dioxide and the sympathoadrenal system in O_2 at high pressure (OHP). Fed. Proc. 14: 81, 1955.

221. Johnson, P. C. and Bean, J. W. Effect of sympathetic blocking agents on the toxic action of O_2 at high pressure. Am. J. Physiol. 188:593–598, 1957.

222. Kabins, S., Molina, C., and Katz, L. Pulmonary vascular effects of serotonin (5-OH-Tryptamine) in dogs and its role in

causing pulmonary edema. Am. J. Physiol. 197: 955–958, 1959.

223. Kann, H. E., Jr., Mengel, C. E., Smith, W., and Horton, B. Oxygen toxicity and vitamin E. Aerosp. Med. 35:840–844, 1964.

224. Kapanci, Y., Weibel, E. R., Kaplan, H. P., and Robinson, R. F. Pathogenesis and reversibility of the pulmonary lesions of oxygen toxicity in monkeys. II. Ultrastructural and morphometric studies. Lab. Invest. 20:101–118, 1969.

225. Kaplan, H. P. Hematologic effects of increased oxygen tensions. In: Proceedings of the Second Annual Conference on Atmospheric Contamination in Confined Spaces. Aerospace Medical Research Laboratories Report AMRL-TR-66-120, pp. 200–222, Wright-Patterson Air Force Base, Ohio, 1966.

226. Kaplan, H. P., Robinson, F. R., Kapanci, Y., and Weibel, E. R. Pathogenesis and reversibility of the pulmonary lesions of oxygen toxicity in monkeys. I. Clinical and light microscopic studies. Lab. Invest. 20: 94–100, 1969.

227. Karsner, H. T. Pathological effects of atmospheres rich in oxygen. J. Exp. Med. 23:149–170, 1916.

228. Karsner, H. T., and Ash, J. E. Further study of the pathological effects of atmospheres rich in oxygen. J. Lab. Clin. Med. 2:254–255, 1916.

229. Kennedy, J. H. Hyperbaric oxygenation and pulmonary damage. The effect of exposure at two atmospheres upon surface activity of lung extracts in the rat. Med. Thoracalis 23:27–35, 1966.

230. Kimball, R. E., Krishna, R., Pierce, T. H., Schwartz, L. W., Mustafa, M. G., and Cross, C. E. Oxygen toxicity: Augmentation of antioxidant defense mechanisms in rat lung. Am. J. Physiol. 230:1425–1431, 1976.

231. King, C. T. G., Williams, E. E., Mego, J. L., and Schaefer, K. E. Adrenal function during prolonged exposure to low concentration of carbon dioxide. Am. J. Physiol. 183:46–52, 1955.

232. Kistler, G. S., Caldwell, P. R. B., and Weibel, E. R. Quantitative electron microscopic studies of murine lung damage after exposure to 98.5% oxygen at ambient pressure: A preliminary report. In: Proceedings of the Third International Conference on Hyperbaric Medicine. (Eds.) Brown, I. W., Jr., and Cox, B. G., NAS/NRC, Publ. 1404, Washington, D.C., 1966.

233. Kistler, G. S., Caldwell, P. R. B., and Weibel, E. R. Development of fine structural damage to alveolar and capillary lining cells in oxygen-poisoned rat lungs. J. Cell Biol. 32:605–628, 1967.

234. Kistler, G. S., Weibel, E. R., and Caldwell, P. R. B. Electron microscopic and morphometric study of rat lungs exposed to 97 percent oxygen at 258 torr (27,000 feet). In: Proceedings of the Second Annual Conference on Atmospheric Contamination in Confined Spaces, Aerospace Medical Research Laboratories Report AMRL-TR-66-120, pp. 147–161, Wright-Patterson Air Force Base, Ohio, 1966.

235. Kydd, G. H. Lung changes resulting from prolonged exposure to 100 percent oxygen at 550 mm Hg. Aerosp. Med. 38:918–923, 1967.

236. Kydd, G. H. Survival of rats exposed to 10 psia of oxygen to further exposure at one atmosphere of oxygen. Aerosp. Med. 39: 739–744, 1968.

237. Kyle, J. D. The effects of 100% oxygen inhalation on adult and newborn rat lungs. S. Med. J. 58:1592, 1965.

238. Laborit, H., Broussolle, B., and Perimond-Trouchet, R. Essais pharmacologiques concernant le mecanisme des convulsions dues a l'oxygene pur en pression chez la souris. J. Physiol. (Paris) 49:953–962, 1957.

239. LaForce, F. M., Mullane, J. F., Boehme, R. F., Kelly, W. J., and Huber, G. L. The effect of pulmonary edema on antibacterial defenses of the lung. J. Lab. Clin. Med. 82:634–648, 1973.

240. LaForce, F. M., Kelly, W., and Huber, G. L. Stimulation of bactericidal activity of alveolar macrophages with surfactant. Clin. Res. 20:579, 1972.

241. Lambertsen, C. J. Physiological effects of oxygen. In: Proceedings of the Second Symposium on Underwater Physiology. (Eds.) Lambertsen, C. J., and Greenbaum, L. J., Jr., pp. 171–187, NAS/NRC Publ. 1181, Washington, D.C., 1963.

242. Lambertsen, C. J. Effects of oxygen at high partial pressure. In: Handbook of Physiology, Section 3: Respiration, Vol. II.

(Eds.) Fenn, W. O., and Rahn, H., pp. 1027–1046, American Physiological Society, Washington, D.C., 1965.

243. Lambertsen, C. J. Oxygen toxicity. In: Fundamentals of Hyperbaric Medicine, pp. 21–32, NAS/NRC Publ. 1298, Washington, D.C., 1966.

244. Lambertsen, C. J. Physiological effects of oxygen inhalation at high partial pressures. In: Fundamentals of Hyperbaric Medicine, pp. 12–20, NAS/NRC Publ. 1298, Washington, D. C., 1966.

245. Lambertsen, C. J., Ewing, J. H., Kough, R. H., Gould, R., and Stroud, M. W., III. Oxygen toxicity. Arterial and internal jugular blood gas composition in man during inhalation of air, 100% O_2 and 2% CO_2 in O_2 at 3.5 atmospheres ambient pressure. J. Appl. Physiol. 8:255–263, 1955.

246. Lambertsen, C. J., Kough, R. H., Cooper, D. Y., Emmel, G. L., Loeschcke, H. H., and Schmidt, C. F. Comparison of relationship of respiratory minute volume to P_{CO_2} and pH of arterial and internal jugular blood in normal man during hyperventilation produced by low concentrations of CO_2 at 1 atmosphere and by O_2 at 3.0 atmospheres. J. Appl. Physiol. 5:803–813, 1953.

247. Laurenzi, G. A., Yin, S., and Guarneri, J. J. Adverse effect of oxygen on tracheal mucus flow. N. Engl. J. Med. 279:333–339, 1968.

248. Lee, C. H., Lyons, J. H., Konisberg, S., Morgan, F., and Moore, F. D. Effects of spontaneous and positive-pressure breathing of ambient air and pure oxygen at one atmosphere pressure on pulmonary surface characteristics. J. Thorac. Cardiovasc. Surg. 53:759–769, 1967.

249. Lee, C. J., Lyons, J. H., and Moore, F. D. Cardiovascular and metabolic responses to spontaneous and positive-pressure breathing of 100 percent oxygen at one atmosphere pressure. J. Thorac. Cardiovasc. Surg. 53: 770–780, 1967.

250. Levens, D., Simmons, G., and Huber, G. Basic physiologic changes and antibacterial defenses in mice exposed to 100% oxygen. Clin. Res. 20:888, 1972.

251. Levine, B. E., and Johnson, R. P. Surface activity of saline extracts from inflated and degassed normal lung. J. Appl. Physiol. 19:333–335, 1964.

252. Levine, B. E., and Johnson, R. P. Effects of atelectasis on pulmonary surfactant and quasi-static lung mechanics. J. Appl. Physiol. 20:859–864, 1965.

253. Lieberman, J. A unified concept and critical review of pulmonary hyaline membrane formation. Am. J. Med. 35:443–449, 1963.

254. Lieberman, J., and Kellogg, F. Hyaline membrane formation and pulmonary plasminogen-activator activity in various strains of mice. Pediatrics 39:75–81, 1967.

255. Lottsfeldt, F. I., Schwartz, S., and Krivit, W. Hyperbaric oxygen, whole-body X-irradiation and cyclophosphamide combination therapy in mouse leukemia L1210. J. Natl. Cancer Inst. 36:37–43, 1966.

256. Loubiere, R., and Pfister, A. Acute and chronic lesions of normobaric hyperoxia in rats. Arch. Anat. Pathol. (Paris) 14:171–175, 1966.

257. Loubiere, R., Pfister, A., Fabre, J., and Violette, C. Recent data on experimental hyperoxia lesions. Rev. Med. Aeronaut. (Paris) 4:8–11, 1965.

258. Lynch, M. J. G. Hyaline membrane disease of lungs. Further observations. J. Pediat. 48:165–179, 1956.

259. Lynch, M. J. G., Mellor, L. D., and Badgery, A. R. Hyaline membrane disease. Its nature and etiology. The poisonous metabolic effects of excess oxygen. Neural control of electrolytes. J. Pediat. 48:602–632, 1956.

260. MacDonald, G. F., Dines, D. E., Sessler, A. D., and Titus, J. L. Prolonged oxygen therapy and mechanical ventilation. Associated pulmonary abnormalities. Minn. Med. 52:1745–1748, 1969.

261. MacEwen, J. D., and Haun, C. C. Oxygen toxicity at near ambient pressures. In: Proceedings of the Second Annual Conference on Atmospheric Contamination in Confined Spaces. Aerospace Medical Research Laboratories Report AMRL-TR-66-120, pp. 65–72, Wright-Patterson Air Force Base, Ohio, 1966.

262. McCarthy, C., Couzens, S., Laguarda, R., Pereira, W., and Huber, G. L. The effect of administration of oxygen on the normal and on the injured lung. Proc. Aerosp. Med. Assoc. 46:9–10, 1975.

263. McCarthy, C., and Huber, G. The dose-dependent nature of pulmonary oxygen toxicity. Chest 64:402–403, 1973.

264. McCord, J., and Fridovich, I. The reduction of cytochrome c by milk xanthine oxidase.

J. Biol. Chem. 243:5753–5760, 1968.
265. McCord, J. M., Keele, B., Jr., and Fridovich, I. An enzyme based theory of obligate anaerobiosis: the physiological function of superoxide dismutase. Proc. Natl. Acad. Sci. U.S.A. 68:1024–1027, 1971.
266. McCurdy, W. C., III, Beller, J. J., Coalson, J. J., and Greenfield, L. J. Pulmonary oxygen toxicity in hemorrhagic shock. Surg. Forum 20:49–50, 1969.
267. McSherry, C. K., and Gilder, H. Pulmonary oxygen toxicity and surfactant. In: Proceedings of the Fourth International Congress on Hyperbaric Medicine. (Eds.) Wada, J., and Iwa, T., pp. 10–15, Igakushoin Ltd., Tokyo, 1970.
268. McSherry, C. K., Panossian, A., Jaeger, V. J., and Veith, F. J. Effects of hyperbaric oxygen, the hyperbaric state and rapid decompression on pulmonary surfactant. J. Surg. Res 8:334–340, 1968.
269. Macklin, C. C. Observations on epicytes of the alveolar wall of the cats' lung and their reactions when stimulated with oxmium tetroxide. Anat. Rec. 70:53, 1938.
270. Macklin, C. C. Silver lineation on surface of pulmonic alveolar walls of mature cat, produced by applying weak silver nitrate solution and exposing to sunrays or photographic developer. J. Thorac. Surg. 7:536, 1938.
271. Macklin, C. C. Residual epithelial cells on pulmonary alveolar walls of mammals. Trans. Roy. Soc. Can. 40:93, 1946.
272. Macklin, C. C. The foam cells of mammalian lungs with special reference to the vacuoloids. In: International Congress of Experimental Cytology, p. 383, Stockholm, 1947.
273. Macklin, C. C. Dust cells in lungs of albino mouse, their structure, relations and mode of action. Lancet 1:432, 1951.
274. Macklin, C. C. The pulmonary alveolar mucoid film and the pneumonocytes. Lancet 2:1099, 1954.
275. Maklari, E., Kellner, M., Kadar, A., Kovach, A. G. B., and Gottsegen, Gy. Studies in experimental pulmonary oedema. I. Pathomechanism of pulmonary oedema induced by hyperoxygenation Acta Med. Acad. Sci. Hung. 23:25–30, 1966.
276. Maklari, E., Kellner, M., Kovach, A. G. B., and Gottsegen, Gy. Studies in experimental pulmonary oedema. II. Effect of dibenzyline on pulmonary oedema induced by hyperoxygenation. Acta Med. Acad. Sci. Hung. 23:31–36, 1966.
277. Marcozzi, G., Messinetti, A., Colombati, M., Mocavero, G., and Zelli, G. P. Oximetric variations and pulmonary lesions induced by inhalation of oxygen concentrations higher than that of the atmosphere. Arch. De Vecchi Anat. Patol. Med. Clin. 32:609–636, 1960.
278. Margolis, G., and Brown, I. W., Jr. Hyperbaric oxygenation: The eye as a limiting factor. Science 151:466–468, 1966.
279. Maritano, M., Cabrai, M., Pattono, R., and Marchiaro, G. Apropos of various vascular and respiratory manifestations with the use of adrenolytic substances during hyperbaric oxygenation in rats. Acta Anesthesiol. Scand. Suppl. 24:343–352, 1966.
280. Marotta, S. F., Hirai, K., and Atkins, G. Adrenocortical secretion in anesthetized dogs during hyperoxia, hypoxia and positive pressure breathing. Proc. Soc. Exp. Biol. Med. 118:922–929, 1965.
281. Marshall, J. R., and Lambertsen, C. J. Interactions of increased PO_2 and P_{CO_2} effects in producing convulsions and death in mice. J. Appl. Physiol. 16:1–7, 1961.
282. Mason, R. J., Huber, G. L., and Vaughan, M. Synthesis of dipalmitoyl lecithin by alveolar macrophages. J. Clin. Invest. 51: 68–73, 1972.
283. Massaro, D., and Massaro, G. Biochemical and anatomical adaptation of the lung to oxygen-induced injury. Fed. Proc. 37: 2485–2488, 1978.
284. Massart, L. Sur une pretendue relation entre l'oxydose et l'acidose gazeuse. C. R. Soc. Biol. 117:265–266, 1934.
285. Matteo, R. S., and Nahas, G. G. Sodium bicarbonate: increase in survival rate of rats inhaling oxygen. Science 141:719–720, 1963.
286. Mengel, C. E., Kann, H. E., Jr., Lewis, A. M., and Horton, B. Mechanisms of *in vivo* hemolysis induced by hyperoxia. Aerosp. Med. 35:857–860, 1964.
287. Menzel, D. B. Toxicity of ozone, oxygen and radiation. Annu. Rev. Pharmacol. 10: 379–394, 1970.
288. Meschan, I., DeArmas, C. R., and Scharyj, M. Adult bronchopulmonary dysplasia: The similarity in roentgen and histopathologic appearance between some cases of oxygen

toxicity, radiation pneumonitis, and post-cytotoxic nonspecific bronchopneumonia. Radiology 92:612–615, 1969.

289. Misra, H., and Fridovich, I. The generation of superoxide radical during the autooxidation of ferredoxins. J. Biol. Chem. 246: 6866–6890, 1971.

290. Morgan, A. P. The pulmonary toxicity of oxygen. Anesthesiology 29:570–579, 1968.

291. Morgan, T. E., Edmunds, L. H., Jr., and Huber, G. L. Biochemical mechanical and morphologic alterations in lungs after pulmonary artery occlusion. Clin. Res. 15:140, 1967.

292. Morgan, T. E., Finley, T. N., Huber, G. L., and Fialkow, H. Alterations in pulmonary surface active lipids during exposure to increased oxygen tension. J. Clin. Invest. 44:1737–1744, 1965.

293. Morgan, T. E., Ulveda, F., Cutler, R. G., and Welch, B. E. Effects on man of prolonged exposure to oxygen at a total pressure of 190 mm Hg. Aerosp. Med. 34:589–592, 1963.

294. Morgan, T. E., Ulvedal, F., Cutler, R. G., B. E. Observations in the SAM two-man cabin simulator. II. Biochemical aspects. Aerosp. Med. 32:591–602, 1961.

295. Motlagh, F. A., Kaufman, S. Z., Giusti, R., Cramer, M., Garzon, A. A., and Karlson, K. E. Electron microscopic appearance and surface tension properties of the lungs ventilated with dry or humid air or oxygen. Surg. Forum 20:219–220, 1969.

296. Mullane, J., Goodenough, S., McCarthy, C., Simmons, G., O'Connell, D., LaForce, M., Laguarda, R., and Huber, G. Alterations in alveolar lining material as the key factor in depression of interpulmonary antibacterial defenses following stress. Chest 66:333, 1974.

297. Nair, C. S. Effects of breathing 100 percent oxygen at ground level on blood clotting time and platelet count. Indian J. Med. Res. 55:123–127, 1967.

298. Nash, G., Blennerhasset, J. B., and Pontoppidan, H. Pulmonary lesions associated with oxygen therapy and artificial ventilation. N. Engl. J. Med. 276:368–374, 1967.

299. Nash, G., Bowen, J. A., and Langlinais, P. C. "Respirator lung": a misnomer. Arch. Pathol. 21:234–240, 1971.

300. Nasseri, M., and Bucherl, E. S. Light and electron-microscopic studies of structural changes in the lung following hyperbaric oxygenation. Virchows Arch. Pathol. Anat. Physiol. Klin. Med. 242:190–198, 1967.

301. Nasseri, M., Kirstaedter, H. J., and Bucherl, E.S. Experimentelle Untersuchungen ueber das Allgemeinverhalten und die morphologischen Veränderungen unter hohem Sauerstoffdruck. Virchows Arch. Pathol. Anat. Physiol. Klin. Med. 341:148–154, 1966.

302. Newman, D., and Naimark, A. Palmitate-14-C uptake by rat lung: Effect of altered gas tensions. Am. J. Physiol. 214:305–312, 1968.

303. Noell, W. K. Effects of high and low oxygen tension on the visual system. In: Environmental Effects on Consciousness. (Ed.) Schaefer, K. E., pp. 3–18, Macmillan, New York, 1962.

304. Northway, W. H., Jr., and Rosan, R. C. Radiographic features of pulmonary oxygen toxicity in the newborn: Bronchopulmonary dysplasia Radiology 91:49–58, 1968.

305. Novelli, G. P., Pagni, E., Pirani, A., Ariano, M., and Pallini, C. La Cosiddetta tossicita dell' ossigeno iperbarico come fenomeno adrenergico. Proposta di interpretazione patogenetica. Acta Anesthesiol. (Padova) 18:801–831, 1967.

306. Ogawa, J., and Saito, H. Hyaline membrane in the lung of premature newborn mammals. Study on the etiological factors. Nagoya Med. J. 7:44, 1961.

307. Ohlsson, W. T. L. Study on oxygen toxicity at atmospheric pressure. Acta Med. Scand. Suppl. 190, 1947.

308. Ozorio de Almeida, A. Recherches sur l'action toxique des haute pressions d'oxygene. C. R. Soc. Biol. 116:1225–1227, 1934.

309. Pagni, E., Palumbo, D., and Cortesini, C. Influenza della medullectomia surrenalica sulla comparsa delle lesioni da ossigeno iperbarico. Boll. Soc. Med. Chir. Pisa 34:1, 1966.

310. Pagni, E., Zampolini, M., and Frullani, F. Blocking of alpha-adrenergic receptors as a method of prevention of lesions from hyperbaric oxygen. Minerva Anesthesiol. 33:49–54, 1967.

311. Paine, J. R., Lynn, D., and Keys, A. Manifestations of oxygen poisoning in dogs confined in atmospheres of 80 to 100

percent oxygen. Am. J. Physiol. 133:406–407, 1941.
312. Paine, J. R., Lynn, D., and Keys, A. Observations on the effects of prolonged administration of high oxygen concentration to dogs. J. Thorac. Cardiovasc. Surg. 11: 151–168, 1941.
313. Parad, R., Simmons, G., Feldman, N., and Huber, G. L. Impairment of adaptive tolerance to oxygen toxicity by systemic immunosuppression. Chest 67:42–43, 1975.
314. Pariente, R., Legrand, M., and Brouet, G. Ultrastructural aspects of the lung in oxygen poisoning at atmospheric pressure in the rat. Presse Med. 77:1073–1076, 1969.
315. Patel, Y. Z., and Gowdey, C. W. Effects of single and repeated exposures to oxygen of high pressure on normal rats. Can. J. Physiol. Pharmacol. 42:245–264, 1964.
316. Paton, W. D. M. Experiments on the convulsant and anesthetic effects of oxygen. Br. J. Pharmacol. Chemother. 29:350–366, 1967.
317. Pattle, R. E. Properties, function and origin of alveolar lining layer. Nature 175:1125, 1955.
318. Pattle, R. E. Properties, function and origin of alveolar lining layer. Proc. R. Soc. (London) 148:217–240, 1958.
319. Pattle, R. E. The cause of the stability of bubbles derived from the lung. Phys. Med. Biol. 5:11–26, 1960.
320. Pattle, R. E. Lung linings. Lancet 2:1298, 1961.
321. Pattle, R. E. The formation of a lining film by fetal lungs. J. Pathol. Bacteriol. 82:333–343, 1961.
322. Pattle, R. E., The lining layer of the lung alveoli. Br. Med. Bull. 19:41, 1963.
323. Pattle, R. E. Surface lining of lung alveoli. Physiol. Rev. 45:48–79, 1965.
324. Pattle, R. E., and Burgess, F. The lung lining film in some pathological conditions. J. Pathol. Bacteriol. 82:315–331, 1961.
325. Pattle, R., Claireaux, A., Davies, P., and Cameron, A. Inability to form a lung-lining film as a cause of the respiratory-distress syndrome in the newborn. Lancet 2:469–473, 1962.
326. Pattle, R. E., and Thomas, L. Lipoprotein composition of the film lining the lung. Nature 189:844, 1961.
327. Patz, A., Eastham, A., Higginbotham, D. H., and Kleh, T. Oxygen studies in retrolental fibroplasia. II. The production of the microscopic changes of retrolental fibroplasia in experimental animals. Am. J. Ophthalmol. 36:1511–1522, 1953.
328. Pautler, S., Cimons, I. M., Cauna, D., Totten, R., and Safar, P. Pulmonary oxygen toxicity at one ata. Acta Anaesthesiol. Scand. Suppl. 24:51–56, 1966.
329. Penrod, D. Lung damage by oxygen using differential catheterization. Fed. Proc. 17: 123, 1958.
330. Penrod, K. E. Nature of pulmonary damage produced by high oxygen pressures. J. Appl. Physiol. 9:1–4, 1956.
331. Pereira, W., Carr, F., Benesova, H., and Huber, G. The effect of experimental prophylactic and therapeutic steroid administration on the course of oxygen toxicity. Chest 68:423–424, 1975.
332. Pfeiffer, C. C., and Gersh, I. The prevention of the convulsions of oxygen poisoning by means of drugs. U.S. Navy NMRI, Project X-192, Report 2, 1944.
333. Pfister, A. M., Fabre, J., Coudert, G., Pingannaud, M., and Miro, L. Normobaric hyperoxia in the rat. Anesth. Analg. Reanim. 24:387–403, 1967.
334. Phillips, L. L., Wyte, S. R., Weeks, D. B., and Soloway, H. B. Fibrinolytic deficit in oxygen intoxicated guinea pigs. Aerosp. Med. 40:744–746, 1969.
335. Pichotka, J. Morphology and morphogenesis of pulmonary changes due to oxygen poisoning. In: German Aviation Medicine in World War II, Vol. I, pp. 515–525, USAF School of Aviation Medicine, 1950.
336. Polgar, G., Antagnoli, W., Ferrigan, L. W., Martin, E. A., and Gregg, W. P. The effect of chronic exposure to 100% oxygen in newborn mice. Am. J. Med. Sci. 252:580–587, 1966.
337. Popovic, V., Gerschman, R., and Gilbert, D. L. Effects of high oxygen pressure on ground squirrels in hypothermia and hibernation. Am. J. Physiol. 206:49–50, 1964.
338. Powney, J., and Addison, C. The properties of detergent solutions. Trans Faraday Soc. 33:1243–1260, 1937.
339. Pratt, P. C. Pulmonary capillary proliferation induced by oxygen inhalation. Am. J. Pathol. 34:1033–1050, 1958.
340. Pratt, P. C. The reaction of the human lung to enriched oxygen atmosphere. Ann. N.Y. Acad. Sci. 121:809–822, 1965.
341. Pratt, P. C. Oxygen toxicity as a factor. J. Trauma 8:854–866, 1968.

342. Puy, R. J. M., Hyde, R. W., Fisher, A. B., Clark, J. M., Dickson, J., and Lambertsen, C. J. Alterations in the pulmonary capillary bed during early O_2 toxicity in man. J. Appl. Physiol. 24:537–543, 1968.

343. Raffin, T. A., Douglas, W., Simon, L. M., Theodore, J., and Robbin, E. Effects of peroxia on ultrastructure and superoxide dismutase (SOD) in cultured type II pneumocytes (type II-P). Clin. Res. 25:422, 1973.

344. Rahn, H., and Farhi, L. E. Gaseous environment and atelectasis. Fed. Proc. 22:1035–1041, 1963.

345. Rehbock, D. J., Oldt, M. R., and Dixon, H. M. Effect of exposure to high oxygen tension on the lungs and heart of the rat. Arch. Pathol. 30:1172–1177, 1940.

346. Reichert, W. Beitrag zum cardiazol-lungenodem. Naunyn-Schmiedebergs Arch. Pharmakol. Exp. Pathol. 197:620–628, 1941.

347. Respiratory Diseases: Task Force on Problems, Research Approaches, Needs. Respiratory Distress Syndromes. DHEW Publ. (NIH) No., 73–432, pp. 165–180, 1973.

348. Riesen, W. H., Gross, A. M., and O'Neill, H. J. Effects of oxygen exposure on the production of surfactant phospholipids in the rabbit lung. Preprints of Annual Scientific Meeting of Aerospace Medical Association, pp. 112–113, 1970.

349. Rister, M., and Baehner, R. L. The alteration of superoxide dismutase, catalase, glutathione peroxidase, and NAD(P)H cytochrome C reductase in guinea pig polymorphoneuclear leukocytes and alveolar macrophages during hypoxia. J. Clin. Invest. 58:1174–1184, 1976.

350. Robertson, W. G., and Farhi, L. E. Rate of lung collapse after airway occlusion on 100% O_2 at various ambient pressures. J. Appl. Physiol. 20:228–232, 1965.

351. Robinson, F. R., Harper, D. T., Jr., Thomas, A. A., and Kaplan, H. P. Proliferative pulmonary lesions in monkeys exposed to high concentrations of oxygen. Aerosp. Med. 38:481–486, 1967.

352. Robinson, F. R., Sopher, R. L., Carter, V. L., and Witchett, C. E. Pulmonary lesions in monkeys exposed to 100% oxygen at 258 and 750 mm Hg pressure. Aerospace Medical Research Laboratories Report AMRL-TR-68-175, pp. 237–249, Wright-Patterson Air Force Base, Ohio, 1968.

353. Robinson, F. R., Sopher, R. L., and Witchett, C. E. Pathology of normobaric oxygen toxicity in primates. Aerosp. Med. 40:879–884, 1969.

354. Robinson, F. R., Thomas, A. A., and Rendon, L. Analysis of pleural effusions from rats exposed to high concentrations of oxygen. Clin. Chem. 14:1066–1073, 1968.

355. Rosenbaum, R. M., Wittner, M., and Lenger, M. Mitochondrial and other ultrastructural changes in great alveolar cells of oxygen-adapted and poisoned rats. Lab. Invest. 20:516–528, 1969.

356. Rosengren, K. Hyaline membrane disease. A radiological investigation in rabbits. Acta Radiol. Suppl. 262:1–64, 1967.

357. Roth, E. M. Selection of space cabin atmospheres. I. Oxygen toxicity. NASA Technical Note D-2008, Washington, D.C., 1963.

358. Rowland, R., and Newman, C. G. H. Pulmonary complications of oxygen therapy. J. Clin. Pathol. 22:192–198, 1969.

359. Rucci, F. S., Satta, U., and Campodonico, A. Effects of adrenal medullectomy on acute intoxication due to O_2 in rats. Acta Anesthesiol. (Padova) 16:265–286, 1965.

360. Sacks, T., Moldow, C. F., Craddock, P. R., Bowers, T. K., and Jacob, H. S. Oxygen radicals mediate endothelial cell damage by complement stimulated granulocytes: An *in vitro* model of immune vascular damage. J. Clin. Invest. 61:1161–1167, 1978.

361. Said, S. I., Avery, M. E., Davis, R. K., Banerjee, C. M., and El-Gohary, M. Pulmonary surface activity in induced pulmonary edema. J. Clin. Invest. 44:458–464, 1965.

362. Saiki, H. Physiological influences of high oxygen respiration in close environmental systems. Ninteenth Congress of the International Astronautical Federation, B183, New York, October 13–19, 1968.

363. Schaffner, F., Felig, P., and Trachtenberg, E. Structure of rat lung after protracted oxygen breathing. Arch. Pathol. 83:99–107, 1967.

364. Schulz, H. Ueber den Gestaltwandel der Mitochondrien in Alveolorepithel unter CO_2 und O_2 Atmung. Naturwissenschaften 9:205, 1956.

365. Sewell, M. T. The phagocytic properties of the alveolar cells of the lung. J. Pathol. Bacteriol. 22:40–55, 1918–1919.

366. Shanklin, D. R. The influence of inter-

mediate oxygen percentages on experimental hyaline membrane disease. S. Med. J. 57:1473, 1964.
367. Shanklin, D. R., and Berman, R. A. The influence of hyperbaric oxygen on hyaline membrane disease in newborn rabbits. S. Med. J. 56:1443, 1963.
368. Shanklin, D. R., and Cunningham, J. J. Vagotomy-oxygen synergism in the pathogenesis of hyaline membrane disease. Am. J. Pathol. 46:27a, 1965.
369. Shanklin, D. R., Cunningham, J. J., Sotelo-Avila, C. and Crussi, F. Oxygen, ascorbic acid, and the lung participation of adrenal ascorbic acid in response to oxygen challenge of the lung. Arch. Pathol. 84:451–459, 1967.
370. Shanklin, D. R., and Lester, E. Role of nitrogen in pulmonary oxygen toxicity. S. Med. J. 61:1338–1339, 1968.
371. Shanklin, D. R., and Sotelo-Avila, C. The effects of components of vagotomy on the lung and the effects of anesthesia on vagotomy-induced lung change. Biol. Neonatorum 11:61–86, 1967.
372. Shanklin, D. R., and Wolfson, S. L. Therapeutic oxygen as a possible cause of pulmonary hemorrhage in premature infants. N. Eng. J. Med. 277:833–837, 1967.
373. Shaw, L. A., Behnke, A. R., and Messer, A. C. The role of carbon dioxide in producing the symptoms of oxygen poisoning. Am. J. Physiol. 108:652–661, 1934.
374. Shilling, C. W., and Adams, B. H. A study of the convulsive seizures caused by breathing oxygen at high pressure. U.S. Naval Med. Bull. 31:112–121, 1933.
375. Simon, L. M., Liu, J., Theodore, J., and Robin, E. D. Effect of hyperoxia, hypoxia and maturation on superoxide dismutase activity in isolated alveolar macrophages. Am. Rev. Resp. Dis. 115:279–284, 1973.
376. Smith, C. W., and Bean, J. W. Adrenal factors in toxic action of O_2 at atmospheric pressure. Fed. Proc. 14:140, 1955.
377. Smith, C. W., and Bean, J. W. Influence of CO_2 on toxicity of O_2 at atmospheric pressure. Am. J. Physiol. 183:662, 1955.
378. Smith, C. W., Bean, J. W., and Bauer, R. Thyroid influence in reactions to O_2 at atmospheric pressure. Am. J. Physiol. 199: 883–888, 1960.
379. Smith, C. W., Bennett, G. A., Heim, J. W., Thomson, R. M., and Drinker, C. K. Morphological changes in the lungs of rats living under compressed air conditions. J. Exp. Med. 56:79–93, 1932.
380. Smith, C. W., Lehan, P. H., and Monks, J. J. Cardiopulmonary manifestations with high O_2 tensions at atmospheric pressure. J. Appl. Physiol. 18:849–853, 1963.
381. Smith, F. J. C., Heim, J. W., Thomson, R. M., and Drinker, C. K. Bodily changes and development of pulmonary resistance in rats living under compressed air conditions. J. Exp. Med. 56:63–78, 1932.
382. Smith, J. L. The pathological effects due to increased oxygen tension in the air breathed. J. Physiol. (London) 24:19–35, 1899.
383. Sornberger, G. C., and Huber, G. L. Models of respiratory clearance. Seminars Resp. Med. 1:274–280, 1980.
384. Stadie, W. C., Riggs, B. C., and Haugaard, N. Oxygen poisoning. Am. J. Med. Sci. 207:84–114, 1944.
385. Stevens, J. B., and Autor, A. P. Oxygen induced synthesis of superoxide dismutase and catalase in pulmonary macrophages of neonatal rats. Lab. Invest. 37:470–478, 1977.
386. Szam, I. Pathogenesis of hyperbaric oxygen intoxication. Experimental studies on the significance of CO_2 and NH_3 accumulation in the central nervous system for the pathogenesis of hyperbaric pulmonary edema. Anesthetist 18:39–43, 1969.
387. Szilagyi, T., Toth, S., Miltenyi, L., and Jona, G. Oxygen poisoning and thyroid function. Acta Physiol. Acad. Sci. Hung. 35:59–61, 1969.
388. Taylor, D. W. Effects of vitamin E deficiency on oxygen toxicity in the rat. J. Physiol. (London) 121:47P, 1953.
389. Taylor, D. W. Effects of high oxygen pressures on adrenalectomized treated and untreated rats. J. Physiol (London) 125:46–47P, 1954.
390. Taylor, D. W. Effects of adrenalectomy on oxygen poisoning in the rat. J. Physiol. (London) 140:23–36, 1958.
391. Taylor, H. J. The effect of breathing oxygen at atmospheric pressure on tissue oxygen and carbon dioxide tensions. J. Physiol. (London) 108:264–269, 1949.
392. Taylor, H. J. The role of carbon dioxide in oxygen poisoning. J. Physiol. (London) 109:272–280, 1949.
393. Tennekoon, G. E. Pulmonary edema due to thiosemicarbazide. J. Pathol. Bacteriol. 67: 341–347, 1954.

394. Thomas, J. J., Jr., Baxter, R. C., and Fenn, W. O. Interactions of oxygen at high pressure and radiation in *Drosphila*. J. Gen. Physiol. 49:537–549, 1966.
395. Thompson, R. E., and Akers, T. K. Influence of sodium pentobarbital on mice poisoned by oxygen. Aerosp. Med. 41: 1025–1027, 1970.
396. Tierney, D., Ayers, L., Herzog, S., and Yang, J. Pentose pathway and production of reduced nicotinamide adenine dinucleotide phosphate. A mechanism that may protect lungs from oxidants. Am. Rev. Resp. dis. 108:1348–1351, 1973.
397. Tiisala, R. Stress reaction to oxygen poisoning in newborn, growing and adult rats. Ann. Pediat. Fenn 5:59–66, 1959.
398. Tiisala, R. Endocrine response to hyperoxia and hypoxia in the adult and newborn rat. An experimental study with radioactive phosphorus. Ann. Acad. Sci. Fenn. Ser. A5 95:1–141, 1962.
399. Treciokas, L. J. The effect of "oxygen" poisoning on alveolar cell mitochondria as revealed by electron microscopy. Aerosp. Med. 30:674–677, 1959.
400. Ulvedal, F. Preliminary observations on testicular function in roosters and mice exposed to increased partial pressure of oxygen. SAM-TR-66-40, USAF School of Aerospace Medicine, 1966.
401. Ulvedal, F., and Roberts, A. J. Endocrine functions in an oxygen atmosphere at reduced total pressure. Aerospace Med. 39:1218–1224, 1968.
402. Van Breemen, V. L., Neustein, N. B., and Burns, P. D. Pulmonary hyaline membranes studies with the electron microscope. Am. J. Pathol. 33, 769–789, 1957.
403. van den Brenk, H. A. S. Lung damage in rats due to radical X-radiation in high-pressure oxygen. Aust. Radiol. 10:375–382, 1966.
404. van den Brenk, H. A. S., and Jamieson, D. Potentiation by anesthetics of brain damage due to breathing high-pressure oxygen in mammals. Nature 194:777–778, 1962.
405. van den Brenk, H. A. S., and Jamieson, D. Pulmonary damage due to high pressure oxygen breathing in rats. 1. Lung weight, histological and radiological studies. Aust. J. Exp. Biol. Med. Sci. 40:37–49, 1962.
406. van den Brenk, H. A. S., and Jamieson, D. Studies of mechanisms of chemical radiation protection *in vivo*. II. Effect of high pressure oxygen on radio-protection *in vivo* and its relationship to "oxygen poisoning." Int. J. Radiat. Biol. 4:379–402, 1962.
407. van den Brenk, H. A. S., and Jamieson, D. Brain damage and paralysis in animals exposed to high pressure oxygen: Pharmacological and biochemical observations. Biochem. Pharmacol. 13:165–182, 1964.
408. Vinogradov, V. N., and Babchinskiy, F. V. Electron microscope observations of changes in the lungs of rats after repeated exposure to pure oxygen. Space Biol. Med. 2:40–45, 1969.
409. Walker, J. B. The involvement of carbon dioxide in the toxicity of oxygen at high pressure. Can. J. Biochem. Physiol. 39: 1803–1809, 1961.
410. Webb, W. R., Lanius, J. W., Aslami, A., and Reynold, R. C. Effects of oxygen tensions on pulmonary surfactant in guinea pigs and rats. J. Am. Med. Assoc. 195:279–280, 1966.
411. Weibel, E. R. Morphometry of the Human Lung. Academic Press, New York, 1963.
412. Weibel, E. R. Discussion. Aerospace Medical Research Laboratories Report AMRL-TR-68-175, p. 246, Wright-Patterson Air Force Base, Ohio, 1968.
413. Weibel, E. R., Lewerenz, M., and Kaplan, H. P. Electron microscopic and morphometric study of rat, monkey and dog lungs exposed to 68% O_2 and 32% N_2 at 258 torr for eight months. Aerospace Medical Research Laboratories Report AMRL-TR-68-175, pp. 189–212, Wright-Patterson Air Force Base, Ohio, 1968.
414. Weir, F. W., Bath, D. W., Yevich, P., and Oberst, F. W. Study of effects of continuous inhalation of high concentrations of oxygen at ambient pressure and temperature. Aerosp. Med. 36:117–120, 1965.
415. Weiss, H. S., Beckman, D. L., and Wright, R. A. Delayed mortality in adult chickens exposed to 1 atmosphere oxygen. Nature 208:1003–1004, 1965.
416. Weiss, H. S., and Wright, R. A. Force feeding on the response of Japanese quail to oxygen toxicity. Comp. Biochem. Physiol. 25:95–106, 1968.
417. Weiss, H. S., Wright, R. A., and Hiatt, E. P. Reaction of the chick to one atmosphere of oxygen. J. Appl. Physiol. 20:1227–1231, 1965.
418. Wittner, M., and Rosenbaum, R. M. Pathophysiology of pulmonary oxygen toxicity.

In: Proceedings of the Third International Conference on Hyperbaric Medicine. (Eds.) Brown, I. W., Jr., and Cox, B. G., pp. 179–188, NAS/NRC Publ. 1404, Washington, D.C. 1966.

419. Whitaker, W., and Lodge, T. The radiological manifestations of pulmonary hypertension in patients with mitral stenosis. J. Fac. Radiol. 5:182, 1954.

420. Wood, C. D., and Perkins, G. F. Factors influencing hypertension and pulmonary edema produced by hyperbaric O_2. Aerosp. Med. 41:869–872, 1970.

421. Wood, C. D., Perkins, G. F., Seager, L. D., and Koerner, T. A. Pulmonary edema resulting from oxygen toxicity in divers. J. Louisiana State Med. Soc. 122:219–221, 1970.

422. Wood, C. D., Seager, L. D., and Perkins, G. Blood pressure changes and pulmonary edema in the rat associated with hyperbaric oxygen. Aerosp. Med. 38:479–481, 1967.

423. Wood, J. D. Development of a strain of rats with greater than normal susceptibility to oxygen poisoning. Can. J. Physiol. Pharmacol. 44:259–265, 1966.

424. Wood, J. D., Stacey, N. E., and Watson, W. J. Pulmonary and central nervous system damage in rats exposed to hyperbaric oxygen and protection therefrom by gamma-aminobutyric acid. Can. J. Physiol. Pharmacol. 43:405–410, 1965.

425. Yamamoto, E., Wittner, M., and Rosenbaum, R. M. Resistance and susceptibility to oxygen toxicity by cell types of the gas-blood barrier of the rat lung. Am. J. Pathol. 59:409–436, 1970.

426. Yarbrough, I. D., Welham, W., Brinton, E. S., and Behnke, A. R. Symptoms of oxygen poisoning and limits of tolerance at rest and at work. U.S. Naval Experimental Diving Unit, Proj. X-337, Sub. No. 62, Report No. 1, 1947.

427. Zal'tsman, G. L., Zinov'eva, I. D., and Kumanichkin, S. D. Increased individual predisposition of a subject to the effect of high partial oxygen pressure. In: The effect of the Gas Medium and Pressure on Body Functions, Collection No. 3. (Ed.) Brestkin, M. P., pp. 287–290, published for NASA, USA and NSF, Washington, D.C. by Israel Program for Scientific Translations. NASA TTF-358, TT 65-50136, 1965.

428. Zal'tsman, G. L., Zinov'eva, I. D., Kumanichkin, S. D., and Turygina, A. V. Experience in the comprehensive study of the conditions of some systems of the human body under increased oxygen pressure. In: The Effect of the Gas Medium and Pressure on Body Functions Collection No. 3. (Ed.) Brestkin, M. P., pp. 148–155, published for NASA, USA and NSF, Washington, D. C. by Israel Program for Scientific Translations, NASA TT F-358, TT 65-50136, 1965.

429. Zeiner, F. N. Sixty-day exposure to artificial atmospheres. Aerosp. Med. 37:492–494, 1966.

15
Oxygen in Closed Environmental Systems

KARL E. SCHAEFER

The normal concentration of oxygen in air is approximately 21%, corresponding to a partial pressure of 0.21 ATA (Atmosphere Absolute) or 159 mm Hg. In closed systems such as pressure chambers, air craft or space craft, the partial pressure of oxygen may become increased and result in oxygen toxicity. On the other hand, there are situations in which a lowering of oxygen partial pressure may occur in all these systems, causing symptoms of anoxia.

Oxygen toxicity is likely to be encountered under the following conditions:

1. For underwater swimmers using closed or semiclosed breathing equipment
2. Oxygen tolerance tests used for the selection of diving personnel in the Navy
3. Saturation diving
4. Use of oxygen breathing to shorten decompression times
5. Recompression for therapeutic purposes
6. During prolonged resuscitation efforts in cases of respiratory failure
7. Prolonged exposure to increased oxygen during flights in spacecraft and aircraft

Hypoxia or anoxia may occur in:

1. Underwater swimmers having equipment problems either due to exhaustion of oxygen cylinders or faulty oxygen–nitrogen mixtures
2. Saturation excursion diving
3. Aircraft operations

Rise in oxygen partial pressure may be due to an increase in the percentage of oxygen in the ambient air or an elevation of ambient pressure, or a combination of both factors. Increases of partial pressure of oxygen at normal atmospheric pressure (1 ATA) produce effects on the respiratory and circulatory system, on erythropoiesis, and endocrine functions, but do not affect the central nervous system (Bean, 1945). Oxygen convulsions that involve the central nervous system, one form of oxygen poisoning, occur only under conditions in which oxygen tensions are in excess of 1 atm (2 ATA).

In Table 1 *threshold values* of increased ambient partial pressure of oxygen causing effects on different organ systems are listed. O_2 levels of inspired air are also given in ATA together with alveolar P_{O_2} values. Dejours et al. (1958) established that the chemoreceptor O_2 drive of respiration (which accounts for 10–15% of the resting ventilation) disappears above 170 mm Hg of alveolar O_2 tension; this corresponds to an

Table 1. *Threshold Effects of Increased Ambient Partial Pressures of Oxygen*

Ambient P_{O_2} (equivalent ATA)	P_{O_2} (mm Hg)	Alveolar P_{O_2} (mm Hg)	Effects	References
0.21	159	110	Normal atmosphere	—
0.30	230	170	Disappearance of chemoreceptor O_2 drive of respiration	Dejours et al. (1958)
0.50	360	300	Change in combined respiratory sensitivity to CO_2 and O_2 Bradycardia effect Transient	Hesser (1962) Stroud (1959); Dripps and Comroe (1947) Meda (1950)
0.50	360	300	Pulmonary limits (revised) established in multiday exposures of men to increased oxygen pressures (saturation diving)	Clark and Lambertsen (1971a)
0.60	460	400	Upper limit (pulmonary) for chronic exposure to increased O_2 at sea level atmosphere and below space craft operations Earlier pulmonary limits predicted on the basis of oxygen tolerance studies	Webb (1962) Clark and Lambertsen (1967)
0.70	532	—	Lung edema after 7 days in rats Death of rat observed after exposure to 0.73 ATA for 4 days	Redding et al. (1975) Smith (1899)
2 ATA	1520	—	Central nervous system limit for oxygen toxicity	Clark and Lambertsen (1971b)

inspiratory P_{O_2} of 230 mm Hg which is equivalent to 30% oxygen in the ambient air. Fifty percent O_2 in nitrogen—corresponding to an inspired P_{O_2} of 360 mm Hg and alveolar P_{O_2} of 300 mm Hg—appears to be the threshold concentration for elicitation of cardiopulmonary effects. Exposure to this mixture was found to produce a reduction in vital capacity after 24 hr (Comroe et al., 1945) and a transitory slowing of the heart rate (Meda, 1950). Another oxygen threshold exists at an alveolar P_{O_2} of 300 mm Hg in regard to the combined respiratory sensitivity to CO_2 and O_2. Breathholding studies under various pressures by Hesser (1962) and Stroud (1959) demonstrated that decreasing the alveolar O_2 tension below 300 mm Hg increases the sensitivity to CO_2.

It is well established that the inhalation of oxygen concentrations above 65%, corresponding to an oxygen tension higher than 494 mm Hg, has deleterious effects in most warm-blooded animals and leads eventually to death due to pulmonary oxygen toxicity. The latter was first discovered by the pathologist J. Lorrain Smith (1899) who observed fatal pneumonia in a rat after 4 days of exposure to 73% O_2 and subsequently described in detail pulmonary changes caused by increased oxygen tensions. Oxygen lung toxicity represents an important clinical problem complicating the care of very ill patients (Huber and Drath 1981).

The toxic effects of oxygen at high pressure (OHP) on the central nervous system were first reported by Paul Bert (1878). In his classic work *La Pression Barométrique* he describes in detail the incidence of convulsions in various animal species exposed to OHP. Behnke et al. (1935–1936) reported a series of exposures of human subjects to hyperbaric oxygen at 3 ATA and 4 ATA. Inhalation of oxygen at a pressure of 3 ATA was found to bring about definite but rapidly reversible changes, such as concentric contraction of the visual field, diminution of visual acuity, dilatation of the pupils, a rise in blood pressure, constriction of the facial vessels, and increased pulse rate. At 4 hr of exposure some subjects noted nausea and sensation of impending collapse. Exposure at 4 ATA terminated after 43 min in one subject having acute syncope and in a second subject after 44 min with convulsions. As a result of these early studies, it was generally accepted that 3 hr exposure to 3 ATA and 30 min to 4 ATA were safe for men at resting conditions.

The development of pulmonary and central nervous system oxygen toxicity depends on partial pressure and the duration of exposure as shown in Fig. 1. There is a latent period prior to the onset of toxicity in both the central nervous system and the lungs. An inspired P_{O_2} of 0.5 ATA was selected as a horizontal asymptote indicating pulmonary limits because multiday exposures of men to oxygen pressures ranging from 0.23 to 0.55 ATA did not produce detectable impairment of pulmonary function (Clark and Lambertsen, 1971a). The horizontal asymptote for the central nervous system curve of oxygen toxicity has been placed at 2 ATA because neurological effects rarely occur in men during oxygen breathing at this pressure (Clark and Lambertsen, 1971b).

Clark and Fisher (1977) pointed out that the rectangular hyperbola describing quantitatively the times of occurrence of oxygen toxicity in the lungs and the central nervous system (Fig. 1) has also been found to apply for oxygen toxicity affecting other functional systems such as inactivation of cellular respiration, nerve conduction block,

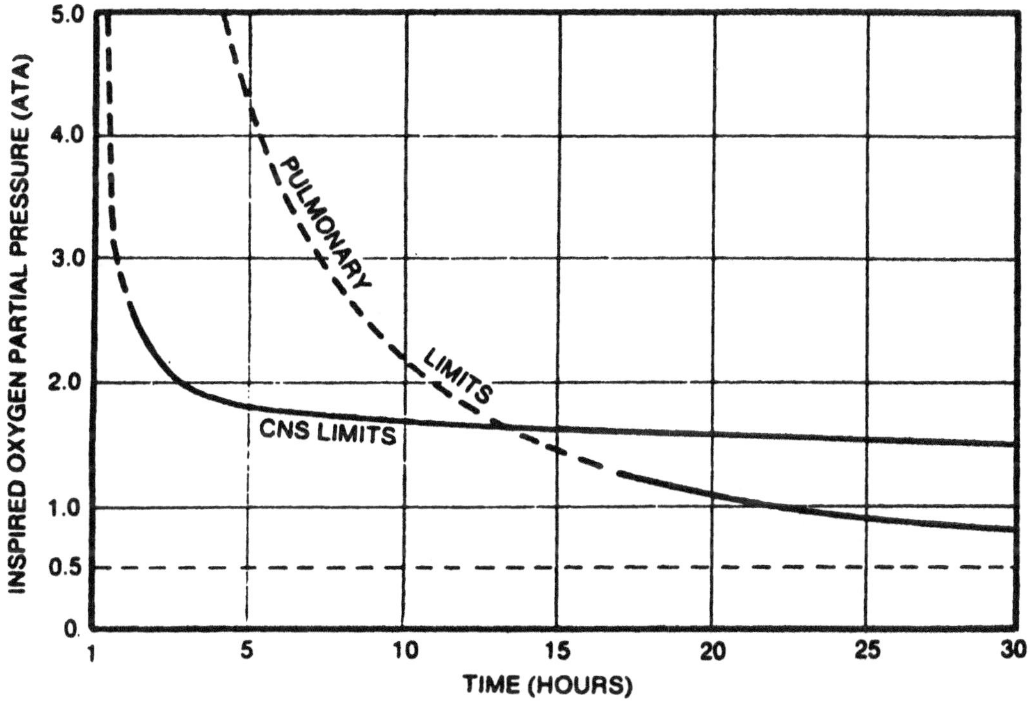

Figure 1. Predicted human pulmonary and CNS tolerance to high pressure oxygen.

and erythrocyte hemolysis. The extensive studies of Donald (1947) which involved more than 2000 exposures of men to pressures exceeding 2 ATA provide the basis of our present knowledge on tolerance of men to oxygen toxicity. One of his important findings was the extremely large variation in susceptibility not only between individuals but also in any one person from day to day. This point is clearly shown in Fig. 2, which illustrates the time of onset of symptoms in an individual who was exposed to 3.1 ATA (70 ft) oxygen on twenty occasions during a 3-month period. Donald also demonstrated that the symptoms of oxygen toxicity occurred earlier when the experiments were carried out in water as compared to those done in the dry chamber. Moreover, the tolerance to OHP was greatly reduced when work was performed during exposure.

Practical use of oxygen breathing in underwater swimming has been considered to be limited to 30 ft (Yarbrough et al., 1947) and 25 ft by Donald (1947). The maximum safe depth for pure oxygen diving is set in the U.S. Navy Diving Manual at 10 m or 33 ft.

I. Clinical Manifestation of Central Nervous System Oxygen Toxicity

Prodromal signs and symptoms listed in order of their percentage of incidence during underwater exposure include vertigo, nausea, lip twitching and other involuntary tremors, convulsions, drowsiness and disorientation, acoustic hallucinations, and paresthesia (Donald, 1947). Symptoms of pulmonary irritation were rarely seen in these experiments, because nervous symptoms forced termination before pulmonary symptoms could develop. However in other studies of oxygen toxicity in underwater swimming, in which 14 out of 50 exposures (28%) were terminated due to toxic manifestations, dys-

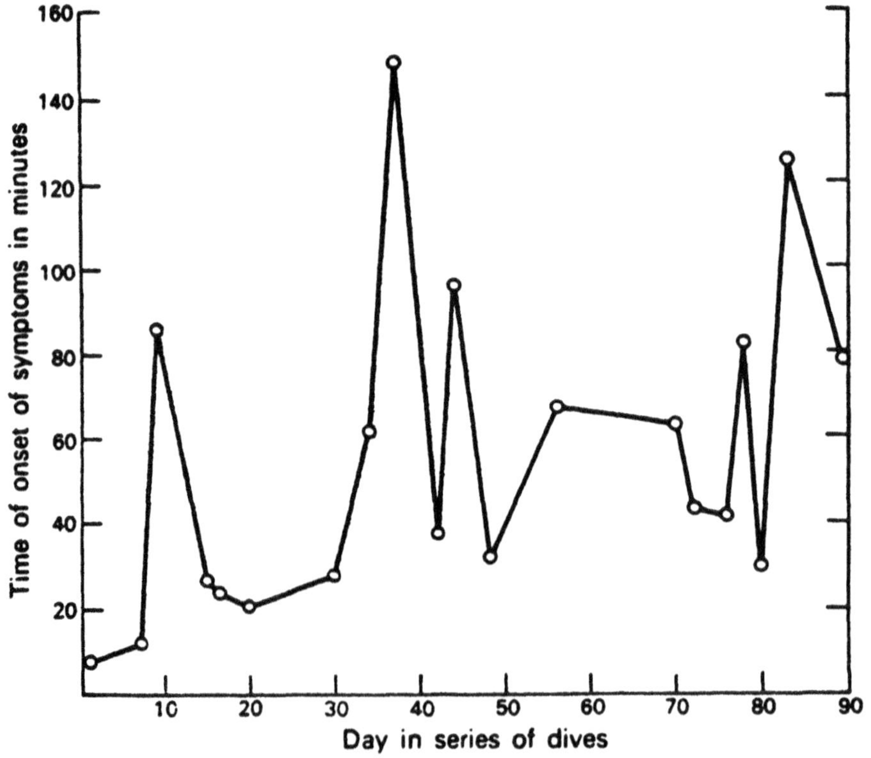

Figure 2. Variation in time and onset of symptoms of oxygen toxicity in the same subject exposed repeatedly to 3.12 ATA (70 ft) over a period of 90 days. The symptom in the vast majority of the dives was lip twitching (Donald, 1947).

pnea was by far the most frequent symptom (Schaefer, 1956). The subjects complaining about dyspnea had a rapid respiratory rate (mostly above 32) and reported a resistance to breathing and inspiratory inhibition.

The differences appear to be related to the different effects of oxygen breathing on respiration in rest and exercise. Under resting conditions, oxygen breathing increases respiratory minute volume at normal atmospheric pressure (Baker and Hitchcock, 1957), and even more at 3.5 atm (Lambertsen et al., 1953). The respiratory rate remains unchanged while tidal volume increases. However, oxygen inhalation causes a reduction in pulmonary ventilation during moderate strenuous exercise (Asmussen and Nielsen, 1947), which has been explained as a predominance of the depressive effect of oxygen on the respiratory center. Since oxygen breathing in underwater swimming produces a strong parasympathicotonic stimulation, as indicated in pulse rate measurements, symptoms of pulmonary irritation and inspiratory inhibition might be caused by excitation of the vagal inhibitory reflex (Schaefer, 1956). The average pulse rate measured at the beginning of 3-min rest periods after each 15 min of underwater swimming showed a consistent decline with continuing exercise, falling even below the initial resting values at the end of 93 min. The cardio-inhibitory action displayed under these conditions led eventually to a fixation of pulse rate, e.g., the differences between exercise and rest disappeared. This sign preceded the development of symptoms in 78% of the cases and can therefore be used as a warning sign of approaching oxygen toxicity, if pulse rate is monitored. Manifestations of symptoms were associated with strong increases in pulse rate and respiratory rate.

Two cardiovascular findings during exposure to increased partial pressure of oxygen at 1 or more atm seem to be well established: the bradycardia which is of vagal origin (Alella and Meda, 1948; Bean and Rottschafer, 1938; Daly and Bondurant, 1962), and the increase noted in the pulse rate and blood pressure if symptoms of oxygen toxicity develop in man (Behnke et al., 1935–1936, 1934–1935; Benedict and Higgins, 1911; Yarbrough et al., 1947). The reported fixation of pulse rate (Schaefer, 1956) appears to denote the end point of the predominate parasympathetic activity and the approaching release of increased sympathetic activity which is associated with symptoms of oxygen toxicity.

Gillen (1966) reported a higher incidence of oxygen convulsions in experienced divers undergoing the oxygen tolerance test (OTT) than those previously found in studies of underwater swimming (Donald, 1947; Yarbrough et al., 1947; Schaefer, 1956). The conditions during the oxygen tolerance test, which involves a 30-min exposure to pure oxygen breathing at an average of 2.82 ATA, may not be comparable to underwater swimming. It was observed that 11 out of 39 divers had their oxygen symptoms during treatment for decompression sickness or air embolism, which may have lowered the threshold for oxygen poisoning.

II. Etiology

Neurophysiological studies of oxygen toxicity demonstrated that electroencephalographic tracings during oxygen convulsions cannot be distinguished from those obtained in grand mal epilepsy (Lennox and Behnke, 1936), and that the seizures appear during or just after a sharp rise in cerebral oxygen tensions as measured with the platinum electrode (Bean, 1961). Investigations of Noell (1955, 1962) on the effect of increased oxygen tension on the visual cell of the rabbit retina demonstrated that oxygen exerts on the visual cell a direct toxic effect which is cumulative. Noell (1962) pointed out that in this cumulative action oxygen poisoning resembles the effects of X-irradiation. This observation supports the theory advanced by Gerschman et al. (1954) that oxygen poisoning and X-irradiation have one common mechanism of action, namely, the formation of free radicals. It has been demonstrated that the lethal effects of X-irradiation are enhanced by high oxygen and reduced by anoxia. Furthermore, agents which give a protection against X-irradiation, such as gluthathione and cysteine, also afford a protection against oxygen poisoning (Gerschman et al., 1954). It is reasonable to assume that oxidizing free radicals such as $HO_2^{\cdot}$, $OH\cdot$, and H_2O_2 would react rapidly and initiate chain reactions.

Under these circumstances the body does require antioxident defenses against uncontrolled oxidations (Gerschman, 1962). Differences in the available antioxidant defenses might account for the wide variety in species and organ susceptibility. The discovery of the superoxide dismutases (McCord and Fridovich, 1969) and the elucidation of their antioxidant function consisting in enzymatic inactivation of superoxide (Fridovich, 1975) contributed greatly in advancing the previously neglected free radical theory of oxygen toxicity. The action of toxic free radicals does not require postulation of an O_2 vasoconstrictor effect on the blood vessels or any specific changes in acid base parameters. The free radical theory offers therefore an explanation for the occurrence of oxygen toxicity under conditions such as exercise at high P_{O_2}, when changes in circulation and arterial and venous P_{CO_2} and P_{O_2} are too small to account for the development of symptoms.

The antioxidant defense mechanisms may become exhausted if too many oxygen free radicals are formed and as a consequence oxidation of essential cellular constituents may occur, such as sulfhydryl-containing proteins.

A variety of metabolic changes produced under conditions of oxygen toxicity have been described and different investigators have attached major significance to different metabolic changes in the development of oxygen toxicity. Britton Chance and coworkers (1965, 1966) have emphasized the inactivation of the energy-linked reverse electron transport pathway. Wood et al. (1969) have accumulated evidence indicating that changes in brain γ-aminobutyric acid (GABA) levels are involved in CNS oxygen toxicity. GABA is known to be a transmitter at central nervous system inhibiting synapses. Hyperoxia has been shown to reduce endogenous output of GABA (Wood et. al., 1969). This decrease is thought to cause convulsions by allowing uncontrolled firing of excitatory nerves. This theory has found further support by the findings of Radomsky and Watson (1973) showing that lithium treatment is effective in inhibiting convulsions and also preventing a decrease in brain GABA which precedes the onset of convulsions in oxygen toxicity.

Exercise (Donald, 1947) and CO_2 inhalation (Bean and Zee, 1966; Marshall and Lambertsen, 1961) have been shown to hasten the onset of oxygen toxicity. Studies of chronic exposure to increased CO_2 levels carried out in rats and guinea pigs demonstrated that the enhancing effect of CO_2 in regard to oxygen toxicity is limited to the acute phase of CO_2 exposure which is associated with a stress effect on the adrenals. During the later phase of CO_2 exposure the stress effect subsides and the onset of oxygen toxicity is delayed (Schaefer, 1974). Animal studies have shown that factors which stimulate metabolism cause in general an enhancement of oxygen toxicity, e.g., epinephrine, norepinephrine, adrenocortical hormones, thyroid hormones, and hyperthermia, while the opposite is true for factors leading to a decrease in metabolism such as hypothyroidism, hypothermia, adrenergic blocking agents, and starvation. Drugs that have been found effective in animal experiments in preventing oxygen toxicity could not be used in man because of existing toxic side effects with the possible exception of disulfiram (Clark and Fisher, 1977).

III. Prevention of CNS Oxygen Toxicity

It has been emphasized by Edmonds et al. (1976) that the only real practical approach for the prevention of oxygen toxicity is to place limits on the exposure. This is especially important for underwater swimming breathing oxygen. The Royal British Navy and Royal Australian Navy limit the depth for a resting dive breathing pure oxygen to 9 m and a working dive to 7 m. The U.S. Navy employes a depth-time of exposure table for the same purpose.

In therapeutic recompression using oxygen, intermittent periods of air breathing are routinely used which shorten the exposure to high levels of oxygen and thereby reduce the occurrence of oxygen toxicity.

IV. Pulmonary Oxygen Toxicity

Pulmonary oxygen toxicity does not play a role in oxygen diving of relatively short duration lasting minutes or hours. However, in saturation diving subjects may be exposed to increased oxygen partial pressures for days and weeks. Under these conditions as well as in oxygen recompression therapy the length of exposure is sufficient to cause pulmonary oxygen toxicity if the ambient oxygen partial pressure is too high.

The limits of oxygen partial pressure to which subjects can safely be exposed for prolonged periods of time have been established in extensive multiday exposures of men. Clark and Lambertsen's earlier prediction (1967) indicated a maximum safe, long-term O_2 pressure of 0.6 ATA. This is in line with the threshold given in the 1962 NASA Life Science Data Book published by Webb. Clark and Lambertsen (1971a) later revised the limit and adopted a more conservative figure of 0.5 ATA. Pulmonary functions studied during shallow habitat air dives which lasted up to 30 days with an average resident O_2 partial pressure of 0.51 and 0.57 ATA did not show any significant changes which would suggest that the earlier prediction of

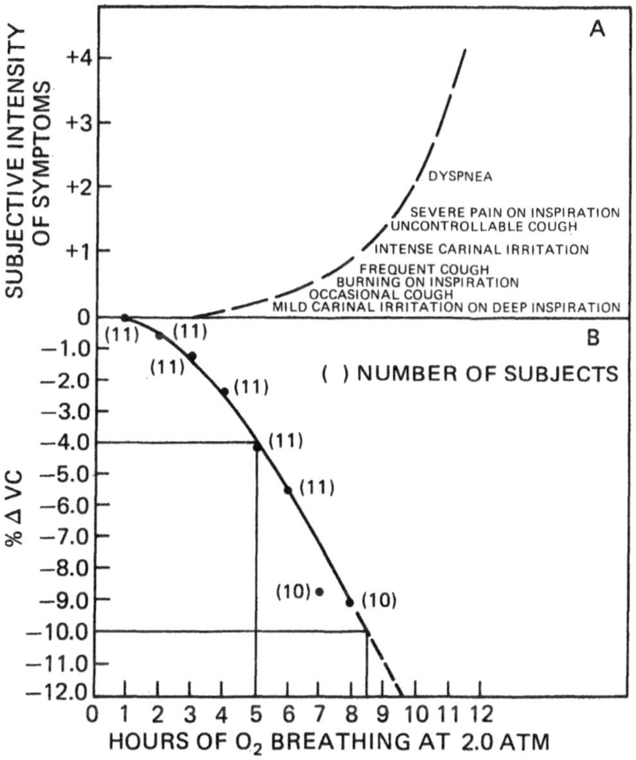

Figure 3. Rate of development of symptoms and decrease in vital capacity during oxygen breathing at 2 ATA. A. Subjective observations of all subjects were combined to construct hypothetical curve. **B.** Curve was derived from vital capacity measurements in same subjects. Reproduced from Clark and Fisher (1977), with permission.

0.6 ATA is more nearly correct than the conservative revision (Dougherty et al., 1978).

V. Clinical Manifestations and Indices of Pulmonary Oxygen Toxicity

Comroe et al. (1945) provided the framework for the establishment of threshold limit values for ambient oxygen levels. They exposed a large number of healthy young subjects to 0.1, 0.5, 0.75, and 1 ATA oxygen for 24 hr. Up to 12 hr, 1 ATA (100%) oxygen could safely be inhaled, but after 24 hr of oxygen breathing symptoms of substernal pain were noted in 86% of the subjects. Exposure of 0.5 ATA for 24 hr induced no symptoms. However, a decrease in vital capacity was found after 24 hr of exposure to both 1 ATA and 0.5 ATA of oxygen.

Fisher et al. (1968) and Clark and Lambertsen (1971b) studied several methods for measuring the rate of development of pulmonary oxygen poisoning in normal men during continuous oxygen breathing at an ambient pressure of 2 ATA. They observed significant changes of the following lung functions at a reversible stage of oxygen poisoning: vital capacity, inspiratory capacity, expiratory reserve volume, inspiratory flow rate, carbon monoxide diffusing capacity, lung compliance, and respiratory rate. Decrease in vital capacity has proved to be a very good indicator of the development of oxygen toxicity in healthy subjects during diving studies and operations.

Figure 3 shows the rate of development of symptoms and the decrease in vital capacity during oxygen breathing at 2 ATA (Clark and Lambertsen 1971b). The sensitivity of vital capacity measurements for monitoring development of pulmonary oxygen toxicity in saturation–excursion diving has recently been demonstrated in an experiment in which subjects were exposed for 8 days to compressed air at 50 fswg (2.52 ATA) with daily 9-hr excursions to 100 fswg (4.03 ATA). The mean oxygen tension was 0.61 ATA. One subject showed a progressive fall in daily forced vital capacity (FVC) measurements beginning with excursion 2. Simultaneously his subjective chest discomfort increased (Dougherty et al., 1978). Figure 4 shows the time course of forced vital capacity in this subject exhibiting an excursion-related decrement of vital capacity and slow recovery at base pressure.

VI. Pulmonary Oxygen Toxicity during Spacecraft and Aircraft Operations

In spacecraft and aircraft operations the ambient atmosphere is less than 1 ATA pressure in contrast to diving operations. Breathing pure oxygen in a rarefied atmosphere facilitates the development of atelectasis in the lungs. Dale and Rahn (1952) found that under one atmosphere pressure the rate of collapse of alveoli, following obstruction of a bronchus, was 60–80 times faster with pure oxygen breathing than with air breathing. Nitrogen in the alveoli acts as a brake in the absorption of gases trapped by an obstruction of a minor airway. Such obstructions can be produced by mucous secretions and by compression during exposure to high gravitational forces.

Occurrences of transitory subsegmental atelectasis in aviators breathing increased concentrations of oxygen while flying high-performance jet aircraft were reported by several investigators (Ernsting, 1960; Green and Burgess, 1962; Langdon and Reynolds, 1961). Atelectasis was detected by chest X-ray after the flight. There is evidence that at least three causal factors play a role in the production of atelectasis under these circumstances. These factors are: (a) acceleration per se; (b) the use of 100% O_2 or increased oxygen concentration as breathing medium; and (c) elevation of the diaphragm and restricted chest cage expansion owing to the use of a G suit (Hyde et al., 1963; McLevy et al., 1962). Histopathological studies of

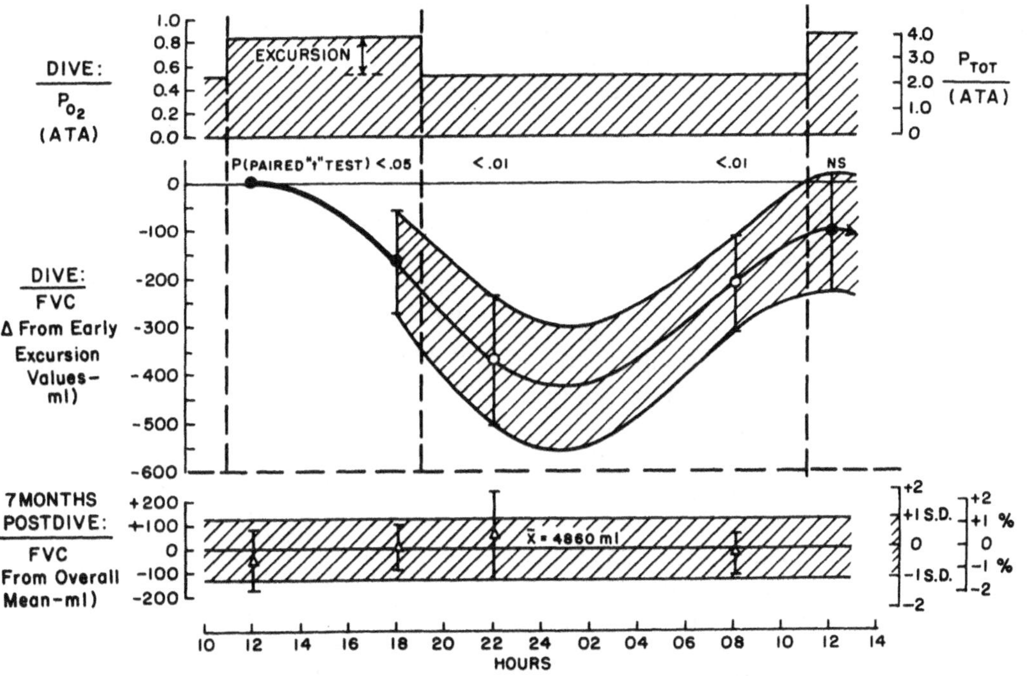

Figure 4. Time course of forced vital capacity (FVC) measurements in subject M. during one day of saturation excursion dive involving 9 hr excursion to 100 fswg (4.03 ATA) from a saturation pressure of 2.52 ATA (50 fswg) breathing compressed air. *Upper part:* dive profile for total pressure P_{O_2}. *Middle part*: changes in forced vital capacity. *Lower part*: 24-hr profile of forced vital capacity measurements 7 months postdive. Reproduced from Dougherty et al. (1978), with permission.

victims of fatal aircraft accidents reported by Gable and Townsend (1962) demonstrated, in five out of 50 cases who had a minimum of 50 hrs of jet flight, clear evidence of atelectasis.

In spacecraft operations the danger of atelectasis has been reduced by the change-over from pure oxygen breathing atmosphere at $\frac{1}{3}$ ATA in the Gemini flights to a 50% nitrogen–50% oxygen mixture at 0.5 ATA in the Apollo flights. Prolonged exposures to partial pressures of oxygen below 0.55 ATA used in space flight have not been found to produce significant lung changes (Clark and Lambertsen, 1971a).

VII. Pathology of Pulmonary Oxygen Toxicity

Acute and chronic phases of pulmonary oxygen toxicity have been described. Acute pulmonary oxygen toxicity caused by exposure to oxygen pressures higher than 0.8 ATA, is characterized by atelectasis, interstitial and alveolar oedema, hemorrhages, hyaline membranes, and swelling and destruction of type 1 alveolar epithelial cells (Kistler et al., 1966; Clark and Lambertsen, 1971a). The histopathological changes observed in acute pulmonary oxygen toxicity are surprisingly similar to those found in acute pulmonary CO_2 toxicity, except that type 2 alveolar epithelial cells are primarily affected (Schaefer et al., 1964). After 72 hr of exposure to 98.5% oxygen the thickness of the air–blood barrier was found to be doubled in rats, indicating a severe impairment of the gas exchange capacity (Kistler et al., 1966). In chronic pulmonary oxygen toxicity, which occurs during prolonged exposure to pressures between 0.5 and 0.8 ATA, hyperplasia of type 2 alveolar epithelial cells develops together with thickening of alveolar walls and arteriolar hypertrophy. An increased rate of collagen synthesis was found during chronic pulmonary oxygen toxicity, which is probably associated with the observed chronic fibrosis of the lungs (Valimaki et al., 1975).

VIII. Etiology for Pulmonary Oxygen Toxicity

The enzymatic changes observed in pulmonary oxygen toxicity are similar to those found in central nervous system oxygen toxicity. Moreover, hyperoxia causes metabolic changes in the lungs such as reductions of oxygen uptake, ATP synthesis and incorporation of radioactive leucine into tissue components by lung homogenates (Gacad and Massaro, 1973; Sanders et al., 1976). A depression of pulmonary serotonin clearance was also found in hyperoxia suggesting impairment of endothelial cell membrane transport (Block and Fisher, 1977). It appears that these changes in intracellular enzymes and endothelial transport occur in acute pulmonary oxygen toxicity after the antioxidant defense mechanisms have been exhausted. This would also explain the markedly higher mortality of older animals exposed to 100% oxygen for prolonged periods of time (Brooksby et al., 1966), since older animals are known to have recuded antioxidant defense mechanisms.

IX. Effects of Hyperoxia on Other Organ Systems

Most of the investigators of oxygen toxicity focused their attention on the effects of hyperoxia on lungs and brain. However it has been shown that 30-day exposure to .33 ATA oxygen, which does not impair pulmonary function in men (Robertson et al., 1964), will produce effects on red cells and endocrine functions (Larkin et al., 1972; Love et al., 1971).

Fisher and Kimzey (1971) have demonstrated that prolonged exposure to increased oxygen levels in spacecraft operations cause abnormal red cell morphology and a decrease in circulating red cell mass. This may be caused by a depression of erythropoiesis induced by the increased oxygen tension, since it is known that the former is closely regulated by the oxygen tension.

The retinal vessels are particularly susceptible to high oxygen pressures. Behnke, et al. 1935–1936) were the first investigators to report on subjective visual impairment of subjects at 3 ATA oxygen. Comroe et al. (1945) found ocular hypotonia in their studies. Noell (1955, 1962) described in detail the metabolic injuries of the visual cell occurring in the outer nuclear layer during breathing of 100% oxygen. Although Noell (1962) believed that a direct toxic effect of oxygen is the cause of the visual impairment, other evidence showing an oxygen-induced decrease in both retinal and uveal blood flow (Saltzman et al., 1965; Trokel, 1965) suggests that the retinal changes in hyperoxia are caused by an intensive vasoconstriction. Visual acutiy changes have also been found in patients following intermittent exposures to 2.5 ATA of oxygen for treatment of osteomyelitis (Hart et al., 1974).

A more recent report on serous otitis media, observed in divers breathing 100% oxygen from semiclosed and closed circuit diving equipment (Strauss, et al., 1973), suggested that this syndrome is caused by the absorption of oxygen from the middle ear. Similar observations had been made previously in aviators exposed to high oxygen during flight.

X. Oxygen Paradox

If a person is in an hypoxic state and suddenly has pure oxygen to breathe, his condition may deteriorate and he may become unconscious for up to 30 sec. This effect of oxygen was first noted in studies involving rapid recompression from simulated altitude. The term oxygen paradox was coined by Ruff and Strughold (1939), Latham (1951) carried out experiments on 52 subjects who were first made anoxic by inhaling air for 4–5 min at 20,000 ft of simulated altitude and then given pure oxygen. Of the 52 subjects, 13 (25%) showed symptoms. After 12 sec of oxygen inhalation, errors in writing tests reached a peak indicating that CNS disturbances are the first manifestations of the oxygen paradox reaction. Cardiovascular responses followed 40 sec later and were associated with muscular incoordination. Somewhat later, signs of impending cardiovascular collapse developed, commensurate with a rapid fall in systolic blood pressure and an increase in peripheral blood flow. The paradox reaction does not occur if oxygen is administered slowly. The neurological manifestations are most likely the consequences of an oxygen-induced impairment of cerebral circulation.

This conclusion is supported by the findings of Miles (1957). In a study of 126 men on oxygen syncope, the threshold for syncope was lowered by incorporating in the test the "stress" conditions frequently encountered in underwater swimming (hyperventilation, postural change, and increased pulmonary pressure). It was found that the syncope rate during oxygen breathing was three times that during air breathing (vasoconstrictor effect of oxygen). However, with the addition of carbon dioxide to the oxygen (vasodilatator effect) no difference was observed.

XI. Hypoxia

For years hypoxia has been the cardinal problem of aviation medicine. Extensive studies have been done on hypoxia and the results have been disseminated in many texts. However, hypoxia can also occur under conditions of diving operations using underwater swimming equipment, and in saturation excursion diving, in chamber operations, when the oxygen supply is for some reason reduced or the gas mixture used has too low an oxygen concentration.

Miles and Mackay (1976) gave an analysis of 200 diving accidents in which they presented the different causes and reported the incidence of accidents due to hypoxia and oxygen poisoning. Out of 200 diving accidents, of which 51 were fatal, 24 were caused by hypoxia (8 fatal) while only 10 were produced by oxygen poisoning (5 fatal). These numbers underscore the danger of hypoxia in underwater swimming and emphasize the need for a thorough training with the equipment used.

In experiments with goats exposed to high pressure with normal ambient oxygen pressure of 159 mm Hg, Chouteau (1971) observed symptoms which he thought were suggestive of hypoxia. This form of hypoxia has been called normoxic hypoxia. The behavioral disorders and paralysis produced by this condition in the goats could be reversed by increasing the partial pressure of oxygen to 191 mm Hg without changing the total absolute pressure.

Similar observations have been made in a saturation dive in which 3 men were exposed to 7 ATA breathing a nitrogen–oxygen mixture which contained 0.23 ATA oxygen (174.8 mm Hg) (Wilson et al., 1977). During the first 85 hr, two out of three divers experienced severe lethargy, slurred speech, difficulty in breathing, and vomiting. The oxygen partial pressure was then raised after 8.5 hr to 0.31 ATA (235 mm Hg) and the symptoms disappeared rather quickly thereafter. One of the divers had during the initial period electrocardiographic changes consisting in a lengthening of the P–R interval (from 0.14 predive to 0.21) the P waves became progressively more flattened and disappeared, resulting in a nodal rhythm. Minor S-T elevations, rSR′ complexes in the right precordial leads, and diphasic T waves appeared during this period, indicating a right ventricular conduction delay. After 8 hr of breathing the increased oxygen concentration the rSR′ complexes and T wave changes had disappeared and the P waves were more demonstrable. In the second diver, premature ventricular contractions had occurred during the initial period on 0.23 ATA oxygen, which disappeared after more oxygen was added.

XII. Summary

In closed environmental systems such as pressure chambers, aircraft, or spacecraft, conditions may occur which result in oxygen toxicity or hypoxia. Threshold values of increased ambient partial pressure of oxygen causing effects on different organ systems have been listed. The clinical manifestations and etiology of central nervous system toxicity have been described. The effects of hyperoxia on other organ systems have been summarized. Also, the incidence of hypoxia in low and high pressure systems has been reviewed.

References

Alella, A., and Meda, E. (1948). Frequenza cardiaca durante la respirazioni di O_2 nell' uomo ed importanza del vago. Boll. Soc. Ital. Biol. Sper. 24:581.

Asmussen, E., and Nielsen, M. (1947). Studies in the regulation of respiration in heavy work. Acta Physiol. Scand. 12:121.

Baker, S. P., and Hitchcock, F. A. (1957). Immediate effects of inhalation of 100% oxygen at one atmosphere on ventilation volume, carbon dioxide output, oxygen consumption and respiratory rate in man. J. Appl. Physiol. 10: 363–366.

Bean, J. W. (1945). Effects of oxygen at increased pressure. Physiol. Rev. 25:1–147.

Bean, J. W., and Rottschafer, G. (1938). Reflexogenic and central structures in oxygen poisoning. J. Physiol. (London) 94:294–306.

Bean, J. W. (1961). Tris buffer, carbon dioxide, and sympathico-adrenal system in reactions to oxygen at high pressure. Am. J. Physiol. 201: 737–739.

Bean, J. W., and Zee, D. (1966). Influence of anesthesia and carbon dioxide on CNS and pulmonary effects of oxygen at high pressure. J. Appl. Physiol. 21:521–526.

Behnke, A. R., Johnson, F. S., Poppen, J. R., and Motley, E. P. (1934–1935). The effect of oxygen on man at pressures from 1 to 4 atmospheres. Am. J. Physiol. 110:565–572.

Behnke, A. R., Forbes, H. S., and Motley, E. P. (1935–1936). Circulatory and visual effects of oxygen at 3 atmospheres pressure. Am. J. Physiol. 114:436–442.

Benedict, F. G., and Higgins, H. L. (1911). Effects on men at rest of breathing oxygen-rich gas mixtures. Am. J. Physiol. 28:1–28.

Bert, P. (1878). La Pression barométrique, recherches de physiologie expérimentale. Paris Masson; 1168 pp. (English translation: Barometric Pressure. Researches in Experimental Physiology, translated by M. A. Hitchcock and F. A. Hitchcock. 1943. Columbus, Ohio: College Book Co., 1055 pp.)

Block, E. R., and Fisher, A. B. (1977). Depression of serotonin clearance by rat lungs during oxygen exposure. J. Appl. Physiol. 42:33–38.

Brooksby, G. A., Dennis, R. L., and Staley, R. W. (1966). Effects of prolonged exposure of rats to increased oxygen pressure. Proceedings of the Third International Conference on Hyperbaric Medicine. Eds., I. W. Brown, and B. G. Cox. Nat. Acad. Sci. Publ. 1404, pp. 208–216.

Chance, B., Jamieson, D., and Coles, H. (1965). Energy-linked pyridine nucleotide reduction: Inhibitory effects of hyperbaric oxygen in vitro in vivo. Nature 206:257–263.

Chance, B., Jamieson, D., and Williamson, J. R. (1966). Control of the oxidation-reduction state of reduced pyridine nucleotides in vivo and in vitro by hyperbaric oxygen. In: Proceedings of the Third International Conference on Hyperbaric Medicine. Eds., I. W. Brown and B. G. Cox. Washington, D. C.: National Academy of Sciences, pp. 15–41.

Chouteau, J. (1971). Respiratory gas exchange in animals during exposure to extreme ambient pressures. In: Proceedings of the Fourth Symposium on Underwater Physiology. Ed., C. J. Lambertsen. New York: Academic Press.

Clark, J. M., and Fisher, A. B. (1977). Oxygen toxicity and extension of tolerance in oxygen therapy. In: Hyperbaric Oxygen Therapy. Eds. J. C. Davis and T. K. Hunt. Bethesda, Maryland: Undersea Medical Society, pp. 61–77.

Clark, J. M., and Lambertsen, C. J. (1967). Pulmonary oxygen tolerance and the rate of development of pulmonary oxygen toxicity in man at two atmospheres inspired oxygen tension. In: Underwater Physiology. Proceedings of the Third Symposium on Underwater Physiology. Ed., C. J. Lambertsen. Baltimore, Maryland: Williams & Wilkins, pp. 439–451.

Clark, J. M., and Lambertsen, C. J. (1971a). Pulmonary oxygen toxicity: A review. Pharmacol. Rev. 23:37–133.

Clark, J. M., and Lambertsen, C. J. (1971b). Rate of development of pulmonary oxygen toxicity in man during oxygen breathing at 2.0 Ata. J. Appl. Physiol. 30:739–752.

Comroe, J. H., Dripps, R. D., Dumke, P. R., and Demming, M. (1945). Oxygen toxicity: The effect of inhalation of high concentrations of oxygen for twenty-four hours on normal men at sea level and at a simulated altitude of 18,000 feet. J. Am. Med. Assoc. 128:710.

Dale, W. A., and Rahn, H. (1952). Rate of gas absorption during atelectasis. Am. J. Physiol. 170:606.

Daly, W., and Bondurant, J. S. (1962). Effects of oxygen breathing on the heart rate, blood pressure, and cardiac index of normal men resting with reactive hyperemia and after atropine. J. Clin. Invest. 41:126.

Dejours, P., Labrousse, Y., Raynaud, J., Girard, F., and Teillac, A. (1958). Stimulus oxygene de la ventilation au repos et au cours de l'exercice musculaire, a basse altitude (50m) chez l'homme. Rev. Fr. Etud. Clin. Biol. 3(2):105.

Dougherty, J. H., Jr., Frayre, R. L., Miller, C. A., and Schaefer, K. E. (1978). Pulmonary function during shallow habitat air dives (Shad I, II, III). Proc. Sixth Symp. on Underwater Physiol. Eds. C. W. Shilling and M. W. Beckett. Bethesda, Maryland: FASEB., pp. 193–204.

Dripps, R. D., and Comroe, J. H., Jr. (1947). The effect of the inhalation of high and low oxygen concentration on respiration, pulse rate, ballistocardiogram and arterial oxygen saturation (oximeter) of normal individuals. Am. J. Physiol. 149:277–291.

Donald, K. W. (1947). Oxygen poisoning in man. I and II. Br. Med. J. 1:667–672, 712–717. See also Admiralty Experimental Diving Unit Report No. 16, 1946.

Edmonds, C., Lowry, C., and Pennefather, J. (1976). Diving and Subaquatique Medicine. Mosman, N.S.W., Australia: Diving Medical Centre Publication.

Ernsting, J. (1960). Some effects of oxygen breathing. Proc. Roy. Soc. Med. 53:96.

Fischer, C. L., and Kimzey, S. L. (1971). Effects of oxygen on blood formation and destruction. Proceedings of the Fourth Symposium on Underwater Physiology. New York: Academic Press.

Fisher, A. B., Hyde, R. W., Puy, R. J. M., Clark, J. M., and Lambertsen, C. J. (1968). Effect of oxygen at 2 atmospheres on the pulmonary mechanics of normal man. J. Appl. Physiol. 24:529–536.

Fridovich, I. (1975). Superoxide dismutases. Ann. Rev. Biochem. 44:147–159.

Gable, W. D., and Townsend, F. M. (1962). Lung morphology of individuals exposed to prolonged intermittent supplemental oxygen. Aerospace Med. 33:1344.

Gacad, G., and Massaro, D. (1973). Hyperoxia: Influence of lung mechanics and protein synthesis. J. Clin. Invest. 52:559–565.

Gerschman, R. (1962). The biological effects of increased oxygen tension. In: Man's Dependence on the Earthly Atmosphere, Ed., K. E. Schaefer. New York: Macmillan, p. 171.

Gerschman, R., Gilbert, D. L., Nye, S. W., Dwyer, P., and Fenn, W. O. (1954). Oxygen poisoning and x-irradiation: A mechanism in common. Science 119:623–626.

Gillen, H. W. (1966). Oxygen convulsions in man. Proceedings of the Third International Conference on Hyperbaric Medicine. Eds., I. W. Brown and B. G. Cox. Nat. Acad. of Sci., Washington, D. C., pp. 217–223.

Green, J. E., and Burgess, E. F. (1962). An investigation into the major factors contributing to post flight chest pain in fighter pilots. Flying Personnel Research Committee. Great Britain Reports No. FTRC-1182.

Hart, G. B., Lee, W. S., Rasmussen, B. D., and O'Reilly, R. R. (1974). Complications of repetitive hyperbaric therapy. In: Proceedings of the Fifth International Hyperbaric Congress, Vol. II. Eds., W. G. Trapp, E. W. Banister, A. J. Davison, and P. A. Trapp. Burnaby, British Columbia: Simon Fraser University, pp. 867–873.

Hesser, C. M. (1962). The Role of Nitrogen in Breath-holding at Increased Pressures. In: Man's Dependence on the Earthly Atmosphere. Ed., K. E. Schaefer. New York: Macmillan Co., p. 327.

Huber, G. L., and Drath, D. B. (1981). Pulmonary oxygen toxicity. This volume.

Hyde, A. S., Pines, J., and Saito, I. (1963). Atelectasis following acceleration: A study of causality. Aerospace Med. 34:150.

Kistler, G. S., Caldwell, P. R. B., and Weibel, E. R. (1966). Quantitative electron microscopic studies of murine lung damage after exposure to 98.5% oxygen at ambient pressure: A preliminary report. Proceedings of the Third International Conference on Hyperbaric Medicine. Eds., I. W. Brown and B. G. Cox. Nat. Acad. Sci. Publ. 1404, pp. 169–178.

Lambertsen, C. J., Kough, R. H., Cooper, D. V., Emmel, G. L., Loeschcke, H. H., and Schmidt, C. F. (1953). Oxygen toxicity. Effects in man of oxygen inhalation at 1 and 3.5 atmospheres upon blood gas transport, cerebral circulation and cerebral metabolism. J. Appl. Physiol. 5: 471–486.

Langdon, D. E., and Reynolds, G. E. (1961). Post flight respiratory symptoms associated with 100% oxygen and G-forces. Aerospace Med. 32:713.

Larkin, E. C., Adams, J. D., Williams, W. T., and Duncan, D. M. (1972). Hematologic responses to hypobaric hyperoxia. Am. J. Physiol. 223: 431–437.

Lennox, W. G., and Behnke, A. R., Jr. (1936). Effect of increased oxygen pressure on the seizures of epilepsy. Arch. Neurol. Psychiat. 35:782.

Love. T. L., Schnure, J. J., Larkin, E. C., Lipman, R. L., and Lecocq, F. R. (1971). Glucose intolerance in man during prolonged exposure to hypobaric-hyperoxic environment. Diabetes 20:282–285.

Latham, F. (1951). The oxygen paradox. The experiments on the effects of oxygen in human anoxia. Lancet 1:77.

Marshall, J. R., and Lambertsen, C. J. (1961). Interactions of increased P_{O_2} and P_{CO_2} effects in producing convulsions and death in mice. J. Appl. Physiol. 16:1–7.

McCord, J. M., and Fridovich, I. (1969). Superoxide dismutase. An enzymic function for erythrocuprein J. Biol. Chem. 224:6049–6055.

McLevy, M., Jaeger, E. A., Stone, R. S., and Doudna, C. T. (1962). Aeroatelectasis: A Respiratory syndrome in aviators. Aerospace Med. 33:987.

Meda, E. I. (1950). Effetti della respirazione di miscele ricche di O_2 sull'apparato cardiovascolare dell'uomo. II. Variazione elettrocardiografiche nell'uomo durante la respirazione di O_2. Boll. Soc. Ital. Biol. Sper. 26:931.

Miles, S. (1957). Oxygen syncope. Med. Res. Council, Royal Naval Personnel Research Commettee, Report No. RNT57-880, UPS 161.

Miles, S., and Mackay, D. E. (1976) Underwater Medicine, 4th Ed. Philadelphia: J. B. Lippincott Co.

Noell, W. K. (1955). Visual cell effects of high oxygen pressures. Fed. Proc. 14:107.

Noell, W. K. (1962). Effects of high and low oxygen tension on the visual system. In: Environmental Effects on Consciousness. Ed., K. E. Schaefer. New York: Macmillan, p. 3.

Radomsky, M. W., and Watson, W. I. (1973). Effect of lithium on acute oxygen toxicity and associated changes in brain-amino butyric acid. Aerospace Med. 44:387–392.

Redding, R. A., Arai, T., Douglas, W. H. J., Tsurupani, H., and Oven, J. (1975). Early changes in lungs of rats exposed to 70% oxygen. J. App. Physiol. 38:136–142.

Robertson, W. G., Hargreaves, J. J., Herlicher, J. E., and Welch, B. E. (1964). Physiologic response to increased oxygen partial pressure. II. Respiratory studies. Aerospace Med. 35: 618–622.

Ruff, S., and Strughold, H. (1939). Grundriss der Luftfahrt Medizin. Leipzig: Barth.

Saltzman, H. A., Hart, L., Sieker, H. O., and Duffy, E. J. (1965). Retinal vascular response to hyperbaric oxygenation. J. Am. Med. Assoc. 191:290.

Sanders, A. P., Gelein, R. S., and Currie, W. D. (1976). The effect of hyperbaric oxygenation on the metabolism of the lung. In: Underwater Physiology V. Proceedings of the Fifth Symposium on Underwater Physiology. Ed., C. J. Lambertsen. Bethesda: FASEB, pp. 483–492.

Schaefer, K. E. (1956). Oxygen toxicity studies in underwater swimming. J. Appl. Physiol. 8: 524–531.

Schaefer, K. E. (1965). Circulatory adaptation to the requirements of life under more than one atmosphere of pressure. In: Handbook of Physiology, Section 2, Circulation, Vol. III. Washington, D. C.: Am. Physiol. Soc. pp. 1843–1873.

Schaefer, K. E. (1974). Chronic hypercapnia and oxygen toxicity at high pressure. Proceedings XXVI International Congress of Physiology, New Dehli, India.

Schaefer, K. E., Avery, M. E., and Bensch, K. (1964). Time course of changes in surface tension and morphology of alveolar epithelial cells in CO_2 induced hyaline membrane disease. J. Clin. Invest. 43:2080–2093.

Smith, J. L. (1899). The pathological effects due to increase of oxygen tension in the air breathed. J. Physiol. 24:29–35.

Strauss, M. B., Lee, W., and Cantrell, R. W. (1973). Serous otitis media in divers breathing 100% oxygen. In: Presented at Annual Scientific Meeting, Aerospace Medical Association, May 7–10.

Stroud, R. C. (1959). Combined ventilatory and

breath-holding evaluation of sensitivity to respiratory gases. J. Appl. Physiol. 14(3):353–356.

Trokel, S. (1965). Effect of respiratory gases upon choroidal hemodynamics. Arch. Ophthalmol. 73:838.

Valimaki, M., Juva, K., and Rantanen, J. (1975). Collagen metabolism in rat lungs during chronic intermittent exposure to oxygen. Aviat. Space Environ. Med. 46:684–690.

Webb, P. (1962). NASA Life Sciences Data Book.

Wilson, J. M., Kligfield, P., Adams, G. M., Harvey, C., and Schaefer, K. E. (1977). Human ECG changes during prolonged hyperbaric exposures breathing N_2–O_2 mixtures. J. Appl. Physiol. 42:614–633.

Wood, J. D., Watson, W. H., and Murray, G. W. (1969). Correlation between decreases in brain gama amino butyric acid levels and susceptibility to convulsions induced by hyperbaric oxygen. J. Neurochem. 16:281–287.

Yarbrough, O. D., Welham, W., Brinton, E. S., and Behnke, A. R. (1947). Symptoms of oxygen poisoning and limits of tolerance at rest and at work. Res. Rept. No. 1, U. S. Naval Exp. Diving Unit, Washington, D. C.

16

Oxygen Tension in the Clinical Situation

I. A. Silver

I. The Historical Background

The importance of hemoglobin oxygenation has long been appreciated and the clinical measurement of it in the 1930s became markedly easier as a result of the work by Millikan (1933) who introduced the photocell oximeter. This instrument allowed not only rapid estimates of oxygen saturation of blood in vitro, but also permitted continuous in vivo measurements by transillumination of the lobe of the ear. Although at that time it was appreciated that hemoglobin saturation was related to oxygen tension and oxygen hemoglobin association/dissociation curves were available, it does not appear to have been widely understood clinically until relatively recently that the oxygen *tension* of the blood is often more important to the survival of the tissue than the oxygen *content*. Furthermore, even among physiologists it was assumed that mixed venous blood oxygen tension could be equated with a rather homogeneous average tissue P_{O_2}.

The importance of oxygen as the respiratory gas had of course been realized since the classic work of Lavoisier (1777) and Priestley (1790) but it was only with the advent of industrially produced oxygen on a large scale that the gas could be applied to the treatment of clinical cases. It was employed initially in the treatment of patients with pneumonia and later in resuscitation treatment of accidental drowning, etc. With the recognition of the need of some pathogenic organisms to grow in anoxic environments and the importance of these in human disease, particularly the activity of gas gangrene bacilli (*Clostridium* species) and the realization that certain diseases were characterized by tissue hypoxia, a slow start was made in the application of oxygen to the treatment of nonrespiratory conditions. It is probable that Starr (1932) was one of the first to apply external oxygen in the treatment of superficial injury, when he introduced the so called "oxygen boot" in which oxygen was pumped around a foot or leg that was suffering from hypoxia, as a result either of vascular disease or injury. A major change in attitude to the clinical appreciation of the importance of tissue oxygenation came with the introduction of methods that enabled the direct measurement of oxygen tension to be carried out relatively simply in animals and patients together with rapid biochemical determination of blood lactate/pyruvate ratios. Although the detection by polarised platinum electrodes of oxygen in aqueous solution had been demonstrated in the late nineteenth century by Danneel (1897–1898) in Nernst's laboratory, the method was not applied to a biological system until Blinks and Skow (1938) used platinum electrodes to measure oxygen generation in suspensions of plant cells. There was then a further gap of 6 years before the classic work of Davies and Brink (1942) who showed that absolute oxygen tension and changes of P_{O_2} could be measured experimentally at the surface of the brain and in the brain substance. Their work was followed and applied to clinical investigations of patients with peripheral vascular disease by Montgomery and Horwitz (1950) and of cancer patients by Cater et al. (1957) using relatively crude open tipped platinum or gold electrodes with separate silver/silver chloride anodes.

II. Oxygen Tension and Radiotherapy

Cater et al. (1957, 1959), Cater and Silver (1960), and Evans and Naylor (1966) studied oxygenation of normal tissues and various tumors in patients undergoing radiotherapy. The interest in measuring oxygen tension before and during radiotherapy is because of its effect on cellular sensitivity to radiation. The work of Gray et al. (1953) demonstrated that below a certain oxygen tension (approximately 20 mm Hg—2.8 kPa) the radiosensitivity of almost all types of cells is progressively reduced until at zero oxygen the radiation tolerance of such cells is approximately twice that of the same cells at an oxygen tension above 3 kPa.

Cater et al. (1957, 1959) first demonstrated and it has frequently been confirmed (see Puffer et al., 1976; Vaupel, 1977) that in contrast to that in normal tissues, the oxygen tension in many malignant growths was extremely low and often below that required for maximum radiosensitivity. They also demonstrated that if patients were given oxygen to breathe at atmospheric pressure, the oxygen tension not only in the normal tissues but also in the tumor tissue would often rise well above the minimum required for maximum radiosensitivity and that one of the effects of "fractionation" of radiotherapy was to provoke increases in local tumor P_{O_2} after each treatment. This work was carried out in collaboration with investigations on the use of oxygen as a clinical radiosensitiser with Mitchell (1957, 1960, 1965). It was later demonstrated that other methods of reducing the radiosensitivity of normal tissue in relation to tumors, which were then in vogue experimentally and proposed for clinical use, such as oxygen washout followed by very rapid irradiation with a linear accelerator, might produce severe hypoxia of the brain but did not cause the oxygen tension within normal tissues such as muscle to fall to the level in tumors and therefore the system was counterproductive (Cater et al., 1963).

It was at one time hoped that substantial improvement in the treatment of hypoxic tumors by radiotherapy would be achieved through the use of hyperbaric oxygen but the results have proved disappointing, except in the case of tumors of the head and neck.

The application of oxygen in the treatment of tumors by radiotherapy has continued and is now routine in many centers, and more recent investigations have refined the approach and also reintroduced the use of hyperthermia as a means of increasing the oxygen tension within a tumor (Bicher, 1980). The sensitivity of tumor cells to high temperatures within the physiological range (41°C) has been known since the observations of Coley (1893) but the explanation of the additive effect of heat and radiation was delayed until Cater et al. (1964) showed that heating could increase tissue P_{O_2} as well as damaging tumor cells directly; this work has been confirmed and extended by Puffer et al. (1978).

III. Tissue P_{O_2} in Surgical Conditions

In a completely different type of investigation the orthopedic surgeon Woodhouse (1961) demonstrated that he could distinguish between viable and nonviable femoral heads in joint surgery by the introduction of an oxygen electrode into the separated femoral head during fracture fixation. On the basis of his findings, he was able to show that if circulation was maintained in a fractured femoral head, as detected by an oxygen electrode, the chances of healing following fracture fixation were very good, whereas if intact circulation could not be demonstrated with the electrode in the isolated bone, the results of surgery were likely to be poor and replacement of the femoral head by a prosthesis was preferable. Much similar work has been carried out in the identification of viable and nonviable tissue following severe trauma or during elective surgery such as radical mastectomy (Niinikoski et al., 1973). In the 1960s Hunt et al. (1965, 1967, 1968, 1969) and later Niinikoski et al. (1972, 1973) and Niinikoski and Hunt (1972) made extensive investigations of oxygen tension in healing wounds and the effects of distant trauma on the oxygen tension in damaged tissue. They were particularly concerned with the incidence of wound infection at high and low oxygen tensions having noticed in experimental animals that those that were kept under high oxygen concentrations (40% oxygen) were almost impossible to infect whereas those that were kept at low levels of oxygen (16% oxygen) developed spontaneous wound infections (Lundgren and Zederfeldt 1969). These observations have been paralleled by clinical experience in high mountain expeditions where infected wounds are a constant problem, and in the hyperbaric chambers maintained under the Mediterranean by Jacques Cousteau where the oxygen tension was high and where even quite severe injuries were found to heal without infection in the absence of antimicrobial treatment. These observations have been extended more recently by the use of hyperbaric oxygen chambers in the treatment of nonhealing wounds (Perrins and Davis, 1977) and of wound infection, particularly where anaerobes have been involved (see Gottlieb, 1977). Another important clinical aspect of tissue oxygenation during healing is the requirement for oxygen in collagen formation and the subsequent effect of efficient or inefficient synthesis on wound strength (see Chvapil et al., 1968; Hunt et al., 1972; Hunt and Pai, 1972; Hunt and Van Winkle, 1979). A simple and ingenious method of measuring subcutaneous or intramuscular P_{O_2} in patients for monitoring of clinical status in local or systemic disturbance was developed by Niinikoski in the form of a Silastic tube tonometer, through which physiological saline could be flushed. The sterile tube was inserted through a trocar and cannula into the appropriate area and analysis of the effluent fluid gave a continuous readout of the conditions in the tissue (see Kivisaari et al., 1975).

IV. Oxygen Toxicity

The clinical use of oxygen requires an appreciation of the dangers of oxygen toxicity (Bean, 1945) which operates at biochemical, cellular, and systemic levels. [For later reviews see Clark and Lambertsen (1971) and Clark and Fisher (1977).] Some of the more immediate effects of excessively high oxygen tensions are seen in the lungs where damage occurs to the epithelium which leads paradoxically to systemic hypoxia, and in brain where the patient may develop convulsions and widespread death of neurons. This effect may be greatly potentiated by some poisons, e.g., paraquat (Fisher et al., 1973), and modified by oxidants and antioxidants (Gilbert, 1963).

One of the most tragic results of failure to appreciate the clinical effects of high systemic oxygen tension, although the dangers were recognized by physiologists, occurred in the 1950s when widespread use of oxygen in premature baby care units led to the blinding of many infants through development of retrolental fibroplasia (Lanman et al., 1954; Patz et al., 1952). This condition arises from initial constriction of the central retinal artery under the influence of high blood P_{O_2} which leads to tissue hypoxia, nerve cell damage, and subsequent fibroblast proliferation. As a result of the identification of the cause of this disease, modifications in the atmosphere supplied in ICUs in pediatric units were carried out and the disease has been more or less controlled except in those cases where high ambient oxygen tensions have been essential for life support.

The problems associated with retrolental fibroplasia and some other conditions have led to the recognition of the paradox that high inspired oxygen concentrations may produce such severe peripheral and central vasoconstriction that the oxygen tension in the tissues may eventually drop to pathological levels.

V. Hyperbaric Oxygen

Indications for the clinical use of hyperbaric oxygen, the associated problems, and the physiological and pathological responses of patients and tissues to very high oxygen tension are the subject of an extensive literature. Many aspects have been reviewed recently (see Davis and Hunt, 1977) and are too extensive to be dealt with here. Hyperbaric oxygen raises the P_{O_2} of all tissues initially but physiological protective mechanisms tend to reduce blood flow to, and therefore to lead to carbon dioxide accumulation in, the tissues. The central nervous system is particularly sensitive to and especially well protected from high blood P_{O_2}, although the vascular mechanisms are of limited value above 1 ATA and long-term high oxygen tension is always harmful. Clinical use of high pressure oxygen is of proved value in a number of conditions, e.g., myocardial infarction, osteomyelitis, maintenance of poor skin grafts, and anaerobic bacterial infections; but its worth is disputed in others, e.g., burns, cerebral edema, and treatment of nonhealing ulcers. There seems little doubt that high P_{O_2} and the absence of P_{O_2} gradients can inhibit growth of new blood vessels, but intermittent high and low P_{O_2} may stimulate growth.

VI. Pediatric Application

Measurement of P_{O_2} in neonates has been reviewed by Strauss et al. (1979) and has been applied intraoperatively in pediatric cases in the clinic (Strauss et al., 1972). Monitoring devices have been developed over the past few years, particularly in the field of pediatric medicine and these include both catheter oxygen electrodes and surface skin monitors (for a recent survey, see *Birth Defects*, 1979, Vol 15). Catheter electrodes in various formats have been produced by many workers, mostly derived from the design of Kimmich and Kreuzer (1969) and several of these have gained commercial acceptance. Surface oxygen monitors have been derived, especially from the work of Lübbers and his group in Dortmund (Huch et al., 1972). The basis of the skin measuring devices has been to produce maximum vasodilation either by the application of some chemical such as nicotinamide to the skin or by warming of the skin with an indwelling heating element around the electrode. When the skin vessels are adequately dilated, oxygen diffuses through the superficial layers at a rate which is in excess of the requirements of the live epithelial layers and therefore the oxygen tension at the surface of the sensor becomes more or less equilibrated with that of the arterial blood. There is some dispute as to the efficiency of these systems but these are matters of detail rather than matters of principle (see Souter and Parker, 1977). The transcutaneous systems originate from the work of Baumberger and Goodfriend (1951) and Evans and Naylor (1967).

VII. Current Practice and Problems

The measurement of oxygen tension in a large number of organs can now routinely be carried out as part of patient monitoring particularly intraoperatively in the operating theater and in intensive care units. The multicathode surface sensor of Kessler and Lübbers (1966), which can be used at the surface of an organ, has made available a powerful clinical tool for use on organs such as muscle, and kidney, and liver, particularly in patients who have suffered multiple trauma and may be in shock.

Many types of catheter electrodes and catheter oximeters (Rybak, 1973) are being used intravenously and intra-arterially for continuous monitoring of patient status but all suffer from the problem of clotting at the sensitive surface. Various methods are being developed to reduce this including the provision of polysulfonate membranes that are similar in structure to a synthetic heparin and have inherent nonstick and anticoagulent properties. Other clinical systems for measurement of tissue oxygenation are the mass spectrometer (Donovan and Myers, 1973) and various optical methods which are either derived from the Millikan (1933) oximeter or record intracellular redox state from fluorescent signals excited by UV irradiation of pyridine nucleotide or flavoproteins (Chance et al., 1978; Jobsis et al., 1972) or depend on infrared (IR) absorbtion spectroscopy. The advantage of the mass spectrometer is that it records from a relatively large volume of tissue and measures O_2, N_2, and CO_2 and anesthetic gases simultaneously; the disadvantage is that insertion of the probe inevitably causes trauma which may affect the reading.

Kunze (1966) and Kunze and Lübbers (1973) have described measurements in human muscle with surface and needle electrodes during investigation of neuromuscular disease and Kessler and Lübbers (1966) designed a multiwire surface electrode which has since become widely used for patient monitoring (Schonleben et al., 1978; Sinagowitz et al., 1978).

The advantage of this system is that histograms of oxygen tension distribution can readily be prepared from the electrode readings and the shape of the histogram indicates with reasonable accuracy the oxygenation state of the tissue or patient under investigation.

In transplant surgery, the status of the organ, after vascular connections have been completed, can be checked immediately by means of surface O_2 readings. In brain surgery, and especially where the edges of infarcts are being examined or where hypoxic areas are being identified prior to vascular surgery, a number of different kinds of oxygen monitor have been introduced. These range from single surface oxygen electrodes through multiwire electrodes to sophisticated optical techniques (Chance et al., 1978; Jobsis, 1977; Jobsis et al. 1972, 1977) Because of the relatively large movement of the human brain due to respiration and vascular changes, it is necessary with the electrode assemblies to introduce some kind of balancing or suspension mechanism, whereby a solid measuring device "floats" on the surface of the tissue and follows its movements without compressing it, and to avoid a movement artifact being introduced. Silver, Chance, and Austin (1980, unpublished data), have introduced a damped Phonoarm type of suspension in which a pivoted counterbalance is used to allow an electrode to touch the brain in a weightless manner and yet to follow the respiratory and vascular movements.

In intensive care units, it is common practice following cerebral vascular accidents to insert plugs through trephine holes in the skull through which intracerebral pressure can be monitored and regulated. O'Connor (personal communication) has proposed that such plugs may serve as an access point for various monitoring devices including oxygen electrodes and this has been carried out on an experimental basis recently.

Since the relatively early days of the introduction of oxygen electrodes into clinical investigation, attempts have been made to identify whether or not muscles are being adequately supplied with blood in suspected cases of peripheral vascular disease. It is frequently difficult to determine whether the blood flow to a leg is mainly being carried in the skin while at the same time the deeper tissues are being rendered ischemic due to peripheral vasoconstriction. Heimstra (1958) carried out extensive investigation on this with simple electrodes and found that he was unable to produce adequately repeatable results for his sytem to be of much use as a diagnostic tool. However, more recently with more sophisticated systems either with surface measurements or by the use of the silastic tonometers (Jussila et al., 1978) or with needle electrodes (Kunze, 1966), the problem has been overcome. The adoption of Lübbers histogram technique has shown that in cases where there is neuromuscular dystrophy or where blood supply is inadequate there is a shift to the left of the average oxygenation in ischemic muscle.

VIII. Future Developments

The practicality of clinical measurement of oxygen tension and the appreciation of its importance in the well being of the patient has led to a large number of developments recently, some of which are beginning to be applied routinely in the clinic and others which are still in an experimental stage.

The ideal oxygen monitor should be a noninvasive device with good resolution which can be used by semiskilled people in the operating theater, intensive care unit, and at bedside. Unfortunately, most of the sensitive devices suffer from "noise" and require skilled handling whereas those which are less sensitive have poor resolution. Recent developments in the semiconductor field have resulted in the production of microsensors that can monitor pH or P_{O_2} and also many ions. These can be made extremely small (of the order of 30–200 μm) but the difficulty is to make them adequately specific and also to eliminate the problems that attend any other kind of implanted sensor, i.e., hemorrhage, protein deposition, and encirclement by inflammatory cells. There is no doubt, however, that such probes on the ends of catheters can be inserted into body cavities, into the blood, etc., and form a useful extension of clinical armoury. Fiber optic systems that incorporate microspheres coated in particular dyes have been produced to measure pH in vivo (Goldstein et al., 1980) and similar devices could be made with a pyrene butyric acid sensor for measuring oxygen. Optical systems are less prone to drift than electrodes and present no electrical hazard.

So far as noninvasive systems are concerned, the two most dramatic are the use of nuclear magnetic resonance (NMR) and IR spectroscopy. The IR system has been developed by Jobsis (1977) who has shown both experimentally and in human volunteers that light in the long IR will penetrate through organs as dense as the skull and that the absorbtion spectra that can be measured on the light emerging through the organ give extensive information on the condition of the tissue ranging from the state of hemoglobin oxygenation to the redox level of the cytochrome aa_3 in mitochondria. While this development is of great promise, it involves the use of extremely sensitive optical equipment and has as yet relatively poor resolution. Its great advantage is that the patient is undisturbed, there is no invasion of tissue and a variety of information can be gained.

The application of NMR to clinical problems is still in the developmental state. Possibly the most clinically useful aspect of the technique is in the detection of ^{31}P where it can give an indication of the amounts of various biologically active substances such as ATP, ADP, and creatine phosphate together with the glycerophosphates and inorganic phosphates within and outside cells. The levels of ATP will indicate the degree of oxygenation of the tissue or at least the degree to which the energy reserve of the tissue is being maintained. NMR does not measure oxygen tension directly but it does measure the all important energy reserve which in most tissues is closely dependent on oxidative phosphorylation and therefore P_{O_2}. Furthermore, with the recent development of the surface coil technique, the resolution of tissue status is now better than 1 cm^3

(Ackerman et al., 1980). This will provide information on areas which have been infarcted in brain, heart, or liver, and will also identify ischemic areas. The disadvantage of the NMR is that it is extremely costly and is a fixed system involving expensive installation and some difficulties in patient access. Even with the very largest magnets now available only a part of the body can be introduced into the machine at any one time, but "whole body" 20-inch magnets are being constructed. The advantage of NMR is that it is noninvasive. It provides relatively good resolution and almost instant readout of tissue energy status. It is clear that machines of this type will have an important part to play in the diagnosis of acute disease and in the identification of chronic lesions but they are unlikely to become widely available on account of their cost. Another disadvantage is the considerable expertise that is required to run the machines and to analyze the data from them, but new "imaging" techniques in which a three-dimensional "picture" of conditions in the tissue can be derived from NMR signals are now available in the development stage.

At the present time the development of intraoperative oxygen measuring devices is proceeding rapidly in a number of centers. The greatest interest has been their association with the extension of microvascular surgery particularly for the relief of cardiac and cerebral atherosclerosis or limb grafting. In stroke patients, Austin et al. (1977) introduced small oxygen electrodes into areas of brain that were suspected of being ischemic and followed what changes in oxygenation occurred when anastomoses were made between extracranial arteries and the middle cerebral arteries supplying the area. They also made optical measurements of cortical metabolism by recording respiratory enzyme redox state. Silver (1980, unpublished data) has developed a more sophisticated system in which a balance arm carries surface electrodes and/or optical devices which can be moved about across the operation area and measure not only the oxygen tension in the brain tissue itself but will also give an indication of arteriovenous differences by measuring the oxygen tension in the arteries and the veins through the walls of the small vessels on the brain surface. A feature of measurements of P_{O_2} on the surface of human brain is that there are small fluctuations that have the same frequency as but are slightly out of phase with respiration. In well-oxygenated tissue with a good blood supply especially if the patient is breathing a 30% oxygen mixture, these fluctuations may be as much as 8 mm Hg; by contrast, in poorly perfused tissue there are no such fluctuations which are due to the difference in oxygen saturation of plasma leaving the lungs at the beginning and end of a respiratory movement and are a good indicator of the effectiveness of the blood supply to the area.

New optical intraoperative methods are being developed from experimental systems to examine the feasibiltiy of using pyridine nucleotide or flavoprotein fluorescence or the reflectance spectra of cytochrome aa_3 from the surface of the human brain for assessment of oxygenation and metabolic state of the tissue (Chance et al., 1978; Jöbsis et al., 1977).

A major difficulty with the optical approach to clinical monitoring has been interference from the absorption spectrum of hemoglobin, which unfortunately coincides rather closely with cytochrome aa_3 and also with the fluorescence signals particularly from flavoprotein. Electronic methods of eliminating the hemoglobin interference are currently being developed. The advantage of the optical methods is that they are completely noninvasive and do not involve the touching of the surface of the brain. They also demonstrate the redox state of the respiratory enzymes and as such are indicative of the metabolic state of the tissue rather than just of the oxygen supply. This particularly applies to the flavoprotein which has a favorable position on the electron transfer chain but the pyridine nucleotide which is easier to measure is, unfortunately, at the substrate end of the chain and will only become reduced when conditions are extremely hypoxic.

IX. Conclusion

The measurement and appreciation of tissue oxygen tension in the clinical situation is clearly very important in both diagnosis and treatment. Some aspects are straightforward but others particularly in the assessment of the desirable or "normal" level of tissue oxygen tension require considerable understanding before use can be made in patient care of the information available. Oxygen is a relatively toxic substance and while an adequate supply for biological energy production is essential, an excess is not only unnecessary but may be actively harmful and lead to oxygen poisoning or excessive reduction in blood flow and buildup of carbon dioxide.

References

Ackerman, J., Grove, T. H., Wong, G. G., Gadian, D. G., and Radda, G. K. (1980). Mapping of metabolites in whole animals by ^{31}P NMR using surface coils. Nature 283:167.

Austin, G., Haugen, G., and LaManna, J. (1977). Cortical oxidative metabolism following microanastomosis for brain ischaemia. In: Oxygen and Physiological Function. Jobsis, F. F., ed. Dallas, Texas: Professional Information Library.

Baumberger, J. P., and Goodfriend, R. G. (1951). Determination of arterial oxygen tension in man by equilibration through intact skin. Fed. Proc. 10:10.

Bean, J. W. (1945). Effects of oxygen at increased pressure. Physiol. Rev. 25:1.

Bicher, H. I. (1980). Tissue oxygenation in normal and hyperthermic conditions. In: Oxygen Transport to Tissue IV. Dora, E. and Kovach, A. G. B., Ed. Budapest: Akademia Kiado.

Birth Defects (1979) Vol. 15, No. 4.

Blinks, L. R., and Skow, R. K. (1938). The time course of photosynthesis as shown by a rapid electrode method. Proc. Natl. Acad. Sci. USA 24:420.

Cater, D. B., Hill, D. W., Lindop. P. J., Nunn, J. F., and Silver, I. A. (1963). Oxygen washout studies in the anaesthetised dog. J. Appl. Physiol. 18:888.

Cater, D. B., Phillips, A. F., and Silver, I. A. (1957). Apparatus and techniques for the measurement for oxidation-reduction potentials, pH and oxygen tension *in vivo*. Proc. Roy. Soc. London, Series B. 146:382.

Cater, D. B., and Silver, I. A. (1960). Quantitative measurements of oxygen tension in the normal tissues and in the tumours of patients before and after radiotherapy. Acta. Radiol. 53: 233.

Cater, D. B., Silver, I. A., and Watkinson, A. (1964). Combined therapy with 220KV roentgen and 10 cm microwave heating in rat hepatoma. Acta. Radiol. (Ther., Phys., Biol.) 2:321.

Cater, D. B., Silver, I. A., and Wilson, G. M. (1959). Apparatus and technique for the quantitative measurement of oxygen tension in living tissues. Proc. Roy. Soc. London, Ser. B. 151: 256.

Chance, B., Barlow, C., Nakase, Y., Takeda, H., Mayevesky, A., Fischetti, R., Graham, N., and Sorge, J. (1978). Heterogeneity of oxygen delivery in normoxic and hypoxic states: A fluorometric study. Am. J. Physiol. 235:H 809.

Chvapil, M., Hurych, J., and Ehrlichova, E. (1968). The influence of various oxygen tensions upon proline hydroxylation and the metabolism of collagenous and non-collagenous proteins in skin slices. Z. Physiol. Chem. 349: 211.

Clark, J. M., and Fisher, A. B. (1977). Oxygen toxicity and extension of the tolerance in oxygen therapy. In Hyperbaric Oxygen Therapy. Davis, J. C. and Hunt, T. K., Ed. Bethesda, Maryland: Undersea Medical Society, Inc.

Clark, J. M., and Lambertsen, C. J. (1971). Pulmonary oxygen toxicity: A review. Pharmacol. Rev. 23:37.

Coley, W. B. (1893). The treatment of malignant tumours by repeated inoculations of erysipelas: With a report of 10 original cases. Am. J. Med. Sci. 105:487.

Danneel, H. (1897-1898) Über den durch diffundierende Gase hervorgerufenen Reststrom. Z. Elecktrochem 4:227.

Davies, P. W., and Brink, F. (1942). Microelectrodes for measuring local oxygen tension in animal tissues. Rev. Sci. Instr. 13:524.

Davis, J. C., and Hunt, T. K. (1977). Hyperbaric Oxygen Therapy. Bethesda, Maryland: Undersea Medical Society Inc.

Donovan, W. E., and Myers, B. (1973). Measurement of tissue gas levels with a mass spectrometer. Adv. Exp. Med. Biol. 37A:67.

Evans, N. T. S., and Naylor, P. F. S. (1966).

Steady state of oxygen tension in human dermis. Respir. Physiol. 2:46.

Evans, N. T. S., and Naylor, P. F. S. (1967). The systemic oxygen supply to the surface of the human skin. Respir. Physiol. 3:38.

Fisher, H. K., Clements, J. A., and Wright, R. R. (1973). Enhancement of oxygen toxicity by the herbicide Paraquat. Am. Rev. Respir. Dis. 107:246.

Gilbert, D. L. (1963). The role of pro-oxidants and anti-oxidants in oxygen toxicity. Radiat. Res. Suppl. 3:44.

Goldstein, S. R., Peterson, J. I., and Fitzgerald, R. V. (1980). A minature fibre optic pH sensor for physiological use. J. Biomech. Eng. 102:141.

Gottlieb, S. F. (1977). Oxygen under pressure and microorganisms. In: Hyperbaric oxygen therapy. Davis, J. C. and Hunt, T. K., Ed. Bethesda, Maryland: Undersea Medical Society Inc.

Gray, L. H., Conger, A. D., Ebert, M., Hornsey, S., and Scott, O. C. A. (1953). Concentration of oxygen dissolved in tissues at the time of irradiation as a factor in radiotherapy. Br. J. Radiol. 26:638.

Heimstra, R. (1958). Lumbale sympathectomie bij perifeer oblitererend arterieel vaatlijden. M. D. Thesis, University of Leiden, Netherlands.

Huch, R., Lübbers, D. W., and Huch, A. (1972). Quantitative continuous measurement of partial oxygen pressure on the skin of adults and new born babies. Pflügers Arch. Ges. Physiol. 337: 185.

Hunt, T. K., and Hutchinson, J. G. P. (1965). Studies on the oxygen tension in healing wounds. In: Wound Healing. Illingworth, C. F., Ed. London: Churchill, 2-68P.

Hunt, T. K., and Pai. M. P. (1972). The effect of ambient oxygen tensions on wound metabolism and collagen synthesis. Surg. Gynecol. Obstet. 135:561.

Hunt, T. K., and Van Winkle, W. (1979). Normal repair. In: Fundamentals of Wound Management. Hunt, T. K. and Dunphy, J. E., Eds. New York: Appleton-Century-Crofts, pp. 2–68.

Hunt, T. K., Twomey, P., Zederfeldt, B., and Dunphy, J. E. (1967). Respiratory gas tensions and pH in healing wounds. Am. J. Surg. 144: 302.

Hunt, T. K., Zederfeldt, B., and Dunphy, J. E. (1968). Role of oxygen tension in healing. J. Surg. (Benares Univ) 4:279.

Hunt, T. K., Zederfeldt, B., and Goldstick, T. K. (1969). Oxygen and healing. Am. J. Surg. 118: 521.

Hunt, T. K., Niinikoski, J., and Zederfeldt, B. (1972). Role of oxygen in repair processes. Acta. Chir. Scand. 138:109.

Jöbsis, F. F. (1977). Non-invasive infrared monitoring of cerebral and myocardial oxygen sufficiency and circulatory parameters. Science 198:1264.

Jöbsis, F. F., O'Connor, M. J., Rosenthal, M., and Van Buren, J. M. (1972). Fluorimetric monitoring of metabolic activity in the intact cerebral cortex. In: Neurophysiology Studied in Man. Somjen, G., Ed. Amsterdam: Excerpta Medica.

Jöbsis, F. F., Keizer, J. H., LaManna, J. C., and Rosenthal, M. (1977). Reflectance spectrophotometry of cytochrome aa_3 *in vivo*. J. Appl. Physiol. 43:858.

Jussila, E., Niinikoski, J., and Inberg, M. V. (1978). Oxygen and carbon dioxide tension in the gastronemicus muscles of patients with lower limb arterial ischaemia. Adv. Exp. Med. Biol. 94:623.

Kessler, M., and Lübbers. D. W. (1966). Aufbau und Anwendungsmöglichkeiten verschiedenen PO_2-Elektroden. Pflügers. Arch. Ges. Physiol. 291:R82.

Kimmich, H. P., and Kreuzer, F. (1969). Catheter PO_2 electrode with low flow dependancy and fast response. In: Oxygen Transport to Tissue. Kreuzer, F., Ed. Progr. Resp. Res. 3:100.

Kivisaari, T., Vihersaari, T., Renvall, S., and Niinikoski, J. (1975). Energy metabolism of experimental wounds at various oxygen environments. Ann. Surg. 181:823.

Kunze, K. (1966). Die lokale kontinuierliche Sauerstoff-druck Messung in der menschlichen Muskulator. Pflügers. Arch. Ges. Physiol. 292: 151.

Kunze, K., and Lübbers, D. W. (1973). Absolute PO_2 measurements with Pt-electrode applying polarizing voltage pulsing. Adv. Exper. Med. Biol. 37A:35.

Lanman, J. T., Guy, L. P., and Dancis, J. (1954). Retrolental fibroplasia and oxygen therapy. J. A.M.A. 155:223.

Lavoisier, A. L. (1777). Oeuvres de Lavoisier. Paris: Imprimerie Imperiale.

Lundgren, C. E. J., and Zederfeldt, B. (1966). Influence of low oxygen pressure on wound healing. Acta. Chir. Scand. 135:555.

Mitchell, J. S. (1957). Some clinical and laboratory studies of chemical radiosensitisers. Acta. Unio. Int. Cancer 13:450.

Mitchell, J. S. (1960). Studies in Radiotherapeutics. Oxford: Blackwells.

Mitchell, J. S. (1965). The Treatment of Cancer. Oxford: Blackwells.

Millikan, G. A. (1933). A simple photoelectric colorimeter. J. Physiol. (London)79:152.

Montgomery, H., and Horwitz, O. (1950). Oxygen tension of tissues by the polarographic method. 1. Introduction: Oxygen tension and bloodflow of the skin of human extremities. J. Clin. Invest. 29:1120.

Niinikoski, J., Heugan, C., and Hunt, T. K. (1972). Oxygen tensions in human wounds. J. Surg. Res. 12:77.

Niinikoski, J., and Hunt, T. K. (1972). Measurement of wound oxygen with implanted silastic tube. Surgery 71:22.

Niinikoski, J., Jussila, P., and Vihersaari, T. (1973). Radical mastectomy wound as a model for studies of human wound metabolism. Am. J. Surg. 126:53.

Patz, A. Hoeck, L. E., and De La Cruz, E. (1952). Studies on effect of high oxygen administration in retrolental fibroplasia. 1. Nursery observations. Am. J. Ophthalmol. 35:1248.

Perrins, D. J., and Davis, J. C. (1977). Enhancement of healing in soft tissue wounds. In Hyperbaric Oxygen Therapy. Davis, J. C. and Hunt, T. K., Ed. Maryland: Undersea Medical Society Inc.

Priestley, J. (1790). Experiments and Observations on Different Kinds of Air. Birmingham, England (3 vols).

Puffer, H. W., Warner, N. E., Schaeffer, L. D., Wetts, R. W., and Bradbury, M. (1967). Preliminary observations of oxygen levels in microcirculation of tumours in C3H mice. Adv. Exp. Med. Biol. 75:605.

Puffer, H. W., Wilson, B. R., and Warner, N. E. (1978). Oxygen levels in C3H mice during hyperthermia: Initial data. Adv. Exp. Med. Biol. 94:377.

Rybak, B. (1973). Catheterizable absolute photometer. Adv. Exp. Med. Biol. 37A:99.

Schonleben, K., Hauss, J. P., Spiegel, U., Bunte, H., and Kessler, M. (1978). Monitoring of tissue PO_2 in patients during intensive care. Adv. Exp. Med. Biol. 94:593.

Sinagowitz, E., Golsong, M., and Halbfass, H. J. (1978). Local tissue PO_2 in kidney surgery and transplantation. Adv. Exp. Med. Biol. 94:721.

Souter, L. P., and Parker, D. (1977). A comparison of transcutaneous and arterial PO_2 in sick neonates. Adv. Exp. Med. Biol. 75:747.

Starr, I., Jr. (1932). On the conservative treatment of gangrene of the feet by a selected temperature, oxygen and dessication. Trans. Assoc. Am. Phys. 47:339.

Strauss, J., Bancalari, E., Feller, R., Gannon, J., Beran, A. V., and Baker, R. (1979). Clinical aspects of continuous neonatal oxygen monitoring. Fed. Proc. 38:2478.

Strauss, J., Beran, A. V., and Baker, R. (1972). Continuous oxygen monitoring of newborn and older infants and of children. J. Appl. Physiol. 33:238.

Vaupel, P. (1977). Hypoxia in neoplastic tissue. Microvasc. Res. 13:399.

Woodhouse, C. F. (1961). An instrument for the measurement of oxygen tension in bone. J. Bone J. Surg. 43A:819.

17

Retinopathy of Prematurity and the Role of Oxygen

ROBERT W. FLOWER AND ARNALL PATZ

The retinopathy of prematurity (retrolental fibroplasia—RLF) has a unique story in the annals of twentieth century medicine. First described by Terry (1) in the early 1940s, RLF became within a decade the largest single cause of child blindness in the United States and, indeed, a greater cause of child blindness than all other conditions combined. The early and mid-1950s saw the incrimination of oxygen as the principal cause of the disease at that time (2–5). Following the identification of oxygen in the etiology of RLF, a drastic curtailment in oxygen usage was instituted in nurseries throughout the world. A dramatic decrease in incidence and severity of this condition then rapidly followed.

In the mid-1960s with the advent of blood gas monitoring capability in the premature nursery, the idiopathic respiratory distress syndrome in the premature infant was documented (6). The severe oxygen deprivation in these infants required increasing amounts of oxygen for their survival. Cases of RLF were noted in the late 1960s and 1970s and a slight but definite increase in incidence over the late 1950 statistics was generally recognized. Most authorities do not associate these cases occurring in recent years with "oxygen overuse" but have suggested that many of the infants who develop RLF with current practices would not have survived some twenty-five years ago to develop the disease.

I. Pathogenesis of the Retinopathy of Prematurity

The pathogenesis of RLF concerns the mechanism of oxygen action on the immature retina, that is, the retina with incomplete vascularization. The process of retinal vascularization is unique when compared to organs throughout the body in that no definitive blood vessels are present in the human retina prior to the fourth month of gestation. The inner layers of the retina, up to the fourth month, are nourished by the embryonic hyaloid vessel system in the vitreous. In the human fetus, starting at approximately four months' gestation, a vanguard of mesenchymal cells emanating from the central hyaloid vessel at the optic disc invade the inner layers of the retina. These endothelial complexes develop into capillaries as vascularization proceeds anteriorly toward the ora serrata in all directions from the optic nerve. As the retinal vascularization progresses, the embryonic hyaloid vessels in the vitreous undergo regression. These developing retinal vessels do not reach the most anterior portion of the retina (ora serrata) nasally until eight months' gestation and the anterior temporal retinal periphery, farthest removed from the optic nerve, is not vascularized until about term for the full-term infant.

The effect of oxygen on the immature retinal vessels can be conveniently divided into a *primary* stage of vascular closure and a *secondary* stage of neovascularization after removal from oxygen. With prolonged breathing of high concentrations of oxygen, the retinal blood vessels of the immature eye are occluded and further normal vascular growth from the disc toward the periphery is halted (primary effect of oxygen). On returning to air breathing, the remaining retinal vessels not permanently occluded undergo neovascular proliferation (secondary effect of oxygen) (see Fig. 1). The new vessels lack structural integrity and hemorrhage is common.

The above primary and secondary effects induced by oxygen are the specific changes of the oxygen pathogenesis of RLF (6). Further progression of the disease is relatively nonspecific and depends on the degree of hemorrhage and amount of traction on the retina that occurs. In many instances the disease progresses so that there is a complete traction detachment of the retina with total or near-total blindness resulting. These eyes have a white membrane representing the totally detached retina behind the crystalline lens. However, the disease may undergo spontaneous regression to a less severe stage during the first few months of life. Usually the permanent cicatricial or scar tissue stage has occurred by 5 to 6 months of age.

It is now appropriate to describe in more detail the specific action of oxygen on the immature retina as observed in the experimental animal model under laboratory conditions.

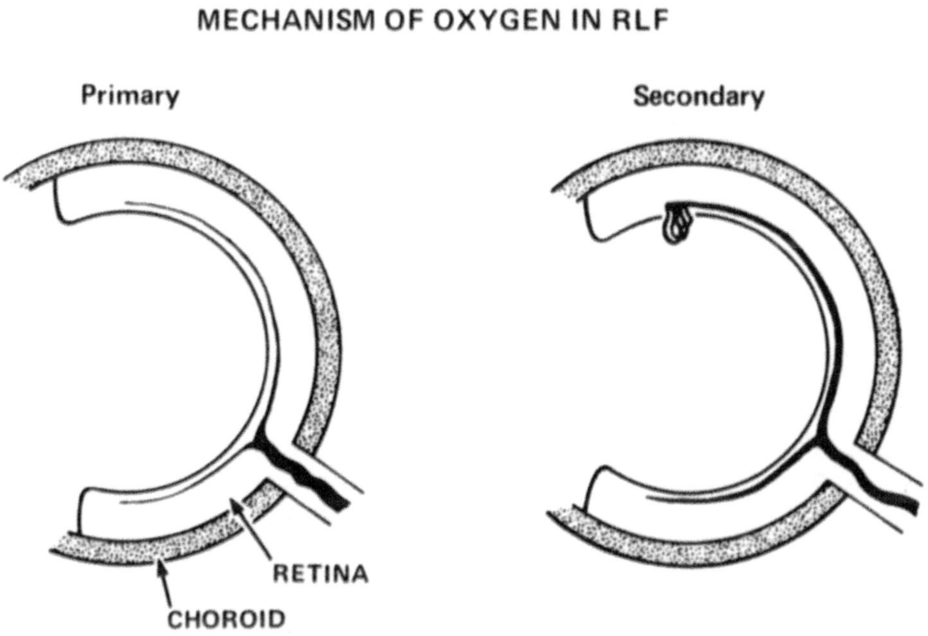

Figure 1. Schematic representation of the mechanism of oxygen in RLF. In the *primary stage* of oxygen exposure marked vasoconstriction occurs. If the oxygen is continued vascular closure starting peripherally results. On removal to air in the *secondary stage* the retinal vessels immediately posterior to the closed vessels undergo proliferation, and the neovascularization erupts through the surface of the retina.

II. Newborn Aminal Model of RLF

Most of the animal experiments have been done with newborn or young kittens and puppies, although Gyllensten and Hellstrom (7) and Gerschman (8) used mice in their studies in the early 1950s. In these species the eyes are relatively immature even in the full-term normal newborn animal. The stage of retinal vascularization in the newborn kitten or puppy is comparable to that observed in the human fetus of approximately 6 to 6½ months gestation (6). Abundant animal experiments conducted in Ashton's laboratory (4,9) and in our laboratory (10) demonstrate that the sensitivity to oxygen damage is directly proportional to the degree of immaturity of the retinal vessels. It is significant that the prematurely born human infant of approximately 6½ to 7 months gestation has part of the nasal retina incompletely vascularized and a large portion of the anterior temporal retina unvascularized. Animal experiments have also shown that when the retina is fully vascularized the damaging effects of oxygen administration are no longer present. Prematurity per se with incomplete vascularization of the retina is fundamental to the development of RLF and is basic to the response of oxygen.

A. Primary Effect of Oxygen: Retinal Vasoconstriction and Vascular Closure

In animal experiments, Ashton has demonstrated that the initial vasoconstriction occurs within the first few minutes after exposure to oxygen. The vessels are reduced approximately 30 to 50% in caliber, then dilate to their original size. Continued oxygen exposure results in progressive but gradual vasoconstriction over the next 4 to 6 hr. At this stage the retinal vasoconstriction is still reversible on removal to air; but if the constriction is sustained, retinal capillary closure results, or "vaso-obliteration," as first described by Ashton (4), occurs. Electron microscopic studies conducted by Ashton and Pedler (11) and subsequently confirmed in our laboratory show marked damage to the endothelial cells lining the capillary bed during the exposure to oxygen. It is not known at the present time whether or not this damage is a direct oxygen cytotoxic effect on the endothelial cells or if the endothelial cells are simply responding to the diminished blood flow produced by the severe vasoconstriction. Animal experiments show that the response to oxygen is dose related and the degree of vascular closure is directly proportional to the concentration and duration of oxygen and inversely proportional to the degree of maturity or development of the retinal vessels at the time of oxygen exposure.

B. Secondary Effect of Oxygen: Retinal Neovascularization

After removal of the experimental subject from sustained hyperoxia to room air, endothelial cells present in remaining retinal capillaries that were not obliterated by the oxygen exposure undergo a marked proliferation. These proliferating endothelial cell complexes break through the internal limiting membrane of the retina to grow on the surface of the retina. Canalization of these endothelial complexes and organization into new vessels occurs and a classical form of retinal neovascularization results. These new capillaries, just as other forms of retinal neovascularization, e.g., those found in diabetic retinopathy, have abnormal permeability and leak molecules the size of fluorescein. Intravenous fluorescein injection is followed by profuse leakage of dye into the vitreous surrounding the new vessels. With further proliferation of these vessels and adherence of the vessels to the posterior cortical vitreous gel, the vessels become elevated on the surface of the retina by the resultant traction and are prone to hemorrhage.

The newborn, or very young kitten or puppy, when placed in oxygenated incubators, with oxygen at ambient pressure, responds as described above. If the kitten or puppy is placed in oxygen at 30 days of age when the retina is fully vascularized, no demonstrable vascular changes result from the oxygen exposure.

In humans and in the animal model, there is a marked predilection for the disease in the temporal periphery. This is especially significant in modern pediatric care where oxygen is not used excessively as was routinely done in the early 1950s. With current pediatric practices the changes are primarily confined, but not always, to the temporal periphery. This observation is consistent with the temporal periphery being the last to become vascularized in normal embryological development of the fetal retina and, therefore, contains the most immature retinal vessel complexes.

III. Monitoring of Blood Oxygen Levels

In view of the continuing trend toward more liberal oxygen administration in premature nurseries, limitations in ability to adequately monitor blood oxygen levels are still a problem in the prevention of RLF. Unfortunately, development of improved monitoring techniques is a most complex challenge. In the first place, infants are placed on oxygen therapy because they present with a life-threatening blood oxygen deficiency, and stabilization of an infant's vital signs is the primary concern of the neonatologist. At best it may be hoped that whatever minimum therapy is required to alleviate that deficiency in each case will be below the maximum safe level for the eye; but at worst, the neonatologist may face the risk of causing ocular damage to save an infant's life.

Even if an ideal, continuous monitoring technique were available, several persistent gaps in understanding the physiology of oxygen transport in the immature eye are encountered. The following is a compilation of what appear to be major obstacles to achieving a better undertanding of RLF. Briefly stated, the most important problems to be considered are the following: (a) It is now known with certainty if the oxygen sensitive retina is responsive to changes in blood oxygen content or to changes in blood P_{aO_2} levels; (b) It is not known if retinal vasoconstriction is an integral part of the pathological process or simply an exaggeration of a normal physiological response to excessively high oxygen levels; (c) The mechanisms by which retinal vasoconstriction and vessel closure are mediated is not understood; (d) The role of the choroidal circulation in RLF, if any, is also unknown.

A. Arterial Blood Oxygen Contents vs. P_{aO_2} *Measurements*

There is a running debate among those who monitor critically ill patients of all ages regarding the merits of monitoring blood oxygen content (i.e., hemoglobin saturation) vs. monitoring arterial blood oxygen levels (i.e., P_{aO_2}). Currently, there is no firm concensus, but of the two, P_{aO_2} levels are more frequently monitored, if for no other reason than the simplicity of the Clark O_2 electrode. In terms of monitoring blood oxygen exclusively for protection of the immature eye, however, there is evidence to suggest that P_{aO_2} level is the most important.

Continuous monitoring of the response of the newborn kitten retinal vasculature to sustained oxygen breathing was observed in our laboratories for periods up to 22 hr (10). Using reduction of blood velocity in the major retinal vessels as an index of vasoconstriction, no changes were observed in any animals whose P_{aO_2} levels were less than 190 torr. In those animals having sustained P_{aO_2} levels greater than 190 torr, both magnitude of blood velocity reduction and speed with which the reduction occurred following onset of oxygen breathing were directly proportional to the magnitude of the sustained P_{aO_2}

level. Hemoglobin saturation of 100% is achieved at a P_{aO_2} of about 90 torr, but vasoconstriciton of the kitten retinal vessels did not occur until more than double that P_{aO_2} level was reached. Therefore, insofar as vasoconstriction in the newborn kitten may be considered a valid model, these data suggest that P_{aO_2} rather than hemoglobin saturation is the important parameter to monitor during oxygen administration.

When extrapolating data from the oxygenated newborn animal with an immature retina to the clinical situation, it should be parenthetically mentioned that there are instances reported in which RLF has been observed in infants who were never placed in high oxygen concentration environments (12,13). These instances are exceptional from the viewpoint that excessive oxygenation means sustained exposure to oxygen concentrations greater than 21% which produce P_{aO_2} levels in excess of 90 to 100 torr. However, from the point of view of the premature infant whose P_{aO_2} *in utero* is in the vicinity of 30 torr, breathing room air alone constitutes excessive oxygenation. This concept moreover suggests the possibility that the immature retina might respond adversely to sudden incremental blood oxygen elevation of a particular magnitude rather than to blood oxygen elevation above a particular absolute level.

B. Retinal Vasoconstriction

Is this response of retinal vasoconstriction to oxygen a pathological process or an exaggerated normal physiological response? Clinical and laboratory studies indicate that retinal vasoconstriction and vascular closure precede retinal neovascularization in RLF; this forms the basis for using vasoconstriction as an indicator of excessive oxygenation. However, even though vascular closure is generally viewed as a necessary first step in the production of RLF, there is no conclusive evidence that this must be the case. Vasoconstriction may be a normal mechanism for regulation of retinal tissue oxygenation *in utero* until the eye is fully developed. The severe vasoconstriction seen in the immature retina with oxygen administration may simply be an exaggeration of the normal physiological response to elevated oxygen levels far outside the range which could occur *in utero*.

Determining the significance of retinal vasoconstriction would be an important step toward discovering the precise mechanism of vascular damage in RLF. To use a simplistic example, if vasoconstriction is a necessary first step in the pathological process, then vessel wall hypoxia might be responsible for subsequent vascular changes. On the other hand, if it were found to be wholly an exaggeration of normal physiological response, it would be more logical to search for some substance diffused from the adjacent choroid in addition to oxygen which causes the vascular changes. Neovascularization can be produced in the animal model, but cicatricial RLF has never been produced. It would be an important step to identify an animal model in which the more advanced stages of RLF are produced that lead to cicatricial changes.

C. Mechanisms of Vasoconstriction

Unless direct evidence to the contrary is found, the mechanism by which retinal vasoconstriction is mediated following the onset of oxygen breathing remains important to understanding the etiology of RLF. Since retinal vessels lack direct nervous innervation, their constriction and dilatation are based upon intravascular resistance-pressure and CO_2 autoregulation. By itself, this fact indicates that any vasotonic control mechanisms must be local, that is, at least within the globe. Ashton and Cook (9) demonstrated in the newborn animal model that vasoconstriction is reversed in a portion of the retina experimentally detached during oxygen breathing. This suggests that an agent emitted from the choroid (oxygen is the prime candidate) is responsible for triggering retinal vasoconstriction.

Consideration must likewise be given to

identifying precisely what are the contractile elements of the retinal vasculature. One or more layers of cells which have the characteristics of smooth muscle cells encircle some retinal arteries and arterioles near the optic disc, but the function of these cells is not clear. In many animal species arteriolar sphincters are found in the precapillary retinal arterioles, but these have not been identified in human retinas. Some investigators have attributed contractile properties to intramural pericytes found in the human vasculature; however, these pericytes are probably undifferentiated primitive endothelial cells. This seems reasonable since the more primitive the cell, the more actively contractile it should be. Many such pericytes acting in unison could create a significant increase in resistance to blood flow through a given segment of the retinal capillary bed, but this too remains hypothesis.

D. Role of the Choroid in RLF

Finally, the role of the choroidal circulation in the etiology of RLF should be investigated. Until the fourth month of gestation, the retina is avascular, deriving its nutrition from the adjacent choroid and the embryonic hyaloid vessels in the vitreous. Complete retinal vascularization does not occur until shortly after birth of the full-term infant. Thus, it seems reasonable that in the immature eye the choroidal circulation might ultimately prove to play an even more important role in controlling retinal tissue oxygenation than it does in the adult eye. Fluorescein dye studies performed in our laboratories showed that in kittens where $P_{a_{O_2}}$ levels were maintained at approximately 475 torr for a period of 4 hr, virtually all of the retinal vessels became threadlike, leaving only the highly reflective disc as a landmark in the fundus. The dye studies also showed that the background choroidal flush was extremely sluggish compared to the rapid choroidal flush seen in the same kitten breathing room air. This strongly suggests that the choroidal circulation, as well as the retinal circulation, can be affected by sustained oxygen breathing.

E. The Role of Antioxidants in RLF

In spite of not knowing answers to the above questions, prevention of the adverse effects of oxygen therapy by administration of antioxidants has been empirically investigated in the laboratory and in the nursery. It is of considerable interest that as far back as 1949, Owens and Owens (14) reported an apparent protective effect of α-tocopherol acetate (vitamin E) on the incidence of RLF in a nursery study. These findings could not be repeated by other observers in the early 1950s and the use of vitamin E as a supplement in the nursery was gradually discontinued. It is significant that Owens and Owens made their recommendation on the known tocopherol deficiency of the premature infant with no concept that it might be serving as an "antioxidant." Indeed, in 1949 the possible role of excess oxygen in the etiology of RLF was not considered and those reports relating to oxygen, which appeared then and in the very early 1950s, suggested that "anoxia" was the primary mechanism if, indeed, oxygen levels had any influence in this disorder. In the mid-1970s Johnson et al. (15) first suggested the use of vitamin E as an antioxidant in the prophylaxis and treatment of RLF. These investigators in their preliminary nursery observations noted an apparent protective effect of vitamin E supplements given to the small, high-risk premature infants. Their preliminary studies have led them to institute a randomized controlled and prospective study on the role of vitamin E in the prevention of RLF and also to further test vitamin E as a possible therapeutic agent once the disease is present.

Phelps and Rosenbaum (16) tested the role of tocopherol on the oxygen-induced retinopathy in kittens. In a controlled study involving young kittens they found that tocopherol administration was beneficial in reducing the severity of oxygen-induced lesions. In a further extension of these studies when the tocopherol was administered after the initial oxygen exposure, they likewise found a slight protective effect, suggesting that the role may be more than simply an antioxidant counteracting the initial primary damage of oxygen. Their findings raise the possibility that vitamin E may theoretically have some role in counteracting the process of abnormal retinal neovascularization. However, further documentation of this concept will be required.

Hall and Hall (17) demonstrated significant quantities of superoxide dismutase in the photoreceptor outer segments. They suggested that superoxide dismutase may inhibit free radical oxidation of polyunsaturated fatty acids, and in light, high O_2 concentrations may activate lipid peroxidation leading to RLF. The possibiility that excess light exposure may theoretically enchance the oxygen effect on the premature retina deserves further study. The role of vitamin E as a protective agent against superoxide radicals will be tested in our laboratory in the experimental RLF model. Studies are now under way to document more precisely the role of tocopherol administration on the experimental model of RLF in our laboratory, and there are now several nurseries involved in a clinical trial to document in the clinical setting its possible protective role against RLF.

We are now investigating the possibility that the prostaglandin system may play a role in regulating retinal vascular status during the perinatal period. Prostaglandins are known to function as local regulators of blood flow. Prostaglandins are known to be formed in response to altered oxygen tensions, and they play a key role in the vascular changes which occur during the course of adaption of the fetal cardiovascular system to the extra-uterine environment. If they are found to play a role in control of the premature infant vascular physiology, it should be possible to manipulate the prostaglandin system pharmacologically in a rational manner and to develop new therapeutic approaches to disorders of the immature retinal vascular system, possibly including RLF.

IV. Summary

The effect of oxygen on the premature retina can be conveniently divided into a *primary stage* of retinal vasoconstriction and vascular closure and a *secondary stage* of retinal neovascularization. The retinopathy produced is directly proportional to the dose, duration, and concentration of oxygen, and inversely proportional to the degree of vascularization of the retina.

The retinal neovascularization produced by oxygen on the incompletely vascularized retina is quite similar to the neovascularization observed in other proliferative retinopathies, such as diabetic retinopathy. In the premature infant, when significant hemorrhage and traction on the retina occur, varying degrees of retinal detachment result causing major impairment of vision.

The precise mechanism of the oxygen damage on these immature retinal vessels is still unknown. The retinal endothelial cells are damaged and ultimately destroyed by the oxygen exposure. The possible role of antioxidants such as tocopherol (vitamin E) in counteracting the oxygen toxic effect on the retinal vessels, as well as the possible role of prostaglandins in regulating vascular status, is under investigation.

References

1. Terry, T. L. (1942). Extreme prematurity and fibroblastic overgrowth of persistent vascular sheath behind each crystalline lens. I. Preliminary reports. Am. J. Ophthalmol. 25:203.
2. Patz, A., Hoeck, L. E., De La Cruz, E. (1952). Studies on the effect of high oxygen administration in retrolental fibroplasia. I. Nursery observations. Am. J. Ophthalmol. 35:1248.

3. Patz, A., Eastham, A., Higginbotham, D. R., et al. (1953). Oxygen studies in retrolental fibroplasia. II. The production of the microscopic changes of retrolental fibroplasia in experimental animals. Am. J. Ophthalmol. 36:1511.
4. Ashton, N., Ward, B., and Serpell, G. (1953). Role of oxygen in the genesis of retrolental fibroplasia: Preliminary report. Br. J. Ophthalmol. 37:513.
5. Kinsey, V. E. (1956). Retrolental fibroplasia. Cooperative study of retrolental fibroplasia and the use of oxygen. Arch. Ophthalmol. 56:481.
6. Patz, A. (1969). Retrolental fibroplasia. Survey Ophthalmol. 14:1.
7. Gyllensten, L. I., and Hellstrom, B. E. (1954). Experimental approach to the pathogenesis of retrolental fibroplasia; changes of eye induced by exposure of newborn mice to concentrated oxygen. Acta Paediat. 43:131.
8. Gerschman, R., Nadig, P. W., Snell, A. C., and Nye, S. W. (1954). Effect of high oxygen concentrations on eyes of newborn mice. Am. J. Physiol. 179:115.
9. Ashton, N., and Cook, C. (1954). Direct observation of the effect of oxygen on developing vessels. Preliminary report. Br. J. Ophthalmol. 38:433.
10. Flower, R. W., and Patz, A. (1971). Oxygen studies in retrolental fibroplasia. IX. The effects of elevated arterial oxygen tension on retinal vascular dynamics in the kitten. Arch. Ophthalmol. 85:197.
11. Ashton, N., and Pedler, C. (1962). Studies on developing retinal vessels. IX. Reaction of endothelial cells to oxygen. Br. J. Ophthalmol. 46:257.
12. Brockhurst, R. J., and Christi, M. I. (1975). Cicatricial retrolental fibroplasia: Its occurrence without oxygen administration in full-term infants. Albrecht von Graefes Arch. Klin. Exp. Ophthamol. 195:113.
13. Foos, R. Y. (1975). Acute retrolental fibroplasia. Albrecht von Graefes Arch. Klin. Exp. Ophthamol. 195:87.
14. Owens, W. C., and Owens, E. U. (1949). Retrolental fibroplasia in premature infants. Am. J. Ophthalmol. 32:1631.
15. Johnson, L., Schaffer, D., and Boggs, T. R., Jr. (1974). The premature infant, Vitamin E deficiency and retrolental fibroplasia. Am. J. Clin. Nutr. 27:1158.
16. Phelps, D. L., and Rosenbaum, A. L. (1977). The role of tocopherol in O_2 induced retinopathy: Kitten model. Pediatrics 59:998.
17. Hall, M. O., and Hall, D. O. (1975). Superoxide dismutase of bovine and frog outer segments. Biochem. Biophys. Res. Commun. 67:1199.

18
Oxygen: An Overall Biological View

DANIEL L. GILBERT

It seems that the oxygen we are breathing today was synthesized in massive stars. When these stars exploded as supernovas, the oxygen was released into the interstellar clouds. Our sun and solar system were born from such a cloud, containing the previously synthesized oxygen (Herbig, 1981).

Almost all the oxygen is in the form of ^{16}O, which is extremely stable. Nuclei which possess the so-called magic numbers of protons or neutrons, i.e., 2, 8, 20, 28, 50, 82, or 126, are very stable. Since ^{16}O contains 8 protons and 8 neutrons, this nuclide contains the double magic number, which accounts for its nuclear stability (Gilbert, 1964a).

Oxygen is a major constituent of the earth's crust (Gilbert, 1981b; Herbig, 1981). During the evolution of the earth, there has been an intimate interrelationship between the biosphere and the atmospheric oxygen. This relationship still exists today (Gilbert, 1981b).

Due to its geochemical and biological importance, it is not surprising that oxygen is currently the third highest volume chemical and the highest volume gas produced commercially in the United States. The major commerical uses of oxygen are for metal manufacturing and fabrication (36%) and for health services (13%) (Storck, 1979). The use of oxygen therapy began shortly after oxygen was discovered (Gilbert, 1981a).

On the other hand, Pasteur discovered that some life is possible without oxygen; indeed, the mere presence of oxygen destroys some living organisms (Gilbert, 1981a). However, oxygen serves as an excellent energy source for living processes; and most organisms use oxygen as a way of life (Gilbert, 1981b). The price that living processes pay for this convenient energy source is oxygen poisoning. Oxygen thus has these dual actions, one promoting life and the other destroying life. Oxygen resembles both the Good Spirit and Evil Spirit in Zoroastrianism, the ancient Persian religion (Gaer, 1956).

Oxygen pressure has been measured in various units (Huber and Drath, 1981). Conversion factors between various pressure units are given in Table 1. The barometric pressure, obtained using the simplified barometric equation (Gilbert, 1981b), equals 1 bar at about 110 m (360 ft) below sea level (Gilbert, 1964b). At sea level, the oxygen pressure equals 158 torr (211 mbar) (Gilbert, 1981b) which corresponds to an oxygen concentration of 8.50 m*M* in the gas phase.

For an aerobic way of life, there is a balance between both these oxygen actions.

Table 1. *Pressure Units*

Pressure unit[a]	Standard atmosphere equivalent	Bar equivalent
Atmosphere absolute (ATA)	1	0.98692
mm Hg (at 0°C)[b]	760	750.06
in Hg (at 0°C)[b]	29.92	29.53
m H_2O (at 3.98°C)[b]	10.33	10.20
ft H_2O (at 3.98°C)[b]	33.90	33.46
m sea water (at 1.03 g/cm^3)[b]	10.0	9.9
ft sea water (at 1.03 g/cm^3)[b]	32.9	32.5
torr	760	750.06
dynes/cm^2 (g cm^{-1} sec^{-2})	1.01325×10^6	1×10^6
bar	1.01325	1
mbar (millibar or mb)	1013.25	1000
Newton/m^2 (N/m^2) (kg m^{-1} sec^{-2})	1.01325×10^5	1×10^5
N/mm^2 (Newton/mm^2)	0.101325	0.1
MN/m^2 (MegaNewton/m^2 or MNewton/m^2)	0.101325	0.1
Pascal (Pa) (kg m^{-1} sec^{-2})	1.01325×10^5	1×10^5
kPa (kiloPascal)	101.325	100
MPa (MegaPascal)	0.101325	0.1
lb/in.^{2} (pounds/in.^{2} or psi)	14.696	14.504

[a]When the pressure units show an increase above sea level atmospheric pressure, 1 ATA must be added. For example, a gas volume under 90 m sea water has a total pressure of 9 ATA plus 1 ATA or 10 ATA (9.8692 bar). Likewise, if a gauge for a pressure chamber indicates 14.696 psi, then the pressure equals 2 ATA (1.97384 bar).

[b]Pressure equals height times density times acceleration due to gravity, where acceleration due to gravity equals 980.665 cm/sec^2. The density of water at 3.98°C is 0.999973 g/cm^3. The density of mercury at 0°C is 13.5955 g/ml $\times$ 0.999973 ml/cm^3 or 13.5951 g/cm^3.

If the oxygen is not sufficient, then the problems of hypoxia are observed. On the other hand, if there is too much oxygen, then oxygen toxicity becomes apparent. A given biological activity will exhibit an optimum oxygen concentration (Fig. 1) balancing these antagonistic effects. In the same organism, the optimum oxygen pressure does not have to be the same for different biological activities. For example, in *Saccharomyces cerevisiae*, the enzymes palmitoyl-CoA desaturase, nonaerobic acetyl-CoA synthetase, and aerobic acetyl-CoA synthetase exhibit optimum specific activities at the oxygen pressures 0.3 to 3 mbar, 3 to 10 mbar, and 210 mbar, respectively (Jahnke and Klein, 1979).

Rosin (1956) writes with tongue in cheek about oxygen as follows: "Very toxic gas and an extreme fire hazard. It is fatal in concentrations of 0.000001 ppm [1×10^{-10}%].... Symptoms resemble very much those of cyanide poisoning.... In higher concentration, e.g., about 20%, the toxic effect is somewhat delayed and it takes about 2.5 billion inhalations [the number of inhalations after 70 years at a rate of 12 inhalations/min equals 0.44×10^9] before death takes place. The reason for the delay is the difference in mechanism of the toxic effect of oxygen in 20% concentration. It apparently contributes to a complex process called aging, of which very little is known, except that it is always fatal.... The first inhalation (occurring at birth) is sufficient to make oxygen addiction permanent.... Oxygen is especially dangerous because it is odorless, colorless, and tasteless, so that its presence cannot be readily detected until it is too late." Campbell (Harrison, 1966), in a similar vein, also wrote about oxygen as follows, "As little as one breath is known to produce a life-long addiction to the gas, which addiction invariably ends in death."

Indeed, as Gerschman (1964) pointed out, oxygen is toxic at all concentrations, even at the lowest oxygen pressures. However, this toxicity can easily be overshadowed by the

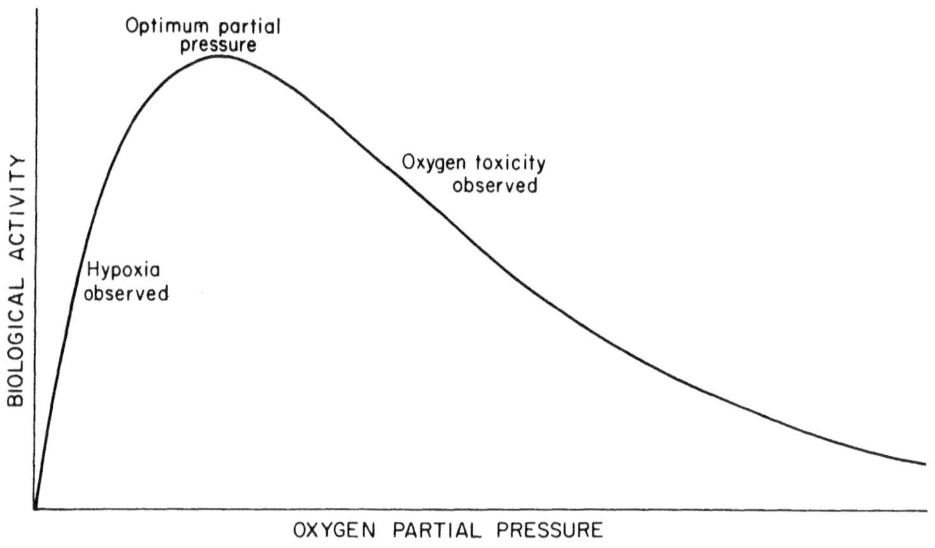

Figure 1. Effect of oxygen on biological activity. Taken from Gilbert (1972).

need of oxygen and therefore experimentally can be overlooked. We have antioxidant mechanisms to defend ourselves from the toxic effect of oxygen (Forman and Fisher, 1981; Fridovich, 1981; Gottlieb, 1981; Kovachich and Haugaard, 1981). If the biosphere could adapt to a lower oxygen concentration, then perhaps the present antioxidant defenses would be better equipped to cope with the toxic effects of oxygen.

I. Hypoxia

There are regions of the world where the oxygen pressure is significantly less than 158 torr, which is the oxygen pressure at sea level. These regions include parts of the ocean and mountains. Indeed, the aerobic part of the biosphere residing in these regions has adapted to this change. For example, diving turtles are very resistant to hypoxia (Hochachka and Guppy, 1977).

At an altitude of 4300 m (14,000 ft), barley and potatoes are grown at Phopagaon, Nepal (Hagen et al, 1961). Reptiles occur at 5500 m (18,000 ft) (Hock, 1964). Humboldt (1849) noted insects at 19,286 ft (5878 m) on Chimborazo in Ecuador. The elevation of Chimborazo is 20,561 ft (6267 m); it is the world's furthest point from the earth's center, being 7057 ft (2151 m) further distant than Mt. Everest (McWhirter, 1979). The flowering plant, *Arenaria muscosa*, has been observed at an altitude of 6218 m (20,400 ft) and salticid spiders were recorded at 6700 m (22,000 ft) (Hutchinson, 1965). Some bacteria have been found at 12,000 m (39,370 ft) (Heath and Williams, 1977).

Humboldt (1849) states that the large condor in the Andes can fly at an altitude about 23,000 ft (7000 m) and that the small humming bird was seen at 14,600 ft (4450 m). The alpine chough, *Pyrrhocorax graculus*, has been seen nesting at 21,500 ft (6500 m). This bird has been spotted at an altitude of 28,000 ft (8500 m) in Tibet. The bar-headed goose, *Anser indicus*, has been seen flying at 30,000 ft (9,000 m) (Swan, 1970). Black and Tenney (1980) observed that this goose can tolerate 10,668 m (35,000 ft) where the oxygen tension is 27 torr without any noticeable behavioral effects. These authors cite a report of a vulture flying at the incredible altitude of 37,000 ft (11,300 m)! Thus, birds can cope with hypoxia remarkably well. Piiper and Scheid (1981) point out that for birds the partial pressure of oxygen in the arterial blood is higher than in the expired air.

Animals which are brought up to high elevations react differently. For example, dogs adapted to the highlands, whereas cats could not adapt (Monge and Monge, 1966).

What of man? It is interesting to compare his adjustment to hypoxia with other forms of life, in order to grasp the biological limits of adaptation. Man has stood on the summit of Mt. Everest (8848 m or 29,028 ft) breathing atmospheric air at a barometric pressure of approximately 250 torr (Pugh, 1964). Mt. Everest, or Chomolongma, meaning mother goddess of the land, has the world's highest elevation (Hagen et al., 1961). A stowaway on an airplane survived flying at 29,000 ft (8840 m) for 5 hr without oxygen (Ward, 1975). Henderson (1938) noted that "Whymper on Chimborazo (20,500) [6250 m] swallowed some potassium chlorate for an extra supply of oxygen." Thus, some men are resistant to hypoxia, and can even survive taking poison at high altitude!

A small but significant proportion of the human population resides at high altitudes where the barometric pressure is significantly below one bar. Above 2500 to 3000 m (8000 to 10,000 ft) has been designated as the level of high altitude regions (Heath and Williams, 1977; Pawson and Jest, 1978). The oxygen pressure in the atmosphere at this altitude is about 70% of the sea level oxygen pressure (Gilbert, 1981b).

Mexico City, Mexico, at an altitude of 6540 ft (2000 m) is the most populous urban area in the world at a moderate altitude (Delury, 1978). Lhasa, China, in Tibet was the highest capital city in the world (12,087 ft or 3684 m). La Paz, the de facto capital of Bolivia, has an elevation of 11,916 ft (3632 m) (McWhirter, 1979). The old salt-trade route between Nepal and Tibet goes through Sanka La, a pass at 5200 m (17,000 ft) (Hagen et al., 1961). A high altitude laboratory is at an elevation of 4540 m (14,900 ft) in Morococha, Peru (Heath and Williams, 1977). Wenchuan, China, is the highest town in the world (16,732 ft or 5100 m) and the highest dwellings in the world are at Basisi, India (19,700 ft or 6000 m) (McWhirter, 1979). At this altitude of 6 km, the ambient oxygen partial pressure is 74 torr (Gilbert, 1981b). Actually, the air passages in the lungs are saturated with water vapor and account for 47 torr regardless of the altitude. Thus, the oxygen pressure in these passages is a constant 10 torr (47 torr $\times$ fraction of O_2 in dry air) less than the ambient oxygen pressure. So at 6 km, the oxygen pressure is 64 torr in the lung passages.

There are three major high altitude areas in the world more or less inhabited by humans (Pawson and Jest, 1978). These are the Tibetan plateau and Himalayan region of Asia (average altitude greater than 4500 m or 14,750 ft), the Andean region of South America (average altitude near 4000 m or 13,000 ft), and the Ethiopian region of Africa (ranges from 2400 to 3700 m or 8000 to 12,000 ft). The highland population of the Andean region is about 17 million (Baker, 1978), whereas in Tibet, the population is 4 to 5 million (Ward, 1975). If we assume that about 25% (Baker, 1978) of the 29 million inhabitants in Ethiopia (Delury, 1978) reside at high altitude, then over 7 million live at high altitudes in that region of the world.

There are other high altitude regions, but the total population from these regions is small. These areas include the Rocky Mountains and Sierra Madre of North America; the Atlas Mountains, Basutoland, and Kilimanjaro of Africa; the Pyrenees and Alps of Europe; and the high mountains of Antarctica (Heath and Williams, 1977). Of the 4219 million people in the world (Delury, 1978), only about 30 million people, or 1% of the world population, reside at high altitude. Therefore, a small but significant part of the human population live under various states of hypoxia, as compared to sea level oxygen pressures.

There are two obvious differences between sea level and highland altitudes. One is a decreased partial pressure of oxygen and the other is a decrease in temperature of 6.5°C per km of altitude up to 10 km (Gilbert, 1964b). Animals are mainly affected by the decrease in the oxygen pressure, whereas plants are mainly affected by the decrease of temperature.

If we believe that biological antioxidant defenses were slowly evolved as the oxygen pressure increased in the earth's atmosphere (Gerschman, 1981; Gilbert, 1981b), then perhaps these defenses would be better equipped to cope with a decreased oxygen pressure. Thus, the high altitude inhabitants would have an advantage over the sea-level residents from the point of view of possessing adequate antioxidant defenses.

On the other hand, the high altitude resident is subjected to the hazards of a greater exposure to radiation (Heath and Williams, 1977). At least part of the biological effects of radiation and of oxygen poisoning are mediated by both free radicals and other activated species according to the Gerschman theory (Gerschman et al., 1954).

At lower oxygen pressures, antioxidant defenses would be better able to cope with the adverse effects of increased radiation (Gerschman, 1962). The stimulation of melanin formation by 290 to 320 nm ultraviolet radiation at high altitude (Little and Hanna, 1978) is perhaps another antioxidant defense mechanism, since melanin is a free radical scavenger (McGinnus and Proctor, 1973; Pirozynski, 1977). Even though the intensity of the 300 nm ultraviolet at 4000 m (13,100 ft) is 2.5 times the corresponding intensity at sea level, there does not seem to be an increase in the incidence of skin cancer at high altitude (Heath and Williams, 1977). These land altitudes are much less than the altitude at which ozone filters out the ultraviolet radiation. The ozone layer lies between 15 and 30 km (50,000 and 100,000 ft) and is much higher at the equator than at the poles (Prinn et al., 1978). It is of interest to note that in spite of this protective radiation filtering action of ozone on the biosphere, it is extremely toxic (Menzel, 1970; Goldstein, 1979). Thus, ozone, another form of oxygen, also has a dual action. When the ozone concentration is high in the upper atmosphere, then it protects the biosphere. On the other hand, when the ozone concentration on the surface of the earth is high, then it destroys the biosphere. Ozone is only beneficial to the biosphere when it is kept in its proper place.

Up to a few kilometers in altitude, the galactic or cosmic radiation doubles for each 2 km. The average galactic radiation is about 3.5 μR/hr or 30 mr/year. At high latitudes, the effect of altitude on galactic radiation is more pronounced. At sea level, the galactic radiation at either pole is 14% greater than at the equator, while at an altitude of 4360 m (14,300 ft), the galactic radiation at the polar regions is 33% greater than at the equator (Eisenbud, 1973). The Andean and Ethiopean highlands are relatively close to the equator, in contrast to the Tibetan highlands at a latitude of 30°N. Thus, the Andean inhabitant receives less radiation than the Tibetan at the same altitude.

The native highlander has superior work performance capacities at high altitude (Buskird, 1978). Lowlanders eventually become acclimatized after a prolonged stay at the high altitudes (Heath and Williams, 1977). There are various physiological and biochemical mechanisms which permit man and other animals to operate under hypoxic conditions (Lenfant, 1973; Ward, 1975; Velásquez, 1976). These mechanisms are difficult to assess since factors such as temperature and nutrition have to be properly controlled (Monge and Whittembury, 1976). Monge and Whittembury (1976) believe that the high hematocrit observed in the native high altitude Andean residents is not a useful mechanism at high altitude. They point out that the high hematocrit produces an undesirable increase in the blood viscosity, which more than offsets the beneficial increase in the oxygen carrying capacity produced by the increase of the red cell mass. A large viscosity increase causes an increased work load of the cardiopulmonary system (Garruto, 1976).

II. Hyperoxia

On earth, there are very few gaseous environments in which the pressure significantly exceeds 1 bar. The deepest exposed depression on earth is the Dead Sea, the surface of which is −1291 ft (−393 m) below sea level (McWhirter, 1979). The calculated oxygen tension here is only 5% greater than at sea level (Gilbert, 1981b). The greatest ocean depth is 36,198 ft (11033 m) (McWhirter, 1979), which corresponds to a pressure of 1100 bar (Table 1). Thus, a gaseous environment would have a tremendous pressure at this depth.

Swimbladders of fish present such a gaseous environment. Some years ago, Haldane and Priestley (1935) wrote, "the living swim-bladder may contain oxygen at a pressure of 100 atmospheres without harm to the cells lining its walls. These cells are apparently 'acclimatized' to the oxygen, just as the cells lining the stomach-wall are acclimatized to hydrochloric acid. It is probable that both the lungs and the rest of the body are capable of acquiring some degree of acclimatization or immunity to the effects of a high pressure of oxygen." Wittenberg et al. (1980) recently analyzed gas contents of swimbladders from rattail fishes at a depth of 3000 m (1000 ft). The analysis showed that 90 to 95% of the gas in the swimbladder is oxygen, which corresponds to 280 bar at this depth. These fish were also caught at a greater depth. Assuming the percent of oxygen in the swimbladder did not change, the corresponding oxygen tension in the swimbladders of these fish would be 460 to 520 bar, i.e., 2,500 times greater than the ambient oxygen tension at sea level! Undoubtedly, there are some extremely effective antioxidant mechanisms at work here. However, exposure of the whole fish to oxygen partial pressures as great as in their swimbladders results in oxygen poisoning (D'Aoust, 1969). Thus, other cells in the fish are sensitive to oxygen toxicity. Blaxter and Tyler (1978) discuss the possibility that oxygen storage is one of the functions of the swimbladder.

J. B. S. Haldane (1948), the son of J. S. Haldane, has described the effects of oxygen poisoning in man very vividly: "At 7 atmospheres the convulsion comes on with little warning.... The clonic convulsions are very violent and in my own case the injury caused by them to my back is still painful after a year. They last for about two minutes and are followed by flaccidity. I wake up into a state of extreme terror in which I may make futile attempts to escape from the steel chamber...." In another description, Haldane (1947) wrote, "One of the naval ratings [trainees] who was being trained in the use of oxygen under water was a boxer. While coming round from a fit he asked 'Who did that?'... The attendant answered 'Oxygen Pete.' Oxygen Pete caught on.... If several people had fits on the same morning people said 'Oxygen Pete's in form today.'" This description can be compared to the account by Acosta of the first documented description of the effects of high altitude hypoxia in going over the Peruvian Andes (Gilbert, 1981a). In crossing the Andes in the same vicinity as Acosta, Gunther (1941) writes also in vivid terms, "The automobile trip from Lima up the Andes... is the most hair-raisingly dramatic I ever took." At the Andean divide, he tried "to walk a step, and then collapse[d] with a crimson roaring in [his] eyes and an exploding blackness in [his] ears from the suddenly realized assault of that incredible tropical altitude."

Effects of both hypoxia and hyperoxia can be quite dramatic! In both cases, the organism must adapt to an oxygen environment which disturbs the delicate balance of oxygen utilization and oxygen poisoning.

It appears that oxygen bubbles produced by *Chara fragilis*, an aquatic plant, in nature inhibit the breeding of mosquitoes (Matheson, 1930). This is an example of natural hyperoxia, caused by the photosynthetic activity of plants (Mauzerall and Piccioni, 1981).

What is meant by high oxygen pressure to a given biological system might be considered low oxygen pressure to another. For example, the exposure of obligate anaerobic organisms to any oxygen is hyperoxic to them (Gottlieb, 1981), but hypoxic to us. Oxygen poisoning to these anaerobes is a natural phenomenon. If they get exposed to the earth's atmosphere polluted with oxygen produced by the "evil" plants (to the anaerobes), they die.

Some distinction is often made between the

effects of oxygen at pressures greater than 1 bar with the effects of oxygen at pressures between 0.2 to 1 bar. Such a distinction is useful only for the techniques used in achieving the oxygen pressure. For the former case, one must pressurize the atmosphere above the normal 1 bar. For the latter case, one only has to change the percent oxygen in the atmosphere. This does not mean that all systems are affected adversely to the same degree at a given oxygen tension. Thus, lung pathology is observed at an oxygen pressure of one bar and convulsions are typically observed at higher oxygen pressures (Dickens, 1962; Huber and Drath, 1981). Bert (1878) showed a century ago that the biological effects of oxygen are due to its partial pressure and not due to the barometric pressure per se. Gerschman (1959) discussed the continuum of oxygen effects as a function of all oxygen tensions.

Survival times of organisms are decreased when they are exposed to an increased oxygen pressure. A linear relation is observed when the logarithm of the survival time is plotted against the logarithm of the oxygen pressure. Williams and Beecher (1944) first observed this relationship. The unicellular Protozoa (Cleveland, 1925) and *Paramecium caudatum* (Gerschman et al., 1958b) have been observed to follow this rectangular hyperbolic function. The survival times of the metazoan forms such as fish, *Seabastodes miniatus* (juvenile) (D'Aoust, 1969), Drosophila (Williams and Beecher, 1944; Fenn et al., 1967), mice (Gerschman et al., 1958a), rats (Jamieson, 1966), and rabbits (Hederer and André, 1940) also follow this relationship. This rectangular hyperbolic relationship also holds for the onset of neurological symptoms as a function of oxygen pressure (Kovachich and Haugaard, 1981).

The adrenal glands are stimulated by high oxygen pressures, as well as by other stress situations (Gerschman and Fenn, 1954). However, in contrast to other types of stress, this stimulation is harmful to the organism (Gerschman, 1964).

Several recent reviews on hyperoxia have been mentioned by Gilbert (1981a). Others include those by Wood (1969), Clark and Lambertsen (1971), Winter and Smith (1972), Haugaard (1974), Balentine (1977), Clark and Fisher (1977), Fridovich (1977), Gottlieb (1977), McCord (1977), Gottlieb (1981), and Kovachich and Haugaard (1981).

III. Normoxia

As mentioned before, oxygen is toxic at all pressures, including normoxic pressures. The ambient oxygen tension of 158 torr in the environment is reduced by a number of barriers before the oxygen gets to the cellular mitochondria, where it is utilized for energy. These barriers from the environment to the cell are discussed by Piiper and Scheid (1981). Rahn (1967) has pointed out that these barriers for oxygen can be coupled to other systems. Thus, for the human, the oxygen tension in the mixed venous blood is only 40 torr. This is one quarter the ambient oxygen tension. Since the solubility coefficient is 1.26 μM O_2/torr for plasma or 1.33 μM O_2/torr for plasma water at 37°C, this oxygen pressure is equivalent to 53 μM (in plasma water) (Altmann and Dittmer, 1971). Wittenberg and Wittenberg (1981) point out that intracellular diffusion is another barrier which must be overcome before the oxygen gets to the mitochondrial sink. This barrier can be partly overcome by oxygen carriers, such as myoglobin in muscle cells and leghemoglobin in the soybean root nodule. These authors believe that intracellular gradients must exist. Thus, at the oxygen sinks, where the oxygen is consumed, the oxygen concentration can be low, perhaps about 5 μM or less. In order to minimize the effects of oxygen poisoning, an oxygen-consuming system should be exposed to the smallest oxygen concentration possible for normal functioning. However, the system should be supplied with a large oxygen capacity transport system. This function is accomplished by myoglobin in the muscle. In the blood, hemoglobin and other blood oxygen carriers possess this function.

At the mitochondrial site within the cell, the oxygen becomes reduced to water releasing energy. Here the cytochrome oxidase is responsible for the dioxygen picking up four hydrogen atoms and forming two water molecules (Chance, 1981). Many destructive and reactive species of oxygen are possible intermediates during this process (Gerschman, 1964). Chance (1981) has shown that the cytochrome oxidase reactions greatly guard against such intermediates, but there are other oxidases. These intermediates include the ionized form of the perhydroxyl free radical or superoxide ion (Chance and Boveris, 1978; Forman and Fisher, 1981; Fridovich, 1981; Kanfer and Turro, 1981), the hydroxyl radical (Chance and Boveris, 1978, Cohen and Cederbaum, 1979, Forman and Fisher, 1981), hydrogen peroxide (Chance and Boveris, 1978; Chance et al., 1979), and possibly singlet oxygen (Singh, 1978; Svingen et al., 1978; Forman and Fisher, 1981; Kanfer and Turro, 1981).

Antioxidant defense mechanisms (Forman and Fisher, 1981) protect the cell against these harmful effects to the cell. One of the most important is superoxide dismutase (Fridovich, 1981). However, Fee (1977) has suggested that one of the superoxide dismutases might have yet another unknown function. Fee and Valentine (1977) conclude that since the superoxide ion does not seem to be sufficiently reactive, it does not seem to be a deleterious species. However, Fridovich (1978) points out that the superoxide ion can give rise to "vastly more effective substances, which are certainly more dangerous." Vitamin E, glutathione peroxidase, and catalase also form part of the antioxidant defense (Forman and Fisher, 1981). Chelators of heavy metals also contribute to the system of antioxidant defenses. They inhibit oxidation (Gilbert et al., 1958) and protect tissue preparations against oxygen toxicity (Kovachich and Haugaard, 1981).

The cells in trees have much shorter lifetimes than the trees themselves (Kramer and Kozlowski, 1960). If we assume that the survival of these cells is at least partly limited by oxygen poisoning, then the turnover of the cells in a tree can permit the survival of the tree to be much longer than its cells. This phenomenon can possibly be viewed as a type of antioxidant defense for trees and thus permit some trees to survive over three millennia (Gilbert, 1972).

Oxidizing free radicals can damage membranes by lipid peroxidation, mediated through chain reactions (Barber and Bernheim, 1967; Wolman, 1975; Forman and Fisher, 1981). These oxidizing reactive intermediates can also inactivate essential sulfhydryl groups (Armstrong and Buchanan, 1978; Kovachich and Haugaard, 1981), and these groups seem to play a role in the functioning of excitable membranes (Shrager, 1977; Zuazaga de Ortiz and del Castillo, 1978). High oxygen pressures can impair membrane function (Stuart et al., 1962; Gilbert and Lowenberg, 1964; Chalazonitis and Arvanitaki, 1970; Cymerman and Gottlieb, 1970; Vaughn, 1971; Gilbert, 1972).

Negative air ions can kill bacteria. This killing effect is inhibited by superoxide dismutase, thus suggesting that the negatively charged superoxide ion is involved (Kellogg et al., 1979). Therefore, negative ion generators for cleaning air may be health hazards.

Antioxidants can also act like pro-oxidants, depending on the system studied. In general, removal of free radicals has an antioxidant effect and addition of free radicals has a pro-oxidant effect. Chain reactions can be initiated by free radical addition and terminated by free radical removal (Gerschman, 1959; Gilbert, 1963).

Excluding the skin, the lung tissue is subjected to the highest oxygen pressure, which is about 104 torr in the human (Altman and Dittmer, 1971). Therefore, one would anticipate finding many antioxidant defenses here (Huber and Drath, 1981). The lung does contain a high concentration of superoxide dismutase, which increases upon exposure to high oxygen pressures (Keele and Schanberg, 1972).

IV. Practical and Clinical Considerations

Small concentrations of free radicals due to an imperfect antioxidant defense could be a factor in the aging process (Gerschman, 1959). Others have also implicated free radicals in the aging process (Harman, 1956, 1968; Tappel, 1968; Emanuel, 1976).

If indeed the free radical concentration determines the damage due to oxygen, then conditions such as increased metabolism should increase the free radical concentration and produce more damage. In support of the Gerschman free radical theory, which states that the effects of X-irradiation and oxygen toxicity are mediated by reactive oxidative intermediates (Gerschman et al., 1954), superoxide dismutase protects against the effects of X-irradiation (Misra and Fridovich, 1976; Petkau, 1978). The effect of radiation can be viewed as mainly being a catalyst for the effects of oxygen by supplying free radicals (Gilbert, 1964a).

Deoxyribonucleic acid, the genetic material, can be degraded under the appropriate conditions when exposed to an increased oxygen tension (Gilbert et al., 1957). High oxygen pressure produces mutagenesis in *Escherichia coli* (Fenn et al., 1957) and chromosome aberrations in chick embryo fibroplasts (Gotzos, 1976). Mutagenesis was also demonstrated at an oxygen pressure of 0.05 bar (Bruyninckx et al., 1978). Carcinogenesis may be due to in part by free radicals (Pryor, 1978), such as the superoxide ion and hydroxyl radical (Greenstock and Ruddock, 1978) and singlet oxygen (Greenstock and Wiebe, 1978).

Rats and mice breathing a saturated vapor of 90% hydrogen peroxide (0.12 m*M*) showed areas of lung edema after 3 days (Oberst et al., 1954). Intraarterial infusion of hydrogen peroxide has caused regression of tumors in patients (Aronoff et al., 1965), as well as removal of lipids and cholesterol from human atheromatous arteries (Finney et al., 1966).

The herbicide paraquat (1,1′-dimethyl-4, 4′-bipyridylium) (methyl viologen) requires oxygen for its herbicidal activity. Photosystem I in the plant (Mauzerall and Piccioni, 1981) reduces the paraquat. The formed paraquat radical is then oxidized back to its original state by molecular oxygen producing the superoxide ion. The damage to the plant is mediated through the superoxide ion and its derivatives. The lung is primarily affected in the human (Smith et al., 1979). It is highly toxic to mammals and it appears to produce lipid peroxidation by enhancing the formation of the superoxide ion (Bus et al., 1974; Smith et al., 1979). Bus et al. (1974) believe that singlet oxygen is involved. Smith et al. (1979) believe that extreme oxidation of the reduced form of nicotinamide adenine dinucleotide phosphate (NADPH) is responsible for its toxicity.

Another instance where oxidizing intermediates play a role is in the defense mechanisms of the organisms against infections. Activation of phagocytic leukocytes results in a respiratory burst with the production of the superoxide ion and hydrogen peroxide (Fridovich, 1981). Possibly singlet oxygen and the hydroxyl radical are also produced (Taubier and Babior, 1978). These oxidizing species then destroy the ingested microorganisms. The cytosol of the leukocytes is protected by superoxide dismutase, glutathione peroxidase, and catalase. Chronic granulomatous disease is characterized by a defect in the superoxide ion generating capacity (Roos and Weening, 1979). Light is emitted in phagocytosis by polymorphonuclear leukocytes, suggesting that singlet oxygen is involved. However, a broad spectrum and not a specific one for the singlet oxygen is obtained. But the use of scavengers for singlet oxygen indicates a possible presence of singlet oxygen. Myeloperoxidase and a halide are involved in the antimicrobial activity (Klebanoff and Rosen, 1979). Thus, the process of phagocytosis involves the generation of destructive, reactive oxidative intermediates. Release of these oxidative intermediates, such as hydrogen peroxide, causes impairment of lymphocyte function (Sagone et al., 1978). Addition of either catalase or superoxide dismutase protects the leukocytes from their own self-inflicted damage (McCord and Wong. 1979).

Since superoxide dismutase possesses anti-inflammatory activity (Huber and Saifer, 1977), the superoxide ion is believed to be involved in the inflammatory process (Fridovich, 1981). The oxidation of arachidonic acid can produce inflammogens, such as some prostaglandins and some oxidizing radicals (Kuehl and Egan, 1980). In addition, the superoxide ion could be released from phagocytes, which are commonly found in inflamed rheumatoid joints (McCord and Wong, 1979). These authors have prepared high molecular weight superoxide dismutases which remain in the circulating blood stream for a prolonged period of time. These modified enzymes show promise in the treatment of inflammatory conditions.

Historically, inflammatory conditions have been treated by either high or low pressures of oxygen. Henshaw in 1664 believed inflammatory diseases could be cured by rarefied air, i.e., a lowered ambient oxygen pressure. Lavoisier in 1785 found that oxygen causes lung inflammation. Fourcroy in 1790 believed that oxygen therapy should be avoided in any inflammatory disease. Also, Gmelin in 1799 thought that oxygen caused some inflammatory fevers. Townsend in 1802 believed that lack of an oxygen source in the food could contribute to inflammatory fevers (Gilbert, 1981a).

There is some suggestion that an inherited deficiency of glucose-6-phosphate dehydrogenase (G-6-PD) in the red blood cell may help overcome a severe malaria infection. There is evidence that infected red cells are extremely sensitive to an oxidative stress. A deficiency of G-6-PD would make the cell more sensitive to such a stress, and the result would be that the cell would be killed with the immature parasite in it (Eaton and Eckman, 1979).

Experimental cataractogenesis of the lens induced by 3-amino-1*H*-1,2,4-triazole increased the hydrogen peroxide by two to three times in the aqueous and vitreous humors. In addition, catalase and superoxide dismutase of the eye tissues were markedly decreased (Bhuyan and Bhuyan, 1979).

The newborn may be subject to the problem of hypoxia, especially during or right after delivery. If the duration and extent of the hypoxia is sufficient, permanent brain damage can occur. The treatment of hypoxic conditions has been the administration of oxygen. An excess of oxygen can lead to the reverse problem of hyperoxia (Hunt, 1976). The newborn is resistant to hyperoxia (Kovachich and Haugaard, 1981). Thus, the newborn mouse can survive 1 bar oxygen for 6 weeks, whereas the adult mouse can survive no longer than 1 week at the same oxygen pressure (Northway et al., 1979). The superoxide dismutase in the lung of newborn rats is increased by the exposure of 1 bar oxygen pressure (Stevens and Autor, 1977). However, there still exists a danger of hyperoxia, even to newborns.

There is a need to know the arterial and tissue oxygen tensions when oxygen administration is indicated, so that the treatment can be carefully controlled (Silver, 1981). This is especially true in high risk situations, e.g., for premature infants (Parker and Soutter, 1975). For example, the premature infant can develop retrolental fibroplasia, a type of blindness, due to an exposure of the infant to high oxygen tensions (Flower and Patz, 1981; Silver, 1981).

Some newborn infants, treated with supplemental oxygen therapy for severe hyaline membrane disease, develop another lung disease called bronchopulmonary dysplasia. Oxygen is thought to be at least a contributing factor in causing this chronic condition, which was first described in 1967 (Northway, 1979). The antioxidant vitamin E seems to have some effect on preventing bronchopulmonary dysplasia (Ehrenkranz et al., 1979).

These studies emphasize the need to know the oxygen tension in the body to assess the quantitative need for oxygen in the clinical situation (Silver, 1981). Of course, the same oxygen data are needed for divers and astronauts (Schaefer, 1981).

Tumor cells are more sensitive to radiation in the presence of oxygen (Silver, 1981). However, in some tumors there are hypoxic cells which make the tumors more radioresistant. Hyperbaric oxygen therapy does not seem to help, presumably due to hypoxic cells existing in such tumors (Stone, 1979). Hyperthermia increases the oxygen tension in such conditions (Silver, 1981).

Hyperbaric oxygen therapy (Davis and Hunt, 1977) has been successfully used in such conditions as decompression sickness, acute carbon monoxide poisoning, and gas gangrene (Kindwall, 1979). However, too often it is used without sufficient cause, reminding us of the claims of those who used compressed air chambers of the nineteenth century (Gilbert, 1981a). There are dangers of oxygen toxicity in this treatment, as well as of decompression sickness, especially when air is used. There was evidently gross negligence in the 1976 accident which claimed five lives in the hyperbaric air chamber at Hannover, Germany (Seemann, 1979). Oxygen administration is not a play toy! It should be carefully controlled. Oxygen therapy should only be used when it is necessary. As Hedley-Whyte and Winter (1967) write, "A great service would be done to patients if oxygen were regarded as a drug—one with great therapeutic utility, but one which will result in toxicity if an overdose is given.... The patient should receive that amount of oxygen which he requires for adequate tissue oxygenation and not more."

V. Summary

Oxygen constitutes a necessity for most of the biosphere. It has provided the energy that has made living organisms so versatile. Stratagems have been invoked by organisms to reduce the hazardous exposures of high concentrations of oxygen, without sacrificing the supply of oxygen, so necessary for an aerobic way of life. However, this energy which nature has bestowed upon oxygen is a mixed blessing. This energy not only permits aerobic life to exist, but in its very nature, it destroys life. The practical applications of this knowledge can lead to efforts to curb the destructive aspects of oxygen by improving on antioxidant defenses. Using superoxide dismutase in some inflammatory diseases is an example. Oxygen possesses a good Dr. Jekyll–evil Mr. Hyde personality. (Stevenson, 1886).

Acknowledgments

Acknowledgment is given to Dr. Claire Gilbert for her extremely valuable criticisms, to Dr. Charles W. Shilling, Undersea Medical Society, Bethesda, Maryland, for his information on hyperbaric medicine, and to Dr. Charles P. Bean, Research and Development Center, General Electric Company, Schenectady, New York for alerting me to the lifespan of cells within trees.

References

Altman, P. L., and Dittmer, D. S. (1971). Respiration and Circulation. Bethesda, Maryland: Fed. Amer. Soc. Exp. Biol.

Armstrong, D. A., and Buchanan, J. D. (1978). Reactions of $O_2^{\overline{\cdot}}$, H_2O_2 and other oxidants with sulfhydryl enzymes. Photochem. Photobiol. 28:743–755.

Aronoff, B. L., Balla, G. A., Finney, J. W., Collier, R. E., and Mallams, J. T. (1965). Regional oxygenation in the diagnosis and management of intra-abdominal and retroperitoneal neoplasms. Cancer 18:1244–1250.

Baker, P. T. (1978). The adaptive fitness of high-altitude populations. In: Baker, P. T. (Ed.). The Biology of High-Altitude Peoples. New York: Cambridge University Press, pp. 317–350.

Balentine, J. D. (1977). Experimental pathology of oxygen toxicity. In: Jöbsis, F. J. (Ed.). Oxygen and Physiological Function. Dallas: Prof. Intern. Library, pp. 311–378.

Barber, A. A., and Bernheim, F. (1967). Lipid peroxidation: Its measurement, occurrence, and significance in animal tissues. Adv. Gerontol. Res. 2:355–403.

Bert, P. (1878). Barometric Pressure. Researches in Experimental Physiology. Trans. by M. A. Hitchcock and F. A. Hitchcock. Columbus, Ohio: College Book Co., 1943.

Bhuyan, D. K., and Bhuyan, K. C. (1979). Mechanism of cataractogenesis induced by 3-amino-1H-1,2,4-triazole. II: Superoxide dismutase of the eye and its role in protecting the ocular lens from oxidative damage by endogenous $O_2^{\bar{}}$, H_2O_2, and/or $OH^{\cdot}$. In: Caughey, W. S. (Ed.). Biochemical and Clinical Aspects of Oxygen. New York: Academic Press, pp. 797–809.

Black, C. P., and Tenney, S. M. (1980). Oxygen transport during progressive hypoxia in high-altitude and sea-level waterfowl. Respir. Physiol. 39:217–239.

Blaxter, J. H. S., and Tytler, P. (1978). Physiology and function of the swimbladder. Adv. Comp. Physiol. Biochem. 7:311–367.

Bruyninckx, W. J., Mason, H. S., and Morse, S. A. (1978). Are physiological oxygen concentrations mutagenic? Nature 274:606–607.

Bus, J. S., Aust, S. D., and Gibson, J. E. (1974). Superoxide- and singlet oxygen-catalyzed lipid peroxidation as a possible mechanism for paraquat (methyl viologen) toxicity. Biochem. Biophys. Res. Commun. 58:749–755.

Buskirk, E. R. (1978). Work capacity of high-altitude natives. In: Baker, P. T. (Ed.). The Biology of High-Altitude Peoples. New York: Cambridge University Press, pp. 173–187.

Chalazonitis, N., and Arvanitaki, A. (1970). Neuromembrane electrogenesis during changes in pO_2, pCO_2, and pH. Adv. Biochem. Psychopharm. 2:245–284.

Chance, B. (1981). The reaction of oxygen with cytochrome oxidase: The role of sequestered intermediates. This volume.

Chance, B., and Boveris, A. (1978). Hyperoxia and hydroperoxide metabolism. In: Robin E. D.(Ed.). Extrapulmonary Manifestations of Respiratory Disease. New York: Marcel Dekker, pp. 185–237.

Chance, B., Sies, H. and Boveris, A. (1979). Hydroperoxide metabolism in mammalian organs. Physiol. Rev. 59:527–605.

Clark, J. M., and Fisher, A. B. (1977). Oxygen toxicity and extension of tolerance in oxygen therapy. In: Davis, J. C., and Hunt, T. K. (Eds.). Hyperbaric Oxygen Therapy. Bethesda, Maryland: Undersea Medical Society, pp. 61–77.

Clark, J. M., and Lambertsen, C. J. (1971). Pulmonary oxygen toxicity: A review. Pharm. Rev. 23:37–133.

Cleveland, L. R. (1925). Toxicity of oxygen for protozoa in vivo and in vitro: Animals defaunated without injury. Biol. Bull. 48:455–468.

Cohen, G., and Cederbaum, A. I. (1979). Chemical evidence for production of hydroxyl radicals during microsomal electron transfer. Science 204:66–68.

Cymerman, A., and Gottlieb, S. F. (1970). Effects of increased oxygen tensions on bioelectric properties of frog sciatic nerves. Aerospace Med. 41:36–39.

D'Aoust, B. G. (1969). Hyperbaric oxygen: Toxicity to fish at pressures present in their swimbladders. Science 163:576–578.

Davis, J. C., and Hunt, T. K. (Eds.) (1977). Hyperbaric Oxygen Therapy. Bethesda, Maryland: Undersea Medical Society.

Delury, G. E. (1978). The World Almanac & Book of Facts. 1979. New York: Newspaper Enterprise Association.

Dickens, F. (1962). The toxic effect of oxygen on nervous tissue. In: Elliott, K. A. C., Page, I. H., and Quastel, J. H. (Eds.). Neurochemistry. The Chemistry of Brain and Nerve. 2nd Ed. Springfield, Illinois: Charles C Thomas, pp. 851–869.

Eaton, J. W., and Eckman, J. R. (1979). Malaria infection and host cell oxidant damage. In: Caughey, W. S. (Ed.). Biochemical and Clinical Aspects of Oxygen. New York: Academic Press, pp. 825–837.

Ehrenkranz, R. A., Ablow, R. C., and Warshaw, J. B. (1979). Prevention of bronchopulmonary dysplasia with vitamin E administration during the acute stages of respiratory syndrome. J. Pediat. 95:873–878.

Eisenbud, M. (1973). Environmental Radioactivity. 2nd Ed. New York: Academic Press.

Emanuel, N. M. (1976). Free radicals and the action of inhibitors of radical processes under pathological states and aging in living organisms and in man. Q. Rev. Biophys. 9:283–308.

Fee, J. A. (1977). Structure-function relationships in superoxide dismutases. In: Michelson, A. M., McCord, J. M., and Fridovich, I. (Eds.). Superoxide and Superoxide Dismutases. New York: Academic Press, pp. 173–192.

Fee, J. A., and Valentine, J. S. (1977). Chemical and physical properties of superoxide. In: Michelson, A. M., McCord, J. M., and Fridovich, I. (Eds.). Superoxide and Superoxide Dismutases. New York: Academic Press, pp. 19–60.

Fenn, W. O., Gerschman, R., Gilbert, D. L., Terwilliger, D. E., and Cothran, F. V. (1957). Mutagenic effects of high oxygen tensions on Escherichia coli. Proc. Nat. Acad. Sci. U.S.A. 43:1027–1032.

Fenn, W. O., Henning, M., and Philpott, M. (1967). Oxygen poisoning in Drosophila. J. Gen. Physiol. 50:1693–1707.

Finney, J. W., Jay, B. E., Race, G. J., Urschel, H. C., Mallams, J. T., and Balla, G. A. (1966). Removal of cholesterol and other lipids from experimental animals and human atheromatous arteries by dilute hydrogen peroxide. Angiology 17:223–228.

Flower, R. W., and Patz, A. (1981). Retinopathy of prematurity and the role of oxygen. This volume.

Forman, H. J., and Fisher, A. B. (1981). Antioxidant defenses. This volume.

Fridovich, I. (1977). Oxygen is toxic! Bioscience 27:462–466.

Fridovich, I. (1978). Superoxide radicals, superoxide dismutases and the aerobic lifestyle. Photochem. Photobiol. 28:733–741.

Fridovich, I. (1981). Superoxide radical and superoxide dismutases. This volume.

Gaer, J. (1956). How the Great Religions Began. Rev. Ed. New York: The New American Library.

Garruto, R. M. (1976). Hematology. In: Baker, P. T., and Little, M. A. Man in the Andes. A Multidisciplinary Study of High-Altitude Quechua. Stroudsburg, Pennsylvania: Dowden, Hutchinson, and Ross, pp. 261–282.

Gerschman, R. (1959). Oxygen effects in biological systems. Symp. Spec. Lect., XXI Intern. Cong. Physiol. Sciences, Buenos Aires, pp. 222–226.

Gerschman, R. (1962). The biological effects of increased oxygen tension. In: Schaefer, K. E. (Ed.). Man's Dependence on the Earthly Atmosphere. New York: Macmillan Co., pp. 171–179.

Gerschman, R. (1964). Biological effects of oxygen. In: Dickens, F. and Neil, E. (Eds.). Oxygen in the Animal Organism. New York: Pergamon Press, Macmillan Co., pp. 475–494.

Gerschman, R. (1981). Historical introduction to the "free radical theory" of oxygen toxicity. This volume.

Gerschman, R., and Fenn, W. O. (1954). Ascorbic acid content of adrenal glands of rat in oxygen poisoning. Am. J. Physiol. 176:6–8.

Gerschman, R., Gilbert, D. L., Nye, S. W., Dwyer, P., and Fenn, W. O. (1954). Oxygen poisoning and x-irradiation: A mechanism in common. Science 119:623–626.

Gerschman, R., Gilbert, D. L., and Caccamise, D. (1958a). Effect of various substances on survival times of mice exposed to different high oxygen tensions. Am. J. Physiol. 192:563–571.

Gerschman, R., Gilbert, D. L., and Frost, J. N. (1958b). Sensitivity of *Paramecium caudatum* to high oxygen tensions and its modification by cobalt and manganese ions. Am. J. Physiol. 192:572–576.

Gilbert, D. L. (1963). The role of pro-oxidants and antioxidants in oxygen toxicity. Radiat. Res. Suppl. 3:44–53.

Gilbert, D. L. (1964a). Atmosphere and evolution. In: Dickens, F., and Neil, E. (Eds.). Oxygen in the Animal Organism. New York: Pergamon Press, Macmillan Co., pp. 641–655.

Gilbert, D. L. (1964b). Cosmic and geophysical aspects of the respiratory gases. In: Fenn, W. O., and Rahn, H. (Eds.). Handbook of Physiology—Section 3: Respiration. Vol. I. Washington, D.C.: American Physiological Society, pp. 153–176.

Gilbert, D. L. (1972). Introduction: Oxygen and life. Anesthesiology 37:100–111.

Gilbert, D. L. (1981a). Perspective on the history of oxygen and life. This volume.

Gilbert, D. L. (1981b). Significance of oxygen on earth. This volume.

Gilbert, D. L., and Lowenberg, W. E. (1964). Influence of high oxygen pressure on the resting membrane potential of frog sartorius muscle. J. Cell. Comp. Physiol. 64:271–278.

Gilbert, D. L., Gerschman, R., Cohen, J., and Sherwood, W. (1957). The influence of high oxygen pressures on the viscosity of solutions of sodium desoxyribonucleic acid and of sodium alginate. J. Am. Chem. Soc. 79:5677–5680.

Gilbert, D. L., Gerschman, R., Ruhm, K. B., and Price, W. E. (1958). The production of hydrogen peroxide by high oxygen pressures. J. Gen. Physiol. 41:989–1003.

Goldstein, B. D. (1979). The pulmonary and extrapulmonary effects of ozone. In: Fitzsimons, D. W. (Ed.). Oxygen Free Radicals and Tissue Damage. Ciba Foundation Symposium 65 (New Series). New York: Elsevier/North-Holland, pp. 295–319.

Gottlieb, S. F. (1977). Oxygen under pressure

and microorganisms. In: Davis, J. C., and Hunt, T. K. (Eds.). Hyperbaric Oxygen Therapy. Bethesda, Maryland: Undersea Medical Society, pp. 79–99.

Gottlieb, S. F. (1981). Oxygen toxicity in unicellular organisms. This volume.

Gotzos, V. (1976). Anomalies mitotiques observées dans les fibroblastes d'embryon de poulet cultivés *in vitro* avec 80% d'oxygène. Acta Anat. 94;520–532.

Greenstock, C. L., and Ruddock, G. W. (1978). Radiation activation of carcinogens and the role of ˙OH and O_2^-. Photochem. Photobiol. 28: 877–880.

Greenstock, C. L., and Wiebe, R. H., (1978). Photosensitized carcinogen degradation and the possible role of singlet oxygen in carcinogen activation. Photochem. Photobiol. 28:863–867.

Gunther, J. (1941). Inside Latin America. New York: Harper and Brothers.

Hagen, T., Wahlen, F. T., and Corti, W. R. (1961). Nepal. The Kingdom in the Himalayas. Berne: Kümmerly and Frey.

Haldane, J. B. S. (1947). Life at high pressures. Science News, No. 4:9–29, July. Penguin Books.

Haldane, J. B. S. (1948). Science Advances. London: George Allen & Unwin Ltd.

Haldane, J. S., and Priestley, J. G. (1935). Respiration. New Ed. (2nd). New Haven, Connecticut: Yale University Press.

Harman, D. (1956). Aging: A theory based on free radical and radiation chemistry. J. Gerontol. 11:298–299.

Harman, D. (1968). Free radical theory of aging: Effect of free radical reaction inhibitors on the mortality rate of male LAF mice. J. Gerontol. 23:476–482.

Harrison, H. (1966). *John W. Campbell*, Collected Editorials from *Analog*. Garden City, New York: Doubleday, p. 226.

Haugaard, N. (1974). The effects of high and low oxygen tensions on metabolism. In: Hayaishi, O. (Ed.). Molecular Oxygen in Biology. Topics in Molecular Oxygen Research. New York: American Elsevier, pp. 163–182.

Heath, D., and Williams, D. R. (1977). Man at High Altitude. The Pathophysiology of Acclimatization and Adaptation. New York: Churchill Livingstone.

Hederer, C., and André, L. (1940). De l'intoxication par les hautes pressions d'oxygène. Bull. Acad. Méd. 123:294–307.

Hedley-Whyte, J., and Winter, P. M. (1967). Oxygen therapy. Clin. Pharm. Ther. 8: 696–737.

Henderson, Y. (1938). Adventures in Respiration. Modes of Asphyxiation and Methods of Resuscitation. Baltimore: Williams & Wilkins Co.

Herbig, G. H. (1981). The origin and astronomical history of terrestrial oxygen. This volume.

Hochachka, P. W., and Guppy, M. (1977). Variation on a theme by Embden, Meyerhof, and Parnas. In: Jöbsis, F. J. (Ed.). Oxygen and Physiological Function. Dallas: Prof. Intern. Library, pp. 292–310.

Hock R. J. (1964). Animals in high altitudes: reptiles and amphibians. In: Dill, D. B., Adolph, E. F., and Wilber, C. G. (Eds.). Handbook of Physiology. Section 4: Adaptation to the Environment. Washington, D.C.: American Physiological Society, pp. 841–842.

Huber, G. L., and Drath, D. B. (1981). Pulmonary oxygen toxicity. This volume.

Huber, W., and Saifer, M. G. P. (1977). Orgotein, the drug version of bovine Cu-Zn superoxide dismutase: I. A summary account of safety and pharmacology in laboratory animals. In: Michelson, A. M., McCord, J. M., and Fridovich, I. (Eds.). Superoxide and Superoxide Dismutases. New York: Academic Press, pp. 517–536.

Humboldt, A. von (1849). Views of Nature: or Contemplations on the Sublime Phenomena of Creation; with Scientific Illustrations. 3rd Ed. (translated from the German by E. C. Ottó and H. G. Bohn). London: Henry G. Bohn 1850. Reprinted New York: Arno Press 1975.

Hunt, J. V. (1976). Environmental risk in fetal and neonatal life and measured infant intelligence. In: Lewis, M. (Ed.). Origins of Intelligence. Infancy and Early Childhood. New York: Plenum, pp. 223–258.

Hutchinson, G. E. (1965). The Ecological Theater and the Evolutionary Play. New Haven, Connecticut: Yale University Press.

Jahnke, L., and Klein, H. P. (1979). Oxygen as a factor in eukaryote evolution: Some effects of low levels of oxygen on *Saccharomyces cerevisiae*. Origins Life 9:329–334.

Jamieson, D. (1966). The effect of anaesthesia and CO_2 on survival time and lung damage in rats exposed to high pressure oxygen. Biochem. Pharmacol. 15:2120–2122.

Kanfer, S., and Turro, N. J. (1981). Reactive forms of oxygen. This volume.

Keele, B. B., Jr., and Schanberg, S. (1972). The

distribution of superoxide dismutase in tissues of normal rats and rats exposed to hyperbaric oxygen. Alabama J. Med. Sci. 9:434–436.

Kellogg, E. W., III, Yost, M. G., Barthakur, N., and Kreuger, A. P. (1979). Superoxide involvement in the bactericidal effects of negative air ions on *Staphylococcus albus*. Nature 281: 400–401.

Kindwall, E. P. (1979). Hyperbaric Oxygen Therapy. Report of the Committee on Hyperbaric Oxygenation. UMS Publication No. 30 CR(HBO) 9-11-79. Bethesda, Maryland: Undersea Medical Society.

Klebanoff, S. J., and Rosen, H. (1979). The role of myeloperoxidase in the microbicidal activity of polymorphonuclear leukocytes. In: Fitzsimmons, D. W. (Ed.). Oxygen Free Radicals and Tissue Damage. Ciba Foundation Symposium 65 (New Series). New York: Elsevier/North-Holland, pp. 263–284.

Kovachich, G. B., and Haugaard, N. (1981). Biochemical aspects of oxygen toxicity in the metazoa. This volume.

Kramer, P. J., and Kozlowski, T. T. (1960). Physiology of Trees. New York: McGraw-Hill.

Kuehl, F. A., Jr., and Egan, R. W. (1980). Prostaglandins, arachidonic acid, and inflammation. Science 210:978–984.

Lenfant, C. (1973). High altitude adaptation in mammals. Am. Zool. 13:447–456.

Little, M. A., and Hanna, J. M. (1978). The responses of high-altitude populations to cold and other stresses. In: Baker, P. T. (Ed.). The Biology of High-Altitude Peoples. New York: Cambridge University Press, pp. 251–298.

Matheson, R. (1930). The utilization of aquatic plants as aids in mosquito control. Am. Nat. 64:56–86.

Mauzerall, D. C., and Piccioni, R. G. (1981). Photosynthetic oxygen production. This volume.

McCord, J. M. (1977). Superoxide and superoxide dismutase in oxygen toxicity. In: Jöbsis, F. J. (Ed.). Oxygen and Physiological Function. Dallas: Prof. Intern. Library, pp. 379–387.

McCord, J. M., and Wong, K. (1979). Phagocyte-produced free radicals: roles in cytotoxicity and inflammation. In: Fitzsimons, D. W. (Ed.). Oxygen Free Radicals and Tissue Damage. Ciba Foundation Symposium 65 (New Series). New York: Elsevier/North-Holland, pp. 343–360.

McGinnus, J., and Proctor, P. (1973). The importance of the fact that melanin is black. J. Theor. Biol. 39:677–678.

McWhirter, N. (1979). Guinness Book of World Records. 17th Ed. New York: Bantam Books.

Menzel, D. B. (1970). Toxicity of ozone, oxygen, and radiation. Annu. Rev. Pharmacol. 10: 379–394.

Misra, H. P., and Fridovich, I. (1976). Superoxide dismutase and the oxygen enhancement of radiation lethality. Arch. Biochem. Biophys. 176:577–581.

Monge [C.], C., and Whittembury, J. (1976). High altitude adaptations in the whole animal. In: Bligh, J., Cloudsley-Thompson, J. L., and MacDonald, A. G. (Eds.). Environmental Physiology of Animals. New York: John Wiley & Sons, pp. 289–308.

Monge M., C., and Monge C., C. (1966). High-Altitude Diseases. Mechanism and Management. Springfield, Illinois: Charles C Thomas.

Northway, W. H. (1979). Observations on bronchopulmonary dysplasia. J. Pediat. 95: 815–818.

Northway, W. H., Petriceks, R., Canty, E., and Bensch, K. G. (1979). Maturation as a factor in pulmonary oxygen toxicity: A preliminary report. J. Pediatrics. 95:859–864.

Oberst, F. W., Comstock, C. C., and Hackley, E. B. (1954). Inhalation therapy of ninety per cent hydrogen peroxide vapor. A.M.A. Arch. Ind. Hyg. Occupat. Med. 10:319–327.

Parker, D., and Soutter, L. P. (1975). *In vivo* monitoring of blood P_{O_2} in new-born infants. In: Payne, J. P., and Hill, D. W. (Eds.). Oxygen Measurements in Biology and Medicine. Boston: Butterworths, pp. 269–283.

Pawson, I. G., and Jest, C. (1978). The high-altitude areas of the world and their cultures. In: Baker, P. T. (Ed.). The Biology of High-Altitude Peoples. New York: Cambridge University Press, pp. 17–45.

Pektau, A. (1978). Radiation protection by superoxide dismutase. Photochem. Photobiol. 28: 765–774.

Piiper, J. and Scheid, P. (1981). Oxygen exchange in the metazoa. This volume.

Pirozynski, K. A. (1977). Melanins as palaeobiological and palaeoenvironmental indicators? Syllogeus 12:59–62.

Prinn, R. G., Alyea, F. N., and Cunnold, D. M. (1978). Photochemistry and dynamics of the ozone layer. Annu. Rev. Earth Planet. Sci. 6: 43–74.

Pryor, W. A. (1978). The formation of free radicals and the consequences of their reactions *in vivo*. Photochem. Photobiol. 28:787–801.

Pugh, L. G. C. E. (1964). Animals in high altitudes: Man above 5,000 meters—mountain exploration. In: Dill, D. B., Adolph, E. F., and Wilber, C. G. (Eds.). Handbook of Physiology. Section 4: Adaptation to the Environment. Washington, D.C.: American Physiological Society, pp. 861–868.

Rahn, H. (1967). Gas transport from the external environment to the cell. In: de Reuck, A. V. S., and Porter, R. (Eds.). Development of the Lung. Ciba Foundation Symposium. Boston: Little, Brown and Co., pp. 3–29.

Roos, D., and Weening, R. S. (1979). Defects in the oxidative killing of microorganisms by phagocytoic leukocytes. In: Fitzsimons, D. W. (Ed.). Oxygen Free Radicals and Tissue Damage. Ciba Foundation Symposium 65 (New Series). New York: Elsevier/North Holland, pp. 225–262.

Rosin, J. (1956). Oxygen *ad absurdum*. Chem. Eng. News 34 (No. 6, Feb. 6):546.

Sagone, A. L., Jr., Kamps, S. and Campbell, R. (1978). The effect of oxidant injury on the lymphoblastic transformation of human lymphocytes. Photochem. Photobiol. 28:909–915.

Schaefer, K. E. (1981). Oxygen in closed environmental systems. This volume.

Seemann, K. E. A. (1979). The hyperbaric air chamber accident in Hannover, Germany: *The final verdict*. Pressure 8 (No. 6):6–8.

Shrager, P. (1977). Slow sodium inactivation in nerve after exposure to sulfhydryl blocking reagents. J. Gen. Physiol. 69:183–202.

Silver, I. A. (1981). Oxygen tension in the clinical situation. This volume.

Singh, A. (1978). Introduction: Interconversion of singlet oxygen and related species. Photochem. Photobiol. 28:429–433.

Smith, L. L., Rose, M. S., and Wyatt, I. (1979). The pathology and biochemistry of paraquat. In: Fitzsimons, D. W. (Ed.). Oxygen Free Radicals and Tissue Damage. Ciba Foundation Symposium 65 (New Series). New York: Elsevier/North-Holland, pp, 321–341.

Stevens, J. B., and Autor, A. P. (1977). Induction of superoxide dismutase by oxygen in neonatal rat lung. J. Biol. Chem. 252:3509–3514.

Stevenson, R. L. (1886). The strange case of Dr. Jekyll and Mr. Hyde. In: Stevenson, R. L. The Great Short Stories of Robert Louis Stevenson. New York: Washington Square Press, 1961, pp. 1–68.

Stone, H. B. (1979). The role of oxygen in the radiation biology of tumors. In: Caughey, W. S. (Ed.). Biochemical and Clinical Aspects of Oxygen. New York: Academic Press, pp. 811–823.

Storck, W. (1979). Industrial gases may take slowdown in stride. Chem. Eng. News 57 (No. 29, July 16):8–11.

Stuart, B., Gerschman, R., and Stannard, J. N. (1962). Effect of high oxygen tension on potassium retentivity and colony formation of Bakers' yeast. J. Gen. Physiol. 45:1019–1030.

Svingen, B. A., O'Neal, F. O., and Aust, S. D. (1978). The role of superoxide and singlet oxygen in lipid peroxidation. Photochem. Photobiol. 28:803–809.

Swan, L. W. (1970). Goose of the Himalayas. Natural History 79 (No. 10):68–75.

Tappel, A. L. (1968). Will antioxidant nutrients slow aging processes? Geriatrics 23(October): 97–105.

Tauber, A. I., and Babior, B. M. (1978). O_2^- and host defense: The production and fate of O_2^- in neutrophils. Photochem. Photobiol. 28:701–709.

Vaughan, G. L. (1971). Oxygen toxicity in a fission yeast. J. Cell. Physiol. 77:363–372.

Velásquez, T. (1976). Pulmonary function and oxygen transport. In: Baker, P. T., and Little, M. A. Man in the Andes. A Multidisciplinary Study of High-Altitude Quechua. Stroudsburg, Pennsylvania: Dowden, Hutchinson, and Ross, pp. 237–260.

Ward, M. (1975). Mountain Medicine. A Clinical Study of Cold and High Altitude. New York: Van Nostrand Reinhold Co.

Williams, C. M., and Beecher, H. K. (1944). Sensitivity of Drosophila to poisoning by oxygen. Am. J. Physiol. 140:566–573.

Winter, P. M., and Smith, G. (1972). The toxicity of oxygen. Anesthesiology 37:210–241.

Wittenberg, J. B., and Wittenberg, B. A. (1981). Facilitated oxygen diffusion by oxygen carriers. This volume.

Wittenberg, J. B., Copeland, D. E., Haedrich, R. L., and Child, J. S. (1980). The swimbladder of deep sea fish: The swimbladder wall is a lipid-rich barrier to oxygen diffusion. J. Mar. Biol. Assoc. U. K. 60:263–276.

Wolman, M. (1975). Biological peroxidation of lipids and membranes. Isr. J. Med. Sci. 11: 1–248.

Wood, J. D. (1969). Oxygen toxicity. In: Bennett, P. B., and Elliott, D. H. (Eds.). The Physiology and Medicine of Diving and Compressed Air Work. Baltimore: Williams & Wilkins, pp. 113–143.
Zuazaga de Ortiz, C., and del Castillo, J. (1978). Induction of excitability in an electrically stable membrane following chemical interference with sulfhydryl and disulfide groups at the cell surface. In: Chalazonitis, N., and Boisson, M. (Eds.). Abnormal Neuronal Discharges. New York: Raven Press, pp. 217–232.

Index

GPSR Compliance

The European Union's (EU) General Product Safety Regulation (GPSR) is a set of rules that requires consumer products to be safe and our obligations to ensure this.

If you have any concerns about our products, you can contact us on ProductSafety@springernature.com

In case Publisher is established outside the EU, the EU authorized representative is:

Springer Nature Customer Service Center GmbH
Europaplatz 3
69115 Heidelberg, Germany

Batch number: 10762209

Printed by Printforce, the Netherlands